RESIDENTIAL BUILDING DESIGN AND CONSTRUCTION

Jack H. Willenbrock, P.E., Ph.D.
Retired: Bernard and Henrietta Hankin Chair
of Residential Building Construction and
Director of the Housing Research Center
The Pennsylvania State University
President: JHW: C/M Consultants

Harvey B. Manbeck, P.E., Ph.D.
Professor, Department of Agricultural and Biological Engineering
The Pennsylvania State University

Michael G. Suchar
Manager, Construction Services Department
Professional Service Industries, Inc.
Charlotte, North Carolina

Prentice Hall
Upper Saddle River, New Jersey 07458

Library of Congress Cataloging-in Publication Data

Willenbrock, Jack H.
   Residential building design and construction / Jack H.
Willenbrock, Harvey B. Manbeck, Michael G. Suchar.
     p.   cm.
   Includes index.
   ISBN: 0-13-375874-5
   1. Dwellings—Design and construction.   I. Manbeck, H.B.
II. Suchar, Michael G.   III. Title
TH4811.W53   1998
690′.837—dc21   97-47688
                                  CIP

Acquisitions Editor: Bill Stenquist
Production Editor: Sharyn Vitrano
Managing Editor: Bayani Mendoza DeLeon
Editor-in-Chief: Marcia Horton
Director of Production and Manufacturing: David W. Riccardi
Cover Designer: Bruce Kenselaar
Manufacturing Buyer: Julia Meehan
Editorial Assistant: Meg Weist

© 1998 by Prentice-Hall, Inc.
Simon & Schuster / A Viacom Company
Upper Saddle River, New Jersey 07458

The author and publisher of this book have used their best efforts in preparing this book. These efforts
include the development, research, and testing of the theories and programs to determine their
effectiveness. The author and publisher make no warranty of any kind, expressed or implied, with
regard to these programs or the documentation contained in this book. The author and publisher shall
not be liable in any event for incidental or consequential damages in connection with, or arising out of,
the furnishing, performance, or use of these programs.

Printed in the United States of America.

10 9 8 7 6 5 4 3 2 1

ISBN 0-13-375874-5

Prentice-Hall International (UK) Limited, *London*
Prentice-Hall of Australia Pty. Limited, *Sydney*
Prentice-Hall Canada Inc., *Toronto*
Prentice-Hall Hispanoamericans, S.A., *Mexico*
Prentice-Hall of India Private Limited, *New Delhi*
Prentice-Hall of Japan, Inc., *Tokyo*
Simon & Schuster Asia Pte. Ltd., *Singapore*
Editora Prentice-Hall do Brasil, Ltda., *Rio de Janeiro*

Dedication

This book is dedicated to the late Bernard Hankin, President of The Hankin Group of Exton, Pennsylvania, his wife Henrietta Hankin, and their three sons, Sam, Robert and Richard. Bernie and his family contributed generously of their personal time and wealth to establish and nurture a program at Penn State which would encourage undergraduate and graduate students in Architectural Engineering, Civil Engineering, and Architecture to consider professional careers in the residential construction industry. "Bernie's Dream" became a reality with the establishment of the initial Bernard Hankin Professorship in Residential Building Construction in 1980, and the eventual expansion of that position into the Bernard and Henrietta Hankin Chair in Residential Building Construction in 1990. This generosity has allowed the program at Penn State to grow and receive national recognition over the last 16 years.

This textbook is also dedicated to two other individuals who have provided the program with both a statewide and a nationwide perspective because of their continued support. The first is J. Roger Glunt, President of Glunt Development Co., Inc. of Turtle Creek, Pennsylvania, who has provided valued counsel from the beginning of the program. Mr. Glunt was the President of the National Association of Home Builders (NAHB) in 1993. The second is David F. Sheppard, Jr., Executive Vice President of the Pennsylvania Builders Association (PBA) of Harrisburg, Pennsylvania, who has provided industry-wide direction and support for the program in Pennsylvania.

Hopefully those who read this book will be beneficiaries of "Bernie's Dream" for many years to come.

Jack H. Willenbrock, P.E., Ph.D.

Contents

Preface

The homebuilding sector is a diverse and significant part of the total construction industry in the United States and a driving force in the national economy. Statistics which define its contribution and describe the number and types of homebuilding firms that operate within it are presented in Chapter 1. Information about the **housing supply** side of the housing supply/demand economic model that is introduced in Chapter 1 provides insight into the evolution, over time, of various housing characteristics for site-built, systems-built, and manufactured housing.

Demographic factors such as the age distribution of the population and household formation rates, as well as the characteristics of different consumer groups, are presented in Chapter 2 as a background for understanding **housing demand**. Consumer preference data gathered from three major studies defines the wants, needs, and demands for various characteristics which homebuyers present to homebuilders and architects when houses are designed.

Chapter 3, which concludes the introduction to the homebuilding industry in the United States, provides additional information about the characteristics of homebuilders and establishes a **management framework** which is appropriate for such firms. An easily understood and original **management growth model** which defines the relationship between **construction output** and **managerial proficiency** in terms of a series of management plateaus and management transition zones, is presented to indicate how homebuilding firms evolve over time.

The regulatory environment within which the homebuilder operates, and often is constrained, is presented in Chapter 4. It is noted that zoning and land development regulations/ordinances are the primary factors that influence the type of subdivisions that can be built. Building codes, with the CABO One and Two Family Dwelling Code™, copyright © 1995[1] (i.e., the 1995 CABO Code) being cited as an example, are the primary regulatory influences that govern the design and construction of houses. The 1995 CABO Code is used as the reference source for a consistent set of technical requirements when the systems in a Case Study House are analyzed in Chapters 9, 10, and 12–14.

A homebuilder and/or architect must blend such technical requirements with the fundamental concepts associated with exterior and interior architectural design and the de-

[1]Used with the permission of the publisher, the International Code Council, Whittier, CA.

mands of the homebuyer during the **house planning phase**. Chapter 5 introduces **exterior architectural design** in terms of a **class**, **type**, and **style** classification system. **Interior architectural design** is introduced in terms of a **use zone** approach. The sensitivity to architectural principles which is created allows a set of working drawings for a specific house to be knowledgeably analyzed in terms of exterior appearance and interior spatial relationships.

The **design phase** of a house includes the development of a **set of contract documents** which typically includes a set of working drawings, material and workmanship specifications, and an agreement (i.e., contract) between the **homebuilder** and the **homebuyer**. Two other important elements, which later guide the decisions made throughout the **construction phase** are the **cost estimate** and the **construction schedule** which the homebuilder prepares. Chapter 6 provides an understanding about the type of information typically provided on each of the working drawings as well as the drafting practices which are used so that a set of working drawings for any house can be analyzed properly. Similar information is provided about the other contract documents, all of which are illustrated in relation to the Case Study House which is presented in Appendices A, B, and C. The illustrated cost estimate and construction schedule for the Case Study House in Chapter 6 provides a detailed introduction to these two important construction management techniques.

The residential construction process is illustrated in Chapter 7 by means of a **series of progress photos** that was taken over two three-month periods as two models of the Case Study House were built in State College, PA. Interspersed among the photos are line drawings which clarify some of the details shown in the photos. This extensively illustrated chapter provides a **transition between** the **qualitative coverage** of the housing industry in Chapters 1–6 and the **technical details** associated with each phase of house design and construction which is found in Chapters 8–14.

Although other structural framing systems are currently used in houses being built in the United States (i.e., light gage structural steel, concrete masonry, poured concrete, structural insulated panels, etc.), the predominant structural framing systems in houses still use **wood in its various forms**. Prior to presenting information about the design of the typical wood structural framing systems, it is necessary to establish a solid understanding about the important characteristics of wood, its various structural properties, the commonly used design methods, and the allowable stress levels for various species and grades of lumber. Chapter 8 provides that understanding.

The structural systems typically used in houses in the United States are described in Chapter 9 within the consistent theme and framework of the 1995 CABO Code, which was developed as a universal building code that could be adopted anywhere in the United States. A number of conventional wood structural framing systems are described and the climatic, geographical, dead load, and live load criteria related to house design are defined. The **design** of both the **foundation system** and the **wood structural floor systems** is covered in great detail and the procedures are illustrated by providing the design calculations for these systems for the Case Study House.

Wall systems and **roof–ceiling systems** are also important parts of the overall structural framing systems in a house. These are described in Chapter 10. With regard to wood wall systems, the design criteria and procedures for load-bearing and non-load-bearing walls, wood wall opening headers, and exterior wall bracing are presented within the context of the 1995 CABO Code and are illustrated for the Case Study House. Walls composed of both **clay masonry units** (i.e., bricks) and **concrete masonry units** are also discussed in terms of the industry requirements and the 1995 CABO Code. The section on wall systems concludes with a presentation of the **interior** and **exterior finish systems** that are typically used in houses.

Typical roof framing systems are then discussed and the design procedures for **ceiling joists**, **roof rafters**, and **roof sheathing** are presented within the context of the 1995 CABO Code and are illustrated for the Case Study House. The section on roof-ceiling systems concludes with a presentation of the **types of roof covering systems** that are typically used in houses. The chapter concludes with a short presentation about the design and construction of chimneys and fireplaces.

Homebuilders and building code officials who do not have either an engineering or architecture degree and are not professionally licensed are limited to using the "structural design by table approach" which is presented in the 1995 CABO Code. This approach, while often conservative, is satisfactory as long as the simplifying assumptions upon which the structural design tables were developed are not violated in the house being analyzed. Move-up and luxury homes often divide the desired living spaces in ways which violate the simplifying assumptions. Chapter 11 illustrates the types of structural design calculations which a **licensed architect or engineer** should perform when a homebuilder is asked to provide structural framing systems for those conditions.

Most homebuilders and building code officials are not as knowledgeable about the important concepts and design procedures related to the **HVAC** (i.e., Heating, Ventilating, and Air Conditioning) **systems**, **plumbing systems**, and **electrical systems** used in houses as they are about the structural systems. This is because they usually rely on the **trade contractors** who specialize in these phases to provide the expertise which is required. Chapters 12, 13, and 14 have been included in order to provide the understanding that is required so that homebuilders (and students learning about the homebuilding process) can make more knowledgeable decisions about the systems that are selected for these phases. Building code officials will also benefit from the information provided in these chapters because they will be able to more knowledgeably inspect the installation of them in houses.

Chapter 12 provides an overview of **HVAC system design**. **Energy consumption** and the basics of **energy movement** in a house are discussed briefly. The key elements in the **residential envelope** (and some of the alternatives that exist with those elements) and the **types of mechanical systems** typically used in houses are then presented. A set of calculations that determine the annual heating performance of the Case Study House is provided in order to illustrate to the student, homebuilder, or building code official how HVAC systems are designed.

Chapter 13 introduces the student, homebuilder, and building code official to the important characteristics of: (1) **the pressurized water supply and distribution systems** and (2) **the sanitary drainage systems** that are typically installed in houses. The design fundamentals that are presented relate directly to the requirements in the 1995 CABO Code. Additionally, these systems are designed for the Case Study House to illustrate the specific design procedures that are involved.

Chapter 14 provides an overview of the important characteristics of the **electrical system** that is typically installed in houses. The design fundamentals that are presented relate directly to the requirements in the 1995 CABO Code. These requirements also directly reflect the requirements in the National Electric Code®,[2] copyright © 1993. Additionally, the electrical system for the Case Study House is designed to illustrate the specific design procedures that are involved.

Jack H. Willenbrock, P.E., Ph.D.

[1]Used with permission of the publisher, The International Code Council, Whittier, CA.

[2]*National Electric Code®* and *NEC®* are registered trademarks of the National Fire Protection Association, Inc., Quincy, MA 02269.

NOTE: Copies of the 1995 CABO One and Two Family Dwelling Code can be obtained from the following model code organization members of the International Code Council Inc.:

BOCA, International, Inc.	ICBO	SBCCI, Inc.
4051 West Flossmoor Road	5360 Workman Mill Road	900 Montclair Road
Country Club Hills, IL 60458-5794	Whittier, CA 90601-2298	Birmingham, AL 35213
(708)799-2300	(800)284-4406	(205)591-1853

Acknowledgments

A program in Residential Building Construction was established in 1980 at The Pennsylvania State University under the auspices of the Departments of Architectural Engineering and Civil and Environmental Engineering. From the beginning, a key component of the program was the two-course sequence of senior undergraduate/graduate residential courses entitled:

CE 433   Residential Subdivision Design and Construction
CE 470   Residential Building Design and Construction

This book represents the final stage of evolution which started with the first set of course notes for AE 470 in 1984. Over the years, many people have contributed to the material which appears in this book. Early assistance with the development of the architectural and mechanical sections of the book, as well as the administration of the AE 470 course, was received from Thomas B. Brown, A.I.A., a former faculty member in the Department of Architectural Engineering who is now retired. His role in later years was filled by Peter Magyar, A.I.A., who is a faculty member in the Department of Architecture.

The HVAC System Design chapter has received contributions over the years from Moses D. F. Ling, P.E., and Grenville K. Yuill, Ph.D., both of whom are faculty members in the Department of Architectural Engineering, and Joseph Staib, a former graduate student in the residential building construction program. The final presentation of the chapter is due to the cooperative work of co-author Michael G. Suchar and H. Edward Carr, Jr., the President of Comfort Home Corporation of Lancaster, PA. The Plumbing System Design chapter received guidance and support from Daniel L. Mattern, P.E., and William P. Bahnfleth, Ph.D., both of whom are faculty members in the Department of Architectural Engineering. Earlier versions of the Electrical System Design chapter were written by Charles N. Claar, P.E., while he served as the Director of the Facilities Management Program at Penn State. The final version of the chapter resulted because of the contributions of Ronald P. Dodson, Jr., P.E., who is a staff electrical engineer with that program.

Several other people who made contributions to this book should also be acknowledged. My co-author, Harvey B. Manbeck, Ph.D., P.E., wrote Chapters 8 and 11 of the book which deal with the properties and strength of wood and the structural analysis of the wood framing system in a house, respectively. The major contribution of my other co-author, Michael G. Suchar, occurred while he was a graduate student in the residential building construction program at Penn State from 1991 to 1993. As a staff engineer with the Housing Research Center at Penn State, he took a loosely organized set of course notes and supportive material for the AE 470 course and converted them into a set of organized course notes which was the immediate predecessor of this book.

Harry J. Burd, the Director of the Centre Region Code Administration Office in State College, PA and an active member of BOCA at the national level, provided valuable guidance and supportive text related to the regulatory environment which interacts with the homebuilding industry. Gopal Ahluwalia, an Assistant Staff Vice President of the Economics, Mortgage Finance, and Housing Policy Division of the National Association of Home Builders, served as a contact with that organization and provided extensive statistical information related to the homebuilding industry. Walter A. Music, currently a Vice President of Toll Brothers Inc. of Huntingdon Valley, PA, who was the first graduate student in the residential building construction program at Penn State, and his successor, Hector A. Dasso, provided the initial research effort which led to the development of the Management Growth Model which is featured in Chapter 3 of this book.

A special word of acknowledgment and thanks is due to Robert E. Poole, the CEO of S&A Custom Built Homes, Inc. of State College, PA for his long-term support and cooperation with the residential building construction program at Penn State. His willingness to share information about his firm and to allow one of his firm's house models to be used as the Case Study House in this book is also very much appreciated.

Several other students, while completing their graduate programs, were associated with the Housing Research Center at Penn State and contributed in various ways to the development of this book. Matt Syal, Ph.D., the first Assistant Director of Operations of the HRC at Penn State and currently an Associate Professor and Director of the Housing Education and Research Center at Michigan State University, also directed a number of housing research projects that provided background material for several chapters. Charles McIntyre, Ph.D., who served as the Assistant Director for Infrastructure Systems and is currently an Assistant Professor in the Department of Civil Engineering and Construction at North Dakota University, provided the initial research effort upon which the Land Development Regulatory Environment section of Chapter 4 and the Design Phase section of Chapter 6 are based. Steven B. Taylor, Ph.D., who also served as an Assistant Director of Operations of the HRC and is currently a Project Leader/Systems Development at the Corporate Research Center of Trus Joist Macmillan, provided valuable encouragement and contributed his knowledge in wood engineering to the development of Chapters 8 and 11 of this book.

This book would not have reached its final state without the continued and professional editorial and typing support that was provided by Ms. Debra Putt. Finally, I would like to thank my wife, Marsha, for her continued encouragement and extreme patience during this entire endeavor. Hopefully the effort put forth and the knowledge contained in this book will guarantee that the families of Peter and Stephanie Willenbrock and their sons Pierce and Jack, Philip and Karen Willenbrock and their children Kelsea and Jacob and our daughter Gretchen will live in quality-built homes during their lives.

1

An Overview of the Homebuilding Industry

INTRODUCTION

The residential building construction industry is highly diversified and offers many challenging career opportunities. It has a multitude of participants which include land developers, engineers/surveyors, architects, Realtors, residential home builders, subcontractors, material and product suppliers, financial institutions, government agencies, and others.

There are homebuilders of all sizes and levels of expertise working in this industry. They range from large multimillion dollar firms, which construct thousands of housing units per year, to homebuilders who construct only a few homes a year. Homebuilders are involved in the construction of single-family residences, townhouses, metropolitan condominiums, luxury apartments, and low-income housing on rural, suburban, and urban sites. In essence, residential building construction is a dynamic and fragmented industry.

Because of its size, the residential building construction industry plays a leading role in the nation's economy, typically accounting for about 3.5% of the GDP[1] (Gross Domestic Product) in the United States each year. Investment in the construction of residential buildings typically represents over 40% of the value of new construction put in place each year.

In order to properly understand the housing design and construction process, there must be an understanding of the context in which that process operates. The beginning of that understanding is provided in this chapter in the form of background information about the homebuilding industry and a historical overview of the housing supply part of the industry's basic supply/demand economic model.

Homebuilder/Homebuyer Interface

A ride through any urban neighborhood or suburban subdivision will indicate the diversity of solutions which homebuilders have found in the past and are continuing to find to meet the needs and desires of homebuyers. Homebuilders, who represent the supply side of the homebuilder/homebuyer interface, can be successful only if they are responsive to the demand side of the interface provided by the homebuyer. Chapters 1 and 2 of this book provide some insight into this supply/demand relationship. Such an understanding places the basic principles related to house design, presented in Chapter 5, in a proper context.

Management Considerations

In the 25 years following World War II, homebuilders did not have to face the rapidly changing social, political, and economic forces that now influence their industry. Today's homebuilders, in addition to being proficient in the construction of the residential units, must also be proficient in financing, marketing, cost control, and scheduling, and they must have the ability to adapt to a rapidly changing construction environment. If the homebuilders of today wish to maintain productive growth-oriented organizations, they can no longer regard themselves as just technically proficient, hammer-in-hand builders; they must view themselves as construction managers who understand the design and construction process and can manage effectively within it.

Unfortunately, many small-to-medium size homebuilders who possess the required technical skills are not aware of the types of management skills they must possess in order to both succeed and, more importantly, to survive in a very cyclical industry. Chapter 3 provides a basic understanding of how the homebuilding process can be managed. The framework for that presentation is a Management Growth Model, which was developed by the Housing Research Center at Penn State[2] to assist homebuilders in improving their managerial performance.

Technological Influences

The last half-century has witnessed some major technological changes in the way that subdivisions are designed and constructed and in the way that houses are built. These changes have allowed homebuilders to provide cost-effective and energy-efficient houses which can be completed in a shorter time period than the houses that were built 10 or 20 years ago.

In concert with these technological changes has been the evolution of the regulations which govern land development and the building codes which establish the minimum standards of design and performance of the houses that are built in the various sections of the United States. Accordingly, Chapter 4 is included in this book to provide some insight into the role of the regulatory process in providing a safe and comfortable living environment.

The Residential Construction Process

The homebuilder's primary role in the residential construction process is to transform the set of house plans and contract documents which describe a house into a finished product which meets the needs of the homebuyer. The development of a set of house plans and contract documents is discussed in Chapter 6. Chapter 7 provides an overview of the entire residential construction process, which begins with site preparation and ends with the interior finishing stage.

Chapters 8 through 14 deal with the various structural systems, HVAC systems, plumbing systems, and electrical systems typically found in a house. They are presented within the context of the 1995 *CABO One and Two Family Dwelling Code*[3] which was developed by the three nationally recognized model codes to specifically apply to residential construction in the United States. Other building codes are also utilized as required.

THE CONTRIBUTION OF THE HOMEBUILDING INDUSTRY TO THE TOTAL U.S. CONSTRUCTION INDUSTRY

The impact which the homebuilding industry has on the overall economy of the United States can best be understood by first establishing the level of contribution which the residential construction industry makes to the total construction industry in the United States. Data published by both the Bureau of the Census of the U.S. Department of Commerce[4] and the National Association of Home Builders (NAHB)[5] can be used to determine that level.

Value of New Construction Put in Place

One of the widely accepted measures that can be used to compare the size of the home-building industry to the size of the total construction industry is the value of the *new construction put in place* over a period of time or for a particular year.

Overall Perspective

Table 1.1 provides such information about both private as well as public construction for 1994 in terms of 1987 dollars. (NOTE: The year 1987 is frequently used as a standard reference year whenever it is desirous to compare data from various years.)

TABLE 1.1   Subcategory Definition of New Construction Put in Place in 1994 (billions of 1987 dollars)

Total New Construction Put in Place		$T = \$415.7$[a]
Private Construction		$307.1 (74% of T)
Private Residential Buildings		$192.2 (46% of T)
Single-family	$124.1	
Multifamily	11.1	
Improvements	56.9	
(additions and alterations)		
Private Non-Residential Buildings		$ 80.4 (20% of T)
Industrial	$ 17.7	
Office	13.8	
Hotels and motels	3.4	
Other commercial	25.9	
Religious	3.0	
Educational	3.7	
Hospitals, other institutional	8.6	
Misc. non-residential bldgs.[b]	4.3	
Private Other Construction		$ 34.5 (8% of T)
Farm	N/A	
Public utilities[c]	N/A	
All other private[d]	2.5	
Public Construction		$108.6 (26% of T)
Public Construction Buildings		$ 42.7 (10% of T)
Housing and redevelopment	$ 3.7	
Industrial	1.2	
Educational	18.7	
Hospital	3.1	
Other[e]	15.9	
Public Construction—Other Structures		$ 65.9 (16% of T)
Highways and streets	$ 35.3	
Military facilities	2.0	
Conservation and development	5.1	
Sewer systems	9.0	
Water supply facilities	5.1	
Miscellaneous[f]	9.4	

[a]Conversion factor: $415.7 billion (1987) = $506.9 billion (1994).
[b]Includes amusement and recreational buildings, bus and airline terminals, animal hospitals and shelters, etc.
[c]Includes telecommunications, railroads, electric light and power, gas, petroleum pipelines
[d]Includes privately owned streets and bridges, parking areas, sewer and water facilities, parks and playgrounds, golf courses, airfields, etc.
[e]Includes general administrative buildings, prisons, police and fire stations, courthouses, passenger terminals, civic centers, postal facilities, etc.
[f]Includes open amusement and recreational facilities, power generating facilities, transit systems, airfields, open parking facilities, etc.

Source: *Construction Review, Quarterly Industry Report,* vol. 41, no. 1, pp. 3–6 (Washington, D.C.: U.S. Department of Commerce, Bureau of the Census, Winter 1995).

4   An Overview of the Homebuilding Industry

Table 1.1 clearly indicates that more privately funded construction than publicly funded construction was completed in the United States in 1994 (74% versus 26%). Construction of private residential buildings accounted for about 46% of all construction put in place in 1994. Within that category, it is evident that *single-family residential buildings* are the largest component of the private residential buildings sector (124.1/192.2 = 65%).

The relative size of the improvements (i.e., remodeling) subcategory of private residential buildings should also be noted; it also is clearly a significant force in the economy.

Historical Record

Table 1.2 expands the data base for the private residential building sector over a period of time. As noted, new single-family housing has been the largest component of the private residential building sector and has risen during the period indicated to a level in 1994 of approximately 65%. Although the percentage devoted to improvements has remained fairly constant at about 30%, there has been a sharp drop in multifamily housing in the period indicated.

Additionally, the construction of private residential buildings has consistently accounted for 39–46% of the total new construction put in place and between 3% and 4% of the GDP of the United States over the period indicated.

TABLE 1.2   Historical Record of Private Residential Construction Put in Place (billions of 1987 dollars)

	1980	1985	1990	1991	1992	1993	1994
Private Residential Buildings (PRB)	128.9	172.3	164.0	141.3	165.1	177.0	192.1
New Single-Family Housing	67.9 52.7%	93.7 54.4%	97.5 59.5%	85.4 60.5%	103.4 62.0%	112.1 63.3%	124.1 64.6%
New Multifamily Housing	21.5 16.7%	21.0 18.0%	17.3 10.5%	13.6 9.6%	11.5 7.0%	9.1 5.1%	11.1 5.8%
Improvements	39.5 30.6%	47.6 27.6%	49.2 30.0%	42.3 30.0%	51.2 31.0%	55.9 31.6%	56.9 29.6%

1. Relationship to Total New Construction Put in Place

	1980	1985	1990	1991	1992	1993	1994
Total New Construction (TNC) Put in Place	328.4	401.9	397.7	360.9	385.8	398.1	415.7
PRB/TNC	39.3%	42.9%	41.2%	39.2%	42.8%	44.5%	46.2%

2. Relationship to the Gross Domestic Product of the United States

	1980	1985	1990	1991	1992	1993	1994
Gross Domestic Product (GDP)	3776.3	4279.8	4897.3	4867.6	4979.3	5134.5	5342.4
PRB/GDP	3.4%	4.0%	3.3%	2.9%	3.3%	3.4%	3.6%

Source: *Housing Market Statistics,* pp. 41 and 48 (Washington, D.C.: National Association of Home Builders, April 1995).

Construction Firm Information

The level of contribution which the residential construction industry makes to the total construction industry in the United States can also be established by examining the construction firms themselves in terms of the number of firms in each category of construction, the number of construction employees involved, etc.

Such information provides a background for a more specific examination of the firms that specialize in the residential homebuilding sector of the construction industry.

Standard Industrial Classification System

The Bureau of the Census, in order to establish a consistent framework for the analysis of all industrial sectors of the U.S. economy, uses its own Standard Industrial Classification (SIC) system. The specific part of that system which applies to the construction industry reports[6] is shown in Table 1.3.

The three major classification categories for the construction industry are: (1) SIC Code 15–Building Construction—General Contractors and Operative Builders; (2) SIC Code 16–Heavy Construction Other Than Building Construction Contractors; and (3) SIC Code 17–Construction—Special Trade Contractors.

Table 1.3 indicates that general building contractors who specialize in residential buildings (SIC Code 152) are further divided into those that focus on either single-family houses (SIC Code 1521) or multi-family buildings (SIC Code 1522). Final reports corresponding to those SIC Codes for the 1992 Census were published in June and April 1995, respectively.[7] Operative builders (SIC Code 1531) are defined by the Bureau of the Census as builders who build on their own account for sale to others. (Investment builders, who build structures on their own account for rental purposes, are classified by the Bureau of the Census as being a part of the Real Estate industry.) *It is important to note that it is common practice to include SIC Codes 1521, 1522, and 1531 as part of the homebuilding industry.*

General building contractors who specialize in the construction of nonresidential buildings are designated by SIC Code 154, which is further divided into SIC Codes 1541 and 1542. Heavy construction general contractors, who are a part of SIC Code 16, are involved primarily in the construction of highways, bridges, and pipelines, as well as power and petro-chemical plants.

The other broad classification which interfaces with both the residential as well as nonresidential construction industries is the **special trade contractor** (SIC Code 17) group.

Historical Perspective

Table 1.4 provides a historical perspective of the number of construction companies in the various SIC Code categories described. Several interesting points should be noted:

1. With regard to the distribution of construction firms in the United States, the highest percentage (approximately 63%) over the period indicated were classified as special trade contractors. Nonresidential building contractor firms and heavy construction firms each represented about 6.4% of the total number of construction firms.

2. The cyclical nature of the homebuilding industry is indicated when the number of homebuilding firms operating in the industry over a period of time is examined. The relative peak years of 1977 and 1992, with 129,245 and 130,827 homebuilding firms, respectively, bracket a trough which reached as low as 93,632 firms in 1982. This reflects the fact that in the early 1980s, a number of firms left the general contracting segment of the industry due to the recessionary period.

3. The percentage of firms in the total construction industry who specialize as general contractors in the homebuilding industry has decreased from its relative high of 26.9%

TABLE 1.3   Standard Industrial Classification System Titles
for the Construction Industry

SIC Code	Industry Titles
15	**BUILDING CONSTRUCTION—GENERAL CONTRACTORS AND OPERATIVE BUILDERS**
152	**General Building Contractors—Residential Buildings**
1521	General Contractors—Single-Family Houses
1522	General Contractors—Residential Buildings, other than Single Family
153	**Operative Builders**
1531	Operative Builders
154	**General Building Contractors—Nonresidential Buildings**
1541	General Contractors—Industrial Buildings and Warehouses
1542	General Contractors—Non-residential Buildings, other than Industrial Buildings and Warehouses
16	**HEAVY CONSTRUCTION OTHER THAN BUILDING CONSTRUCTION—CONTRACTORS**
161	**Highway and Street Construction, Except Elevated Highways**
1611	Highway and Street Construction Contractors, Except Elevated Highways
162	**Heavy Construction, Except Highway and Street Construction**
1622	Bridge, Tunnel, and Elevated Highway Construction Contractors
1623	Water, Sewer, Pipeline, and Communications and Power Line Construction Contractors
1629	Heavy Construction Contractors, Not Elsewhere Classified
17	**CONSTRUCTION—SPECIAL TRADE CONTRACTORS**
171	**Plumbing, Heating, and Air-Conditioning Special Trade Contractors**
1711	Plumbing, Heating, and Air-Conditioning Special Trade Contractors
172	**Painting and Paper Hanging Special Trade Contractors**
1721	Painting and Paper Hanging Special Trade Contractors
173	**Electrical Work Special Trade Contractors**
1731	Electrical Work Special Trade Contractors
174	**Masonry, Stone Work, Tile Setting, and Plastering Special Trade Contractors**
1741	Masonry, Stone Setting, and Other Stone Work Special Trade Contractors
1742	Plastering, Drywall, Acoustical, and Insulation Work Special Trade Contractors
1743	Terrazzo, Tile, Marble, and Mosaic Work Special Trade Contractors
175	**Carpentry and Floor Work Special Trade Contractors**
1751	Carpentry Work Special Trade Contractors
1752	Floor Laying and Other Floor Work Special Trade Contractors, Not Elsewhere Classified
176	**Roofing, Siding, and Sheet Metal Work Special Trade Contractors**
1761	Roofing, Siding, and Sheet Metal Work Special Trade Contractors
177	**Concrete Work Special Trade Contractors**
1771	Concrete Work Special Trade Contractors
178	**Water Well Drilling Special Trade Contractors**
1781	Water Well Drilling Special Trade Contractors
179	**Miscellaneous Special Trade Contractors**
1791	Structural Steel Erection Special Trade Contractors
1793	Glass and Glazing Work Special Trade Contractors
1794	Excavation Work Special Trade Contractors
1795	Wrecking and Demolition Work Special Trade Contractors
1796	Installation or Erection of Building Equipment, Special Trade Contractors, Not Elsewhere Classified
1799	Special Trade Contractors, Not Elsewhere Classified

Source: 1992 Census of Construction Industries, General Contractors—Single-Family Houses: Industry 1521, CC92-I-1, p. B-1 (Washington, D.C.: U.S. Department of Commerce, Bureau of the Census, June 1995).

in 1977 to a level of 22.8% in 1992. At the same time, the percentage of special trade contractors has increased from its 1977 level of 59.9% to a level of 64.2% in 1992. In terms of the number of special trade contractor firms, there were 80,000 more in 1992 than there were in 1977.

4. Although not all of the special trade contractors work in the homebuilding industry, these statistics probably reflect the increasing dependence which homebuilders

TABLE 1.4   Historical Record of Number of Construction Industry Establishments with an Employee on the Payroll

SIC Code	Industry	Year[a]			
		1977	1982	1987	1992
	Residential (Total)	129,245	93,632	119,287	130,827
	Percent of Grand Total	26.9%	20.5%	21.9%	22.8%
1521	SF General Contractors	100,993	72,061	90,378	107,284
1522	MF General Contractors	4,775	7,570	8,140	6,500
1531	Merchant (Operative) Bldr.	23,447	14,001	20,769	17,043
	Nonresidential (Total)	26,726	29,547	38,351	37,337
	Percent of Grand Total	5.6%	6.5%	7.0%	6.5%
1541	Industrial Bldgs. & Warehouses	8,259	7,435	7,014	7,727
1542	Other Non-residential[b]	18,467	22,112	31,337	29,610
16	**Heavy Construction**	31,296	28,187	36,599	37,348
	Percent of Grand Total	6.5%	6.2%	6.7%	6.5%
	Special Trade Contractors (Total)	287,674	299,410	342,023	367,765
	Percent of Grand Total	59.9%	65.6%	62.8%	64.2%
171	Plumbing, HVAC	56,435	60,243	69,556	75,364
172	Painting, Paper Hanging, Decor.	27,369	24,779	29,867	32,114
173	Electrical Work	36,764	39,563	49,436	53,992
174	Masonry, Plastering, Tile Set.	45,451	40,460	46,182	47,698
175	Carpentry, Flooring	33,357	37,438	44,183	48,267
176	Roofing, Sheet Metal	20,577	21,152	25,673	27,695
177	Concrete Work	16,974	19,986	23,422	26,141
178	Water Well Drilling	4,305	3,551	3,414	3,730
179	Miscellaneous Special Trade	46,442	52,238	50,290	52,764
6552	**Subdividers and Developers**	5,078	5,925	7,995	NA
	Percent of Grand Total	1.1%	1.3%	1.5%	—
	GRAND TOTAL	480,019	456,702	544,215	573,277

[a]Dates of five-year Periodic Census of Construction Industries.
[b]These are general building contractors of nonresidential buildings of which the most important construction activities are: (1) office and bank buildings, (2) stores, restaurants, public garages, and automotive services.

Source: *Five-Year Periodic Census of Construction Industries.* Compilation of data provided by the Economic, Mortgage Finance, and Housing Policy Division of NAHB.

have placed on subcontractors over the last 15 years. A clear shift has occurred in the homebuilding industry because many homebuilders no longer complete the major portion of the construction of the house with their own forces; instead, they view themselves as **construction managers** of a large number of subcontractors. This phenomenon has created a major shift in the managerial aspects of homebuilding.

5. The number of homebuilding firms that specialized in multifamily housing is considerably smaller than the number that built single-family houses over the period of time examined.

Focus on Remodeling

A specific focus on the homebuilding sector of the construction industry in 1992 is provided in Table 1.5 in terms of the number of homebuilders who either are involved in only new construction, or remodeling, or both.

In 1992 the largest percentage of homebuilders (82%) focused their attention on the construction of single-family homes, while only 5% focused on just multifamily housing. Because of the definition of Operative Builders, it is difficult to determine how they are focused.

TABLE 1.5   Builder Establishments by Type of Construction—1992

| | Type of Construction | | | | |
	New Construction Only	Remodeling Only	Both New & Remodeling	Total	% of T
General Contractor Single-Family (SIC 1521)	22,490	39,626	44,981	107,097	82%
General Contractor Multifamily (SIC 1522)	1,169	3,442	1,883	6,494	5%
Operative Builders (SIC 1531)	12,614	251	4,183	17,048	13%
All Residential	36,273	43,319	51,047	$T = 130,639$	100%
Percentage of T (Grand Total)	28%	33%	39%	100%	

Source: Gopal Ahulwalia, "A Decade of Remodeling," *Housing Economics—The National Association of Home Builders,* May 1995, p. 12.

Additionally, it should be noted that the percentage of firms that focus their primary attention on remodeling (33%) exceeds the percentage that focus on only new construction. The largest percentage of firms (39%) provide a dual focus on both new construction and remodeling.

THE SUPPLY/DEMAND MODEL

Additional insight about the homebuilding industry can be obtained by: (1) examining the diversity of products which this industry has provided, and continues to provide, to homebuyers, and (2) defining, in some organized fashion, the stated needs and desires of these homebuyers.

The supply/demand model which is presented by the National Association of Homebuilders (NAHB) in its report entitled *The Future of Homebuilding: 1992–1994 and Beyond*[8] provides a good framework for such an analysis. The fundamental assumption upon which that model was based is that:

$$\frac{\text{The Supply of}}{\text{New Housing Units}} = \frac{\text{The Demand for}}{\text{New Housing Units}}$$

Figure 1.1 illustrates the variables which appear on each side of the equation.

Supply-Side Orientation

The homebuilding industry, in order to remain economically viable, must continue to supply the correct number of housing units that society needs each year. This **Housing Supply** component, as described in the remainder of this chapter, is met from a strictly economic

Figure 1.1
NAHB
Supply/Demand
Model

Source: The Future of Homebuilding: 1992–1994 and Beyond, (Washington, DC: National Association of Home Builders, 1992), p. 8.

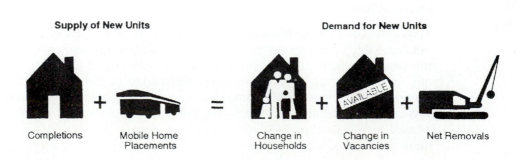

Supply of New Units   Demand for New Units

Completions   Mobile Home Placements   Change in Households   Change in Vacancies   Net Removals

point of view by the combination of single-family, multifamily, and mobile homes that are built each year. From a demographic and business point of view, however, it is equally important to provide the correct mix of housing types and styles in order to meet the different needs of each age group within the population.

Demand-Side Orientation

One of the basic assumptions that is made in the supply/demand model is that demographic forces play a major role in establishing both the level, as well as the nature, of the **Housing Demand** Component. Another assumption is that most of the new home demand can be accounted for by the number of households that are formed each year. In those periods when household formation is high, there is also a greater demand for new housing.

The vacancy variable on the demand side of the equation serves as a short-term cushion during times when the industry's production rate of new houses is out of balance with the homebuyer's demand rate. In an overzealous building environment, for instance, the unsold inventory of houses typically increases until a level is reached which dampens additional speculative building activity. A number of additional factors also influence the vacancy variable.

The final variable on the right side of the equation is net removals, which represents the net effect of the changes in the existing number of housing units due to intentional demolition of dilapidated structures, the effects of natural disasters, and various use–conversion activities.

The focus on the supply side of the equation in this chapter provides a broad historical perspective of the output of the homebuilding industry in the last 20 years. Some of the trends which have developed are documented. The demand-side emphasis in Chapter 2 analyzes the expressed housing needs and desires of various age groups.

HOUSING SUPPLY

Overview of Completed Privately Owned Housing

Information about the supply of housing in the United States can be obtained from the U.S. Bureau of the Census, Series C-25 Annual Report, entitled *Characteristics of New Housing*.[9] Table 1.6 provides data from that report on the number of privately owned housing units that were completed in 1990 and 1993.

It should be noted that information about one-family housing is provided in terms of its two primary components, (1) housing units built for sale and (2) housing units for owner occupancy, and a minor component which represents one-family rental units. Multifamily housing information is divided into "for sale" and "for rent" components, with the latter being the larger of the two.

One-Family versus Multifamily Housing Units

Table 1.6 clearly indicates that the majority of the privately owned houses that are built in the United States are one-family units. In 1993, this category accounted for 87% of all of the houses that were built. This is a considerable increase from 1990, when only 74% were one-family units because the multifamily sector was much stronger. This large increase occurred at the expense of multifamily housing, which dropped from 26% of the total in 1990 to 13% in 1993.

Both townhouses and apartment houses are usually classified as multifamily structures. It is important to note, however, that a segment of the townhouse sector is classi-

TABLE 1.6   Completed Privately Owned Housing Units

	1990		1993	
	No. of Units	% of Grand Total	No. of Units	% of Grand Total
One-Family Completed Housing Units[a] (in 000s)				
Housing units built for sale	594	46%	642	54%
Housing units for owner				
occupancy on owner's land				
Contractor built[b]	199	15%	216	18%
Owner built[c]	147	11%	159	13%
Housing units built for rent	26	2%	23	2%
Total one-family housing units	966[d]	74%	1039[e]	87%
Multifamily Completed Housing Units				
Housing units built for sale	76	6%	44	4%
Housing units built for rent	266	20%	109	9%
Total multifamily housing units	342	26%	153	13%
Grand Total, Privately Owned Housing Units Completed	1,308	100%	1,193	100%

[a]Includes one-family detached houses and townhouses attached together, in groups of two or more, which are partitioned from each other by a ground-to-roof wall.
[b]Contractor-built: houses built for owner occupancy, on owner's land, with construction under the supervision of a single general contractor.
[c]Owner-built: houses built for owner occupancy, on owner's land, under the supervision of the owner acting as his or her own general contractor.
[d]Grand total of one-family houses sold in 1990 was 534,000.
[e]Grand total of one-family houses sold in 1993 was 666,000.

Source: *Characteristics of New Housing: 1993*, C25/93A, p. 3 (Washington, D.C.: U.S. Department of Commerce, Bureau of the Census).

fied by the Bureau of the Census as one-family houses. As noted: "In addition to single-family detached houses, the definition of single-family houses also includes those townhouses attached together in groups of two or more which are partitioned from each other by a ground-to-roof wall. Also, these units must not share heating/air-conditioning systems or interstructural public utilities, such as water supply, power supply, or sewage disposal lines."[10]

In 1993, 2% of the one-family completed housing units were "built for rent" (Table 1.6). Attached townhouses, fitting this definition, represent a large share of those units.

One-Family for Sale versus Custom Homes

The two largest categories of one-family housing units in Table 1.6 are: (1) housing units for sale and (2) housing units for owner occupancy on owner's land. This is an important division because it separates the housing units that are built for sale by homebuilders to the general public from **custom-built** houses that are built directly for owner occupancy on the owner's land. As noted, the total percentage of one-family custom-built houses was 26% in 1990 and 31% in 1993.

Unsold Inventory

Footnotes (d) and (e) of Table 1.6 indicate an important fact about the homebuilding industry, namely, that not all houses completed in a given year are sold. In 1990, for instance, 966,000 one-family housing units were built, but only 534,000 (i.e., 55%) were sold. This percentage was 69% in 1993. This comparison indicates a problem for the homebuilding industry that directly relates to the supply/demand model shown in Figure

1.1. One of the forces that clearly affects the health of the industry is the level of unsold inventory of new houses that exists at any particular point in time.

One-Family Completed Housing Units

The basic data provided by the U.S. Bureau of the Census in its C-25 report series has routinely been compiled and analyzed by the Department of Economics and Housing Policy of NAHB for a number of years. The results of this analysis are published in NAHB's monthly *Housing Economics* reports. Table 1.7, which was developed to provide a 20-year perspective about the "one-family completed housing units" portion of Table 1.6, and other tables presented in this chapter rely heavily on the previously noted sources for their data. One of the first things to note in Table 1.7 is the extreme variability in the yearly number of houses completed over the 20-year period—clear evidence of the cyclical nature of the homebuilding industry in the United States. Some of the changes in the physical characteristics of houses that have occurred are discussed next.

TABLE 1.7   Characteristics of Completed New One-Family Housing: 1974–1994

	1974	1978	1981	1986	1990	1993	1994
One-Family Completions							
(000s)	940	1369	819	1120	966	1039	1157
Type of House							
1 story	65%	61%	61%	51%	46%	48%	49%
2 stories or more[a]	25	28	32	44	49	48	47
Split level or other	10	11	7	5	5	4	4
Square Feet of Floor Area							
Average	1,695 sf	1,755 sf	1,720 sf	1,825 sf	2,080 sf	2,095 sf	2,100 sf
Median	1,560	1,655	1,550	1,660	1,905	1,945	1,940
Square Feet of Floor Area							
Under 1,200 sf	24%	18%	24%	17%	11%	9%	9%
1,200–1,500 sf	↑	28	29	30	22	21	21
1,600–1,999 sf	63	24	20	21	22	23	24
2,000–2,399 sf	↓	16	12	14	17	18	18
2,400–2,999 sf	13	14	15	11	16	16	15
3,000 sf or more	↓	↓	↓	7	13	13	13
Lot Size							
Average	NA	18,760 sf	NA	15,840 sf	14,680 sf	13,440 sf	NA
Median[b]	NA	9,790	8,650 sf	8,960	10,000	9,600	NA
Foundation							
Full or partial basement	45%	42%	33%	37%	38%	40%	39%
Slab	36	40	47	45	40	40	41
Crawl space or other	19	18	20	18	22	20	20
Garage							
No carport or garage	22%	25%	25%	21%	16%	14%	13%
Carport or one-car garage	16	23	22	19	12	9	9
Two- or more-car garage	52	52	53	60	72	77	78
Exterior Wall Material							
Brick	35%	30%	29%	20%	18%	21%	21%
Wood	NA	40	44	43	39	31	27
Aluminum	NA	11	8	9	5	5	4
Stucco	NA	13	11	13	18	14	15
Vinyl	NA	NA	NA	NA	NA	25	28
Other[c]	NA	7	8	14	20	4	5

NA = data not available

[a]Includes a small number of houses with $1\frac{1}{2}$, $2\frac{1}{2}$, or 3 stories.

[b]Denotes number of new homes sold; not available for homes built on owner's lot.

[c]Includes cinderblock, stone, and other types; data prior to 1992 include vinyl siding.

Source: *Characteristics of New Housing: 1993*, C25/93A (Washington, D.C.: U.S. Department of Commerce, Bureau of the Census); and information provided by the Economics, Mortgage Finance, and Housing Policy Division of NAHB.

Type of House

Over the last 20 years there has been a clear shift of preference from one-story to two-story houses. The 65%/25% split which favored the one-story house in 1974 had become approximately equal by 1994. During the same period, the popularity of split-level and other types of houses fell from 10% to 4%.

Floor Area Analysis No. 1

A comparison of the houses that were built in 1974 and 1994, in terms of both the average and median measure, clearly indicates the increase in floor area that has occurred in the last 20 years. Both of these measures have added approximately 25%, or 400 sq ft, to the living space of houses over the last 20 years.

Lot Size

While the square footage of houses was expanding, the size of the lot upon which they were being built was decreasing. This probably was due to the steadily increasing cost of land over the 20-year period. The net result of accommodating an increased number of bedrooms, bathrooms, etc., on a lot of equal or smaller size probably is one of the primary driving forces behind the shift to two-story houses.

Floor Area Analysis No. 2

Table 1.7 also illustrates the changes that have occurred with regard to the various floor size categories. The share of new houses with 2,400 sq ft or more was fairly constant at a level of about 14% until the early 1980s, when it began to rise fairly rapidly to a level of about 28% in 1994. The proportion of new houses of 3000 sq ft or more, a category that was introduced into the database in 1986, has almost doubled, from 7% to a level of 13% in 1994. It should be noted that during the same period, the market share of smaller-size houses of less than 1,200 sq ft, the type that typically is purchased by first-time homebuyers, was reduced from 24% in 1974 to 9% in 1994. This is probably because many first-time homebuyers no longer felt that houses of less than 1,200 sq ft met their needs.

Cost Considerations

There is, of course, a direct correlation between the floor area provided in a house and its cost. An appreciation for the order of magnitude of the sales price of houses and the related price/sq ft (with the value of the improved lot deducted) for the period between 1989 and 1993 is presented in Tables 1.8 and 1.9, respectively. As indicated, this period was one of relative stability in terms of both of these cost factors.

TABLE 1.8   Price Distribution of Sold Houses

	1989	1990	1991	1992	1993
Number of Houses Sold (000)	650	534	509	610	666
Under $70,000[a]	13%	12%	11%	9%	7%
$70,000–$79,999	8	9	9	7	6
$80,000–$99,999	16	16	17	19	17
$100,000–$119,999	12	12	13	13	14
$120,000–$149,999	16	17	17	18	20
$150,000–$199,999	15	16	16	16	18
$200,000–$249,999	8	8	7	8	8
$250,000+	12	12	11	10	9
Average Sales Price	$148,800	$149,800	$147,200	$144,100	$147,700
Median Sales Price	$120,000	$122,900	$120,000	$121,500	$126,500

[a]The sales price is the price agreed upon between the purchaser and seller at the time the first sales contract is signed or a deposit is made for the house.

Source: *Characteristics of New Housing: 1993*, C25/93A, p. 40 (Washington, D.C.: U.S. Department of Commerce, Bureau of the Census).

TABLE 1.9   Price/Square Foot Distribution of Sold Houses

	1989	1990	1991	1992	1993
Number of Houses Sold (000)	650	534	509	610	666
Under $35/SF[a]	14%	13%	11%	10%	6%
$35.00–$39.99	13	14	12	10	8
$40.00–$44.99	16	15	15	13	13
$45.00–$49.99	13	14	14	15	14
$50.00–$54.99	11	11	13	14	13
$55.00–$59.99	8	9	9	11	13
$60.00–$64.99	5	6	8	7	10
$65.00–$69.99	4	5	5	6	7
$70.00+	14	13	12	14	15
Average Price per SF	$53.25	$53.05	$53.05	$53.85	$55.95
Median Price per SF	$47.20	$47.85	$49.15	$50.35	$53.20

[a]Price per square foot of floor area excludes the value of the improved lot.
Source: *Characteristics of New Housing: 1993*, C25/93A, p. 43 (Washington, D.C.: U.S. Department of Commerce, Bureau of the Census).

Type of Foundation

One of the primary structural elements of a house is the foundation. Both time and cost of construction, as well as the ease with which living space can be expanded, are influenced by this factor. As indicated in Table 1.7, in the last 20 years there has been a stand-off between the full or partial basement and the slab-on-grade approach, with each accounting for approximately 40% of the completed housing on a national scale. It is important to note that regional differences must be considered with regard to this factor. In the Northeast, for instance, the majority of homebuyers "expect" houses to have basements, while homebuyers in certain sections of the South recognize that expansive soil conditions preclude the use of basements.

Garage

The overall layout of a house is influenced by the number and size of the spaces that are included in the garage. As indicated in Table 1.7, there appears to have been a consistent shift over the last 20 years from the basic one-car garage or carport option to two- or more-car garages (from 52% of the houses completed in 1974 to 78% in 1994). It is believed that one of the reasons for the popularity of larger garages is the increased share of women in the workforce, and hence the prevalence of two- (and often three-) car families.

Type of Exterior Wall Material

As noted in Table 1.7, the most common types of exterior wall materials in 1978 (limited data was available for 1974) were brick and wood and wood products. Both of these, along with aluminum, have declined in popularity since then, most likely in favor of vinyl siding, which had a 28% market share by 1994 (statistics on vinyl siding were not collected until 1992). The share captured by wood may continue to decline if the supply of timber available for residential use declines, and if the price of wood rises higher, in the future.

Interior Features

Data is also collected routinely by the Bureau of the Census about several interior features of completed houses. Some of this data is summarized in Table 1.10 and discussed next.

TABLE 1.10   Interior Features of Completed New Housing: 1974–1994

	1974	1978	1981	1986	1990	1993	1994
One-Family Completions (000s)	940	1369	819	1120	966	1039	1157
Bathrooms							
2.5 or more baths	21%	25%	24%	33%	45%	48%	49%
3 or more baths	NA	NA	NA	NA	18	14	15
Bedrooms							
4 or more bedrooms	23%	24%	20%	20%	29%	30%	30%
Fireplace							
1 or more fireplaces	49%	64%	54%	63%	66%	63%	64%
Air Conditioning							
Central a.c. installed	48%	58%	65%	69%	76%	78%	79%
Type of Heating Fuel							
Gas	NA	37%	41%	47%	59%	66%	67%
Electricity	NA	52	50	44	33	29	29
Oil	NA	8	7	5	5	3	3
Other types or none	NA	3	2	4	3	2	1

NA = data not available

Source: *Characteristics of New Housing: 1993*, C25/93A (Washington, D.C.: U.S. Department of Commerce, Bureau of the Census); and information provided by the Economics, Mortgage Financing, and Housing Policy Division of NAHB.

BATHROOMS AND BEDROOMS

The growth trend related to the floor area of houses is reflected clearly in the historical record related to bathrooms and bedrooms in the last 20 years. The percentage of houses with 2.5 or more baths held steady until the early 1980s, and has virtually doubled since then [from 24% in 1981 to 49% in 1994; (Table 1.10)]. Data collected after 1986 indicate the increasing influence of the upscale buyer because a significant percentage of houses now contain three or more bathrooms.

During the late 1970s and early 1980s, only about one in five completed houses in a particular year were designed with four or more bedrooms. That figure rose to about 30% by 1990 and has remained relatively steady at that level.

FIREPLACES

The inclusion of one or more fireplaces occurred in 64% of the completed houses in 1994, a level which was recorded as a previous high in 1978. It appears that after a decline to 54% in the early 1980s, acceptance of fireplaces as desired amenities in houses has gradually increased to its present level over a 10-year period. It is, however, an amenity which could be viewed as expendable in times of economic difficulty.

CENTRAL AIR CONDITIONING

In 1974, as noted in Table 1.10, 48% of the houses that were built included central air conditioning. The popularity of this option has increased steadily since then, to the point where 79% of the houses included it in 1994. As might be expected, however, the popularity of this option varies in different geographical areas of the U.S.

TYPE OF HEATING FUEL

In the last 20 years, the popularity of high-energy efficiency in houses has fluctuated as the nation has focused on, and then appeared to ignore, the long-term availability of various fuels for heating houses. Gas appears to be the fuel of choice, with a two-thirds share in 1994 (Table 1.10). This almost two-fold gain since 1978 has been offset by a steady decrease in the use of electricity as a primary heating source. In addition, oil has lost over 50% of the market share it held in 1978. It should be noted that the cost and availability of the various fuels is dependent, to a certain extent, on the geographical area of the United States in which the house is being built.

OTHER INTERIOR FEATURES

This presentation certainly has not touched on all of the important trends that relate to the interior configuration of the house. Other aspects will be discussed when customer preferences are presented in Chapter 2.

Multifamily Completed Housing Units

An analysis of the trends in the multifamily sector over the last 20 years is more complicated than the analysis of one-family housing. One of the reasons, as can be seen in Table 1.11, is because the majority of multifamily housing (74% in 1994) is built with a rental objective in mind. As a result, investors in such properties are influenced greatly by federal government taxation policies and the availability of credit provided by lending institutions. These outside financing influences have helped to create a cyclical environment for this sector of the housing industry.

The Effect of Outside Influence

There has been a steady decline in multifamily housing for the period indicated, with an incredible drop from 788,000 housing unit completions in 1974 to 185,000 in 1994 (Table 1.11). In between, a peak of 636,000 units in 1986 occurred. In 1974 the ratio of multifamily housing production to one-family housing production was 84%. This comparative percentage had decreased to 16% by 1994. The NAHB publication, entitled *The Future of Homebuilding: 1992–1994 and Beyond,*[11] provides an in-depth coverage of the outside influences that caused these fluctuations.

Floor Area Analysis

The size of multifamily units declined in the late 1970s, but then began a relatively steady increase during the 1980s and early 1990s (Table 1.11). The standoff between the two cat-

TABLE 1.11   Characteristics of Completed New Multifamily Housing: 1974–1994

	1974	1978	1981	1986	1990	1993	1994
Multifamily Completions (000s)	788	498	447	636	342	153	185
Multifamily/One Family	788/940=	498/1369=	447/819=	636/1120=	342/966=	153/1039=	185/1157=
Completions	84%	36%	55%	57%	36%	15%	16%
Purpose of Construction							
For rent	76%	82%	63%	79%	78%	71%	74%
For sale	24	18	37	21	22	29	26
Square Feet of Floor Area							
Average	1,021 sf	902 sf	980 sf	911 sf	1,005 sf	1,065 sf	1,040 sf
Median	922 sf	863 sf	930 sf	876 sf	955 sf	1,005 sf	1,010 sf
Less than 1,000	NA	71%	60%	69%	56%	49%	49%
More than 1,000	NA	29%	40%	31%	44%	51%	51%
Percentage with Two or More Bathrooms	24%	20%	32%	36%	44%	45%	44%
Number of Bedrooms							
One bedroom	NA	39%	34%	37%	33%	24%	27%
Two bedrooms	49%	52	55	53	54	57	55
Three or more bedrooms	10	9	11	7	11	17	15
Percent with Air Conditioning	86%	79%	86%	83%	82%	82%	87%

NA = data not available

Source: *Characteristics of New Housing: 1993,* C25/93A (Washington, D.C.: U.S. Department of Commerce, Bureau of the Census); and information provided by the Economics, Mortgage Finance, and Housing Policy Division of NAHB.

egories indicated for 1994 (49% of the units were less than 1000 sq ft, 51% were greater than 1000 sq ft) shows that the needs of both the low-end affordable housing customer and the high-end luxury housing customer appear to be addressed equally at present.

Interior Features

BATHROOMS AND BEDROOMS

As noted in Table 1.11, it was unusual to have two or more baths in a completed multi-family unit in 1974 (24%). Over the following 20 years, however, that percentage has gradually increased to 44% in 1994. During the same time period, one-bedroom units have declined from 39% in 1978 to 27% in 1994, while the percentage of two-bedroom units has stayed fairly constant at about 55%.

CENTRAL AIR CONDITIONING

Most people who either own or rent multifamily units apparently expect central air conditioning as one of the interior features (Table 1.11). It is probably safe to say that multifamily housing without this feature would suffer during the resale process.

Alternatives to Site-Built Housing

The supply side of the basic housing model shown in Fig. 1.1 is not complete until the "mobile home placements" variable is also considered. This is a particularly important component of housing supply when the broad concept of "affordable housing" is examined. The mobile home category (also called manufactured home) should not be considered by itself, however, because it is only one of the alternatives to site-built housing that are available to the homebuyer.

Two of the alternatives which are either partially or fully built off-site and then transported to their foundations are: (1) **systems-built housing** (also called factory-produced housing or in-plant-built housing) and (2) **manufactured housing** (i.e., mobile homes).

One of the basic differences between the two categories is the level of influence that local building codes have on each. The design and construction of a systems-built house essentially is governed by the same building code that applies in the locality where its site-built (also called stick-built) alternative is being built. Mobile homes, on the other hand, comply with a set of building standards that are administered by the U.S. Department of Housing and Urban Development (HUD) and, therefore, preempt the jurisdiction of the local building code.

A second basic difference is that manufactured homes must be built on a permanent chassis which consists of steel beams that run the length of the structure and transmit the load to wheels which allow the manufactured home to be truly "mobile." This capability typically is not needed once the manufactured home reaches its destination because the wheels then are often removed in favor of a more permanent foundation.

A third basic difference is that the U.S. Bureau of the Census does not differentiate between site-built houses and systems-built houses when it tabulates the number of housing completions as presented in Tables 1.6–1.11. As a result, it is very difficult to determine an accurate estimate of the systems-built housing contribution to the overall output of the housing industry each year. Separate statistics are, however, collected and tabulated. Without knowledge of these separate statistics, a full picture of the housing industry would not be possible.

Additional information about the systems-built housing alternative can be obtained in a report entitled *A Consumer Guide to Factory-Built Housing,* which was issued by the Housing Research Center at Penn State in 1991.[12]

Systems-Built Housing

INTRODUCTION

Almost all houses less than 25 years old contain some components that were produced in factories. What distinguishes a systems-built house from its traditional site-built alternative is the degree of prefabrication that is performed in the factory. The category typically includes: (1) precut houses, (2) panelized houses, (3) modular (i.e., sectional) houses, and (4) log houses.

PRECUT HOUSES

A **precut house** package contains components that are precut in a factory to the correct length, but delivered to the building site unassembled. It is the systems-built alternative that has the lowest percentage of preassembly completed in the factory.

PANELIZED HOUSES

A **panelized house** is constructed using factory-produced wall panels 8 ft high and ranging from 4 ft to 40 ft in length. Sometimes the doors and windows are factory installed in the panels; at other times they are installed at the site. Panelized houses generally come in two versions: open- and closed-wall panels. Open-wall panels contain the studs and exterior sheathing, as well as rough frame openings for doors and windows. Plumbing, wiring, insulation, and interior sheathing are installed at the building site.

   Closed-wall panels are shipped from the factory as complete wall systems with exterior sheathing, utilities, and interior wall and finish material already in place. Code compliance of the closed-wall panels is verified at the factory before shipment.

MODULAR HOUSES

A **modular house** is constructed using three-dimensional structures, or "modules," which are 90–95% complete (including interior and exterior sheathing, utility lines, interior partitions, and stairs) when they come off the assembly line. They are generally shipped on a truck bed in two or more sections (12 ft to 16 ft wide, up to 60 ft long) to a permanent site and erected on a conventional foundation on the lot. There they are connected to electrical, water, and other utility lines.

   Modular homes must comply with the same state and local building codes as the conventional stick-built houses. They typically are certified as meeting those requirements by a third-party inspection agency before leaving the factory.

LOG HOUSES

Log houses are provided in the form of packages which utilize either prepared logs or solid log walls as the primary structural members. They are heavily marketed in a variety of styles which take advantage of the beauty of either hand-hewn or modern machine-milled logs. The industry is customer driven and will custom design packages to meet the needs of a broad spectrum of homebuyers.

USE OF SYSTEMS-BUILT HOUSING

Statistics describing the systems-built alternatives are difficult to obtain. One source of information is an extensive survey of homebuilders which NAHB conducted in both 1987 and 1994,[13] in part to estimate the number of precut, panelized, or modular housing units started by builders. The results of that survey and related U.S. Bureau of the Census estimates are compared in Table 1.12.

TABLE 1.12   Percent Share of In-Plant Built Housing Starts:
Comparison of NAHB and U.S. Bureau of the Census Data

	1986[a]	1992[b]	1993[b]	1993[c]
Precut	3.0%	1.0%	1.0%	3.3%
Panelized	7.0%	2.0%	1.0%	6.7%
Modular	2.4%	4.0%	3.0%	3.3%

[a]NAHB 1987 Survey of Builders.
[b]Estimates from the U.S. Bureau of the Census.
[c]NAHB 1994 Survey of Builders.

Source: Information provided by the Economics, Mortgage Finance, and Housing Policy Division of NAHB.

Mobile Homes[a]

Mobile homes became a viable alternative to conventional housing after World War II. As noted by Dan Mercer, in his article entitled "The Market for Mobile Homes,"[14] production of mobile homes steadily increased from 944,000 units during the 1950s to 2.06 million in the 1960s, and finally to a highwater mark of 3.65 million units in the 1970s. Mobile home shipments slowed during the 1980s (to 2.47 million units), and have now reached the levels shown in Table 1.13.

Mercer also notes that even though there are a large number of mobile homes shipped each year, their share of residential *investment* is small compared to the contribution of conventional homes. This is because mobile homes truly are an affordable alternative, as indicated in Table 1.13 by the relatively low median sales price for such homes between 1989 and 1993. Mercer cautions that even though the original purchase price is low, a problem with mobile homes is that they do not appreciate in value as site-built housing usually does.

TABLE 1.13   Supply and Characteristics of Mobile Homes

	1989	1990	1991	1992	1993
Completed Privately Owned Housing Units (000s)[a]	1,423	1,308	1,091	1,158	1,193
Completed Mobile Homes (total; 000s)	203	195	174	212	242
Single-wide	107	104	95	114	127
Double-wide	94	89	78	96	112
Mobile Homes Completed/Completed Privately Owned Housing Units	14%	15%	16%	18%	20%
Median Floor Area					
Total	1,120 sf	1,120 sf	1,125 sf	1,215 sf	1,235 sf
Single-Wide	980	980	985	985	1,060
Double-Wide	1,455	1,455	1,460	1,460	1,565
Median Sales Price					
Total	$24,000	$24,400	$24,800	$25,500	$27,700
Single-Wide	18,600	18,800	19,000	19,900	21,300
Double-Wide	34,400	35,000	35,700	35,900	38,400

[a]Completed mobile home total is *not* included in the total of completed privately owned housing units.

Source: *Characteristics of New Housing: 1993,* C25/93A (Washington, D.C.: U.S. Department of Commerce, Bureau of the Census), and Dan Mercer, "The Market for Mobile Homes," *Housing Economics—The National Association of Home Builders,* January 1995.

[a]For consistency with the references cited in this textbook, the term "mobile home" will be used instead of the term "manufactured housing."

SIZE

Mobile homes are available in single-wide or double-wide versions (Table 1.13). Single-wide versions, which arrive on the placement site as one unit, range in width from 8 ft to 18 ft and contain a median living area of approximately 1,000 sq ft. Double-wide units, which contain a median living area of approximately 1,500 sq ft, are shipped as two units which are then joined at the placement site. Double-wides, which contain more amenities, had a median sales price of $35,880 during the 1989 to 1993 period, which was roughly $16,000 higher than the median sales price of $19,520 for single-wides.

SOCIAL AND LEGISLATIVE ISSUES

One of the problems which the mobile home industry typically faces whenever an attempt is made to secure zoning for a new mobile home park is the resistance raised by the local community to such authorization. Even though a mobile home park provides an affordable housing alternative for families in the lower-income brackets, many communities view such parks as potential eyesores.

CHAPTER SUMMARY

The contribution of the homebuilding industry to the total U.S. construction industry was analyzed in Chapter 1. Attention is directed to two possible measures of that contribution: (1) the *value of new construction put in place* and (2) the *number and type of homebuilding firms* that operate in the industry.

The concept of a *housing supply/demand model* is introduced in order to provide a suitable framework for describing: (1) the diversity of housing products which the homebuilding industry provides, and (2) the demands of homebuyers which have resulted in that diversity. The *supply side* of that model was discussed in Chapter 1. Included in the supply side are *systems-built housing* and *mobile homes.*

Extensive information about the evolution of various characteristics of one-family and multifamily completed housing units is provided in order to create a context for the presentation of housing design guidelines in Chapter 5. Information of a similar nature, from the demand-side viewpoint, is provided in Chapter 2.

An understanding of the historical context in which houses have been designed and built is important for the homebuilder, the architect, and the homebuyer. A word of caution is necessary, however, because records of the past rarely can be extrapolated directly into the future. Forces appearing in the 1990s related to downsizing, more efficient use of the natural resources of land and materials, etc., may reverse some of the indicated trends which resulted in larger, more expensive housing instead of smaller, more affordable housing.

Additionally the design of a house involves a much more complicated process than merely selecting a certain house size, a given number of bedrooms and bathrooms, and the type of exterior wall material. Satisfaction of the homebuyer will be achieved only if the type of information presented in this chapter is used as a starting point for the interaction between an experienced, sensitive architect or homebuilder and the homebuyer.

NOTES

[1] *Housing Market Statistics,* pp. 41 and 48 (Washington, D.C.: National Association of Home Builders, April 1995).

[2] Jack H. Willenbrock, *Management Guidelines for Growth Oriented Homebuilding Firms,* HRC Research Series: Report No. 37 (University Park, PA: The Pennsylvania State University, 1994).

[3] *CABO One and Two Family Dwelling Code, 1995 Edition* (Falls Church, VA: The Council of American Building Officials, 1995).

[4] *1992 Census of Construction Industries,* Final Industry Series, CC92-I-1 to -27 (Washington, D.C.: U.S. Department of Commerce, Bureau of the Census, April–August 1995); and *Construction Review, Quarterly*

Industry Report, vol. 41, no. 1 (Washington, D.C.: U.S. Department of Commerce, Bureau of the Census, Winter 1995).

[5]*Housing Market Statistics,* National Association of Home Builders.

[6]*1992 Census of Construction Industries,* Bureau of the Census, April to August 1995.

[7]*1992 Census of Construction Industries, General Contractors—Single-Family Houses: Industry 1521,* CC92-I-1 (Washington, D.C.: U.S. Department of Commerce, Bureau of the Census, June 1995); and *1992 Census of Construction Industries, General Contractors—Residential Dwellings, Other Than Single-Family Houses: Industry 1522,* CC92-I-2 (Washington, D.C.: U.S. Department of Commerce, Bureau of the Census, April 1995).

[8]*The Future of Home Building: 1992–1994 and Beyond,* p. 8 (Washington, D.C.: National Association of Home Builders, 1992).

[9]*Characteristics of New Housing: 1993,* C25/93A (Washington, D.C.: U.S. Department of Commerce, Bureau of the Census, June 1994).

[10]Ibid., p. B-2.

[11]*The Future of Home Building,* NAHB, Ch. 3–5.

[12]Phyllis A. Barner and Don Prowler, *A Consumer Guide to Factory-Built Housing* (University Park, PA: The Pennsylvania State University, 1991).

[13]*1987 and 1994 Home Builder Industry Survey* (Washington, D.C.: National Association of Home Builders).

[14]Dan Mercer, "The Market for Mobile Homes," *Housing Economics—The National Association of Home Builders,* January 1995, pp. 15–18.

BIBLIOGRAPHY

1987 and 1994 Home Builder Industry Survey. Washington, D.C.: National Association of Home Builders.

1992 Census of Construction Industries, Final Industry Series, CC92-I-1 to -27. Washington, D.C.: U.S. Department of Commerce, Bureau of the Census, April–August 1995.

1992 Census of Construction Industries, General Contractors—Single-Family Houses: Industry 1521, CC92-I-1. Washington, D.C.: U.S. Department of Commerce, Bureau of the Census, June 1995.

Ahulwalia, G., "A Decade of Remodeling," *Housing Economics—The National Association of Home Builders,* May 1995.

Barner, P.A. and D. Prowler, *A Consumer Guide to Factory-Built Housing.* University Park, PA: The Pennsylvania State University, 1991.

CABO One and Two Family Dwelling Code, 1995 Edition. Falls Church, VA: The Council of American Building Officials, 1995.

Characteristics of New Housing: 1993, C25/93A. Washington, D.C.: U.S. Department of Commerce, Bureau of the Census.

Construction Review, Quarterly Industry Report, vol. 41, no. 1. Washington, D.C.: U.S. Department of Commerce, Bureau of the Census, Winter 1995.

The Future of Home Building: 1992–1994 and Beyond. Washington, D.C.: National Association of Home Builders, 1992.

Housing Market Statistics. Washington, D.C.: National Association of Home Builders, April 1995.

Mercer, D., "The Market for Mobile Homes," *Housing Economics—The National Association of Home Builders,* January 1995.

Willenbrock, J.H., *Management Guidelines for Growth Oriented Homebuilding Firms,* HRC Research Series: Report No. 37. University Park, PA: The Pennsylvania State University, 1994.

2

Housing Demand

INTRODUCTION

The supply-side information presented in Chapter 1 provides only one-half of the story about the physical characteristics of the houses that have been built in the past and are presently being built in the United States. The other half relates to the demand side of the housing model in terms of what homebuyers demand of the homebuilding industry as a solution to their wants and needs.

CHANGING HOUSING NEEDS

Housing demand, in whatever form it is sought or received, dictates to homebuilders/ developers what should be built, where it should be built, and in what quantity and price range so that the needs of *most* of the prospective homebuyers are met. The needs of the portion of the population that can afford to purchase the type of housing it desires are met directly by such marketplace forces. The housing needs of the disadvantaged and the poor also are met in the marketplace, but often require social subsidies that are provided by local, state, and federal agencies. The primary focus in this chapter is on the homebuyers who purchase their own homes directly.

In assessing the needs of these homebuyers, it must be recognized that the housing needs of people change as they pass through the various stages of their lives. In the United States, each stage is typically accompanied by selling an existing house and buying another. In other countries, houses belong to the same families for generations.

Growth Stages of Housing Decisions

For most people who move away from home either to attend school or begin their first jobs, first-stage housing needs are met by renting an apartment or a house. In the second stage, a transition from being a *renter* to being a *first-time homebuyer* usually occurs. This transition is made by some people while they are still single and unattached, and by others when they get married and enter the "household formation" stage of their lives.

The second stage can involve either the purchase of an existing house or the construction of a new house. Most likely because of economic constraints, **entry-level housing** will be purchased to meet the immediate needs of the first-time homebuyers and possibly their young children.

On one or more occasions in the future, some of the homebuyers again enter the marketplace as **move-up buyers.** At this point, the homebuilder is dealing with homebuyers who are much more knowledgeable about their wants and desires, particularly as related to the children that most of them now have. For a fortunate small percentage of move-up buyers, these stages lead through the **custom-home** stage to the **luxury-home** stage.

Knowledge of the percentages of homebuyers in each stage is extremely important to homebuilders because it allows them to design and market a product line that meets the diverse needs at each stage. Table 2.1 presents such information.

Downsizing Stage of Housing Decisions

After passing through all of the move-up stages that a particular household has desired, a time is reached when downsizing decisions are often made. This sometimes occurs either when the last adult child leaves home and the parents become **empty-nesters** or at the time of retirement of the major wage earner. Downsizing decisions focus on houses that are easier to maintain and manage so that individual goals and interests can be pursued. Such decisions may result in a move away from the family "homestead" to an active retirement community in another part of the country.

As the household approaches the **advanced elderly stage,** there is a greater likelihood that chronic health problems will develop. The last stage of downsizing, therefore, occurs when the final home is sold and people move into an assisted living, congregate care, or nursing home facility. For those individuals who moved to an active retirement community, this stage might include moving back from their retirement location to be closer to their adult children.

DEMOGRAPHIC PREDICTION OF NEW HOUSE DEMAND

As noted in Chapter 1 when the housing model in Figure 1.1 was discussed, demographic forces control both the level as well as the nature of new housing demand. Two of these forces are: (1) the size of the population in each age group in that year or period of time;

TABLE 2.1   Percent of First-Time and Move-Up Buyers by Year

Single-Family Homes	First-Time Buyer	Second-Time Buyer	Third-Time Buyer	Fourth (or More)-Time Buyer
Detached				
1976	32.2%	35.0%	15.3%	14.5%
1978	30.4	37.4	18.5	13.8
1981	37.6	32.9	16.4	13.2
1986	35.3	30.5	17.7	16.4
1990	34.6	29.8	16.5	19.1
1993	35.2	28.7	18.6	17.5
Attached				
1989	36.2%	22.7%	15.5%	25.6%
1990	47.3	19.9	15.9	17.0
1991	39.8	23.8	14.5	21.9
1993	49.2	20.2	12.0	18.6

Source: *Profile of the New Home Buyer*, Exhibit 6-1, p. 44 (Washington, D.C.: National Association of Home Builders, 1992); and information provided by the Economics, Mortgage Finance, and Housing Policy Division of NAHB.

and (2) the rate at which people in each of those age groups form households. Household growth, in turn, historically accounts for most new home demand.

Age Distribution of the Population

According to the NAHB publication entitled *The Future of Home Building: 1992–1994 and Beyond,*[1] a prediction of age distribution in the population can be made if information about (1) past births, (2) prospective deaths, and (3) immigration is combined properly.

Past Births

The pattern of births in the United States is well known and, as indicated in Fig. 2.1, is represented as a series of baby booms and busts. After reaching a low value in 1933, the number of births each year began to rise, grew rapidly during World War II, and increased more dramatically after the war. From 1946 to 1964 the birth rate was so high that the period became known as the great "baby boom." The peak year was 1957, when 4.308 million babies were born. Thus, in 1996, for instance, the adults who were at the "leading edge" of the baby boom reached 50 years of age and those at the "trailing edge" reached 32 years of age.

Prospective Deaths

David D'Alessandris, in his article entitled "Demographics and Household Formation,"[2] noted that the death rate is fairly constant, but is changing in two ways: the rate is decreasing for older Americans due to improved health, but is increasing for younger Americans as a result of AIDS. According to D'Alessandris, the death rates have changed slowly since the 1989 Census Bureau forecast. As a result, they can be predicted fairly accurately.

Immigration

The variable that has changed dramatically in the recent past is immigration. According to the NAHB,[3] slightly more than one-fourth of the net increase in U.S. population during the 1980s was attributable to net foreign immigration. The NAHB predicts that in the 1990s, as national population growth slows and the numerical restrictions on immigration are relaxed, net immigration could account for more than a third of the net population increase in the United States.

Figure 2.1
Historical Record
of Live Births in
the United States
Source: National
Center for Health
Statistics, U.S.
Department of
Health and Human
Services.

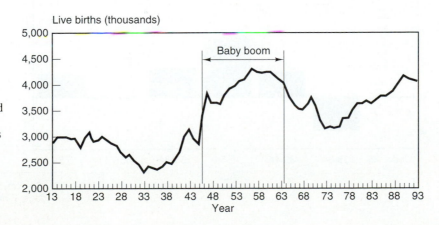

Household Formations

Headship Rates

Household formations are typically measured by the **headship rate**—the percentage of the population in each age group who are heads of households. U.S. Bureau of the Census statistics, as noted by NAHB,[4] indicated that there are large differences in headship rates for different age groups.

Figure 2.2 indicates the predicted change in households for each age group in the period from 1995–2000. A similar analysis could be made for any period in the future.

Significance of Household Formation Information for Homebuilders

The information presented in Fig. 2.2 is very significant to the homebuilding industry at the national level because it allows predictions to be made about the types of housing that will be demanded by each type of homebuyer. Similar information at either the state or local level would be equally significant to homebuilders who operate in a specific geographical area.

Prediction of Demand by Type of Homeowner

As noted earlier, households go through a well-recognized set of homebuying stages. When information of the type provided in Fig. 2.2 is associated with these stages, predictions about new house demand (such as those provided by the publication entitled *The State of the Nation's Housing—1993,* published by the Joint Center for Housing Studies of Harvard University) can be made:

> With the aging of the baby boom, the number of households aged 35–54 will grow rapidly in the decade ahead. This group is moving into the peak earning years; it is also well into the childrearing phase of the lifecycle, when families tend to consume the most housing. With some 23 million (53%) of all baby-boom households already owning homes, tradeup demand will be extremely strong. . . . The increasing number of affluent elderly households also has important implications for housing investment in the 1990s. Elderly households have high rates of homeownership and low rates of mobility; most of these households are therefore unlikely to move to new homes or even make major investments in their present homes. Elderly households do, however, spend relatively large amounts on maintenance and repair, and thus will add to the overall growth of this segment of the market in the

Figure 2.2
Predicted Change
in Households
1995–2000
(millions)
Source: The State of the Nation's Housing, 1993 (Boston, MA: Joint Center for Housing Studies of Harvard University, 1994).

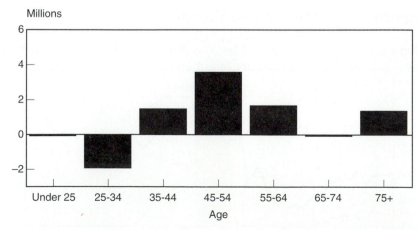

> years ahead. . . . Moreover, the outlook for the first-time buyer market is promising. Having been shut out of homeownership in the 1980s by high housing costs, more young households have had time to accumulate the required down payment and to earn the income to qualify for a mortgage. The movement of new immigrant households into the economic mainstream should also buoy the first-time buyer market.[5]

Predictions of this type, particularly if presented in a "regional market" format, can assist homebuilders in providing a product line which will appeal to their customers. Another source of information about the needs and demands of homebuyers can be obtained through consumer research.

CONSUMER RESEARCH PREDICTION OF NEW HOUSE DEMAND

Consumer research studies have become almost a necessity if individual builders are to remain competitive in their regional markets. The development of such studies presents many market research challenges because it is often difficult for consumers to identify the features in a house that they desire if these features are not already within the realm of their past experience. As a result, these studies are often expensive to design and implement.

Non-Proprietary Consumer Research Studies

The reports generated by studies initiated by individual homebuilders are usually proprietary and cannot serve as references for this chapter. Fortunately, several organizations do publish the results of the consumer research studies which they co-sponsor as a service to homebuilders. Three of these are:

1. *Builder's* combined 14th Annual Home Buyer Survey (1991) and Focus Group Analysis (1991).

2. *Professional Builder's* 21st annual consumer survey entitled *1995 Consumer New Home Survey* (1995).

3. NAHB's homebuyer preference study entitled *What First-Time Home Buyers Want* (1991).

Taken together, the findings in these studies can provide homebuilders with an excellent perspective about the wants, needs, and demands of the consumers who will buy the houses which they design and build.

Segmentation of Homebuyer Groups

There are a number of approaches that can be used when the buyers of new homes are segmented for consumer research purposes. Some of the commonly used approaches are:

By Consumer Group
- First-time buyer
- Move-up buyer
- •
- •
- •
- Retiree

By Household Category

- Single male (or female)
- Couple without children
- •
- •
- •
- Unrelated individuals

By Annual Income

- Less than $35,000
- $35,000–$44,999
- •
- •
- •
- Over $200,000

An emphasis is placed on the *consumer group* approach in the following sections of this chapter because it provides the most meaningful type of information related to the housing characteristics that are being demanded.

Consumer Research Study #1

Builder, in its July 1992 issue entitled "Welcome to a Buyer's Market,"[6] reported on two comprehensive consumer research studies which were co-sponsored by *Builder* and other organizations involved in the homebuilding industry. One was *Builder's* 14th Annual Home Buyer Survey, which included 1,346 new-home shoppers who visited model homes throughout the United States and responded to two-part questionnaires.[7] The other related study was conducted using 12 focus groups dispersed throughout the United States representing various consumer groups who were on the traffic lists of 11 co-sponsoring homebuilders.[8]

Description of the Consumer Groups

The four consumer groups that were defined for the *Builder* studies were:

1. first-time homebuyers,
2. move-up homebuyers,
3. luxury homebuyers, and
4. singles and couples.

The information which was obtained in the study about these four consumer groups is summarized in Table 2.2 and is discussed next.

First-Time Homebuyers[9]

INTRODUCTION

As noted earlier, interaction with the homebuilding industry typically begins when a shift from renter status to homeownership status is contemplated. The *Builder* study found that the average age of the first-time buyer is 30 years and the average household income is $54,000. For 59% of the households, this level is achieved by two wage earners (Table 2.2).

At this stage, 75% of the first-time buyers are either still single or are married without children. Table 2.2 indicates that 75% of these first-time homebuyers expressed an interest in purchasing single-family detached houses. Some of these same people, as well

TABLE 2.2   Description of Various Homebuyer Groups

Characteristic	All New Homebuyers	First-Time Buyers	Move-Up Buyers	Luxury Buyers	Singles/ Couples
		Who They Are			
Average Age	40	30	36	40	41
Household Income	$72,000	$54,000	$71,000	$104,000	$70,000
One income	37%	41%	30%	33%	42%
Two or more incomes	63%	59%	70%	67%	58%
Household Type					
Singles	25%	44%	10%	16%	41%
Couples with children	46	25	66	58	—
Couples without children	30	31	21	26	59
Number in Household					
One	10%	18%	6%	5%	20%
Two	42	56	26	36	72
Three	19	12	25	22	5
Four	20	9	30	24	3
Five or more	9	3	13	13	—
		What They Want			
Type of House[a]	All Buyers	First-Time Buyers	Move-Up Buyers	Luxury Buyers	Singles/ Couples
Single-family detached	82%	75%	92%	89%	74%
Patio	13	18	9	11	20
Townhouse	15	24	8	11	23
Mid-/high-rise condo	5	8	1	5	8

[a]More than one response allowed.

Source: *Builder,* vol. 15, no. 8, pp. 55, 81, 113, 147, July 1992, ©Hanley-Wood, Inc.

as some of the others questioned, also considered purchasing either patio homes/townhouses (43%) or mid-/high-rise condos (8%).

GENERAL DESCRIPTION

The study found that a large diversity of needs must be satisfied for this consumer group. Singles and couples without children are not as willing as others to travel long distances to escape urban areas. They prefer to be closer to their workplace and to cultural opportunities. Couples with children, on the other hand, are hoping to find safe neighborhoods where their children can find playmates, and where there are organized recreational facilities, nearby shopping, and a good school system.

In the price range of the first-time buyer, the large-volume builders cannot customize each house design. The desire for status-symbol design which reflects individuality must be balanced against the cost effectiveness of uniformity of design.

Perhaps the overriding concern of first-time homebuyers is whether the house they are buying will have appreciated resale value in three to five years. They typically view their purchase as a short-term investment. Thus it is more important from a resale viewpoint that the style of the house reflect time-tested, American architecture (perhaps with a porch, a large eat-in kitchen, etc.) rather than their own sense of individuality.

Move-Up Homebuyers[10]

INTRODUCTION

Move-up buyers are a key demographic group for the homebuilding industry. Table 2.1 indicates that approximately 65% of all single-family homebuyers come from this group.

Because of prior ownership experience, and the positive and negative aspects associated with it, move-up buyers have a better idea about what they want and need, and they are better prepared to judge the quality of new home construction.

The average age of the move-up homebuyer is 36, and the average two-person household income has risen to $71,000 (Table 2.2). Agewise, this group is part of the leading edge of the baby boomer generation. Only 10% of these households are composed of singles. Most are entrenched in the demands of raising a family; two-thirds of these homebuyers are couples with children. Single-family detached housing is the clear choice of 92% of these homebuyers (Table 2.2).

GENERAL DESCRIPTION

The major drawing force which leads move-up homebuyers to move is probably the desire to raise their children in the best possible environment; that goal often guides what, when, where, and why they want to move. Move-up homebuyers already have established roots in their existing communities. As a result, there are many financial and psychological factors that tempt move-up homebuyers to stay in their existing houses. There often is little urgency to move unless employment change also accompanies the move. As a result, move-up buyers can be more particular and also are less likely than first-time buyers to buy on impulse.

Luxury Homebuyer[11]

INTRODUCTION

Luxury homebuyers typically deal with custom homebuilders because they can afford to incorporate their desires for individuality and creativeness into the architectural design and construction process. The average luxury homebuyer is only four years older than the average move-up buyer, is family oriented (58% have children), and has an average two wage-earner household income which has risen to $104,000 (Table 2.2). Associated with that increase of $35,000 in income is a basic difference in terms of class; typically, luxury homeowners are professional and business leaders. In terms of the type of housing desired, they mirror the move-up buyer by clearly desiring single-family detached housing (89%).

GENERAL DESCRIPTION

Because of the upward mobility factor, the location in which luxury homebuyers want to build must be convenient to work and they most likely will pay a premium to accomplish that objective. They typically buy in the best communities so the fear of living next to neighbors who "don't fit" their social class is no longer significant.

Separation and zoning of living spaces in the house is important because of the desire for privacy, particularly if there are teenage children in the household. Although they want larger, more expensive houses (3,000+ sq ft, $350,000+), they are not particularly interested in oversized rooms; they typically reserve volume space for public settings such as the living room and family room. Rooms for family and private functions should be comfortable, but not overwhelming.

Singles and Couples[12]

INTRODUCTION

These two homebuyer groups are sometimes joined together because of their one common characteristic: they do not have children. They are the age clusters that bracket the "family" stage of life directly before (i.e., singles or newlyweds) or directly after (i.e., empty nesters) the childbearing and childrearing years. Fifty-nine percent are married couples without children; 35% are never-marrieds, divorced, or widowed and living alone; and 6% are unrelated people sharing living quarters (Table 2.2).

Because of the clustering effect in this homebuyer group, the average age of 41 is deceptive. The average household income of $70,000 compares favorably with the average income of $72,000 for all new homebuyers. The majority of households prefer single-family detached housing (74%), although 43% of them also indicate an interest in either patio or townhouses, which is higher than the percentage for all new homebuyers. Only 8% preferred either mid- or high-rise condominiums.

GENERAL DESCRIPTION

As noted, this group either leads or trails the family-oriented bulk of the baby boom population. It was found, however, that they share many of the baby boomers' values. They all appear to want houses that, in some fashion, reflect their need for self-expression and self-fulfillment. The young childless professionals view their first house as an opportunity to build equity, take advantage of the tax break associated with home ownership in the United States, and as a stepping stone to the next house so resale value is critical. The empty nesters, on the other hand, have amassed equity and the advantages of homeownership. They are buying for the long haul.

For both age clusters, high-density or attached housing with repetitive facades are not exciting. They both appear to want a detached family house: a reminder of their childhood homes on the inside and the ambience of an adult-oriented community on the outside. Since the presence of children is not a factor, the singles/couples homebuyer group prefers an open floor plan design with a formal dining room and a large kitchen to support entertaining.

New Housing Demands of Homebuyer Groups[13]

The four homebuyer groups used in the consumer research studies co-sponsored by *Builder* provide a convenient starting point for presenting new housing demand information in a systematic fashion.

EXPECTATIONS

Table 2.2 indicates that each of the groups overwhelmingly indicated the desire to purchase a single-family detached house. Table 2.3 provides some additional insight into the expectations of each group with regard to some of the specific characteristics of these single-family detached houses.

As expected, the price and floor area increase as a shift is made from first-time homebuyer to luxury homebuyer. The singles/couples group tends to behave like the move-up group but does not need as much living space. A two-level house clearly is the most popular, although all groups except first-time homebuyers also have a slight interest in three levels. First-time homebuyers and the singles/couples group tend to favor three bedrooms and either two or two and a half baths, while some luxury homebuyers and, to a lesser degree, some move-up buyers, prefer five bedroom/four or more bathroom houses.

EXPECTED STANDARD FEATURES

The results of the consumer research study also provide some interesting information about the type of features each homebuyer group wants (Table 2.4). In addition, information is provided about the percentage of homebuyers in each group that felt that the listed feature should be standard in the house they were purchasing.

A homebuilder who is targeting a particular homebuyer segment can use information such as that provided in Table 2.4 to improve the attractiveness of the firm's houses. Although a feature-by-feature analysis will not be provided here, a few comments of a general nature can be made. First-time homebuyers do not appear to have outlandish expectations about the features they expect; features such as built-in microwave ovens, walk-in closets, etc. appear to be practical requests. Based upon experience, move-up homebuyers are not willing to settle for the minimum level of features; they demand

TABLE 2.3   Housing Characteristic Expectations of Various Homebuyer Groups

Characteristic	All Buyers	First-Time Buyers	Move-Up Buyers	Luxury Buyers	Singles/ Couples
Price range	$160,000– 209,000	$117,000– 153,000	$142,000– 183,000	$307,000– 415,000	$150,000– 196,000
Monthly payment	$1,113– 1,479	$945–1,246	$1,076– 1,438	$1,820– 2,445	$1,070– 1,457
Square footage	2,290	1,883	2,368	2,918	2,108
Maximum commute time	34 minutes	34 minutes	35 minutes	36 minutes	34 minutes
Number of levels					
One	33%	38%	22%	19%	39%
Two	62	58	71	72	55
Three	6	—	7	9	6
Number of bedrooms					
Two	9%	14%	—	—	14%
Three	40	61	27%	26%	54%
Four	45	23	64	53	30%
Five or more	6	—	5	18	—
Number of bathrooms					
1.5	4%	9%	—	—	6%
2.0	27	41	19%	10%	34
2.5	43	42	50	34	40
3.0	15	6	17	27	14
3.5	9	—	12	22	6
4 or more	1	—	—	6	—

Source: *Builder,* vol. 15, no. 8, pp. 58, 84, 117, and 151, July 1992, © Hanley-Wood, Inc.

enhanced living features such as specialized living and storage spaces, upgraded appliances and bath fixtures, and additional rooms for a home office and game room.

Luxury homebuyers can afford to pay for all of the features they want and, as indicated in Table 2.4, they expect high-end features to be standard and they express a desire for specialized rooms. The singles/couples group, without children to consider, are able to seek self-expression and self-fulfillment. As a result, they care more about features than extra space.

Consumer Research Study #2[14]

This 21st Annual Consumer Survey of *Professional Builder* was conducted in order to determine the attitudes, opinions, and preferences of consumers regarding newly constructed housing. When the data are sorted by consumer group, it is evident, as shown next, that the groupings used by *Professional Builder* overlap those used by *Builder* in its study.

Homebuyer Groups

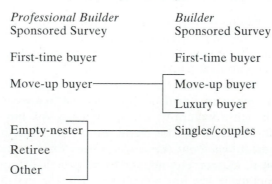

Professional Builder Sponsored Survey	*Builder* Sponsored Survey
First-time buyer	First-time buyer
Move-up buyer	Move-up buyer
	Luxury buyer
Empty-nester	Singles/couples
Retiree	
Other	

TABLE 2.4   Top Features Expected by Various Homebuyer Groups

First-Time Homebuyer		Move-Up Homebuyer	
Feature	Percent[a]	Feature	Percent[a]
High ceilings	86%	Separate tub and shower	89%
Bay windows	82%	Walk-in pantry	78%
Refrigerator	75%	Grilltop range	75%
Fireplace in family room	74%	His-and-her closets	62%
Ceiling fan in kitchen	72%	Matching cabinet/ appliance fronts	62%
Separate laundry room	70%	Double oven	58%
Built-in microwave oven	69%	Formal dining room	56%
Ceramic tile in master bath	68%	Storage room	55%
Stainless steel kitchen sink	67%	Front porch	52%
Single oven	66%	Bookshelves	51%
Separate tub and shower	65%	French doors in family room	48%
Walk-in closets	63%	Attic	39%
Walk-in pantry	54%	Hardwood entry flooring	33%
		Home office	36%
		Game room	30%

Luxury Homebuyer		Singles/Couples Homebuyer	
Feature	Percent[a]	Feature	Percent[a]
Separate tub/shower	93%	Fireplace	86%
Walk-in pantry	84%	High ceilings	84%
Double vanity, master bath	82%	Bay window	80%
Matching cabinet and appliance fronts	79%	Microwave	80%
Ceramic tile, master bath	75%	Separate tub and shower	76%
French doors	66%	Ceiling fan	73%
Double entry door	63%	Two-car garage	69%
Ceramic tile, kitchen	62%	Skylight	68%
Hot water dispenser	61%	Grilltop range	63%
Water purifier	61%	Walk-in pantry	54%
Rear deck	59%	Water purifier	52%
Three-car garage	53%	His-and-her closets	51%
Office	52%	Hot water dispenser	51%
Library	42%	Double oven	49%
Marble entry flooring	40%	Spice rack	38%
Sunroom	39%	Ceramic tile, entry floor	35%

[a]The percentage of homebuyers who feel that the indicated feature should be standard.

Source: *Builder,* vol. 15, no. 8, pp. 59, 85, 117, and 151, July 1992, © Hanley-Wood, Inc.

As a result of the grouping selected by *Professional Builder,* the move-up buyer class is broader and does not differentiate as well between the needs of the early-stage move-up buyers and the unique needs of the luxury homebuyer. Conversely, the *Professional Builder* classification more definitively separates the needs of the empty-nester from the retiree. It also avoids the problems created by the catch-all singles/couples grouping used by *Builder.* Selected results of the *Professional Builder* study are presented next. The presentation does not, however, provide an exhaustive analysis of the data that is available in the final report of the study.[15] Therefore, the interested reader is encouraged to obtain a copy of the entire report in order to fully assess the robust database of information that is available.

Type of House

As the decision-making process of the homebuyer progresses, the issue of the desired type of house must be addressed. With regard to a simple choice between (1) a single-family house on its own lot and (2) an attached house such as a townhouse or condominium, all categories of homebuyers in the study favored the *single-family house* with the following percentages:

First-time buyer	88%
Move-up buyer	97%
Empty-nester	84%
Retiree	87%

Table 2.5 presents the responses to a related question which broadened the selection menu for housing type. The data reinforces the response to the simple choice question, while also indicating the preference for a house built by a "production builder" rather than a "custom builder." Only the retirees clearly preferred the custom-built option, although that option was selected by 39% of the move-up buyers and 31% of the empty-nesters. The move-up buyer result is somewhat surprising because it is usually assumed that they prefer the personal attention which a custom builder can provide. It is also interesting to note that the system-built option came in third with all buyer categories except the first-time buyer.

Description of Desired Housing

These results clearly indicate the popularity of one-family detached housing in the United States. The specific characteristics of one-family detached houses that are desired by each type of homebuyer were addressed in a series of questions in the survey. The results summarized in Table 2.6 provide a composite description of the desired characteristics for each type of homebuyer. Several interesting observations can be made:

- Three of the homebuyer groups (i.e., first-time, empty-nester, and retiree) indicated a preference for a single-level house when given the options of (1) single-level, (2) split-level (bi-level, tri-level), (3) 1½ story, and (4) 2 story or more. It is

TABLE 2.5   Selection of Preferable Type of House for Each Homebuyer Category

Type of House You Would Purchase (percentage of responses)	First-Time Buyer	Move-Up Buyer	Empty-Nester	Retiree
A detached production house	49%	45%	33%	26%
One-of-a-kind custom house	28	39	31	38
A modular, panelized, or precut package house	6	5	14	14
A mobile home	5	4	6	8
Townhouse	9	2	12	3
Specialty house	2	4	—	1
Duplex	2	1	—	3
An apartment-like condominium in a low-rise building	1	1	4	6
An apartment-like condominium in a high-rise building	—	—	—	1

Source: 1995 Consumer New Home Survey, (Des Plaines, IL: *Professional Builder,* 1995), p. 50.

TABLE 2.6 Desired Characteristics of New Housing

Characteristics of Desired Housing	First-Time Buyer	Move-Up Buyer	Empty-Nester	Retiree
Type	single level	2 story or more	single level	single level
Architectural Style	traditional	traditional	contemporary	traditional
Exterior Finish on Front of House	brick	brick	brick	brick
Parking Facilities	2-car garage	2-car garage	2-car garage	2-car garage
Average Number of Bedrooms	3	4	3	3
Location of Master Bedroom	1st floor (61%)	1st floor (61%)	1st floor (90%)	1st floor (87%)
Average Number of Bathrooms	2.2	2.4	2.1	2.1
Average Size of Desired House	1,811 sf	2,304 sf	1,676 sf	1,885 sf
Average Price of Desired House	$107,000	$167,000	$114,000	$138,000

Source: 1995 Consumer New Home Survey, pp. 52, 61, 64, 74, 76, 78, 80, 118, and 121 (Des Plaines, IL: *Professional Builder*, 1995); and *Professional Builder*, vol. 60, no. 11, pp. 126, 128, 130, and 132, July 1995.

interesting to note that this result contradicts the Consumer Research Study #1 results shown in Table 2.3.

- A rather large selection of architectural styles was provided to the potential homebuyers. Included were: (1) Contemporary, (2) Victorian, (3) Prairie/ Craftsman, (4) Traditional, (5) Colonial, (6) Tudor, (7) Spanish, and (8) Georgian. Only the empty-nester category chose a contemporary architectural style; all the rest of the homebuyer types selected the traditional style.

- All four homebuyer types indicated a preference for brick siding on the front of the house when given a choice between: (1) brick, (2) aluminum/metal siding, (3) stucco, (4) wood siding, (5) stone, (6) concrete block, (7) wood shakes/shingles, (8) vinyl siding, (9) plywood siding, and (10) hardboard siding. Since brick is the most expensive exterior finish option, it may be one of the first features that is changed when cost considerations are taken into account.

Interior Design Features

Another important aspect of house design relates to the interior architectural features which transform the space into a comfortable, efficient solution to the living needs of each of the homebuyer categories. The desired features were determined in the survey by asking the potential homebuyers to rate a list of interior design features for the price and size range of the house they desired. The rating was on a scale of 1 to 5, with 5 being the most desirable and 1 being the least desirable.

The average responses to each of the choices by the homebuyer groups are tabulated in Table 2.7. The results obtained appear to be relatively consistent across all homebuyer groups for most of the features. The only features which separate the first-time buyer and the move-up buyer from the empty-nester and retiree are: (1) interior lofts and balconies, (2) space for a full-time home office, (3) finished basement, (4) three-car garage, and (5) bonus room over garage.

It probably can be assumed that any interior architectural feature that was rated near 4.0 or above would help "sell" the house if included in a house design. Based on this viewpoint, the features that should definitely be incorporated into houses are: (1) covered entry, (2) porch, (3) patio, (4) storage space in garage, (5) pantry, (6) linen closet, (7) two closets in the master suite, and (8) space for family entertainment.

It is more difficult to identify those features that clearly do not enhance the attractiveness of a particular house design. If one assumes that any interior architectural feature

TABLE 2.7   Average Responses to Interior Architectural Design Features
(scale: 1 = least desirable, 5 = most desirable)

Interior Architectural Design Feature	First-Time Buyer	Move-Up Buyer	Empty-Nester	Retiree
Raised, but flat ceilings in major living areas for volume effect	3.1	3.3	2.7	3.2
Cathedral, vaulted-type ceilings in major living areas for volume effect	3.4	3.6	2.8	3.5
Interior lofts and balconies	2.6	2.5	1.9	1.9
Sun room or greenhouse-type room	2.7	2.7	2.8	2.7
"Garden" bathroom (i.e., bathroom plan that includes private landscaped outdoor space)	2.3	2.3	2.1	2.3
Home entertainment center or "theater" for TV, recorders, stereo, etc.	3.0	3.0	2.5	2.8
Home fitness center for exercise equipment	2.6	2.5	2.2	2.0
Inclusion of more built-in furniture (dressers, cabinets, shelves, etc.) to free up more floor space	3.1	3.0	2.8	3.2
Covered entries	3.8	3.9	4.0	4.0
Porch	4.0	4.0	4.1	4.1
Patio	3.8	4.0	4.2	4.0
Storage space in garage	3.9	4.1	4.3	4.3
Pantry	4.2	4.4	4.2	4.3
Linen closet	4.4	4.5	4.5	4.5
Two closets in the master suite	3.7	3.8	3.9	3.7
Dramatic window treatments	3.0	3.2	2.7	2.9
Space for formal entertaining	2.9	3.2	2.5	2.9
Space for family entertainment	4.0	4.2	3.4	3.7
Space for full-time home office	2.9	2.9	2.4	2.5
Space for a part-time home office and/or room for adult hobbies	3.2	3.2	2.8	3.0
Private living space for mother, father, or other relative	2.3	2.3	2.5	2.3
Oversized master bath	3.6	3.8	3.1	3.3
Finished basement	2.7	2.8	2.2	2.3
Three-car garage	2.5	3.1	2.2	2.2
Bonus room over garage	2.6	2.8	2.1	2.1

Source: 1995 Consumer New Home Survey, pp. 184–240 (Des Plaines, IL: *Professional Builder,* 1995).

that is generally rated near 2.0 or below meets that definition, then only (1) garden bathroom and (2) private living space for mother, father, or other relative are identified.

Consumer Research Study #3[16]

The final study which is briefly mentioned represents a unique attempt by the Economics, Mortgage Finance, and Housing Policy Division of NAHB to focus on the demands of the first-time homebuyer. The population from which the data were obtained were people who were presently renting. Responses were received from 1,360 renters.

The median age of the potential first-time homebuyer respondents was 31.4 years, the median income was $42,000, and 47% of them were married couples. Among the households which planned to move in the next two years, about one-third planned to buy, one-third planned to rent, and one-third were not sure. For those renters who planned to buy a house in the next two years, about one-fourth preferred a new house, 17% preferred a new house built to their specifications, and 48% preferred an existing house.

Gopal Ahluwalia, in his article entitled "What First-Time Homebuyers Want,"[17] presents a composite listing of the preferences expressed by the renters who wanted to

TABLE 2.8 Wish List of First-Time Homebuyers

- Single-family detached home
- Suburban location
- 1,900 sq ft of finished area
- Two-story home with brick front
- Walk-out basement
- 3 bedrooms
- 2 bathrooms
- 2-car garage
- Master bedroom that is separated from the other bedrooms
- Walk-in closet in the master bedroom
- Master bath with a closet, vanity with one bowl, and a toilet separated from the bathtub
- Kitchen with a walk-in pantry, microwave oven, vinyl flooring, bay window, and large countertop
- Family room/great room that is visually open to the kitchen/family room arrangement
- $88,000 sales price

Source: Gopal Ahluwalia, "What First-Time Homebuyers Want," *Housing Economics—The National Association of Home Builders,* October 1993, p. 6.

purchase their first house (Table 2.8). The projected sales price of $88,000 in Table 2.8 appears to be unrealistically low, perhaps because renters are not very familiar with the costs of new or existing housing.

CHAPTER SUMMARY

A short section entitled "Changing Housing Needs," which highlights the typical life cycle relationship between the homebuyer and the homebuilding industry, sets the stage for the presentation in this chapter of the **demand** side of the **housing supply/demand model.** Knowledge about housing demand, which has been found to be a function of (1) the age distribution of the population and (2) household formation rates, is important for homebuilders because it allows predictions to be made about the types and features of housing that will be demanded by various types of homebuyers.

The characteristics of four consumer groups (first-time homebuyers, move-up homebuyers, luxury homebuyers, and singles and couples) were described in this chapter and their housing needs, as determined in a study sponsored by *Builder* in 1991, were presented. Information obtained by two other studies supplements the *Builder* presentation. The first study, sponsored by *Professional Builder,* indicated the criteria which influence the purchase decision of homebuyers and the exterior and interior design features which first-time buyers, move-up buyers, empty-nesters, and retirees wanted in the houses they had or were going to purchase. The final study, which was conducted for the Economics, Mortgage Finance, and Housing Policy Division of NAHB, focused on first-time homebuyers.

An understanding of the wants, needs, and demands of homebuyers is important for both the homebuilder and the architect as a housing product line is being designed for a particular market. An appropriate incorporation of some of the features presented, either as a standard part of the product line or as additional features, enhances the attractiveness of the various models.

A word of caution, however, is necessary with regard to housing demand. The design of a house for a particular homebuyer is a complicated combination of *push–pull* forces.

An experienced architect or homebuilder must react appropriately to the *push* forces created by the homebuyer who wants certain characteristics and features incorporated into the design of the house. An appropriate reaction might be to disagree with the suggestions if they do not enhance the architectural design of the house. The responsibility of the architect or homebuilder is also to *pull* the homebuyer into agreeing to the incorporation of innovative and appropriate characteristics and features which will improve either the performance or the appearance of the house.

In either case, the information presented in this chapter can serve as the beginning point of such a push–pull process of design.

NOTES

[1] *The Future of Home Building: 1992–1994 and Beyond* (Washington, D.C.: The National Association of Home Builders, 1992), pp. 11–14.

[2] David D'Alessandris, "Demographics and Household Formation," *Housing Economics—The National Association of Home Builders,* May 1994, p. 5.

[3] *The Future of Home Building,* p. 13.

[4] *The Future of Home Building,* p. 17.

[5] *The State of the Nation's Housing, 1993.* Boston, MA: The Joint Center for Housing Studies of Harvard University, 1994, pp. 20–21.

[6] *Builder,* vol. 15, no. 8, July 1992.

[7] Ibid., p. 16.

[8] Ibid.

[9] June Fletcher, "First Time," *Builder,* vol. 15, no. 8, July 1992, pp. 55–61.

[10] June Fletcher, "Move Up," *Builder,* vol. 15, no. 8, July 1992, pp. 81–90.

[11] June Fletcher, "Luxury," *Builder,* vol. 15, no. 8, July 1992, pp. 113–119.

[12] June Fletcher, "Singles/Couples," *Builder,* vol. 15, no. 8, July 1992, pp. 147–153.

[13] June Fletcher, "First-Time," pp. 55–61; "Move-Up," pp. 81–90; "Luxury," pp. 113–119; "Singles/Couples," pp. 147–153.

[14] Andrea LaFreniere, "21st Annual Consumer Survey: How to Build What New Home Buyers Seek," *Professional Builder,* vol. 60, no. 11, July 1995, pp. 123–135.

[15] 1995 Consumer New Home Survey. Des Plaines, IL: *Professional Builder Magazine,* 1995.

[16] *What First-Time Home Buyers Want.* Washington, D.C.: The National Association of Home Builders, 1993.

[17] Gopal Ahluwalia, "What First-Time Homebuyers Want," *Housing Economics—The National Association of Home Builders*, Oct. 1993, pp. 5–7.

BIBLIOGRAPHY

1995 Consumer New Home Survey. Des Plaines, IL: *Professional Builder Magazine,* 1995.

Ahluwalia, Gopal, "What First-Time Homebuyers Want," *Housing Economics—The National Association of Home Builders,* October 1993, pp. 5–7.

Belsky, Eric, "Home Building: The Five-Year Outlook," *Housing Economics—The National Association of Home Builders,* January 1995, pp. 5–9.

Builder Magazine, vol. 15, no. 8, July 1992.

D'Alessandris, David., "Demographics and Household Formation," *Housing Economics—The National Association of Home Builders,* May 1994, pp. 5–8.

Fletcher, June, "First Time," *Builder,* vol. 15, no. 8, pp. 55–61, July 1992.

Fletcher, June, "Luxury," *Builder,* vol. 15, no. 8, pp. 81–90, July 1992.

Fletcher, June, "Move-Up," *Builder,* vol. 15, no. 8, pp. 113–119, July 1992.

Fletcher, June, "Singles/Couples," *Builder,* vol. 15, no. 8, pp. 147–153, July 1992.

The Future of Home Building: 1992–1994 and Beyond. Washington, D.C.: The National Association of Home Builders, 1992.

LaFreniere, Andrea, "21st Annual Consumer Survey: How to Build What New Home Buyers Seek," *Professional Builder,* vol. 60, no. 11, pp. 123–135, July 1995.

Professional Builder Magazine, vol. 60, no. 11, July 1995.

Profile of the New Home Buyer. Washington, D.C.: The National Association of Home Builders, 1992.

The State of the Nation's Housing, 1993. Boston, MA: The Joint Center for Housing Studies of Harvard University, 1994

What First-Time Home Buyers Want. Washington, D.C.: The National Association of Home Builders, 1993.

3

Management of the Homebuilding Process

INTRODUCTION

The supply/demand model of the homebuilding industry, which is introduced in Chapter 1 and continued in Chapter 2, provides insight into: (1) the size of the homebuilding industry in relation to the total construction industry, (2) the types of houses that are supplied by homebuilders, and (3) the types of housing demands which the homebuying public places on homebuilders. The information presented does not, however, provide an insight into the homebuilding firms themselves in terms of size, production volume, market share, type of ownership, organizational aspects, etc. That type of information is provided in the first part of this chapter, prior to addressing the issue of how homebuilders can effectively manage the homebuilding process.

HOMEBUILDER ANALYSIS

The National Association of Homebuilders (NAHB), in order to obtain more detailed information about homebuilders than can be provided by Bureau of Census data, has sponsored numerous surveys of the homebuilding industry since the 1960s. The latest of these, "The Homebuilder Industry Survey,"[1] was published in 1994. Results from these surveys, and other NAHB documents and reports, provided the background for the following discussion.

Size Considerations

The homebuilding industry is highly fragmented and diverse, and homebuilders range in size from large multimillion dollar firms, which construct thousands of housing units per year, to homebuilding firms that construct only a few houses per year. It is logical, therefore, to focus on the issue of size as one approach to understanding more about homebuilding firms.

Housing Starts by Size of Homebuilding Firm

Data from the NAHB surveys provide some significant information about the relationship between the size of homebuilding firms and the number of housing starts generated by each

TABLE 3.1   Distribution of NAHB Builders by Number of Starts
(percent of builders in each size category)

	1969	1978	1982	1990	1992	1993	1994
Small Builders (1–25 units)							
Fewer than 11 units	32	46	54	53	59	61	55
11 to 25 units	26	27	19	22	20	20	24
Subtotal	58	73	73	75	79	81	79
Medium Builders							
26–100 units	30	20	18	16	14	13	14
Large Builders (101+ units)							
101–500 units	10	6	8	7	6	5	NA
500 and more units	1	1	2	2	2	1	NA
Subtotal	11	7	10	9	8	6	7

Source: Gopal Ahluwalia, "Home Builders and Their Companies," *Housing Economics—The National Association of Home Builders,* July 1995, p. 10.

size group. Table 3.1, for instance, indicates that the percentage of NAHB member firms that build 10 houses or less per year rose, but not always steadily, from 32% in 1969 to 55% in 1994. In 1994, 79% of NAHB member firms were in the **small builder** category (25 or less starts per year). During the same period, both the **medium builder** and **large builder** categories dropped to their present levels of 14% and 7%, respectively.

Although it is recognized that not all homebuilders in the United States belong to NAHB, the data suggests that the majority of the homebuilders in the United States are small entrepreneurs that build less than 25 houses per year.

Market Share by Size of Homebuilding Firm

It is interesting to note that even though most homebuilders are in the small-volume category, the percentage of the houses built in any year by the large builders accounts for a major portion of the production. This fact is illustrated in Table 3.2, which indicates that the largest homebuilder members of NAHB represented only 7% of the homebuilding firms in

TABLE 3.2.   NAHB Builders by
Size and Number of Starts in
1992 and 1994

Percentage of Units by Builder Size—1992		
	Percent of	
Number of Units	Builders	Starts
25 or less	79	13
26–100	14	18
101 and over	8	69

Percentage of Units by Builder Size—1994		
	Percent of	
Number of Units	Builders	Starts
25 or less	79	14
26–100	14	16
101 and over	7	70

Source: Information provided by Gopal Ahluwalia, Department of Economics, The National Association of Home Builders, June 1996.

1994 but built 70% of the total number of houses constructed that year. Similarly, small builders, even though they represented 79% of the membership, built only 14% of all of the houses.

Market Focus by Size of Homebuilding Firm

Small-volume homebuilders, according to Ahluwalia,[2] traditionally build a high percentage of large custom homes at the upper end of the housing market, while large-volume builders build a higher percentage of smaller houses and houses at the middle to low end of the market.

This assertion is supported by the 1993 data in Table 3.3, which categorizes each size of builder (i.e., small, medium, or large) according to the percentages of houses that were built in certain price ranges. The percentage of homebuilders who built houses that cost more than $250,000 in 1993 ranged from 19% for small-volume homebuilders to only 5% for large-volume homebuilders. Additionally, only 14% of the houses built by small-volume homebuilders cost less than $100,000, while 29% of the production of large-volume homebuilders fell into that class. Information about the median cost of houses for each size of homebuilder also indicates that the small-volume homebuilders favor the high-end housing product.

Analysis of the "Giant" Builders

In Tables 3.1–3.3, all homebuilders that started over 100 housing units per year were classified as **large homebuilders.** An alternative approach is to rank order the largest homebuilders in terms of an appropriate indicator of size such as revenue, housing starts, housing closings, etc., and then establish a threshold above which every homebuilder is considered to be either "giant" or "large" in size.

Professional Builder,[3] in its annual April issue, establishes the threshold at 400 and considers every homebuilder above that level to be part of "The Giant 400" group. *Builder,*[4] on the other hand, uses a threshold of 100 to establish its "Builder 100" group in its annual May issue.

HISTORICAL RECORD

Ahluwalia has provided the historical record of the percentage of houses started by the giant homebuilders, as shown in Table 3.4, for the period from 1974 to 1992. The percentage of

TABLE 3.3   Price of Homes Started in 1993 (percent built in each price range)

| | Number of Units Started by NAHB Member Homebuilders | | | |
	Small Volume (<25)	Medium Volume (25–99)	Large Volume (100+)	Total
<$75,000	4%	10%	7%	7%
$75,000 to $99,999	10	19	22	19
$100,000 to $149,999	22	31	33	30
$150,000 to $199,999	36	19	22	24
$200,000 to $249,999	9	11	10	10
$250,000 to $499,999	14	8	5	8
$500,000 and over	5	2	—	2
Median	$169,444	$133,890	$140,000	

NOTE: Prices are for single-family detached houses, including land.

Source: Gopal Ahluwalia, "Home Builders and Their Companies," *Housing Economics—The National Association of Home Builders,* July 1995, p. 11.

TABLE 3.4   Share of Homes
Started by Top 400, Top 100, and
Top 50 Builders

| | Total Starts | | |
Year	Top 400	Top 100	Top 50
1974	21%	16%	12%
1977	15	10	8
1980	20	14	11
1982[a]	21	16	13
1986	22	15	11
1989	24	16	12
1990	23	15	11
1991	21	14	11
1992	18	13	10

[a]1982 based on 344 builders.

Source: Gopal Ahluwalia, "Changing Industry Structure," *Housing Economics—The National Association of Home Builders,* February 1994, p. 7.

total housing starts has ranged from 15% to 24% for the *Top 400* homebuilders, and from 10% to 16% for the *Top 100* homebuilders.

Although the ranges indicate a rather significant "market capture" by the "housing giants," it should be recalled from Table 3.2, for instance, that during 1992 69% of the housing starts were credited to homebuilders who built over 100 housing units per year. Reference to Table 3.4 indicates that in 1992, the Top 400 started only 18% of the houses. This left approximately 51% of the houses to be built by homebuilders who built more than 100 housing units per year but were not large enough to be included in the Top 400 group.

HOUSING STARTS IN 1995

"The 29th Annual Report of Housing's Giants" appeared in the April 1996 issue of *Professional Builder.*[5] The report noted that the Giant 400 homebuilding firms collectively *started* 268,700 for-sale and rental housing units, or 19.9% of the 1.347 million total housing starts in the United States in 1995.

Professional Builder rank orders the Giant 400 in terms of housing revenues (which do not include revenues due to property management, land sales, and non-residential building and remodeling). Table 3.5 provides detailed information about the five largest homebuilders during 1995, as well as several others at various checkpoint levels, based on the housing revenue criterion.

The largest homebuilder was Pulte Home Corp., which *closed* 12,456 houses in 1995 with total housing revenues of $1,936 million. As is typical of the very large homebuilders, Pulte Home Corp. is dispersed geographically throughout the United States (it sells 17% of its homes in the North, 30% in the South, 34% in the Central, and 18% in the West).

If it is assumed that Pulte Home Corp. started the same number of houses that it closed in 1995, then the largest homebuilder in the United States captured less than 1% of the housing starts (12,456/1,347,000 = .009) in the total U.S. market. This dramatic statistic clearly indicates the extreme fragmentation of the homebuilding industry.

Also of interest in Table 3.5 is the fact that the tenth largest homebuilder (Del Webb Corp.) builds only 35% of the annual production of Pulte Home Corp. This relative production drops off dramatically to 14% when the 100th largest homebuilder (Miles Homes Inc.) is considered. The 400th largest homebuilder (Lazenby Construction) would just have qualified as a large builder in Tables 3.1–3.3.

TABLE 3.5   Giant Homebuilders, 1995

Rank	Company	Home Office Location	1995 Closings	Housing Revenue (Millions)
1	Pulte Home Corp.	Bloomfield Hills, MI	12,456	$1,936
2	Centex Corp.	Dallas, TX	11,790	1,922
3	Ryland Group Inc.	Columbia, MD	8,950	1,460
4	Fleetwood Enterprises Inc[a]	Riverside, CA	68,974	1,431
5	Kaufman & Broad Home Corp	Los Angeles, CA	8,182	1,328
10	Del Webb Corp.	Phoenix, AZ	4,316	764
50	Brisben Cos.	Cincinnati, OH	1,553	193
100	Miles Homes Inc.	Plymouth, MN	1,706	100
200	Hammonds Homes Inc.	Houston, TX	327	49
300	Dunmore Homes Inc.	Fair Oaks, CA	230	28
400	Lazenby Construction	Fayetteville, AR	108	7

[a]Mobile home manufacturer.

Source: "The Giant 400," *Professional Builder,* vol. 61, no. 6, pp. 100–140, April 1996.

Additional Characteristics of Homebuilders

Homebuilders come from all walks of life. The individuals that participate in this industry have backgrounds covering business, engineering, banking, carpentry, etc. Very few of these individuals are specifically trained to cope with the wide variety of problems and situations that a participant in the homebuilding industry must face on a daily basis.

The term "homebuilder" is very loosely applied because anyone who constructs houses can be called a homebuilder. Individuals who build multi-story, inner-city apartments are called homebuilders, but so are carpenters who build only one or two entry-level single-family detached houses a year. Individuals can be called homebuilders if they are involved in the entire life cycle of a residential development (from the conceptual planning stage to the sale and closing of the last unit), but so can those who purchase previously developed lots and construct the houses on them.

In order to obtain some additional insight into the characteristics of a typical homebuilder, NAHB sponsored a Home Builder Industry Survey,[6] which was mailed out to NAHB member companies in April 1994. It had a total of 1,857 responses, so it provides a relatively comprehensive database about the homebuilding industry.

Type of Ownership

Table 3.6 presents the ownership information obtained in the NAHB survey in terms of both the number of housing units built in 1993 and the dollar volume of residential activity. It is interesting to note that a large percentage of small homebuilders is incorporated. As expected, the popularity of the sole proprietorship form of ownership rapidly decreases in favor of the corporation form as the size of the homebuilding firm increases.

Description of Top Management

The continuity of ownership in homebuilding firms is indicated in Table 3.7. Approximately three out of every four homebuilding firms were started, and are continuing to operate, as first-generation builders. This is particularly interesting with regard to the larger homebuilding firms because it appears that a majority of them were able to grow to their present size in one generation.

TABLE 3.6   Type of Firm Ownership in 1993 (percent of respondents)

	Number of Housing Units Started		
	Less than 25	25–99	100 or more
Sole Proprietorship	27%	5%	2%
Partnership	5	5	11
Corporation—regular	33	39	42
Subchapter (S) Corporation	33	48	39
Subsidiary or division of another firm	1	1	5
Other	1	2	2
Respondents	1,133	243	103

	Dollar Volume of Residential Construction Activity			
	Less than $1 million	$1–$4.9 million	$5–$9.9 million	$10 million or more
Sole Proprietorship	41%	11%	4%	2%
Partnership	6	3	6	8
Corporation—regular	22	44	45	40
Subchapter (S) Corporation	30	39	42	43
Subsidiary or division of another firm	—	1	1	5
Other	1	2	1	2
Respondents	707	629	149	130

Source: Home Builder Industry Survey (Washington, D.C.: The National Association of Home Builders, April 1994); information provided by the Economics, Mortgage Finance, and Housing Policy Division of NAHB.

TABLE 3.7   First-Generation Homebuilders in 1993 (percent of respondents)

	Number of Housing Units Started		
	Less than 25	25–99	100 or more
Head of firm is a first-generation builder	72%	70%	77%
Head of firm is not a first-generation builder	28%	30%	23%
Respondents	1,142	250	102

	Dollar Volume of Residential Construction Activity			
	Less than $1 million	$1–$4.9 million	$5–$9.9 million	$10 million or more
Head of firm is a first-generation builder	74%	71%	74%	70%
Head of firm is not a first-generation builder	26%	29%	26%	30%
Respondents	709	638	152	132

Source: Home Builder Industry Survey (Washington, D.C.: The National Association of Home Builders, April 1994); information provided by the Economics, Mortgage Finance, and Housing Policy Division of NAHB.

TABLE 3.8   Phases of the Housing Project that are Subcontracted

	Always	Sometimes	Never
Foundations	81%	13%	6%
Framing	59%	20%	21%
Roofing	73%	19%	8%
Wood flooring	69%	17%	14%
Carpeting	95%	3%	2%
Drywall	85%	12%	3%
Paneling	63%	14%	24%
Fireplace	83%	12%	5%
Plumbing	95%	3%	2%
Electric wiring	93%	4%	3%
Bathroom	76%	13%	11%
Kitchen cabinets	71%	17%	13%
Kitchen countertops	75%	16%	9%
Brick work	93%	5%	2%
Exterior siding	63%	23%	15%
Interior doors	57%	16%	27%
Painting	74%	20%	6%
Doors and windows	64%	14%	22%
Concrete flatwork	78%	17%	6%

Source: Home Builder Industry Survey (Washington, D.C.: The National Association of Home Builders, April 1994); information provided by the Economics, Mortgage Finance, and Housing Policy Division of NAHB.

Supply-Side Information

Table 3.8 indicates that the average homebuilder subcontracts most of the construction work on a house. This data supports the information presented in Chapter 1 about the impact of special trade contractors on the homebuilding industry and the fact that most homebuilders have assumed the role of CONSTRUCTION MANAGERS.

A MANAGEMENT FRAMEWORK FOR HOMEBUILDERS

It is clear that the majority of the homebuilders in the United States are in the small-to-medium-size category. Many of these homebuilders either: (1) are not aware of the types of management skills they must possess in order to both succeed and, more importantly, to survive in a very cyclical industry; or (2) do not have the time, or the inclination, to address the management of their firms in a systematic way.

In order to properly analyze how firms within the residential construction industry are, or should be, managed, an adequate framework of management theory must first be established which is appropriate to an industry that predominately consists of small-to-medium-size entrepreneurs (Table 3.1). The approach presented in this section of the chapter accomplishes this objective by drawing upon three credible sources of management theory that can be related to the unique situation of a homebuilder.

The Evolution of Leadership[7]

Louis A. Allen provides an excellent analysis of the relationship between a leader and the group which is being led. He contends that the nature of the group being led is not static; after formation it does not continue in its original form. The first driving force that causes groups to change occurs "because members of the group want more than what the group presently provides." Until personal satisfaction is enjoyed by most group members, there will be continuous pressure exerted for change. The second force is due to the realization by group members that there is an unequal distribution of the satisfactions that are derived from being a part of the group.

The Natural Leadership Stage

Allen feels that when an individual first assumes a leadership position in an "immature" organization, the individual exhibits a tendency toward intuitive action since he/she is dealing with individuals who give a higher priority to their own needs rather than to the needs and objectives of the group as a whole. According to Allen, such a situation usually requires a **natural leader** who displays some, or all, of the following characteristics:

- Promotion of personal interests
- Technical specialty emphasis
- Centralized decision making
- Intuitive action
- Personalized organization
- One-way communication
- Control by inspection

These characteristics provide a close parallel between a natural leader and a typical small-to-medium-size homebuilder. Large homebuilders, on the other hand, tend to be management leaders.

The Transition Stage—Inadequacy of Natural Leadership

As an organization grows, its members become more confident and technically proficient. In a young (i.e., immature) organization the members often are willing to accept the authoritarian leadership of a natural leader, but as the group matures and better understands the objectives of the organization, they demand more freedom to make and implement their own decisions.

At this critical transitional stage, one reason why a normal evolution of leadership does not proceed is because most successful natural leaders tend to be very resistant to change. Since the success of the firm has been due to the natural leader's own effort and ability, it is difficult to accept the fact that the firm has outgrown him or her, or, at the very least, the style of leadership currently being practiced.

Allen has identified the following symptoms of dissatisfaction that typically develop:

- Dissatisfaction of employees
- Decreased innovation
- Organization emphasis
- Increased expense for lower productivity
- Creation of bureaucracy

The Management Leadership Stage

As previously noted, as a group begins to gain confidence and proficiency, its members begin to feel the need for a new form of leadership. At this point, the group will continue to be effective only if the leadership mode changes. Allen contends that a shift must occur to **management leadership,** which has the following seven characteristics:

- Domination of group objectives
- Management emphasis
- Decentralized decision making

- Logical action
- Rational organization
- Two-way communication
- Control by exception

Importance to Homebuilders

An understanding of the three stages of leadership can allow homebuilders to make important changes as their firms evolve. In a situation of growth, or where the original founder was a first-generation homebuilder (Table 3.7) and is now close to retirement, it is important to realize that not wanting to "let go" of the *natural leadership* phase and hence move into the *management leadership* phase can seriously threaten the survival of the firm.

Managerial Work[8]

An understanding of the nature of managerial work was provided by Henry Mintzberg, whose research identified the work characteristics common to managers and the roles which managers typically fulfill.

Common Work Characteristics of Managers

Based upon direct observations of managers, Mintzberg identified the following six basic work characteristics of managers regardless of their level in the corporate structure or the industry within which they work:

- Unrelenting pace—Managers work long hours with virtually no break in the constant stream of activities and duties they must, or want to, fulfill.
- Brevity, variety, and fragmentation—Managers spend brief periods of time on each activity and must rapidly change their frame of mind and attitude during the day.
- Preference for live action—Managers prefer specific and concrete activities rather than general activities such as long-range planning.
- Attraction to the verbal media—Mintzberg found that the five types of media that managers predominantly use are mail, telephone, unscheduled meetings, scheduled meetings, and tours.
- Communication link—Managers serve as the fundamental communication link between the organization and outsiders.
- Control of activities—Managers try to control their activities and use them to their advantage.

A Manager's Working Roles

Mintzberg, through direct observation of managers, also identified the roles that managers typically perform. He defined a role as one that is observable, and is common to all managers at all levels in an organization.

A manager, by the very nature of the position held, has formal authority and status granted by the organization. As a result of this authority and status, Mintzberg classified managerial activities into the following three roles:[9]

   1. Interpersonal Roles
   —Figurehead
   —Leader
   —Liaison

      2. Information Roles
         —Monitor
         —Disseminator
         —Spokesperson
      3. Decisional Roles
         —Entrepreneur
         —Disturbance Handler
         —Resource Allocator
         —Negotiator

Importance to Homebuilders

Many small-to-medium-size homebuilders have major problems with time; there is never enough of it because of their unrelenting pace and the need to constantly change focus from dealing with the homebuyers, making subcontractor calls in the evening, etc. Often they are unable to properly carry out all of the roles presented by Mintzberg, particularly if they also get involved in professional organizations. Recognition of the environment in which these small-to-medium-size homebuilders operate and the types of roles that must be performed should provide the basis for changing priorities if that is deemed necessary in order to succeed and survive.

A Systems Approach to Management[10]

A system, by definition, is an assemblage of connected or interdependent things, facts, or principles that form a complex, unified scheme. A management system is a combination of external and internal environmental constraints, management theory, and people that interact to produce goods and services.

An Input/Output Transformation Model

The Koontz, O'Donnell, and Weihrich systems approach to management is presented in Fig. 3.1. An enterprise is assumed to receive **inputs** and then **transform** them into **outputs** for use by the environment.

    The inputs from the external environment consist of the human, capital, management, and technology resources needed to produce the end products. The external environment also has goal inputs placed by employees, stockholders, communities, etc. There are also external variables such as market opportunities and material or distribution constraints.

    The manager must be cognizant of the multitude of often incompatible inputs and be able to transform then "in an effective and efficient manner" to produce outputs. The transformation process has been organized by Koontz et al. into an operational system which is categorized into the following five functions of management: (1) **Planning,** (2) **Organizing,** (3) **Staffing,** (4) **Leading,** and (5) **Controlling.**

    The outputs of the enterprise come in many forms, such as products, services, profits, satisfaction, and goal integration. Satisfaction must be an output if the enterprise hopes to elicit further contributions from its members.

    Re-energizing the system comes when the enterprise is successful in its endeavors and has a surplus, such as profits, that it can redirect back into its operations.

Importance to Homebuilders

The work of Koontz et al. has provided a systems framework which includes five management functions which can be used by homebuilders to systematically manage their firms. For the small-to-medium-sized homebuilder, the task of managing in terms of these functions

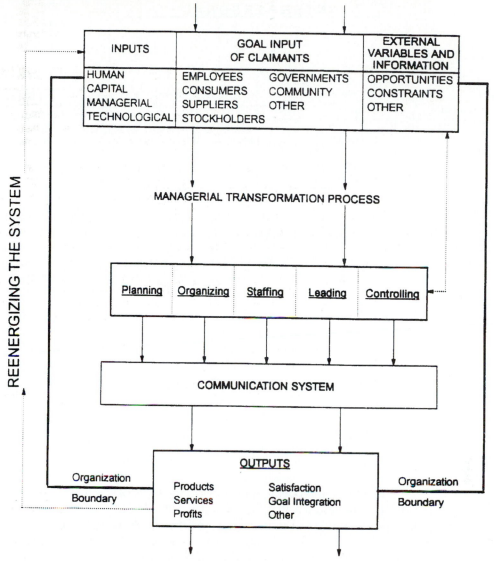

Figure 3.1

A Systems Approach to Management

Source: Harold Koontz, Cyril O'Donnell, and Heinz Weihrich, *Management*, 8th ed. (New York: The McGraw-Hill Companies, 1994), p. 19. Reproduced with the permission of the McGraw-Hill Companies.

can appear to be insurmountable. It goes without saying, however, that these functions must be performed by all managers in all sizes and varieties of organizations.

A small-to-medium-sized homebuilder cannot take advantage of the economics of scale available to the "Top 400" homebuilders, for instance, who often address each of the management functions within separate departments. As a result, smaller homebuilders are forced to approach management in one of two ways: (1) they can choose to practice all of the functions of management in a superficial manner, thus risking mediocrity and never becoming proficient at anything, or (2) they can selectively emphasize the most important function and, as a result, influence the other functions in a satisfactory manner.

It is believed that the **keystone function** which should receive the greatest emphasis is *planning* because it is the catalyst that drives the other functions (Fig. 3.2). Proper planning is not only the keystone to a complete management philosophy; it also alleviates the undesirable consequences of the typical homebuilder's crisis management approach. Crisis man-

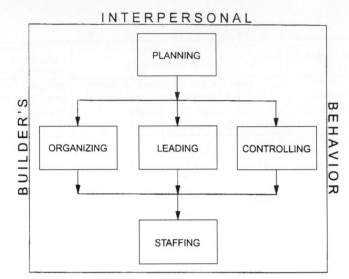

Figure 3.2
The Hierarchy of Management Functions
Source: Jack H. Willenbrock and Walter A. Music, *Managerial Aspects of Residential Construction* (University Park, PA: The Pennsylvania State University, 1985), p. 96.

agement is exhibited through making instantaneous decisions that would have been better decided with forethought. Too often, due consideration is not given to the long-term ramifications resulting from crisis management decisions.

The builder's interpersonal behavior provides a frame around the five functions of management (Fig. 3.2). Interpersonal behavior describes those ill-defined activities and relationships that one person has with another. These are reflected in the tone of voice, facial expressions, words that are used, mannerisms, and sensitivity to others' needs and motivations. Each of these behavior characteristics, when taken collectively, has a direct bearing on the relationships that are established between the homebuilder, the community, government organizations, financiers, customers, the development team, and the builder's employees.

Interpersonal behavior pervades all aspects of the management of the homebuilder's firm. It is imperative that homebuilders understand and be aware of the importance that interpersonal relationships can have on the success of their business.

THE HOUSING INDUSTRY SYSTEM[11]

A generalized "Systems Approach to Management" is presented in Figure 3.1. The equivalent system which takes into account the specific environment in which a homebuilder/developer operates is shown in Fig. 3.3.

The housing industry system consists of two distinct, yet interdependent, parts. This symbiotic relationship can be described as the interaction between: (1) society's requirement for quality affordable housing, and (2) the builder/developer's perception and subsequent response to those needs.

Society's Housing Requirements

The residential housing process consists of a complex interaction of many factors and people that affect the transformation of a parcel of land from one use to another, or a given building lot to a completed house. The intent of this transformation process is to create a profitable venture for the homebuilder/developer as well as an acceptable and beneficial product for society. Society's housing requirements can be described in terms of: (1) the housing market, (2) government controls, and (3) financial parameters (Fig. 3.3).

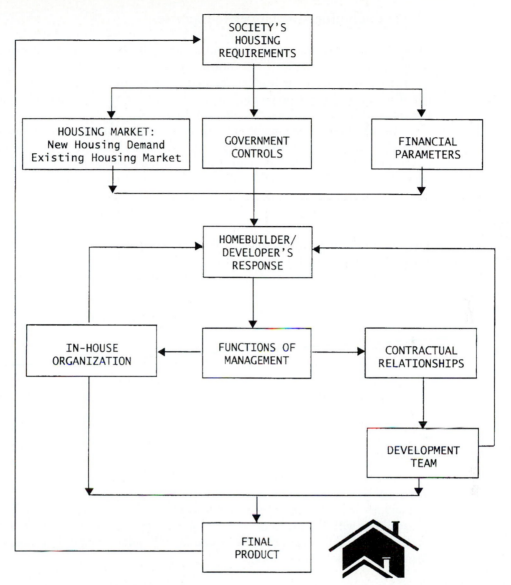

Figure 3.3   A Systems Approach to the Housing Industry
Source: Jack H. Willenbrock and Walter A. Music, *Managerial Aspects of Residential Construction* (University Park, PA: The Pennsylvania State University, 1985), p. 82.

The **housing market,** which combines all of the societal requirements, is a delicate balance between the demand for new housing and the availability of existing housing. It dictates to the homebuilder/developers where to build, what to build, how many units to build, and what price to charge.

Government controls represent a collective and concerted effort by society to regulate its social structure in a manner that is responsive to its needs and desires. With regard to residential housing, these controls exist at the federal, state, and local levels with the maximum effect occurring at the local level in the form of subdivision regulations and building codes.

The homebuilding industry, unlike most other industries, is extremely sensitive to the **financial parameters** operating in the marketplace because of its dependence upon construction loans and its close association with the availability of home mortgage loans to potential homebuyers. Both of these are closely associated with the interest rates that lending institutions charge for the use of money.

The Homebuilder/Developer's Response

Homebuilders/developers are the central actors in the development process because their actions determine the direction a housing project will take. These individuals, by reason of their risk position, are the leaders, the first and final arbiters, and the ultimate decision makers. They make the go–no go decisions, and are the ones who must ultimately balance their concepts with the realities of the marketplace.[12]

In addition to managing an "in-house organization," the homebuilder/developer, through a series of professional and contractual relationships, must manage a development team which typically consists of an attorney, a land planner, a civil engineer, a land surveyor, a landscape architect, an architect, a financier, a realtor, and subcontractors, depending, of course, on the magnitude of the development.

The five functions of management which are used in the transformation process shown in Figure 3.1 (i.e., planning, organizing, staffing, leading, and controlling) provide what is believed to be an adequate framework around which the homebuilder/developer can implement the management responsibilities.

EVOLUTION OF MANAGEMENT IN RESIDENTIAL CONSTRUCTION[13]

Table 3.1 and the descriptive material presented earlier in this chapter indicate the broad range of homebuilding firms that operate within the industry. In terms of growth, a crucial set of questions that often arises is: What characteristics will be required if a firm desires to grow from a small-volume homebuilder, building between 1 and 25 units per year, to a much larger volume homebuilder? Are there commonly held managerial characteristics among homebuilders at similar annual construction volumes? Also, are these managerial characteristics distinctly different from those found in firms at dissimilar annual construction volumes?

Answers to questions such as these provide valuable information to a homebuilder wishing either to expand or contract operations. These answers can be provided within the framework of a **Management Growth Model** which assumes that the homebuilder is not constrained with regard to growth by the external environment, that is, society's requirements. Constraint occurs because of the homebuilder's inability to manage more than the firm's present volume of work.

The Evolutionary Growth Process

It is hypothesized that the growth process for a homebuilding firm consists of the various stages of management evolution shown in Fig. 3.4. The abscissa defining this process is labeled **Construction Output.** The ordinate, labeled **Managerial Proficiency,** implies that there are different levels of management skills, techniques, and awareness that are required to achieve different levels of output (measured in terms of the number of units built per year, annual dollar volume, etc.).

Construction Output

It is difficult to categorize homebuilders according to any single common output parameter. Output often is categorized in terms of the types of housing units constructed, sales volume per year, or number of housing units constructed per year. These categories, however, are not mutually exclusive but rather interdependent.

For the purpose of this discussion, **construction output** is considered to be a generic term used to reflect various levels of production which can be applied uniformly to all homebuilders.

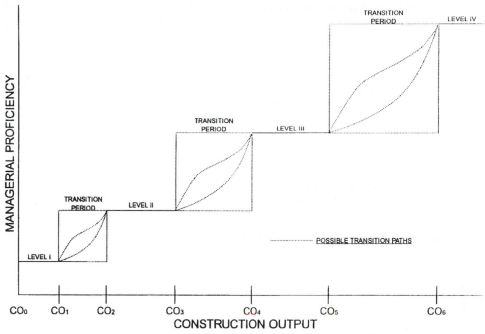

Figure 3.4
Stages in the Evolution of Management in Homebuilding Firms
Source: Jack H. Willenbrock and Walter A. Music, *Managerial Aspects of Residential Construction* (University Park, PA: The Pennsylvania State University, 1985), p. 104.

Factors Influencing Construction Output

In order to grow, it is assumed that the homebuilder must increase output in order to increase profits (assuming that the homebuilder is currently managing a cost-efficient operation). It is proposed that an increase in output is a function of two internal variables. The first variable represents the **effort** that must be expended in order to increase volume, and the second can be represented by the term **managerial proficiency.** Effort can be defined as the amount of work required to increase the number of units constructed at the same level of managerial proficiency. Managerial proficiency is defined as the set of management characteristics, when refined and developed, which also enable a builder to increase construction output.

Effort simply is the brute force method of increasing volume and, hopefully, profits. It entails working longer hours, with more workers or subcontractors, more days per week, with bigger hammers, and exponentially increasing frustration. Effort is limited by the endurance possessed by the homebuilder. Managerial proficiency represents doing the work more efficiently with moderate increases in the energy expended. An increase in managerial proficiency is assumed to be limited only by the homebuilder's training, expertise, and willingness to be innovative.

Managerial Proficiency

Managerial proficiency can be defined in terms of four characteristics that are readily identifiable within any homebuilding firm, regardless of the number of units constructed. It is hypothesized that individually each of the characteristics goes through an evolutionary process which, when considered collectively, determine the evolution of managerial proficiency. These characteristics are (1) management systems, (2) management techniques, (3) educational attitudes, and (4) organizational structure.

MANAGEMENT SYSTEMS

Management systems provide the tools that are used to transform external data, such as market analyses, into a workable set of parameters and objectives which allow construction activity to occur. The management systems also include the tools which manipulate internal information in order to control costs, develop schedules, perform accounting tasks, and forecast the future.

Homebuilding firms require a set of management subsystems that are appropriate to their current level of construction output. Management systems can be divided into the following ten key subsystems that are critical to the effective and efficient management of homebuilding firms:

Business Planning
Office Management
Estimating
Cost Accounting
Cost Control
Scheduling
Subcontracting
Quality Management
Customer Service
Safety

MANAGEMENT TECHNIQUES

Management techniques are those techniques that can be associated directly with the way homebuilders personally manage their firms. The first of these is the relationship between the amount of time a homebuilder spends performing *managing work* compared to the amount of time that is spent performing the *operating work*. Another characteristic is the amount of personal control a homebuilder exerts on the construction process. Is the homebuilder constantly present on the projects, continually making all the decisions, or does the homebuilder simply monitor the progress of each project on a weekly or biweekly basis, thus allowing subordinates to make decisions concerning their own individual areas of responsibility?

A third management technique is associated with the relative amount of intuitive versus deductive decision making practiced by the homebuilder. In other words, how many decisions are based on "guesses and feelings" as opposed to those based on data collection and analyses?

EDUCATIONAL ATTITUDE

Experience indicates that there is a wide variation in the attitude of homebuilders toward the importance of specialized and advanced education in the homebuilding industry. Most small homebuilders, for instance, probably feel that practical "hands-on" experience is more critical to the success of the firm than employing people with specialized education. Such an attitude seriously affects the homebuilder's ability to successfully and efficiently expand the firm's operations into new and more challenging areas.

ORGANIZATIONAL STRUCTURE

A homebuilder's **organizational structure** defines the business relationship that employees have with each other, the responsibilities associated with each position, the amount of authority and accountability associated with the assigned responsibilities, and the specialization of each position within the firm.

It can be assumed that a higher level of managerial proficiency can be attained in an organization with specialized functions rather than in an organization in which the homebuilder assumes the "jack-of-all-trades" approach. It is not to be implied that an extensive organization is required in order to be successful; it is merely proposed that a homebuilder cannot personally perform all of the tasks and still expect to achieve high levels of construction output.

Management Plateaus/Transitional Periods

There are four management plateau levels which are each separated by transition periods (Fig. 3.4). The importance of these aspects of the Management Growth Model is discussed in the following sections.

Management Plateaus

The four **management plateaus** (Levels I to IV) shown in Fig. 3.4 represent the range of construction outputs that can be achieved at the various levels of managerial proficiency. These levels are "stability zones," where construction output matches the effort exerted and the management proficiency achieved. These are also the periods during which the homebuilder must consciously plan the modifications to current management practices if the firm is to successfully experience the subsequent transition period to the next plateau level in the least disruptive way possible and also achieve the next desired increase in construction output.

The Transitional Period

The transitional period occurs when a homebuilder makes a conscious decision to increase construction output by increasing the firm's managerial proficiency. This increase is the result of an iterative process of learning through personal experience and also adapting the most valuable ideas of other homebuilders in similar situations. Each time an attempt is made to increase output in an efficient and profitable manner, lessons are learned that make the next attempt less traumatic.

The transition period often results in an uncomfortable experience for the individuals within the firm. It generally causes interpersonal conflicts because of new expectations and higher anticipated efficiencies. Employees are usually being asked to step out of their personal "comfort zones" and meet new challenges.

Characteristics of Management Plateaus

The four components of management proficiency—management systems, management techniques, educational attitudes, and organizational structure—have unique developmental characteristics at the four levels shown in Fig. 3.4. For example, management techniques probably evolve from a fairly crude and unstructured process in Level I to a sophisticated set of management principles in Level IV. When the management techniques component of managerial proficiency is displayed at each level in conjunction with each of the other components at that level, an evolutionary process for managerial proficiency results.

Managerial Proficiency Level I

A homebuilder functioning at this managerial proficiency level typically is a craftsperson or an individual with little or no experience in the management of the homebuilding process. The organization is an extension of the homebuilder's personality. All functions of management are probably centralized and authoritative in nature.

Managerial Proficiency Level II

A Level II homebuilding firm continues to strongly reflect the personality of the homebuilder. Control still is authoritative in nature but is less centralized due to the increased construction output. A need for structured management systems has been realized and they are, for the most part, manually implemented. For example, accounting and payroll might be recorded manually in ledger books.

Managerial Proficiency Level III

A Level III organization has evolved into a well-defined group of specialists in functional areas such as planning, accounting, and construction supervision. The homebuilder has relinquished control of routine activities and is primarily concerned with long-range or strategic planning. A structured management information system is used extensively. The firm now has a "corporate identity" that is not nearly as dependent on the idiosyncrasies of the homebuilder.

Managerial Proficiency Level IV

Homebuilders operating at the fourth level typically do not focus their major attention on the daily operations of the firm; these concerns are the responsibility of subordinates. There is generally a great deal of depth in the organization, and personnel are hired for their particular skills and experience. It generally can be said that these organizations are the "General Motors®" of the homebuilding industry. They are represented by the upper levels in the "Giant 400" or "Top 100" listings discussed earlier in this chapter.

Management Plateaus for Business Planning

Although each of the four components of managerial proficiency (i.e., management systems, management techniques, educational attitudes, and organizational structure) are important, the one which is probably most useful to the homebuilder is the "management systems" component. If homebuilders compare the management systems presently implemented in their firms with the characteristics of the same management system which a Level I, Level II, etc., builder **should have**, then it is fairly simple to identify the current managerial level of the homebuilder.

As noted earlier, ten key management subsystems are considered to be critical for effective and efficient management of homebuilding firms. These subsystems can be classified into three major categories, as follows:

1. Planning Systems
   a. Business planning
   b. Office management
2. Control Systems
   a. Estimating
   b. Cost accounting
   c. Cost control
   d. Scheduling
   e. Subcontracting
3. Service Systems
   a. Quality management
   b. Customer service
   c. Safety

A graphical representation of one of the management subsystems, **business planning**, is presented in Fig. 3.5. Information of the type shown in Fig. 3.5 will allow homebuilders to answer questions such as "Where am I now?" and "What do I need to do in order to improve and grow?" The answers will enable homebuilders to evaluate their current management systems. Details of the levels will also transmit information about the management systems of homebuilders who currently are at a higher level of managerial proficiency. Homebuilders who wish to improve their management proficiency, as well as their ability to increase construction output, will understand what is required to achieve the next higher level. It is further anticipated that this insight will provide homebuilders with an understanding of how to successfully run their companies for their own benefit, for the benefit of their employees, and for the benefit of all those who will need housing in the future.

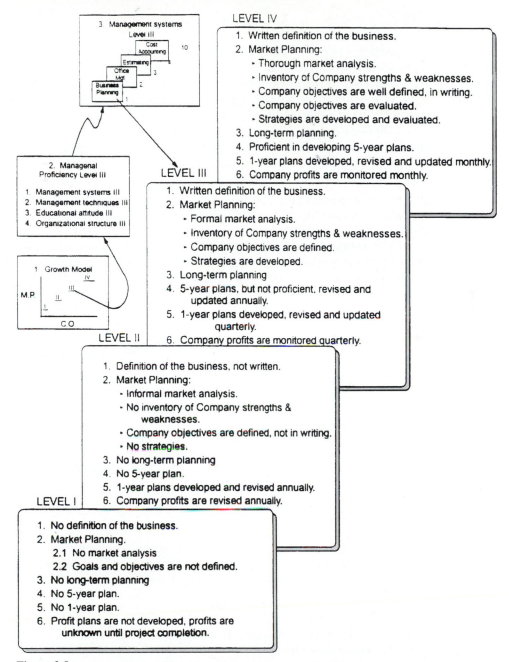

Figure 3.5

Classification of Management Systems: Business Planning

Source: Jack H. Willenbrock, *Management Guidelines for Growth Oriented Homebuilding Firms*, HRC Research Series: Report No. 37 (University Park, PA: The Housing Research Center at The Pennsylvania State University, 1994), p. 60.

As changes to the various management subsystems are made, it is felt that the home-builder's current approaches to the management techniques, educational attitudes, and organizational structural components of managerial proficiency will also naturally be modified.

Two publications of the Housing Research Center at Penn State provide a full development of all of the management systems in a format similar to Fig. 3.5.[14] The interested reader is referred to these publications for additional information.

CHAPTER SUMMARY

The primary objective of this chapter was to provide guidance about the basic concepts of management which can increase the effectiveness of homebuilders as they apply their technical expertise to the homebuilding process. The chapter began with a homebuilder analysis which established the appropriate context for the management framework which was presented later in the chapter.

The information presented indicated the large diversity of firms that operate within the homebuilding industry. Although the majority of the firms can be defined appropriately as **small homebuilders**, there also are a number of **giant** firms that collectively build a majority of the houses in the United States each year. Important market share and size information for both types of homebuilders was provided. Additional characteristics which relate to the types of ownership of homebuilding firms and the characteristics of top management in these firms were also provided.

The **management framework** for homebuilders which was presented in this chapter drew upon the work of three credible sources of management theory which can be related to the unique situation of a homebuilder that was defined in the beginning of the chapter. This framework requires an understanding of: (1) the evolutionary stages of leadership which span from the natural leadership stage to the management leadership stage, (2) the type and characteristics of managerial work typically performed by managers, and (3) a systems approach to management which transforms inputs into the organization into appropriate outputs. After this management framework was established, it was related to the **housing industry system** within which a homebuilder must operate.

The final section of this chapter presented a **Management Growth Model** which can be used to describe the various stages of growth which can be experienced by a homebuilding firm as its (1) management systems, (2) management techniques, (3) educational attitude, and (4) organizational structure evolve over time. This conceptual model can be used by a homebuilder to determine which management level is most appropriate for the firm. An example of how the model can be applied to the **business planning** function was also presented.

NOTES

[1] *Home Builder Industry Survey* (Washington, D.C.: The National Association of Home Builders, April 1994).

[2] Gopal Ahluwalia, "Changing Industry Structure," *Housing Economics—The National Association of Home Builders,* February 1994, p. 6.

[3] *Professional Builder,* vol. 61, no. 6, April 1996.

[4] *Builder,* vol. 19, no. 5, May 1996.

[5] "The Giant 400," *Professional Builder,* vol. 61, no. 6, pp. 100–140, April 1996.

[6] Home Builder Industry Survey, April 1994.

[7] Louis A. Allen, *The Management Profession* (New York: McGraw Hill, 1964); and *Professional Management: New Concepts and Proven Practices* (Berkshire, England: McGraw Hill (U.K.) Limited, 1973).

[8] Henry Mintzberg, *The Nature of Managerial Work* (Englewood Cliffs, N.J.: Prentice Hall, 1980).

[9] Ibid., p. 59.

[10] Harold Koontz, Cyril O'Donnell, and Heinz Weihrich, *Management,* 8th ed. (New York: McGraw Hill, 1984).

[11] Jack H. Willenbrock and Walter A. Music, *Managerial Aspects of Residential Construction* (University Park PA: The Pennsylvania State University, 1985), pp. 81–92.

[12] Paul W. O'Mara, Allan Borut, and Frank H. Spink, Jr., *Residential Development Handbook,* 3rd printing (Washington, D.C.: The Urban Institute, 1978), p. 6.

[13] Willenbrock and Music, *Managerial Aspects,* pp. 102–122.

[14] Ibid.; and Jack H. Willenbrock, *Management Guidelines for Growth Oriented Homebuilding Firms,* HRC Research Series: Report No. 37 (University Park, PA: The Housing Research Center at The Pennsylvania State University, 1994).

BIBLIOGRAPHY

Ahluwalia, Gopal, "Changing Industry Structure," *Housing Economics—The National Association of Home Builders,* February 1994.

Ahluwalia, Gopal, "Home Builders and Their Companies," *Housing Economics—The National Association of Home Builders,* July 1995.

Allen, Louis A., *The Management Profession.*   New York: McGraw Hill, 1964.

Allen, Louis A., *Professional Management: New Concepts and Proven Practices.*   Berkshire, England: McGraw Hill (U.K.) Limited, 1973.

Builder, vol. 19, no. 5, May 1996.

"The Giant 400," *Professional Builder Magazine,* vol. 60, no. 6, April 1995.

The Future of Home Building: 1992–1994 and Beyond.   Washington, D.C.: The National Association of Home Builders, 1992.

Home Builder Industry Survey.   Washington, D.C.: The National Association of Home Builders, April 1994.

Koontz, Harold, Cyril O'Donnell, and Heinz Weihrich, *Management,* 8th ed.   New York: McGraw Hill, 1984.

Mintzberg, Henry, *The Nature of Managerial Work.*   Englewood Cliffs, NJ: Prentice Hall, 1980.

O'Mara, Paul W., Allan Borut, and Frank H. Spink, Jr., *Residential Development Handbook,* 3rd printing. Washington, D.C.: The Urban Institute, 1978.

Professional Builder, vol. 61, no. 6, April 1996.

Willenbrock, Jack H., *Management Guidelines for Growth Oriented Homebuilding Firms,* HRC Research Series: Report No. 37 (University Park, PA: The Housing Research Center at The Pennsylvania State University, 1994).

Willenbrock, Jack H. and Walter A. Music, *Managerial Aspects of Residential Construction.*   University Park, PA: The Pennsylvania State University, 1985.

4

The Regulatory Environment

INTRODUCTION

The primary role of the homebuilder/developer, as suggested in Fig. 3.3, is to transform society's housing requirements (both in terms of land and houses) into final products which homebuyers will purchase. These requirements most directly affect the homebuilder/developer as: (1) land development regulations and zoning ordinances, and (2) building codes.

The regulatory environment which results from these two mechanisms is described in this chapter.

THE LAND DEVELOPMENT REGULATORY ENVIRONMENT

Residential land development, in most areas of the United States, is governed by a set of procedures established by local governmental authorities which guide homebuilders/developers in the preparation and submission of land development plans. Typically, these plans (1) must consider the existing characteristics of the site, (2) must utilize appropriate engineering design principles and improvement standards, and (3) are subject to review and approval by the local governmental authorities to ensure compliance with existing zoning ordinances, land use plans, regulations, and standards.

A Land Development Model

The effect which government regulations have on residential land development can be best understood in relation to an appropriate land development model. McIntyre et al.,[1] for example, have divided the land development process into three phases:

> Phase I—Preliminary Project Planning
> Phase II—Project Design and Review
> Phase III—Construction and Delivery

Fig. 4.1 indicates the interrelationship between these three phases.

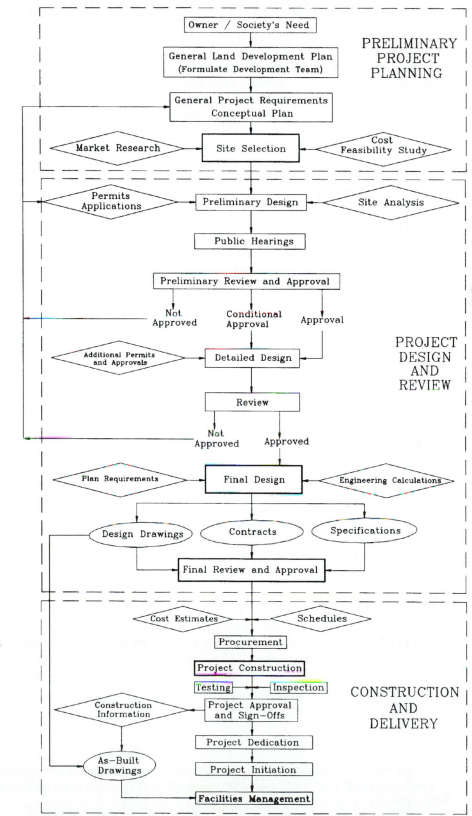

Figure 4.1

A Residential Land Development Model

Source: Charles McIntyre, M. Kevin Parfitt, and Jack H. Willenbrock, *A Residential Land Development Model*, HRC Research Series: Report No. 28 (University Park, PA: The Housing Research Center at The Pennsylvania State University, August 1994), Fig. 4.4, p. 69.

Preliminary Project Planning

Preliminary project planning begins by defining the **need** for a project. The need could arise because the builder/developer is looking for a location on which to place a predescribed housing product or because the builder/developer already owns the land and is looking for the appropriate housing product to put on it. In either case, the homebuilder/developer identifies the need and usually relates it to a for-profit opportunity.

At the early stages, several possible land parcels (sites) are probably considered. Developers must be sensitive to the characteristics of each of the sites and the adjacent neighborhoods, as well as site topography, geology, hydrology, vegetation, and climate. The decision about which of the alternative sites will be selected for the preliminary design phase relies on the information that is gathered from both market research and cost feasibility studies. It is also contingent upon the availability of the land for each site.

MARKET RESEARCH

Market research assists the homebuilder/developer in identifying the target market for the project, thus providing a clearer definition of the general project requirements. Information usually is gathered about:

1. Existing types of housing on the market
2. The status of the resale market
3. The important local competition
4. The economic forces influencing the project
5. The demographics of the affected area

COST FEASIBILITY STUDY

A **cost feasibility study** helps to determine the costs associated with the development of each of the parcels of land being considered. The study should strive to maximize the economical use of the land, while not compromising the natural character of the site, the adjacent neighborhood, or the community as a whole. The bottom line of the study is that it provides the cost information necessary to establish the price of individual parcels or lots.

Although detailed design information is not available at this point in the process, an attempt should be made to establish reasonable costs for each item associated with the alternative sites. Brewer and Alter, for instance, suggest that the following cost factors should be considered:[2]

1. Land
2. Pavement
3. Sidewalks
4. Storm Sewers
5. Sanitary Sewers
6. Utilities (gas, electric, water)
7. Engineering/Surveying Fees
8. Legal Fees
9. Finances

Additional cost factors that must be considered for each alternative site include any impact fees associated with a particular site, as well as taxes, overhead costs, and desired profit margins.

The combination of market research and cost feasibility studies assists the homebuilder/developer in making a decision about which of the set of possible alternative sites should be selected.

Project Design and Review

Once a site is selected, initial applications and permits must be filed with appropriate regulatory agencies. A detailed site analysis is associated with this process. A **preliminary design** is then completed. This design typically includes: the road and lot layout, access to adjoining streets or properties, drainage considerations, and the availability of sewer, water, and utilities. After this preliminary design is subjected to public hearings and preliminary reviews by the local regulatory agencies, modifications are almost always required prior to the issuance of a preliminary approval.

The same basic sequence is followed during the detailed design, review, and approval phase. A **detailed design** requires extensive engineering calculations, detailed design drawings, and usually a project narrative complete with the necessary permits and applications.

A **final design,** complete with contract documents and specifications as well as any desired property restrictions (also called deed restrictions or restrictive covenants), concludes the project design and review phase. In addition, cost estimates and schedules are usually prepared by the engineering firm and are used as project guidelines when contractor bids are evaluated.

Construction and Delivery

The initial phase of construction begins with the procurement of the necessary contractor or contractors. Resources, both human and material, must be supplied at the site. Materials and equipment for construction activities may be supplied either by the contractors or the homebuilder/developer.

Inspection and testing ensures that a satisfactory level of quality will be achieved on the project. In-house inspection, or inspection by independent testing laboratory personnel, for **quality control** purposes is often provided by the contractors. These testing programs often are supplemented by **quality assurance** testing by inspectors representing the local municipality or other jurisdiction with authority over the project to verify that the construction meets or exceeds their specification requirements.

Once the entire residential land development project, or a designated phase of the project, is completed and judged to be in conformance with the required regulations and standards, the easements, utilities, streets, etc. are usually dedicated to the local municipality or other appropriate jurisdiction. The construction of houses can then begin as the individual lots are sold either directly to homebuyers or to contractors that build scattered lot speculative housing.

The municipality or the provider of a utility (water and sewer, for instance) has the continued responsibility of providing and maintaining the service. Eventually all systems need maintenance, repair, or replacement: roads will have to be resurfaced, sewer and storm water pipes and manholes will have to be cleaned, and parks and playgrounds will have to be maintained and upgraded. **Facilities management**, thus, is an integral part of the overall residential and development process.

The Project Design and Review Phase

Although the homebuilder/developer continually interacts with the local governmental agencies throughout the three phases shown in Figure 4.1, it is clear that the maximum influence of governmental regulations is felt during the project design and review phase.

The development of the preliminary design and the transformation of that design into the final design are the two major milestones of the project design and review phase. As indicated in Fig. 4.1, this transformation occurs as an iterative process which includes several regulatory agency review and approval steps interspersed with design modification steps.

Three of the factors which provide the framework within which the preliminary design of the selected site can proceed are:

- Zoning
- Land development regulations
- General and specific existing site conditions

Zoning

Zoning is the most common system of legal constraints in the United States which influences subdivision and site planning. Most states have passed enabling legislation, which authorizes zoning at the local governmental level. Zoning ordinances—enacted at the county, city, or town level—enable local governmental entities to plan for the type, number, and location of the various land uses within their geographical boundaries.

PURPOSE OF ZONING

Zoning is defined as the application of land-use controls to protect the rights of the individual property owner, as well as the rights of others in the community. Zoning is formulated and controlled by the residents of the community. Its function is to control how the environment and public services of the community are used. It also requires the segregation of incompatible land uses into separate districts for the protection of the public's health, safety, and welfare, which includes property values. Zoning can be thought of as a collective property right which maintains the neighborhood and business environment and the right to the public services which complement that environment.

The main purpose of zoning is to achieve the "best" possible use for any given parcel of land. Zoning provides the enforcement and implementation mechanism for the comprehensive land use plan of an area which defines desired future land use. Sound planning for land use includes social factors, economic conditions, and the physical characteristics of the land. In this context, the homebuilder/developer must stipulate the size and types of housing units that will be constructed in a subdivision. Houses may be required to have a minimum square footage of interior space and must blend in with other structures in the neighborhood with respect to exterior, building height, and setback distances, with the intention of protecting the rights of current and future property owners.

CONVENTIONAL ZONING ORDINANCES

The types of controls that typically are specified in conventional zoning ordinances include: (1) use regulations, including primary and accessory uses; (2) lot requirements (density control), including size, width, and coverage; (3) setback requirements for front, side, and rear yards; and (4) maximum height restrictions.

Use is controlled by dividing the jurisdictional area into districts and then permitting only certain activities to occur in each district. Most zoning ordinances create: (1) agricultural (or rural), (2) residential, (3) commercial, and (4) industrial districts. An official zoning map is usually issued to define the geographical area for each zoning district. As an example, Table 4.1 lists the zoning districts for Ferguson Township, located in Centre County, Pennsylvania. Figure 4.2 illustrates a portion of a typical local zoning map which is consistent with the districts defined in Table 4.1.

Most communities have a zoning officer who is empowered to enforce the regulations. Larger municipalities may have an entire zoning office or department. The decisions of the zoning officer, or the performance standards of the ordinance, may be appealed. In general, three basic tests determine the legality of a zoning regulation:

- It must promote the health, safety, and welfare of the public.
- It cannot be arbitrary and capricious.
- It must be reasonable and consistent.

TABLE 4.1 Zoning Districts in the
Zoning Ordinance of the Township of
Ferguson, Centre County, Pennsylvania

Districts	Map Symbol
Rural Districts	
Rural Agricultural District	RA
Rural Residential District	RR
Residential Districts	
Single-Family Residential District	R-1
Suburban Single-Family Residential District	R-1B
Two-Family Residential District	R-2
Townhouse Residential District	R-3
Multi-Family Residential District	R-4
Village District	V
Mobile Home Park District	MHP
Planned Residential Development District	PRD
Commercial Districts	
Office Commercial District	OC
General Commercial District	C
Industrial Districts	
General Industrial District	I
Light Industry, Research and Development District	IRD

Source: Zoning Ordinances 224 and 506, p. 453, (Ferguson Township, PA, Township of Ferguson: Centre County, Oct. 1991).

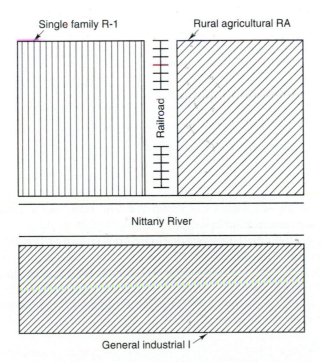

Figure 4.2
A Portion of a
Typical Official
Zoning Map

ZONING SYMBOLS AND PERFORMANCE STANDARDS

In order to properly interpret the requirements in a zoning ordinance, the homebuilder/developer must understand the types of symbols that are typically used on land development plans and drawings. Some of these symbols, as indicated in Fig. 4.3, include:

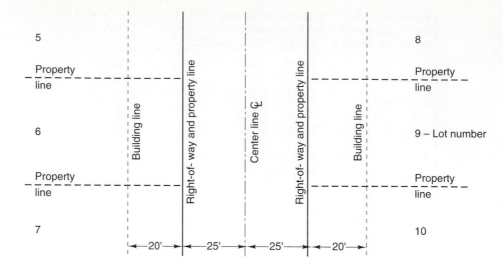

Figure 4.3
Typical Symbols
on Land
Development
Drawings

- R-O-W (right of way): The portion of the subdivision that is allocated for public use. This area, which is maintained by the municipality, contains the street and the utility lines.
- C/L (center line): Refers to the center line of the R-O-W and usually, but not always, the center line of the street.
- P/L (property line): Designates the ownership line of a parcel of property. The property line towards the street is a common line with the R-O-W.
- B/L (building line): Marks the limits of a building location on a given property. The B/L is often called the setback line or setback distance.
- Lot number: References or identifies a lot within a subdivision.[3]

At the very least, a local zoning ordinance will typically specify the following performance standards:

1. Minimal front, rear, and side yard setbacks
2. Maximum building height limits
3. Off-street parking requirements

These minimum and maximum values provide the basis for the maximum building envelope or maximum building shell on a site. The definition of these values that are used in the Zoning Ordinance of the Township of Ferguson is shown in Fig. 4.4. A three-dimensional interpretation is shown in Fig. 4.5.

The actual lot size and yard setback requirements for each acceptable type of use within a designated zoning district are usually presented in a separate table in the zoning ordinance. The one that applies to the R-1 District designation in the Zoning Ordinance of the Township of Ferguson is shown in Table 4.2.

ZONING CHANGES

One of the first questions that must be answered during the site analysis phase is whether the parcel of land that has been selected for residential development is zoned for the use which the homebuilder/developer intends. If it is not, and the homebuilder/developer still wants to pursue the issue, then it will be necessary to appear before the local zoning board to request a change of zoning. The three typical approaches that can be used are to request: (1) a variance, (2) a special exception, or (3) an amendment. As noted by Pisani and Pisani:

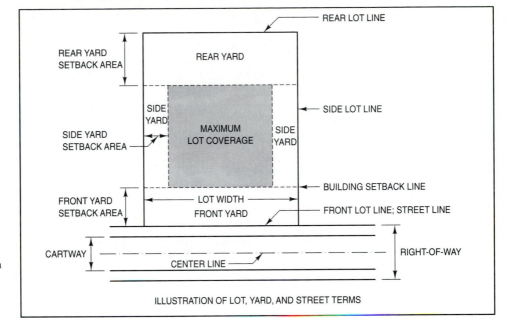

Figure 4.4
Illustration of
Lot, Yard, and
Street Terms
Source: Zoning
Ordinance of the
Township of
Ferguson (Ferguson
Township, Centre
County, PA,
October 1991),
p. 548.1.

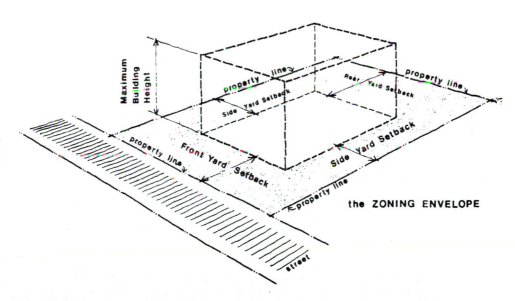

Figure 4.5
Three-
Dimensional
Representation of
Euclidean Zoning
Setbacks
Source: R. Gene
Brooks, *Site
Planning—
Environment,
Process and
Development*
(Englewood Cliffs,
NJ: Prentice Hall,
1988), p. 268.

A **variance** is a request to change a portion of the zoning requirements for the property without changing the zoning itself. . . . The applicant for a variance must demonstrate a hardship if the zoning ordinance is strictly applied.

A **special exception** (also known as a "conditional permit") allows the use of land for a purpose different than that permitted in the zoning ordinance. The ordinance typically lists those uses that are considered special cases. . . . The difference between a variance and a special exception is that no hardship or unique characteristics need be shown to receive a special exception . . .

An **amendment** is the formal process which partially or wholly changes a zoning district by changing the zoning ordinance itself. Only the governing body can make these changes, either voluntarily or through court action brought by the developer.[4]

TABLE 4.2 Criteria Standards for the R-1 District in Ferguson Township

Permitted Uses — Land and structures may be used for only the following situations.	Lot Requirements — The following lot requirements shall be met for each primary use.				Yard Setback Requirements — The following yard requirements shall be met.			
	Minimum Size	Minimum Width	Maximum Coverage	Maximum Impervious Coverage	Front Yard	Side Yard	Rear Yard	Maximum Height
PERMITTED USES								
1. Single-family detached dwellings, with off-site sewer service	10,000 sq ft	80 ft at the building setback line; 50 ft at the street line			20 ft on local and collector streets; 50 ft on arterial streets	10 ft	30 ft	
2. Single-family detached dwellings, with on-site sewer service	1 acre	100 ft at the building setback line; 75 ft at the street line	30%	50%				40 ft
3. Churches and other places of worship, parish houses and convents	1 acre	150 ft at the building setback line; 100 ft at the street line						
4. Public and private nursery, kindergarten, elementary and secondary schools					50 ft	50 ft	75 ft	
5. Public park and recreational areas	—	—						
6. Personal care boarding homes	8,500 sq ft	80 ft at the building setback line; 50 ft at the street line			30 ft on local and collector streets; 50 ft on arterial streets	10 ft	30 ft	

ACCESSORY USES
7. Home occupations
8. Customary uses accessory to the above; essential services

See Primary Use above to which it is accessory

(Added to by *Ordinance 237*, September 7, 1982).

Source: Zoning Ordinances 224 and 506, p. 474 (Ferguson Township, PA, Township of Ferguson: Centre County PA, Oct. 1991).

Land Development Regulations

Most local governmental agencies consider that the **zoning ordinance** and **land development regulations** are the two primary documents which define how land development will proceed in their jurisdiction. The basic difference between these documents is that the zoning ordinance defines how the land can be used, while the land development regulations specify the minimum standards that must be applied during the application, design, approval, and construction phases of the project. After a site has "passed" an analysis of its proper zoning classification, it must then be analyzed in terms of the specific requirements which appear in the land development regulation.

Although the contents of each set of land development regulations differ, most of them probably contain information about:

1. The plan submittal and review procedure.

2. The information that is required on the preliminary as well as the final set of plans.

3. The specific design standards for each phase of the land development project.

4. The procedures that are involved in the administration and enforcement of the land development regulation.

THE PLAN SUBMITTAL AND REVIEW PROCEDURE

Information about the local government's plan submittal and review procedure is important to the homebuilder/developer during the site analysis stage because it indicates: (1) the number of steps in the process, (2) the number of different levels of review that are required, and (3) the anticipated timetable which begins once the preliminary plans are submitted.

One of the problems with land development regulations in the United States is that unified or model land development regulations do not exist. Since they are usually local documents, the jurisdiction of these regulations in some states stops at the boundaries of the local governmental agencies or at the city limits. As a result, land development regulations often change, sometimes dramatically, as one crosses a municipal or county line.

In order to address this issue, the NAHB Research Center, under a contract sponsored by the U.S. Department of Housing and Urban Development, developed a proposed Model Land Development Standard and accompanying Model State Enabling Legislation.[5] This effort represented the first attempt in the United States to standardize land development regulations and the land development process. The proposed land development approval process which was included in the report is shown in Fig. 4.6.

INFORMATION REQUIRED ON THE SET OF PLANS

This section of the land development regulations establishes the requirements for the format as well as the content of the sets of plans that are submitted in both the preliminary and final submittal stages.

SPECIFIC DESIGN STANDARDS FOR EACH PHASE OF THE LAND DEVELOPMENT PROJECT

The engineering analysis and design of the various phases of the typical land development project usually include technical requirements which govern:

- Street widths
- Construction of pavement
- Water supply
- Fire hydrant location
- Sanitary sewers
- Storm sewers

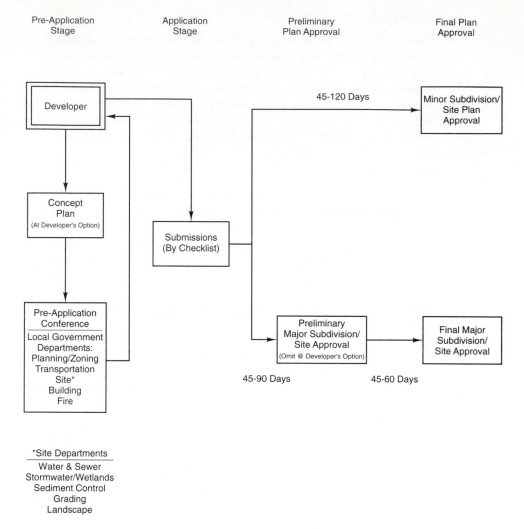

Figure 4.6
Simplified Land
Development
Approval
Process—
Subdivision
and/or Site Plan

*Source: Proposed
Model Land
Development
Standards and
Accompanying
Model State
Enabling Legis-
lation, 1993 edition
(Washington, DC:
U.S. HUD, 1993),
p. A-4.*

ADMINISTRATIVE AND ENFORCEMENT PROCEDURES

This section of the land development regulations typically indicates how the entire land development process is administered and the enforcement options that are available to the local governmental agency in the event that there is noncompliance with any of the procedures in the ordinance.

General and Specific Existing Site Conditions

As the engineering and design process proceeds, increased attention is focused on both the **general and specific existing site conditions.** On-site information is integrated with the proposed improvements that are necessary to obtain a total understanding of the scope of the project.

Some of the existing conditions related to the site which influence the final design are discussed in the following sections.

CLIMATE

Each site has a characteristic climate which is typical of the region within which it is located. **Climate** is expressed in terms of annual precipitation, number of possible sun days, relative humidity, and prevailing wind direction. Climate can influence the orientation of structures, heating and air-conditioning costs, building material selection, and landscape plantings.

TOPOGRAPHY

Topography defines the slope gradient of a site expressed as the relationship of vertical feet of elevation over horizontal feet of distance. Topography influences the location of houses on their lots as well as the placement of roads, pathways and utilities. Slopes of 8–10% usually represent the maximum values for vehicular or pedestrian traffic. Slopes of greater than 15% are usually prohibitive for most uses. Slopes of less than 2% may require additional grading to ensure adequate drainage. Topographical maps (Fig. 4.7), based upon information obtained from U.S. Geological Survey Maps, can be used to identify areas of concern when developing a site plan.

GEOLOGY

A wide variety of rock formations and soils can affect even a relatively small site. Specific **geological conditions** can be a liability or an asset. For example, granite bedrock close to the surface may increase excavation costs, limestone or dolomite substrate may be subjected to sinkhole formation, etc. Geological maps [Fig. 4.8(a)] derived from information provided by the U.S. Soil Conservation Service can be used to interpret site characteristics and potential.

SOIL

Site-specific **soil types** must be documented in order to evaluate the permeability, erodibility, bearing strength, and agricultural value of the land surface. This information can be extracted from soil surveys provided by the U.S. Soil Conservation Service [Fig. 4.8(b)]; however, the engineering properties of the soils should also be evaluated and analyzed through site visitations and investigations.

It is also extremely important to check the specific condition of the soil where a house will be located. One of the main reasons for foundation failures and cracking is poor soil conditions. If the soil has been disturbed previously, it should be removed and replaced with fill that is compacted using the proper equipment in order to transform the soil into a system which is firm and dense enough to carry the weight of the house. Other alternatives in this situation include excavating deeper until undisturbed soil is encountered and designing

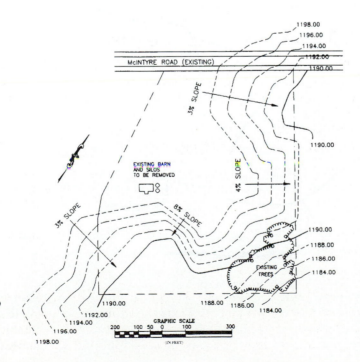

Figure 4.7
Topographic Map of Existing Site Conditions at Hankin Estates

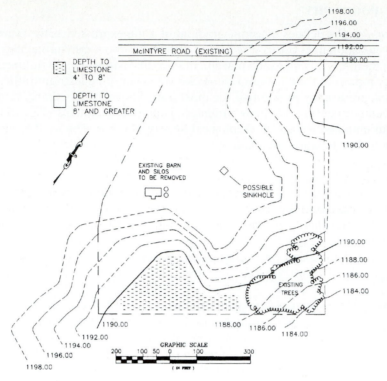

(a.) Geological Map

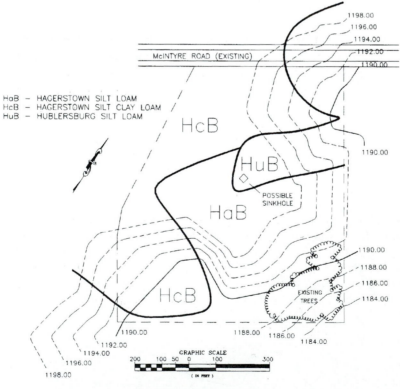

(b.) Soils Map

Figure 4.8
Existing Geology
and Soils
Information of
Hankin Estates

a special foundation. Any soil that has not been disturbed previously should be tested for its load-bearing capacity. Since the entire weight of the building is supported by the ground, the characteristics of the soil will determine how much weight it can support.

HYDROLOGY

In general, **hydrology** refers to the movement of water above or below the land surface. Even though high market value is placed upon property adjacent to open bodies of water, caution is warranted due to periodic flooding considerations and flood plain regulations.

Wetlands also require special attention. Increased environmental awareness of these sensitive areas has occurred because wildlife habitat is sometimes destroyed when developments are built without due consideration being given to this issue. Environmentally sensitive land development provides the best solution for everyone. Stricter governmental regulations have been imposed on many areas to ensure that such solutions are achieved. As a result, areas that have been defined as wetlands often carry restrictions against a number of proposed land activities.

The location of the water table is another important factor that must be considered during the design and construction phases of a project. A water table depth of less than 8 ft may pose problems for utility and foundation excavation. Permanent cellar sump pumps may also be required in certain locations.

EXTERNAL NOISE, ODORS, AND VIEWS

Noise pollution from adjacent sites should be documented during the site investigation phase. In addition, offensive, as well as pleasant, odors should be investigated. A project can also be greatly enhanced by attractive views and its value generally reduced if it is adjacent to a visually undesirable area.

SOLID AND HAZARDOUS WASTE

The presence of any waste material on a site must be investigated. Chemical or hazardous wastes pose potential liability problems for developers and builders. Undocumented hazardous waste sites may be uncovered during the site analysis, but often are discovered only after site construction begins.

ACCESS

Site access must be evaluated to determine the need for an access road. Not only is the initial access to the site important during construction, but an adequate means of access/egress to and from the site is also critical during the occupancy stage (Fig. 4.9).

UTILITIES

The utilities that are available to the site must be determined early in the planning stages of subdivision development. If sewer lines to a central treatment plant are not accessible, a septic tank and drainfield or some other acceptable alternative must be installed and maintained. In addition, if a centralized water supply is not available to the site, a deep well provided with a pump will be required. When selecting a site, it is important to consider how much the initial and annual maintenance costs of such installations increase the original land costs.

LEGAL ISSUES

Present zoning, covenants, and legal restrictions on the proposed parcel of land must be researched. Existing deeds and titles should be reviewed by an attorney, at which time a full title search should be performed to determine the legal limits of the parcel and any imposed restrictions. Deed restrictions or covenants may allow only a certain type or size of house to be built on the land. Contact with local planning boards or review agencies is suggested in order to determine any additional covenants or restrictions that may apply to the parcel of land being considered.

Figure 4.9
Proposed Access
Analysis of
Hankin Estates

THE BUILDING CODE REGULATORY ENVIRONMENT

The design and construction of houses are typically regulated at the local governmental level through the use of building codes. Not all areas of the United States are covered by building codes since the local governmental level must first adopt a particular building code by ordinance before it can be used.

Background

The history of building codes reaches far back in time. Building codes as we know them today did not evolve in any orderly manner. In fact, construction regulations were often non-existent in many societies or probably were formulated only as a reaction to a particular disaster. As specific construction products and techniques proved to be capable of withstanding specific natural forces, thereby providing a degree of safety, they became accepted as approved methods of construction. These approved methods correspondingly became the foundation of modern building codes.

Building Regulations in the United States—1648 to Present[6]

Building regulations in the United States probably started as early as 1648, when wooden chimneys were no longer permitted in New York City and inspectors of chimneys were appointed. A fire district was created in 1766 that required all buildings within the district to be constructed of stone or brick with roofs made of tile or slate. The first real building code was established in New York City in 1862, and it provided for building construction with specific exit requirements. Regulations that followed included plumbing in 1880, elevators in 1883, height of non-fireproof buildings in 1885, and precautions for workmen's safety in 1896.

The Chicago fire of 1871 (which lasted two days, killed 250 persons, and destroyed 17,000 buildings) prompted that city to enact a building code and fire prevention ordinance in 1875. In 1892, a new building code for Boston contained limits on stresses for wood, steel, cast iron, and masonry. Professional and scientific contributions to an understanding of how building materials and systems functioned were made after the American Society of Civil Engineers was formed in 1852, the American Society for Testing Materials was formed in 1902, and the National Board of Fire Underwriters was formed in 1905.

By the early 1900s, the regulations in building codes had become very extensive in an attempt to encompass the increasing number of different building materials. Several model code organizations which developed building code documents were founded during this period. The divergence of new materials and opinions, plus public demand for uniformity in codes and lower construction costs, forced the United States Senate to establish the Department of Commerce Building Code Committee in 1921. The committee, composed of seven nationally recognized architects and engineers, was formed to make recommendations regarding the development of building codes which were based upon scientific data; to write codes that assured uniformity of practice; and to provide a basic starting point for the preparation of local building codes.

Types of Codes

Today's model building code regulations are the product of historical events and the contributions of university, private, industrial, and governmental laboratories; professional and technical societies; and code-writing organizations whose members study, coordinate, and adapt for use the vast amount of data obtained in the laboratory and in the field.

The two basic types of codes that have evolved are **specification codes** and **performance codes.** Each of these incorporates into their documents **industry standards** which have been developed by some of the more than 500 standards-writing organizations in the United States.

SPECIFICATION CODE

A specification-type code typically presents a fixed concept or solution for a particular situation and allows little or no provisions for alternatives or innovation. This type of code describes in detail the types of materials and methods to be used in achieving the required result. New materials and methods of construction are often denied, thus at times increasing expenses without providing effective benefits. An example of a specification code stipulation is: "Steel columns placed in exterior walls shall be protected *by approved masonry not less than four inches thick."*

PERFORMANCE CODE

A performance code, unlike the rigid requirements of a specification code, makes statements in terms of the measurable performance desired. This allows for the acceptance of new materials and methods of construction that have been tested and proven to be acceptable. Performance codes are not always entirely free of specification code stipulations, but they do help to stimulate competitive innovations by allowing for alternatives. An example of a performance code stipulation is: "Steel columns placed in exterior walls shall be protected *with an approved fire resistance rating of four hours."* The specification code states what the columns shall be protected with, while the performance code states only the desired end result.

INCORPORATION OF REFERENCE STANDARDS

A **standard** is a common language that promotes the flow of goods between buyer and seller and protects the general welfare. The word *standard*[7] can be used in a building code in a number of ways, such as: (1) a standard definition,(2) a standard recommended practice, (3) standard test methods, (4) standard classification, and (5) standard specification. The source of the standard also can vary; thus either a (1) company standard, (2) industry standard, (3) government standard, or (4) full consensus standard might be the appropriate one to use for a particular application.

As building codes have evolved over the years, it has become common practice to incorporate various standards into the codes by reference in order to reduce the size of the particular code. Such incorporation, from a legal standpoint, is often considered to be equivalent to including the entire standard directly in the building code.

Organizations that write standards that have been included in building codes as they have evolved include:

1. American National Standards Institute (ANSI)
2. American Society of Mechanical Engineers (ASME)
3. American Society of Testing and Materials (ASTM)
4. National Fire Protection Association (NFPA)
5. American Society of Heating, Refrigerating and Air Conditioning Engineers (ASHRAE)

Model Building Codes in the United States

As a result of the evolution of building codes in the United States, there are now three major model building codes that regulate both commercial and residential construction in the geographical regions shown in Fig. 4.10. These codes are generally considered to be performance codes, even though they contain many standard specification codes either by inclusion within the code text, by reference in the Appendix, or by local amendments. In this manner, building codes allow designers and contractors a large amount of freedom with regard to the selection of materials and the method of construction, respectively. In addition, there is a model building code which has been jointly developed by the three model building code organizations specifically for the residential sector of the construction industry. This code is designed to be more of a specification code, since it is used by many people who do not have the training to perform an independent engineering analysis of each unique situation. Each of these is discussed in the following sections.

BOCA (Building Officials and Code Administrators International, Inc.)

BOCA, founded in 1915, is the oldest professional association of construction code officials in the United States; presently it is headquartered in Country Club Hills, Illinois. BOCA publishes a complete model code package centered around *The BOCA National Building*

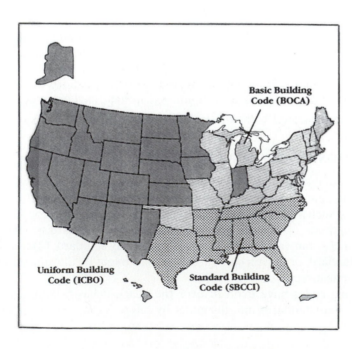

Figure 4.10
Approximate
Areas of Model
Code Influence
Source:
Understanding
Building Codes and
Standards in the
United States
(Washington, DC:
The National
Association of
Home Builders,
1984), p. 29.

Code, which is updated every three years.[8] BOCA's jurisdiction primarily covers the midwestern and eastern regions of the United States (Fig. 4.10).

ICBO (International Conference of Building Officials)

ICBO was founded in 1922, is headquartered in Whittier, California, and publishes the *Uniform Building Code*[9] on a three-year cycle as part of its package of supporting codes. ICBO has a very large region of coverage since it essentially influences the entire western half of the United States (Fig. 4.10).

SBCCI (The Southern Building Code Congress International)

SBCCI primarily provides its services and develops its building code documents for ten states in the southeastern part of the United States (Fig. 4.10). It was founded in 1940, is headquartered in Birmingham, Alabama, and publishes the *Standard Building Code*[10] and a related series of supportive codes.

CABO (Council of American Building Officials)

In most jurisdictions in the United States, building construction is regulated at either the local or state level by building and related codes. Although a small number of municipalities (mostly large cities) write and revise their own building codes, most *adopt, by ordinance,* one of the previously mentioned model building codes. A city in Colorado, for instance, may decide to adopt ICBO's Uniform Building Code, and its supporting codes, as the governing code for both commercial and housing construction. An alternative to using one of the three model building codes for both types of buildings became available in 1971, when the forerunner of what is now the 1995 edition of the *CABO One and Two Family Dwelling Code* (CABO–OTFDC; hereinafter referred to as the 1995 CABO Code)[11]* was first published.

The Council of American Building Officials (CABO) was incorporated in 1972 as the culmination of approximately 15 years of effort by the leadership of BOCA, ICBO, and SBCCI to coordinate areas of code enforcement and address code-related issues at the national level. One of the major activities of CABO, which is the umbrella organization for the three model code organizations, is publishing the CABO Code on a three-year cycle. Some of the other activities of CABO include:

1. Publishing the Model Energy Code (MEC)[12] on behalf of the three model code organizations.
2. Administering the Building Officials Certification Program to enhance the skills of professionals in the code enforcement field.
3. Administering the National Evaluation System (NES), which is a national program for evaluating new or innovative building materials, products, and systems.

The most direct impact of CABO on the housing industry, however, is through the CABO Code. This code presents the latest code concepts as they appear in the three model codes in a format which directly applies to the construction, prefabrication, alteration, repair, use, occupancy, and maintenance of one- and two-family dwellings. The technical maturation of the CABO Code over the last 25 years and the broad recognition and acceptance which it has received now make it possible for a city in Colorado, for instance, to adapt ICBO's Uniform Building Code for commercial construction and the CABO Code for housing construction.

---

*Sections and figures from this Code are cited for instructional purposes only, and the treatment of the Code provisions illustrated in this chapter are not intended as a substitute for a complete and thorough understanding of all building code provisions to which any particular project may be subject.

CABO One and Two Family Dwelling Code

The CABO Code, as noted in the preface to the 1995 edition,[13] standardizes requirements for detached one- and two-family dwellings (of not more than three stories) by using a compilation of data from the national model codes. Since the CABO Code probably has the best chance of being recognized as a building code for housing that can be applied throughout the United States, it is the one that receives the primary focus in the following discussion. The 1995 CABO Code is also the one that is the primary reference of choice for the remainder of this textbook.

Code Adoption

The process of adopting a building code varies throughout the United States. An individual state, through its legislative process, may adopt, on a statewide basis, a particular model building code as written, amend certain parts of the code, or write its own code. Also, through enabling legislation, states may grant authority to their local governmental bodies to adopt and enforce building codes which they select.

The particular format of the ordinance adopted by a local governmental body is usually developed by the legal department of the organization or by someone like a municipal solicitor. A sample adopting ordinance for the 1995 CABO Code is shown in Fig. 4.11.

The adopted model building code will often contain local amendments that address special concerns in the community. In the broad sense, amendments usually refer to the action of adding new text to the model code, altering the model code text with words or sentences, or deleting portions of the model code text.

Scope of the CABO Code

The primary sections of the 1995 CABO Code illustrate the scope of coverage of the code. These sections include:

- **I.** Administration   (1 chapter)
- **II.** Building   (9 chapters)
- **III.** Mechanical   (18 chapters)
- **IV.** Plumbing   (10 chapters)
- **V.** Electrical   (8 chapters)
- **VI.** Appendices   (5 appendices)

The scope of the 1995 CABO Code is defined in Sections 102 and 103 of "Chapter 1: Administration," as follows:

Figure 4.11
Sample of the CABO Code Adoption Ordinance

Source: Reproduced from p. 5 of the 1995 edition of the *CABO One and Two Family Dwelling Code*,™ copyright © 1995, with the permission of the publisher, The International Codes Council, Whittier, CA.

ADOPTING ORDINANCE
CABO ONE AND TWO FAMILY DWELLING CODE

ORDINANCE NO. _____

An ordinance regulating the fabrication, erection, construction, enlargement, alteration, repair, location and use of detached one- and two-family dwellings, their appurtenances and accessory structures in the jurisdiction of _____; and providing for the issuance of permits therefore providing penalties for the violation thereof, and repealing all ordinances and parts of ordinances in conflict therewith.

Be it ordained by the _____ of the jurisdiction of _____ as follows:

Section 102—Purpose

102.1 Minimum Standards. The purpose of this code is to provide minimum standards for the protection of life, limb, health, property, environment and for the safety and welfare of the consumer, general public, and the owners and occupants of residential buildings regulated by this code.

Section 103—Scope

103.1 Application. The provisions of the code apply to the construction, addition, prefabrication, alteration, repair, use, occupancy, and maintenance of **detached one- and two-family dwellings and one-family townhouses not more than three stories in height,** and their accessory structures. Compliance with the requirements of this code may be considered as prima facie evidence of compliance with the locally adopted code.[14]

Administration of the CABO Code

After a local governmental body adopts the CABO Code, it becomes necessary for it to appoint someone to administer and enforce the code. This individual, called the **building official,** should have the necessary technical training and experience to qualify for certification under the CABO-administered "Building Officials Certification Program." In small jurisdictions, the code enforcement office may consist of only one individual; in larger towns and cities, a complete code enforcement organization may be staffed.

The primary functions of the code enforcement process are: (1) review and approval of plans, (2) issuance of building permits, (3) inspection during the construction phase, and (4) issuance of the occupancy permit.

PLAN REVIEW AND APPROVAL

Before a building permit can be issued to authorize the beginning of the construction phase, the building plans and specifications, perhaps accompanied by a completed application form similar to the one shown in Fig. 4.12, must be submitted to and approved by the Code Enforcement Office. The specific requirements for this transmittal are covered in Section 112 of the 1995 CABO Code.

ISSUANCE OF BUILDING PERMIT

Construction of the house can begin once a building permit, similar to the one shown in Fig. 4.13, is issued and displayed at the job site. The type of work which requires a permit is covered in Section 111.1 of the 1995 CABO Code.

The requirement to display the building permit is specified in Section 116 of the 1995 CABO Code. The issued building permit is considered to be null and void if the homebuilder does not begin work within six months of the date of issuance or if the work is suspended or abandoned for six months after it starts. The 1995 CABO Code has provisions for the extension or renewal of a building permit.

INSPECTION DURING THE CONSTRUCTION PHASE

Figure 4.13 indicates that an inspection card is included with the building permit so that the representative of the Code Enforcement Office can "sign off" at each of the indicated stages. The minimum number of inspections required is specified in Section 113 of the 1995 CABO Code:

Section 113—Inspection

113.1 Foundation inspection. Commonly made after poles or piers are set or trenches or basement areas are excavated and forms erected and any required reinforcing steel is in place and prior to the placing of concrete. The foundation inspection shall include excavations for thickened slabs intended for the support of bearing walls, partitions, structural supports, or equipment and special requirements for wood foundations.

113.1.2 Plumbing, mechanical and electrical. Rough inspection: Commonly made prior to covering or concealment, before fixtures are set, and prior to framing inspection.

CENTRE REGION CODE ADMINISTRATION
Fraser Plaza, Suite #4, 131 South Fraser Street
State College, PA 16801
Telephone: 814-231-3056

Building Permit No.
Zoning Permit No.
Department of Labor and Industry No.
Worker's Comp. Ins. No.

APPLICATION FOR BUILDING PERMIT

APPLICATION REQUIREMENTS: Documents to be submitted with an application for –

NEW SINGLE FAMILY BUILDINGS – Zoning, Water and Sewer Permits and **Two** Sets of Plans
NEW COMMERCIAL BUILDINGS – Zoning, Water and Sewer Permits, **Two** Sets of Plans and Labor & Industry Approval
BUILDING ADDITIONS – Zoning Permit, **Two** Sets of Plans and May Need Water and/or Sewer Permits and/or Labor & Industry
          Approval
OTHER WORK – **Two** Sets of Plans and May Need Zoning, Water, and Sewer Permits.

LOCATION OF PROPOSED WORK OR IMPROVEMENT

Municipality _____
Tax Parcel No. _____
Number and Street _____
Rural Directions _____
_____
_____

TYPE AND COST OF WORK OR IMPROVEMENT

Type of Improvement
1 ☐ New building
2 ☐ Addition
3 ☐ Alteration
4 ☐ Repair, replacement
5 ☐ Demolition
6 ☐ Electrical (only)
7 ☐ Plumbing (only)
8 ☐ Sprinkler System (only)

Describe Work:
_____
_____
_____
_____
_____
_____
_____

Declared Cost (Omit cents)

$ _____

Dimensions
Height in feet .... _____
Number of stories .... _____
Total square feet of all floor areas
(inc. garage & basement) based on
exterior dimensions ...... _____

Type of sewage disposal
☐ Public or private company
☐ Private (septic tank, etc.)
Type of water supply
☐ Public or private company
☐ Private (well, cistern)

IDENTIFICATION

	Name	Mailing address - *number, street, city, and state*	Phone no.
1. Owner			
2. Contractor			
3. Architect			

AFFIDAVIT

I hereby certify that I am the owner in fee or the authorized agent of the owner in fee of the property upon which the work authorized by the permit sought will be performed. All work will be performed in accordance with all applicable laws of the Commonwealth of Pennsylvania and this jurisdiction.

Signature of owner or authorized agent	Address	Application date

Figure 4.12
Application for a Building Permit
Source: Information provided by the Centre Region Code Administration Office, State College, PA.

SITE PLAN – DIMENSION TO BE FILLED IN BY APPLICANT.

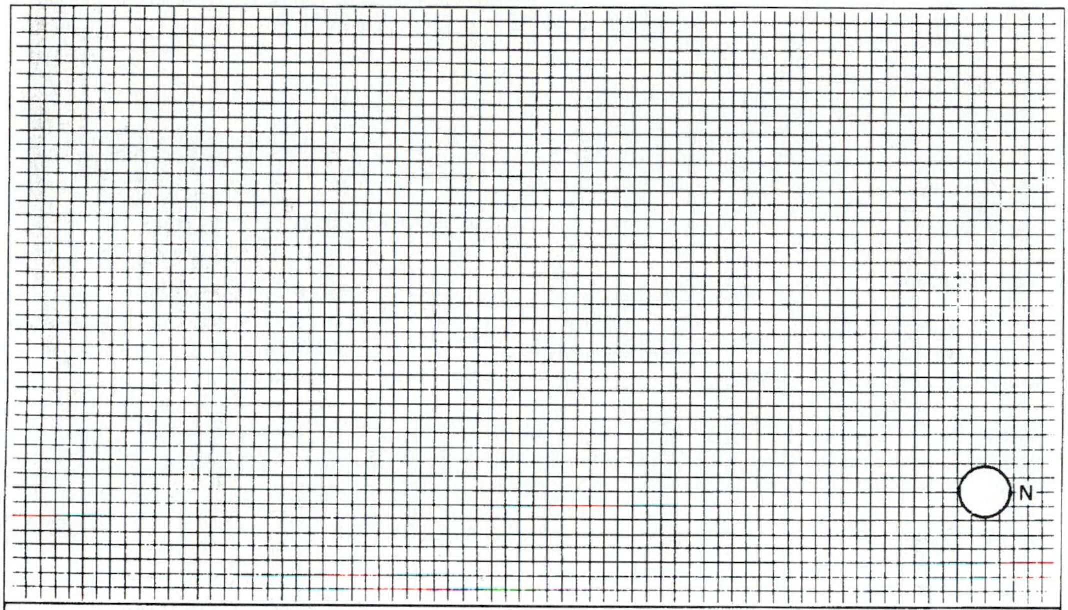

N

ZONING PLAN EXAMINER'S NOTES

Zone	Lot Square Footage		Percent Coverage	Permit No.

Set Backs	Required	Provided	Front (place √)	Number of off-street parking spaces
North				1 Enclosed .... _____
East				2 Outdoors .... _____
South				Date
West				Permit issued _____ 19 ____
Notes:				Approved _____

OCCUPANCY INFORMATION

Type of Construction _____   Use Group _____

	Number of Units	Maximum Occupancy Load	Maximum Live Loads lbs. per sq. ft.		Number of Units	Maximum Occupancy Load	Maximum Live Loads lbs. per sq. ft.
BASEMENT				SEVENTH FLOOR			
FIRST FLOOR				EIGHTH FLOOR			
SECOND FLOOR				NINTH FLOOR			
THIRD FLOOR				TENTH FLOOR			
FOURTH FLOOR				OTHER			
FIFTH FLOOR				ROOF			
SIXTH FLOOR							

VALIDATION

Building
Permit Number _____   Date
                                       Permit Issued _____ 19 ____

Permit Fee $ _____   Approved _____

Figure 4.12 (cont.)

113.1.3 Frame and masonry inspection. Commonly made after the roof, masonry, all framing, firestopping, draftstopping and bracing are in place and after the plumbing, mechanical and electrical rough inspections are approved.

113.1.4 Lath and/or wallboard inspection. Commonly made after all lathing and/or wallboard interior is in place, but before any plaster is applied, or before wallboard joints and fasteners are taped and finished.

CENTRE REGION CODE ADMINISTRATION
Fraser Plaza, Suite #4, 131 South Fraser St.
State College, PA 16801
Telephone: 814-231-3056

BUILDING PERMIT

DATE: PERMIT #:

ADDRESS:
MUNICIPALITY: TAX PARCEL #:

TYPE OF WORK:

CONSTRUCTION TYPE: USE GROUP: STORIES: HT: ft   SQft:
NO. OF UNITS: PARKING SPACES: USE ZONE: PERMIT #:
SET BACKS (FRONT) NORTH: EAST: SOUTH: WEST:

OWNER: ADD:
CONTR: ADD:
ARCH: ADD:

LABOR AND INDUSTRY #:   ESTIMATED COST: FEE:

NOTES:
I/S:
LIC:
                                        _____
                                                CODE OFFICIAL

Office Copy

POST THIS PERMIT AT WORKSITE

NOTICE call 231-3056

FOR INSPECTIONS

APPROVED PLANS MUST BE RETAINED ON JOB AND THIS PERMIT
POSTED UNTIL FINAL INSPECTION HAS BEEN MADE. BUILDING SHALL
NOT BE OCCUPIED UNTIL FINAL INSPECTION HAS BEEN MADE.

STRUCTURE MUST BE CONSTRUCTED IN
ACCORDANCE WITH STATE ENERGY
CONSERVATION ACT.

INSPECTOR MUST SIGN ALL SPACES PERTAINING TO THIS JOB.

	Inspector	Date
1 - Footer inspection before pouring		
2 - Foundation prior to backfill		
3 - Water service inspection		
4 - Plumbing under slab inspection		
5 - Rough plumbing inspection		
6 - Rough mechanical inspection		
7 - Rough electrical inspection		
8 - Fireplace inspection		
9 - Framing members inspection		
10 - Electrical service inspection		
11 - Chimney inspection		
12 - Woodburner inspection		
13 - Demolition		
14 - Fire Suppression/Protection		
15 - Final electrical inspection		
*16 - Final zoning inspection for occupancy		
*17 - Building sewer or septic system inspection		
*18 - Final inspection		

*MUST BE APPROVED PRIOR TO OCCUPYING THE STRUCTURE

Figure 4.13
Building Permit
Source: Information provided by the Centre Region Code Administration Office, State College, PA.

113.1.5 Other inspections. In addition to the called inspections above, the building department may make or require any other inspections to ascertain compliance with this code and other laws enforced by the building department.

113.1.6 Final inspection. Commonly made after the building is completed and ready for occupancy.[15]

There are also provisions in the 1995 CABO Code for those situations where the code inspector deems that the work is being performed improperly. In such cases, the **police powers** delegated to the building official by the local governmental body become operative because either a **violation notice** or a **stop work order** can be issued.

ISSUANCE OF THE OCCUPANCY PERMIT

After the final inspection has been completed successfully, the building official will typically issue an *Occupancy Permit,* similar to the one shown in Fig. 4.14. This permit allows for the connection of utility service to the property so that it can be occupied legally according to the health, safety, and energy codes.

CENTRE REGION CODE ADMINISTRATION

CERTIFICATE OF USE AND OCCUPANCY

HAS BEEN INSPECTED AND THE FOLLOWING OCCUPANCY IS HEREBY AUTHORIZED.

	Number of Units	Maximum Occupancy Load	Maximum Live Loads lbs. per sq. ft.		Number of Units	Maximum Occupancy Load	Maximum Live Loads lbs. per sq. ft.

This certificate is required to be posted and permanently maintained in a conspicuous space at or close to the entrance of the building or structure (except Owner Occupied Structures of Use Group R-3 and R-4).

....
Director of Code Administration

Figure 4.14
Certificate of Use and Occupancy
Source: Information provided by the Centre Region Code Administration Office, State College, PA.

COST IMPLICATIONS OF REGULATIONS

There are many effects on the housing industry because of the land development regulations and building codes that are discussed in this chapter. Many of them are positive because collectively they ensure that the public is properly protected and that the overall quality of land development and housing in the United States satisfactorily meets the needs of the majority of the population.

It should be noted, however, that there is a cost impact associated with the implementation of these regulations. In fact, some of these regulations have been identified as creating regulatory barriers to the goal of providing affordable housing in the United States.

The issue of the cost impacts of these regulations on housing is not considered to be within the scope of the presentation in this chapter. The interested reader is, therefore, referred to a study commissioned by President Bush in 1989, entitled *Not in My Back Yard: Removing Barriers to Affordable Housing,*[16] which focuses on this important issue.

CHAPTER SUMMARY

This chapter has reviewed the regulatory environment in which residential land development projects are completed and individual houses are built. **Zoning and land development regulations/ordinances** are the primary factors that influence the type of subdivisions that can be built. **Building codes**, with the CABO One and Two Family Dwelling Code being cited as an example, are the primary influences that govern the design and construction of houses. In both instances, the primary regulations are promulgated and administered at the local governmental level.

NOTES

[1]Charles McIntyre, M. Kevin Parfitt, and Jack H. Willenbrock, *A Residential Land Development Model,* HRC Research Series: Report No. 28 (University Park, PA: The Housing Research Center at The Pennsylvania State University, August 1994), pp. 58–77.

[2]William E. Brewer and Charles P. Alter, *The Complete Manual of Land Planning and Development* (Englewood Cliffs, NJ: Prentice Hall, 1988), pp. 153–154.

[3]Brewer and Alter, *The Complete Manual of Land Planning,* pp. 7–8.

[4]Ralph R. Pisani and Robert L. Pisani, *Investing in Land—How to be a Successful Developer* (New York City, NY: Wiley, 1989), pp. 127–128.

[5]*Proposed Model Land Development Standards and Accompanying Model State Enabling Legislation, 1993 Edition* (Washington, D.C.: U.S. HUD, 1993).

[6]Harry Burd, "Introduction to Building Codes," and Jack H. Willenbrock, Thomas B. Brown and Harry Burd, *Participant Manual for the Residential Building Code Training Program for Pennsylvania Code Officials and Inspectors, Volume I,* HRC Research Series: Report No. 17 (University Park, PA: The Housing Research Center at The Pennsylvania State University, 1991), pp. 1.6–1.9.

[7]*Understanding Building Codes and Standards in the United States* (Washington, D.C.: National Association of Home Builders, 1984), pp. 13–14.

[8]*The BOCA National Building Code, 1993* (Country Club Hills, IL: Building Code Officials and Code Administrators International, Inc., 1993).

[9]*Uniform Building Code, 1994* (Whittier, CA: International Conference of Building Officials, 1994).

[10]*Standard Building Code, 1994* (Birmingham, AL: Southern Building Code Congress International, 1994).

[11]*CABO One and Two Family Dwelling Code (OTFDC), 1995 Edition.* Falls Church, VA: The Council of American Building Officials, 1995.

[12]*CABO Model Energy Code, 1995 Edition.* (Falls Church, VA: The Council of American Building Officials, 1995).

[13]*1995 CABO Code,* p. iii.

[14]Ibid., p. 1.

[15]Ibid, p. 3.

[16]*Not in My Back Yard: Removing Barriers to Affordable Housing.* Report to President Bush and Secretary Kemp. (Washington, D.C.: Advisory Commission on Regulatory Barriers to Affordable Housing, 1991).

BIBLIOGRAPHY

The BOCA National Building Code, 1993. Country Club Hills, IL: Building Code Officials and Code Administrators International, Inc., 1993.

Brewer, William E. and Charles P. Alter, *The Complete Manual of Land Planning and Development.* Englewood Cliffs, NJ: Prentice-Hall, Inc., 1988.

Brooks, R. Gene, *Site Planning—Environment, Process and Development.* Englewood Cliffs, NJ: Prentice Hall, 1988.

Burd, Harry, "Introduction to Building Codes," and Jack H. Willenbrock, Thomas B. Brown, and Harry Burd, *Participant Manual for the Residential Building Code Training Program for Pennsylvania Code Officials and Inspectors, Volume I,* pp. 1.1–1.17. HRC Research Series, Report No. 15. University Park: The Housing Research Center at The Pennsylvania State University, 1991.

CABO One and Two Family Dwelling Code, 1995 Edition. Falls Church, VA: The Council of American Building Officials, 1995.

CABO Model Energy Code, 1995 Edition. Falls Church, VA: The Council of American Building Officials, 1995.

McIntyre, Charles, M. Kevin Parfitt, and Jack H. Willenbrock, *A Residential Land Development Model,* HRC Research Series: Report No. 28. University Park, PA: The Housing Research Center at The Pennsylvania State University, August 1994.

Not in My Back Yard: Removing Barriers to Affordable Housing. Report to President Bush and Secretary Kemp. Washington, D.C.: Advisory Commission on Regulatory Barriers to Affordable Housing, 1991.

Pisani, Ralph R. and Robert L. Pisani, *Investing in Land—How to be a Successful Developer.* New York City, NY: Wiley, 1989.

Proposed Model Land Development Standards and Accompanying Model State Enabling Legislation, 1993 Edition. Washington, D.C.: U.S. HUD, 1993.

Standard Building Code, 1994. Birmingham, AL: Southern Building Code Congress International, 1994.

Understanding Building Codes and Standards in the United States. Washington, D.C.: The National Association of Home Builders, 1984.

Uniform Building Code, 1994. Whittier, CA: International Conference of Building Code Officials, 1994.

Zoning Ordinances 224 and 506 of the Township of Ferguson. Ferguson Township, PA, October 1991.

5

Housing Design Guidelines

INTRODUCTION

Houses are designed and built to meet the needs of the homebuyers who live in them. Although the needs of each consumer group may be different, all homebuyers go through a similar process leading to the selection of a house that meets their needs, some in an organized, systematic fashion and others in a series of random and often chaotic experiences.

Homebuyers may retain the services of an architect (an individual who has passed his profession's licensing examination) or they may choose to work directly with a homebuilder. These professionals help to refine the ideas and concepts which the homebuyers may have gathered as a result of: (1) their own life experiences in previous houses, (2) research into periodicals and books which feature house plans and deal with housing design issues, (3) visits to model houses in existing housing developments, and (4) conversations with friends and acquaintances.

At some point in the interaction with an architect or homebuilder, a list of requirements is developed which becomes the basis for transforming the dream into a reality. These requirements typically include both the **exterior architectural design** and the **interior architectural design** features.

This chapter introduces some of the important concepts and guidelines associated with each of these design areas. It is recognized, however, that architecture cannot be separated simply into its exterior and interior aspects; their integration is the art that architects spend a lifetime perfecting.

The regulatory requirements of "Chapter 3: Building Planning" of the 1995 CABO Code[1]* which affect these design areas are included at appropriate points in the presentation.

---

*Sections, figures, and tables from this Code are cited only for instructional purposes, and the treatment of the Code provisions contained in the design solutions illustrated in this chapter are not intended as a substitute for a complete and thorough understanding of all building code provisions to which any particular project may be subject.

EXTERIOR ARCHITECTURAL DESIGN

One of the most important attributes of a house is its exterior appearance. Each type of homebuyer appears to value this attribute somewhat differently. The first-time homebuyers mentioned in Chapter 2 appear to want the style of their houses to reflect *time-tested American architecture* rather than their own desires for individuality. They typically view their first house as a short-term investment and they do not want exterior appearance to jeopardize its resale potential. Luxury homebuyers value exterior appearance for another reason: they want it to reflect their economic and social status and convey the resultant message of success to others. As a result, it is this group that is most likely to retain the services of an architect to ensure that their dreams are transformed into an attractive, satisfying reality.

There does not appear to be a universally accepted system of describing and classifying houses in the United States. As a result, problems of misunderstanding occur at all levels of the housing industry, and it is difficult to reach an agreement on what is recognized as *time-tested American architecture*. The **CTS system**,[2] which was developed for the real estate industry, provides one possible framework which can be used to address this classification system. In the CTS system, a particular house can be described in terms of the attributes of: (1) **housing class**, (2) **housing type**, and (3) **housing style**.

Introduction to the CTS System

Housing Class

The attribute of **housing class** in the CTS system identifies the number of families that can be accommodated in a housing unit and whether the house is detached or shares a common party wall with other housing units. The nine basic classes which are used, along with their abbreviations, are shown in Table 5.1.

In this system, the definition of one-family detached housing is quite clear, but the one-family party wall class can be interpreted in several ways. It can apply, for instance, to: (1) a one-family townhouse unit, (2) one of the units in a four-plex or quadro (a structure with four housing units built together), or (3) a single housing unit in a 20-story condominium high-rise building. In addition, a limitation of the current system is that it recognizes individual housing units in apartment houses only if they are sold individually in condominium or cooperative forms of ownership.[3]

TABLE 5.1 Description of Housing Classes

Code	Description	Abbreviation
1	One-family, detached	1 FAM D
2	Two-family, detached	2 FAM D
3	Three-family, detached	3 FAM D
4	Four-family, detached	4 FAM D
5	One-family, party wall	1 FAM PW
6	Two-family, party wall	2 FAM PW
7	Three-family, party wall	3 FAM PW
8	Four-family, party wall	4 FAM PW
9	Other	OTHER

Source: Henry S. Harrison, *Houses: The Illustrated Guide to Construction, Design and Systems, 2nd Ed.* (Chicago, IL: Real Estate Education Company, 1992), p. 399.

Used with permission: *Houses*, Henry S. Harrison

TABLE 5.2   Description of Housing Types

Code	Description	Abbreviation
1	One-story	1 STORY
2	One-and-a-half story	1-1/2 STORY
3	Two-story	2 STORY
4	Two-and-a-half story	2-1/2 STORY
5	Three-or-more stories	3 STORY
6	Bi-level	BI-LEVEL
6	Raised ranch	R RANCH
6	Split entry	SPLT ENT
7	Split-level	SPLT LEV
8	Mansion	MANSION
9	Other	OTHER

Source: Henry S. Harrison, *Houses: The Illustrated Guide to Construction, Design and Systems, 2nd Ed.* (Chicago, IL: Real Estate Education Company, 1992), p. 400.

Housing Type

The second attribute in the CTS system is the **housing type**. The nine basic types of housing listed in Table 5.2 probably include most of the ones that are of interest to the typical home-buyer. The advantages and disadvantages of each type are discussed later in this chapter.

Housing Style

Style is the attribute of houses that is the most difficult to define because there are so many ways in which the exterior architectural features of houses can be classified. A separate section of this chapter is devoted to a discussion of housing style.

The CTS system, at its highest level of classification, provides a division into the following nine **style groups**:[4]

1. Colonial American
2. English
3. French
4. Swiss
5. Latin
6. Oriental
7. 19th Century American
8. Early 20th Century American
9. Post-World War II American

Both historical time periods and countries of origin are mixed as broad-based style identifiers. Each style grouping is further subdivided so that the exterior architectural features of a house can be described more accurately (Table 5.3).

A **Prairie House**, for instance, appears in the Early 20th-Century American style grouping. Using the CTS system approach, a house designed according to the Prairie House style, being built as a single-family, detached, one-story house, could be described as:

	Class	Type	Style
NARRATIVE	One-family, detached	one-story	Prairie House
ABBREVIATION	1 FAM D	1 STORY	PRAIRIE
COMPUTER	1	1	801

TABLE 5.3 Style Classification: CTS System

Code	Description	Abbrev.	Code	Description	Abbrev.
100	**COLONIAL AMERICAN**	**COL AMER**	**700**	**19th CENTURY AMERICAN**	**19th CTY**
101	Federal	FEDERAL	701	Early Gothic Revival	E GOTH
102	New England Farmhouse	NE FARM	702	Egyptian Revival	EGYPT
103	Adams	ADAMS CO	703	Roman Tuscan Mode	RO TUSC
104	Cape Cod	CAPE COD	704	Octagon House	OCTAGON
105	Cape Ann	CAPE ANN	705	High Victorian Gothic	HI GOTH
106	Garrison Colonial	GARR CO	706	High Victorian Italianate	VIC ITAL
107	New England	N E COL	707	American Mansard	MANSARD
108	Dutch	DUTCH CO	707	Second Empire	2nd EMP
109	Saltbox	SALTBOX	708	Stick Style	STICK
109	Catslide	CATSLIDE	708	Carpenter Gothic	C GOTH
110	Pennsylvania Dutch	PENN DUT	709	Eastlake	EAST L
110	PA German Farmhouse	GER FARM	710	Shingle Style	SHINGLE
111	Classic	CLASSIC	711	Romanesque	ROMAN
111	Neoclassical	NEO CLASS	712	Queen Anne	Q ANNE
112	Greek Revival	GREEK	713	Brownstone	BROWN S
113	Southern Colonial	SOUTH CO	713	Brick Row House	BR ROW
114	Front Gable New England	F GAB NE	713	Eastern Town House	E TOWN
114	Charleston	CHARLES	714	Western Row House	WEST ROW
114	English Colonial	ENG COL	714	Western Town House	W TOWN
115	Log Cabin	LOG CAB	715	Monterey	MONTEREY
			716	Western Stick	W STICK
200	**ENGLISH**	**ENGLISH**	717	Mission Style	MISSION
201	Cotswald Cottage	COTSCOT			
202	Elizabethan	ELIZ	**800**	**EARLY 20th CENTURY**	**EARLY-20C**
202	Half Timber	HALFTIM	801	**AMERICAN**	
203	Masonry Tudor	M TUDOR	802	Prairie House	PRAIRIE
203	Jacobean	JACOBEAN	803	Bungalow	BUNGALOW
204	Williamsburg	WILLIAMS	803	Pueblo	PUEBLO
204	Early Georgian	E GEORG	804	Adobe	ADOBE
205	Regency	REGENCY	805	International Style	INTERNAT
206	Georgian	GEORGE	806	California Bungalow	CAL BUNG
207	Tudor	TUDOR	807	Shotgun	SHOTGUN
			808	Foursquare	F SQUARE
300	**FRENCH**	**FRENCH**	808	Art Deco	A DECO
301	French Farmhouse	FR FARM		Art Moderne	A MOD
302	French Provincial	FR PROV			
303	French Normandy	FR NORM	**900**		
304	Creole	CREOLE	901	**POST-WORLD WAR II**	**POST WW2**
304	Louisiana	LOUISIA	902	**AMERICAN**	
304	New Orleans	NEW OR	902	California Ranch	C RANCH
			903	Northwestern	NORTH W
400	**SWISS**	**SWISS**	903	Puget Sound	P SOUND
401	Swiss Chalet	SWISS CH	904	Functional Modern	FUN MOD
			905	Contemporary	CONTEMP
500	**LATIN**	**LATIN**	906	Solar House	SOLAR
501	Spanish Villa	SP VILLA	907	"A" Frame	A FRAME
502	Italian Villa	IT VILLA	909	Mobile Home	MOBILE
			909	Plastic House	PLASTIC
600	**ORIENTAL**	**ORIENT**	910	Contemporary Rustic	C RUSTIC
601	Japanese	JAPAN		California Contemporary	CAL CONTEMP
				Postmodern	P MODERN

Source: Henry S. Harrison, *Houses: The Illustrated Guide to Construction, Design and Systems, 2nd Ed.*
(Chicago, IL: Real Estate Education Company, 1992), pp. 401–403.

This introduction to the CTS system provides the framework for the in-depth discussion of Class–Type–Style which follows.

Classes of Houses

Overview

The important differences between a detached house, a town/row house, and a semi-detached house are shown in Figure 5.1. From a realtor's standpoint, knowing if a particular house is detached or attached probably is sufficient information about the class of the house.

A cost estimator might feel, however, that the nine classes in Table 5.1 are too globally defined because they do not differentiate between the following **cost classes**: (1) Economy Class, (2) Average Class, (3) Custom Class, and (4) Luxury Class. Some of the basic differences between these cost classes are explained and illustrated for the one and one-half story house shown in Fig. 5.2.

Detached House
This category of residence is a free-standing separate building with or without an attached garage. It has four complete walls.

Town/Row House
This category of residence has a number of attached units made up of inner units and end units. The units are joined by common walls. The inner units have only two exterior walls. The common walls are fireproof. The end units have three walls and a common wall.

Semi-Detached House
This category of residence has two living units side-by-side. The common wall is a fireproof wall. Semi-detached residences can be treated as a row house with two end units.

Figure 5.1
Classes of Houses
Source: *Means Residential Cost Data 1995*, p 9. Copyright R.S. Means Co., Inc., Kingston, MA, 617–585–7880, all rights reserved.

(a) One and one-half story
economy model

(b) One and one-half story
average model

(c) One and one-half story
custom model

(d) One and one-half story
luxury model

Figure 5.2
Cost Classes for a One and One-Half Story House
Source: *Means Residential Cost Data 1995*, pp. 13, 25, 45, and 61. Copyright R.S. Means Co., Inc., Kingston, MA, 617–585–7880, all rights reserved.

Economy Class[5]

An **economy class residence** is usually mass-produced from stock plans. The materials and workmanship are sufficient only to satisfy minimum building code requirements. Low construction cost is more important than distinctive features. Design is seldom other than square or rectangular.

Average Class[6]

An **average class residence** is simple in design and is built from standard designer plans. Material and workmanship are average, but often exceed the minimum building code requirements. Frequently, there are special features that give the residence some distinctive characteristics.

Custom Class[7]

A **custom class residence** is usually built from a designer's plans which have been modified to give the building a distinction of design. Material and workmanship are generally above average with obvious attention given to construction details. Construction normally exceeds minimum building code requirements.

Luxury Class[8]

A **luxury class residence** is built from an architect's plan for a specific owner. It is unique in design and workmanship. There are many special features, and construction usually exceeds all building code requirements. It is obvious that primary attention is placed on the owner's comfort and pleasure. Construction is often supervised directly by an architect.

Types of Houses

Each type of house (see Fig. 5.3) provides advantages, as well as disadvantages, to each homebuyer group and for each particular set of site conditions (climate, topography, soil and geological characteristics, and the configuration and orientation of the lot). Some of the housing types achieved popularity in the past because they met the changing needs of society either for a particular period of time or in a particular region of the United States.

One-Story House

This type of house, sometimes called a **ranch** or **rambler**, confines all of the living/social, private/sleeping, and working/service functions to one floor unless a basement is also provided. Roofs are usually low-pitched because the attic serves only as a limited storage area.

Among the advantages of a one-story house are the elimination of stair-climbing hazards for young children and the elderly and ease of exterior maintenance because of low height. One of the primary disadvantages is the additional land cost associated with spreading out the building footprint on a larger-sized lot in order to achieve the same living area of a two-story house.

Construction costs per square foot of living area and heating and cooling costs also tend to be higher for a one-story house. Interference problems between members of a household are also potentially higher, particularly in smaller entry-level versions of the one-story house.

One and One-Half Story House

Additional private/sleeping space can be obtained easily if the low-pitched roof of the ranch-type house is changed to a high-pitched roof, resulting in a one and one-half story house. The actual living area that is gained generally is limited to the area with a ceiling height of 6 ft or more (typically 50–90% of the ground floor).

Although the *primary* living/social, private/sleeping, and working/service functions are still carried out on the first floor (and possibly in the basement) of a one and one-half story house, it is now possible to separate some of the sleeping areas by utilizing the second floor. The cost per square foot of living space should be less than for a one-story house since the foundation and roof areas have not been increased greatly to accommodate the additional living space.

A flight of stairs, which takes away living space from the first floor and creates a potential hazard for children and the elderly, must, of course, be added. If the pitch of the roof is not sufficient, the rooms on the second floor often can appear to be cramped and have limited headroom. Insufficient roof insulation and second-floor ventilation can result in the second-floor space being hot in the summer and cold in the winter.

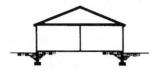

(a) One-story house

(e) Three-story house

(b) One and one-half story house

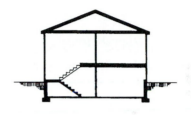

(f) Bi-level house

(c) Two-story house

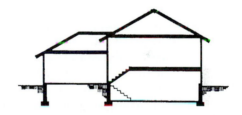

Figure 5.3
Types of Houses
Source: Means Residential Cost Data 1995, pp. 6–7. Copyright R.S. Means Co., Inc., Kingston, MA, 617–585–7880, all rights reserved.

(g) Tri-level house

(d) Two and one-half story house

NOTE: Basements can be added to Houses (a) to (e) at an additional cost

Two-Story House

Many of the problems associated with a one and one-half story house are eliminated when a full two-story house is selected. It is now possible to completely separate the private/sleeping functions from the other functions by placing all of the sleeping areas on the second floor. Alternatively, the master bedroom can be retained on the first floor and the bedrooms for the children and guests can be located on the second floor.

Many of the traditional styles of houses in the United States are based upon a two-story house configuration, resulting in this being a widely recognized and accepted type of housing. Separation of functions in two-story houses, however, does require an increased use of stairs with the same inherent problems noted for one and one-half story houses. Maintenance of the exterior of the house also is more of a problem if an extensive amount of painting of windows and walls is required. The use of siding material that does not need continual maintenance somewhat alleviates this problem.

Two and One-Half and Three-Story Houses

As the need for additional living space becomes an issue, either because of the size of the household or because it adds to the flexibility and graciousness of lifestyle, the alternatives of two and one-half and three-story houses can be considered. Some of the traditional architectural styles require these larger-size house configurations. Custom and luxury homebuyers often recognize, and are willing to accept, the increased costs involved with heating and maintenance when they purchase two and one-half and three-story houses. These additional costs usually can be justified if all of the additional space is used effectively. As the number of living levels increases, it is obvious that there is a continuous need to climb stairs in order to fully utilize this additional space.

In the case of the three-story house, the possibility of a multifamily living situation or room for a professional office exists if the space is not needed for extra sleeping, recreational, and storage rooms.

Bi-Level House

A bi-level house (also called a raised ranch, split-entry, and split-foyer) has two living levels, one which is about 4 ft below grade and one which is about 4 ft above grade. Both of these approximately equal-in-size areas have full ceiling heights. Upon entry into the house, a person must either go up or down a half-flight of stairs in order to be on a living area.

Living/social and working/service functions are typically carried out on both levels since the lower level receives natural light and is not designed to serve as a basement. Private/ sleeping space is usually provided on the upper level.

In addition to the need to always climb or descend stairs to reach a living area, there also are problems associated with raising the basic ranch-type house 4 ft out of the ground. The part that is below ground can be damp, musty, and cold if proper good practice standards are not followed when the exterior basement walls are insulated and water proofed. Also, the resulting part of the house that is above ground and exposed often detracts from the exterior appearance of the house if it has not been finished properly to "blend in" with the exterior finish on the rest of the house.

Tri-Level House

A tri-level house, also called a split-level house, has three levels of living area: one at grade level and the other two at 4 ft above and 4 ft below the grade level, respectively. In such a house, the private/sleeping areas can be separated from the rest of the house by placing these areas at the highest living level.

A tri-level house is ideally suited for a sloped lot. Its advantage over a bi-level house is that entry into the house is at a living area level so it is often unnecessary to either climb or descend a set of stairs to enter the kitchen, dining room, and living room. Separation of these living/social and working/service functions from the remainder of the house ensures that there is good interior circulation throughout the house.

This type of house is definitely not appropriate for level lots because the adjustments that must be made to accommodate such an installation often result in unattractive exterior features. A considerable amount of stair climbing is required to reach the levels.

From an energy standpoint, bedrooms at the highest level can become hot in the summer due to the natural convection currents that are established in a tri-level house. Rooms at the lowest levels can be cold. These problems can be alleviated somewhat if the house is properly ventilated and insulated. A structural framing system that is more complicated than the basic platform framing system used for one- or two-story houses is required because of the different levels.

Styles of Houses

Overview

Closely associated with the homebuyer's decision about the house type is probably a decision about the architectural style of the house. Invariably, as house plan books are studied and visits are made to existing housing developments, questions are raised about the style of particular houses. These questions usually refer to the predominant architectural, structural, and ornamental features which are generally recognized as belonging to a particular **architectural style**.

Architectural style, as indicated by the CTS classification system defined in Table 5.3, has both cultural and historical roots. Homebuyers who take a "purist" approach commission their houses to be designed and built in strict accordance with the recognized attributes of a particular historical architectural style. The majority of homebuyers typically settle for something less than perfection by selecting houses which are eclectic in style because they contain characteristics that have their roots in a number of different architectural styles.

Classification of Architectural Style

Since the typical homebuilder, or for that matter the typical homebuyer, does not have an architectural background, both parties certainly can benefit from an increased knowledge of the more prevalent architectural styles of houses. A number of well-illustrated books can provide that knowledge. For instance, *A Field Guide to American Houses,* by Virginia and Lee McAlester, provides a historical overview of the evolution of architectural style.[9]

Other excellent references which include an illustrated classification of the architectural styles of houses have been written by Henry S. Harrison,[10] Lester Walker,[11] Mary Mix Foley,[12] Carol Rifkind,[13] and John Milnes Baker.[14] Each of these works proposes a somewhat different classification system. Some follow the mixed cultural/historical approach used in the CTS system (Table 5.3). Others trace the evolution of the architectural style of houses through a series of historical periods.

Although a comparative analysis of the historical periods used in the previous references is beyond the scope of this chapter, the one proposed by McAlester and McAlester (Table 5.4) provides a representative framework of the important architectural styles which were prevalent during the various historical periods indicated. A number of the architectural styles which appear in Table 5.3 also appear in Table 5.4 in a more organized historical context.

Regional Approach to Architectural Housing Styles

Jim Kemp's somewhat different approach focuses on the regional influences which provided the historical roots for the architectural styles of houses currently being built in the United States.[15] Kemp considers regional architectural styles as **vernacular** because they: (1) used materials that were abundant in their regions, (2) were adaptable to local geographic and climatic conditions, and (3) reflected the unique history of the region in which they were built. He notes:

> Though distinctly different, these types of houses have much in common. Each reflects its locale in terms of building methods and materials, climate control, and exterior styling. In New England, for example, houses were built with high-peaked roofs so that snow would slide easily to the ground instead of accumulating atop the structure. The massive walls of the New Mexico abode, sometimes two feet thick, shielded families from the strong heat of the sun during the day. On cool desert nights, the walls re-radiated the solar gain indoors as space heating. . . .
>
> The development of regional house-styles also was governed by the size of building plots. Some of the oldest residences are farmhouses, which were the nerve centers of

TABLE 5.4   Prevalent Architectural Housing
Styles During Various Historical Periods

Historical Period	Styles of Houses
Colonial Houses (1600–1820)	Postmedieval English Dutch Colonial French Colonial Spanish Colonial Georgian Adam Early Classical Revival
Romantic Houses (1820–1880)	Greek Revival Gothic Revival Italianate Exotic Revivals Octagon
Folk Houses	Native American (up to 1900) Pre-Railroad (before 1850–1890) National (after 1850–1890)
Victorian Houses (1860–1900)	Second Empire Stick Queen Anne Shingle Richardson Romanesque Folk Victorian

Eclectic Houses (1880–1940)	Anglo-American, English, and French Period Houses
	    Colonial Revival
	    Neoclassical
	    Tudor
	    Chateauesque
	    Beaux Arts
	    French Eclectic
	Mediterranean Period Houses
	    Italian Renaissance
	    Mission
	    Spanish Eclectic
	    Monterey
	    Pueblo Revival
	Modern Houses
	    Prairie
	    Craftsman
	    Modernistic
	    International

American Houses (since 1940)	Modern Neoeclectic Contemporary Folk

extensive agrarian operations. As the owners' need for space grew, farmhouses often were expanded again and again by succeeding generations, creating structures that seem to meander across the landscape. Later, as great urban centers developed, local architecture evolved to make the most of the smaller lots available to city dwellers. In Charleston, South Carolina, the response was a prototype town house called the Charleston Single. Instead of facing the street, the Charleston Single was oriented to one side of the lot to exploit the deep, narrow building sites.[16]

Six specific regions of the United States provide the highest level of classification for Kemp's System (Table 5.5). In addition, a **National Style** category is used for those architectural housing styles that he feels are more correctly associated with the United States as a whole.

It is beyond the scope of this chapter to provide an illustrative discussion of each of the diverse housing styles listed in Table 5.5. The highlighted styles in Table 5.5 are presented, however, to start the architectural housing style recognition process for homebuilders and homebuyers who are interested in how exterior architecture features are selected and how they are "fit together" in some pleasing and integrated fashion. Although Kemp's classification system was used in the selection process, the illustrative examples that are used were developed by Harrison,[17] who provides similar examples for many of the housing styles listed in Table 5.3.

House Shape and Roof Line Considerations

One of the basic features of a house which will be associated with a particular architectural housing style is its form or shape. Another is the shape and slope of the roof. Each of these characteristics is briefly introduced to provide some background information for understanding the styles of houses which follow.

HOUSE SHAPES

McAlester and McAlester provides guidance with regard to house shape by noting:

> House form is not endlessly varied. Instead, a few fundamental shapes, and relatively minor variations on them, tend to be used again and again through a range of changing architectural styles. . . . In theory, ground plans and elevations can be varied to form an infinite number of house shapes. In practice, there are only a relatively few common variants of both plan and elevation. These combine to make several fundamental families of house shapes that dominate American domestic architecture.[18]

Both simple-plan and compound-plan families of shapes are shown in Fig. 5.4. Simple-plan families result when individual square or rectangular room-sized units are combined into larger squares or rectangles. Compound-plan families are created when the rectangles and squares of simple plans are combined into shapes representing the letters L, T, U, or H.

ROOF SHAPES

The roof shapes for traditionally styled houses, as shown in Fig. 5.5, can be represented by the following families: (1) **Gabled**, where the triangular upward extension of either side walls or the front walls of the house provide the end supports for the two sloping roof planes; (2) **Hipped**, where four sloping surfaces form the roof; and (3) **Flat**, where a single or slightly sloped surface defines the roof.

In this presentation, gambrel roofs are considered to be part of the Gabled family, and mansard roofs are considered to be members of the Hipped family.

ROOF PITCH AND ROOF SLOPE

As indicated in Fig. 5.5, the appearance of the house will change as the roof pitch or slope is changed from a low slope to a steep slope. The three key variables associated with the roof pitch and roof slope are: (1) the rise, (2) the run, and (3) the span (Fig. 5.6).

TABLE 5.5   A Portfolio
of Regional Styles

Region	Style
New England	Cape Cod Cottage **Colonial** Saltbox Nantucket Cottage Shingle Style Farmhouse Colonial Garrison Barn
Mid Atlantic	**New York City Brownstone** New York City Townhouse Camp Style Dutch Farmhouse Flemish House German Stone House
The South	Plantation House Charleston Single Dogtrot Cajun Cottage French Colonial Shotgun Southern Rural Miami Mediterranean
The Southwest	**Adobe** Rocky Mountain Desert Style Native American Influence
The Midwest	Ohio Farmhouse Prairie School
The West	Bay Area Style San Francisco Victorian California Bungalow Western Farmhouse Sod House Hispanic Tradition
The National Styles	Log House Greek Revival Cottage **Victorian Styles** Romanesque A-Frame American Post Modern Energy Efficient Architecture

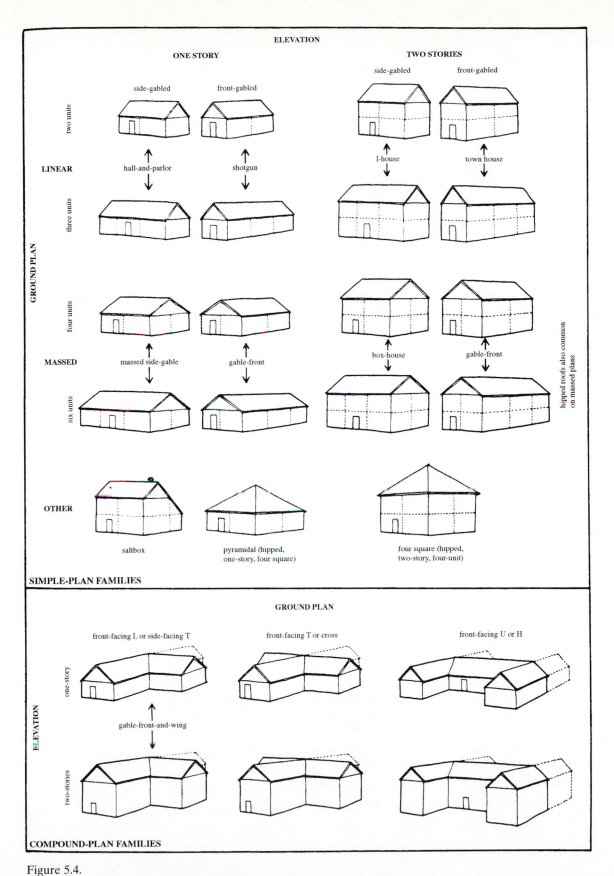

Figure 5.4.

Characteristic Families of Housing Shapes

Source: From p. 27 of *A Field Guide to American Houses* by Virginia and Lee McAlester. Copyright © 1984 by Virginia Savage and Lee McAlester. Reprinted by permission of Alfred A. Knopf, Inc.

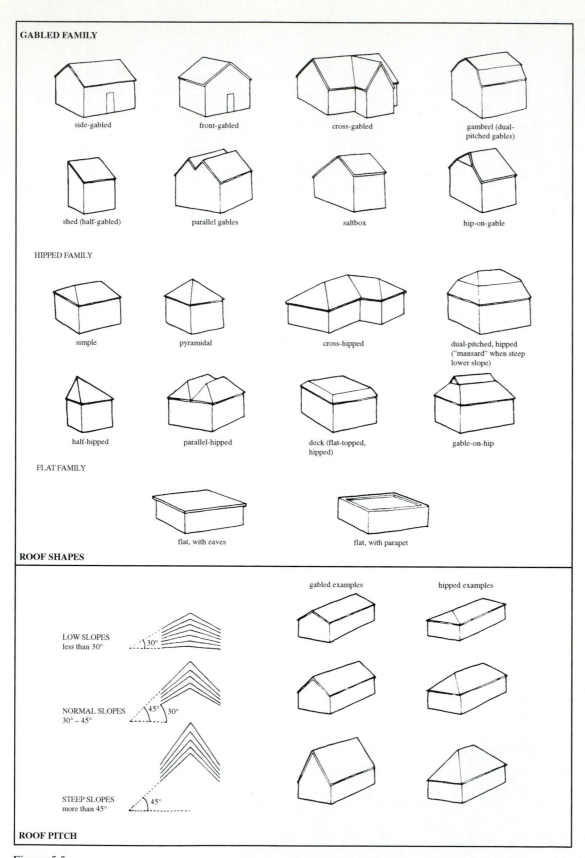

Figure 5.5

Families of Roofing Shapes

Source: From p. 43 of *A Field Guide to American Houses* by Virginia and Lee McAlester. Copyright © 1984 by Virginia Savage and Lee McAlester. Reprinted by permission of Alfred A. Knopf, Inc.

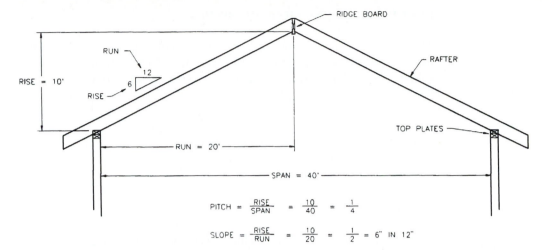

Figure 5.6
Definition of
Roof Pitch and
Roof Slope

Roof pitch is expressed as the value of the vertical rise over the value of the span of the roof (the distance between the outside walls of the house). Thus, the pitch of the roof illustrated in Fig. 5.6 is:

$$Pitch = Rise/Span = 10/40 = 1/4$$

Roof slope is expressed as the value of the vertical rise over the value of the run of the roof rafter. It is usually represented as inches of rise per 12 in. of run. Thus, the slope of the roof illustrated in Fig. 5.6 is:

$$Slope = Rise/Run = 10/20 = 1/2 \text{ or } 6 \text{ in } 12$$

A very low-sloped roof will have a substantially smaller rise than run (i.e., 2 in 12); a medium-sloped roof will range up to about 6 in 12, while a roof having a 45% slope (12 in 12) is considered to be a steep roof.

Examples of Architectural Styles

Four very different architectural styles are illustrated to indicate the diversity in housing styles that has evolved over the years in the United States.

NEW ENGLAND COLONIAL

The **New England Colonial** house shown in Fig. 5.7 has seen many revivals of popularity over the years. Examples of it can be found in many communities in the United States.

Figure 5.7
A New England
Colonial House
Source: Henry S.
Harrison, *Houses:
The Illustrated
Guide to
Construction,
Design and Systems,
2nd Ed.* (Chicago,
IL: Real Estate
Education
Company, 1992),
p. 113.

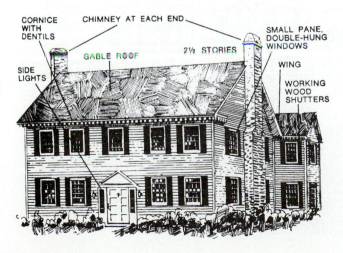

KEY DISTINGUISHING CHARACTERISTICS

The New England Colonial is a two and one-half story, generally symmetrical, square or rectangular boxlike house with side or rear wings. The traditional material is narrow clapboard siding. The roof usually is the gable type covered with shingles.

OTHER DISTINGUISHING CHARACTERISTICS

Originals had chimneys at each end for heating.
Modern versions have only one chimney at the end or in the center.
Windows are the double-hung type with small glass panes.
Shutters are the same size as the windows.
Central entrance door often has sidelights and a fanlight.
Elaborate cornice with dentils.
First floor has a central hallway running from the front to the rear.
Bedrooms are on the second floor.

HISTORY

These large, roomy houses evolved from the Cape Cod style.[19]

MID-ATLANTIC: NEW YORK CITY BROWNSTONE

The **brownstone** house met the limited space availability situation that existed in most American cities in the latter half of the 19th Century where lots were extremely narrow and deep (Fig. 5.8). Entry through ornate double doors usually occurred on the second floor,

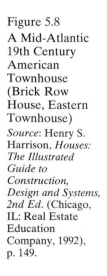

Figure 5.8
A Mid-Atlantic
19th Century
American
Townhouse
(Brick Row
House, Eastern
Townhouse)
Source: Henry S.
Harrison, *Houses:
The Illustrated
Guide to
Construction,
Design and Systems,
2nd Ed.* (Chicago,
IL: Real Estate
Education
Company, 1992),
p. 149.

which was reached by an exterior staircase. When used as a single-family residence, the public rooms typically were on the second floor with bedrooms on the third level and the kitchen in the basement.

KEY DISTINGUISHING CHARACTERISTICS

These houses usually cover an entire city block. Most are four or five stories with a stoop leading up to the first floor. They have common side walls with the house on either side.

OTHER DISTINGUISHING CHARACTERISTICS

Built of brick.
Often faced or trimmed with chocolate sandstone called brownstone.
Flat roof.
Simple double-hung windows.

HISTORY

Brownstones became popular in the late 1800s in New York and other large eastern cities. They tended to develop their individual characteristics in each different city.[20]

THE SOUTHWEST: ADOBE HOUSE

Adobe is a building product that was commonly made and used in the Southwest from a blend of soil, water, and straw that was cast in wooden frames as bricks that dried in the sun. Adobe houses were built in many styles and typically had massive walls whose exterior was coated with a protective layer of finish. Peeled logs were used in the ceiling (Fig. 5.9).

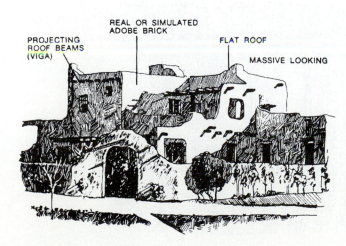

Figure 5.9
An Adobe
(Pueblo) House
Source: Henry S. Harrison, *Houses: The Illustrated Guide to Construction, Design and Systems, 2nd Ed.* (Chicago, IL: Real Estate Education Company, 1992), p. 156.

KEY DISTINGUISHING CHARACTERISTICS

This house uses adobe brick or some other material that is made to look like adobe brick. The characteristic projecting roof beams are called viga.

OTHER DISTINGUISHING CHARACTERISTICS

Houses are massive looking.
Roofs are flat.
Long rainwater gutters called "canales."

HISTORY

Based on the design of the houses of the Indians of New Mexico and Arizona, this became a popular style throughout the Southwest in the early 1900s.[21]

THE NATIONAL STYLE: VICTORIAN

The **Victorian** era, from which the style received its name, spanned from the crowning of Queen Victoria in England in 1837 until her death in 1901. The diversity of housing styles that flourished during this long period of time precludes the selection of one "Victorian style."

Kemp has observed:

> This period was an era of innovation and revival in architecture and the decorative arts. The Victorian Age began with a revival of the Medieval Gothic style and ended with the emergence of Art Nouveau. In between, a number of historic styles were revived and reinterpreted and new house types were introduced in architecture: Elizabethan, Italianate, Tuscan, the Octagon house, French Second Empire, the Stick and Eastlake styles, Victorian Queen Anne, Richardson Romanesque, Sullivan Romanesque, and the Shingle Style.[22]

An example of one of these styles, the High Victorian Italianate, is shown in Fig. 5.10.

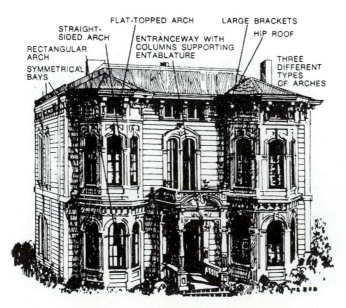

Figure 5.10
A High Victorian Italianate House
Source: Henry S. Harrison, *Houses: The Illustrated Guide to Construction, Design and Systems, 2nd Ed.* (Chicago, IL: Real Estate Education Company, 1992), p. 142.

KEY DISTINGUISHING CHARACTERISTICS

The use of three different kinds of window arches is the primary distinguishing characteristic of this style. They are the straight-sided arch, the flat-topped arch, and the rectangular arch.

OTHER DISTINGUISHING CHARACTERISTICS

Square shape.
Symmetrical bays in the front.
Hip roof.
Small chimneys protruding through the roof in irregular locations.
Entranceway has columns supporting an entablature.
Cornice supported with oversized brackets.
Exterior appears to be cut up and ornate.

HISTORY

The style seems to have come to this country from Italy by way of England, where the style was popular in the early 1800s. The house pictured is in Portland, Oregon, and was built by Mark Morris in 1882.[23]

INTERIOR ARCHITECTURAL DESIGN

When the homebuyer discusses the program requirements with either an architect or a homebuilder, decisions are made concurrently about both the exterior architectural appearance of the house and the interior architectural design. Decisions about the exterior architectural appearance can be structured around either the CTS system or some modification of it. Interior architectural design begins by recognizing that the interior of a house is usually divided into a number of **use zones**.

Once this concept is understood, the room requirements, in terms of both type and size, can be addressed. Decisions about the room requirements usually address the following types of space demands and needs:

Living Area	Laundry
Dining Area	Storage
Kitchen	Basement
Utility Room	Garage
Family Area	Porch
Bedrooms	Emergency Shelter
Hobby Room	Terrace
Entertainment Center	Garden
Bathrooms	Fireplace[24]

When a reasonable consensus has been reached about the room requirements, the process of developing a floor plan for the house can proceed. As Weidhaas notes:

> There are two major considerations in the design of the floor plan of a house. First, each room must be designed so that it is pleasant, functional, and economical; and second, the

rooms must be placed in a correct relationship to one another. This second consideration may be likened to working out a jigsaw puzzle in which the pieces may change in shape and size, giving various solutions. Some of these solutions will be better than others, but in the design of a house, the overall plan can be no better than the design of the individual rooms.[25]

The development of the layout and size of the floor plan which satisfies the homebuyer is an iterative process which goes through the typical steps of conceptual design, preliminary design, and final design. Other important considerations of spatial organization also influence the final design of the house. These include factors such as artistic or creative expression; site conditions such as views, slopes, trees and vegetation, breezes, etc.; interaction of interior and exterior spaces; or strict conformance to a particular architectural style. Limitations imposed by materials or construction methods or sequences also can affect the final design decisions.

Use Zone Analysis

The analysis of a set of plans for a particular house, or the evolution of the design of a floor plan which finally meets the needs of a specific homebuyer, can proceed effectively if the common **use zones** of a house are recognized. These use zones, and related entrances to the house, consist of:

Use Zones	Entrances (circulation zones)
Living/social	Family entrance
Private/sleeping	Guest entrance
Working/service	
Circulation[26]	

The opportunity to reach an optimal relationship between these use zones increases as the number of stories in the house and the size of its floor plan increase. Special attention must be given to the circulation zone, which typically includes the entrance to the house and the halls and stairs. This zone should reflect the patterns of traffic that are expected from the various members of the household so that there is convenient and efficient circulation.

Use Zones for a One-Story House

Figure 5.11 provides an example of the type of use zone relationships that exist in a one-story house, an option which provides only a limited number of ways that the various use zones can be separated from each other. The guest entrance, with an adjacent closet for the storage of garments, leads directly into the living/social zone, which in this case consists of the living room/dining area.

The working/service zone, which is the kitchen, is isolated from the guest entrance, has direct access to the living room/ dining area, and is also used as the family entrance to the house from the garage. A terrace/deck could be added adjacent to the living room/dining area to provide direct access to the outdoors. The private/sleeping zone, which consists of three bedrooms and one bathroom, is separated on the right side of the house. The limited floor plan flexibility in a one-story house means that the privacy of the bedrooms is disturbed whenever guests or family members use the bathroom.

Use Zones for a Two-Story House

The advantage of the two-story configuration is reflected in the fact that the private/sleeping zone can be completely separated from the working/service and the living/social zones (Fig. 5.12). The kitchen on the first floor is the primary working/service zone and is conveniently placed adjacent to the dining room. A terrace/deck also could be added in this house adjacent to the living room to provide direct access to the outdoors. In this house, the living/social zone has been divided into two parts by the main circulation zone on the first

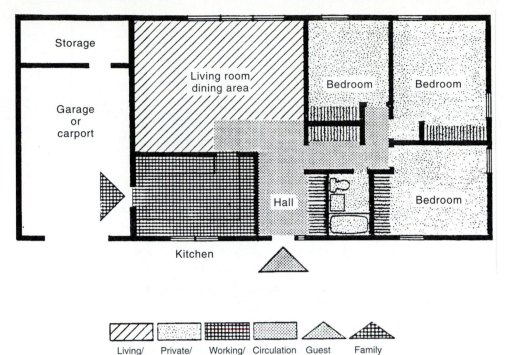

Figure 5.11
Zoning: One-Story Ranch House

Source: Henry S. Harrison, *Houses: The Illustrated Guide to Construction, Design and Systems, 2nd Ed*. (Chicago, IL: Real Estate Education Company, 1992), p. 42.

floor. Thus, guests can be served dinner while other family members, perhaps children, can use the living room without disturbing the diners.

One of the problems with the floor plan shown in Fig. 5.12 is that none of the bedrooms has a private bathroom. Everyone must use the one bathroom on the second floor, which is placed at the head of the stairs and, therefore, does not provide sufficient privacy from the first floor.

Both Figs. 5.11 and 5.12 illustrate the basic differences between the guest entrance, which should lead into the center of the house, and the family entrance, which ideally should lead from the garage into the kitchen or to a circulation zone that leads to the kitchen. If such an entrance is provided, then groceries can be carried from the automobile to the kitchen without being exposed to the weather. In each of these cases, it is also possible for a family member of the household to move from the work/service zone to the private/sleeping zone without passing directly through the living/social zone.

Common Floor Plan Deficiencies

The preceding discussion of use zones has highlighted some of the deficiencies that occur in a set of floor plans as an attempt is made by the architect or homebuilder to meet all of the stated needs of the homebuyer while working under either physical or budgetary constraints. Harrison has provided the following list of additional deficiencies:

1. Front door entering directly into the living room.

2. No front-hall closet.

3. No direct access from the front door to the kitchen, bathroom, and bedrooms without passing through other rooms.

4. Rear door not convenient to the kitchen and difficult to reach from the street, driveway, and garage.

5. No comfortable space for the family to eat in or near the kitchen.

6. A separate dining area or room is not easily reached from the kitchen.

7. Stairway between levels is off a room rather than off a hallway or foyer.

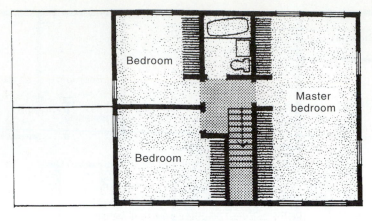

Second floor

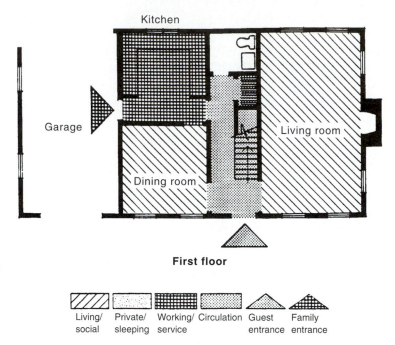

First floor

Figure 5.12
Zoning: Two-
Story House
Source: Henry S.
Harrison, *Houses:
The Illustrated
Guide to
Construction,
Design and Systems,
2nd Ed*. (Chicago,
IL: Real Estate
Education
Company, 1992),
pp. 43–44.

| Living/ social | Private/ sleeping | Working/ service | Circulation | Guest entrance | Family entrance |

8. Bedrooms and bathrooms located so they are visible from the living room or foyer.

9. Walls between the bedrooms not soundproof (best way to accomplish this is to have them separated by a bathroom or closets).

10. Recreation room or family room poorly located.

11. No access to the basement from outside the house.

12. Outdoor living areas not accessible from the kitchen.

13. Walls cut up by doors and windows, making it difficult to place the furniture around the room.[27]

Analysis of Individual Rooms and Spaces

The individual rooms and spaces within the living/social, private/sleeping, working/service, and circulation use zones each serve a different function for the members of a household. The size and shape that are recommended for each are directly related to that function.

Chapter 3 of the 1995 CABO Code provides the following **minimum** room size requirements in Section 304:

Section 304
Room Sizes

304.1 Minimum area. Every dwelling unit shall have at least one habitable room which shall not have less than 150 sq ft (13.9 m^2) of floor area. Other habitable rooms shall have an area of not less than 70 sq ft (6.50 m^2). Every kitchen shall not have less than 50 sq ft (4.64 m^2) of floor area. Habitable rooms, except kitchens, shall not be less than 7 feet (2134 mm) in any horizontal dimension.[28]

In addition, the following requirements appear in Sections 306 and 307:

Section 306
Sanitation

306.1 Toilet facilities. Every dwelling unit shall be provided with a water closet, lavatory, and a bathtub or shower.

306.2 Kitchen. Each dwelling unit shall be provided with a kitchen area and every kitchen area shall be provided with a sink of approved nonabsorbent material.[29]

Section 307
Toilet, Bath and Shower Spaces

307.1 Privacy required. Every water closet, bathtub, or shower required by this code shall be installed in a room which will afford privacy to the occupant.[30]

These *minimum* size requirements can be used as design guidelines for affordable housing for lower-income homebuyers. Clearly, another source of information is required in order to establish some realistic room size and shape recommendations for houses that are designed for homebuyers whose needs and demands exceed these minimum requirements. Weidhaas notes that there is not a true average, or standard, or even ideal size or shape for each room or space. Sizes and shapes of rooms should be selected to meet the requirements of function, aesthetics, and economy.[31] He does, however, provide a starting point for size and shape selection by presenting some minimum and average room size guidelines (Table 5.6).

Additional guidance regarding the interior architectural design considerations for some of the individual rooms and spaces in a typical house is provided.

TABLE 5.6   Size and Shape Guidelines for Individual Rooms and Spaces

Room	Minimum (inside size)	Average
Living room	12 ft × 16 ft	14 ft × 20 ft
Dining room	10 ft × 12 ft	12 ft × 13 ft
Dining area	7 ft × 9 ft	9 ft × 9 ft
Bedroom (master)	11.5 ft × 12.5 ft	12 ft × 14 ft
Bedroom (other)	9 ft × 11 ft	10 ft × 12 ft
Kitchen	7 ft × 10 ft	8 ft × 12 ft
Bathroom	5 ft × 7 ft	5.5 ft × 8 ft
Utility (no basement)	7 ft × 8 ft	8 ft × 11 ft
Hall width	3 ft	3.5 ft
Closet	2 ft × 2 ft	2 ft × 4 ft
Garage (single)	9.5 ft × 19 ft	12 ft × 20 ft
Garage (double)	18 ft × 19 ft	20 ft × 20 ft

Source: Ernest R. Weidhaas, *Architectural Drafting and Construction*, 4th ed., p. 143 (Boston, MA: Allyn & Bacon, 1989).

Living Room

The **living room**, traditionally, is the place where guests are entertained and therefore typically is located in the front of the house near the main entrance. Closet space generally is provided adjacent to the living room to store overcoats, umbrellas, etc. The emergence of the **family room** as the focal point for family activities in many households has diminished the function and status of the living room as a room to be *lived in.*

A living room will generally have one or more focal points or centers of attraction. Furniture can be grouped around a picture window, entertainment center, fireplace, or television. Harrison notes:

> The ideal conversation area is a circle approximately 10 ft in diameter. Furniture groups should be arranged with this in mind. The family and/or guests are more comfortable when seated within this range for most people like to congregate in small groups whether in a large or a small room. Within the 10-ft diameter circle, they can see each other easily and communicate comfortably without shouting.[32]

When the traditional pieces of furniture (i.e., couch, two comfortable chairs, end tables, coffee table) are considered in conjunction with the concept of a 10-ft circle, it often is suggested that the ideal width of the living room is 14 ft. A **minimum** size living room which does not meet that criteria, and therefore appears to be cramped, is shown in Fig. 5.13.

Family Room

While the living room traditionally has been used to entertain guests in a more formal setting, the less formal **family room** is a center for recreational activities, meeting friends, and a good place for children to play. Family rooms often are located in the basement of older houses. For new houses, it is felt that the family room should be near the kitchen and away from the private/sleeping zone. If the family room is near the rear of the house, there is more privacy and it is likely to be near both the kitchen and patio/outdoor entertainment area.

Figure 5.13
A Minimum Size
Living Room
Source: Henry S. Harrison, *Houses: The Illustrated Guide to Construction, Design and Systems, 2nd Ed.* (Chicago, IL: Real Estate Education Company, 1992), p. 54.

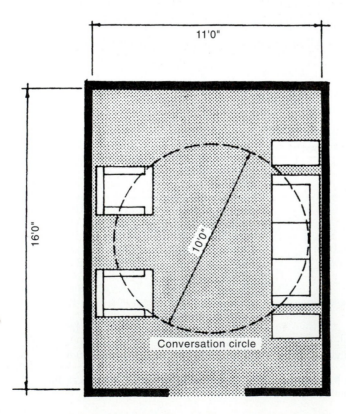

Table 5.6 does not provide any size and shape guidance for family rooms. Such information for the family room is provided by Harrison, as follows:

> A good size for a small house is approximately 12×16 ft to 13×18 ft, but, whatever the room area, the smallest wall dimension should be no shorter than 10.5 linear ft.[33]

Dining Room

The basic purpose of the **dining room** is to provide a special space for eating. Acceptable alternatives to the dining room in many houses are dining areas set up at one end of the family room, living room, or kitchen. Many homes now contain an additional dining area which consists of an outdoor patio or deck with a grill. Wherever it is situated, access from the kitchen service area to the dining table should be short and direct.

Bedrooms

The number of **bedrooms** desired by the homebuyer depends upon the size and stage of the household, the desire for flexibility to convert a bedroom into some other use such as an office or a hobby room, and the frequency of overnight guests. The use zone analysis indicated the desirability to isolate the bedrooms from the rest of the house. A related decision must be made about providing a separation between the master bedroom and the bedrooms used by the children and by guests.

The first priority when determining the size of a bedroom is to provide adequate space for the bed (or beds), the other bedroom furniture, and closet space for clothes. The standard bed sizes that can be used as a starting point in room size determination are:

Twin bed	39 in. × 75 in.
Full (standard)	54 in. × 75 in.
Queen bed	60 in. × 80 in.
King bed	76 in. × 80 in.

If the clearances indicated in Fig. 5.14 are considered and a 2-ft closet depth is assumed, it is clear that the average 10×12 ft bedroom size indicated in Table 5.6 has already been exceeded.

Bedrooms are no longer viewed just as sleeping areas; they are also considered to be hiding places for relaxation and privacy. Whenever possible, the sizes of the children's bedrooms should be large enough to accommodate and display personal games, books, hobbies,

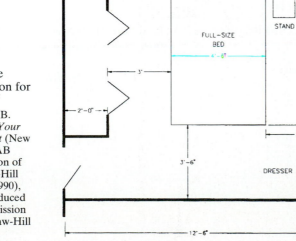

Figure 5.14

Minimal Size Determination for a Bedroom

Source: Gene B. Williams, *Be Your Own Architect* (New York, NY: TAB Books, Division of The McGraw-Hill Companies, 1990), p. 108. Reproduced with the permission of The McGraw-Hill Companies.

and collectible treasures. In addition, the master bedroom in larger houses now often includes a sitting area which can be used for reading and lazy contemplation, a place to unwind and relax after a day at work. Also adding to the size of the master bedroom is the integral bathroom, which can include dual lavatories, large dressing and storage areas, spas, whirlpools, steam baths, and saunas.

The presence of natural light and ventilation also is important in bedrooms, both from a health as well as a fire safety standpoint. Chapter 3 of the 1995 CABO Code provides the following requirements for windows in bedrooms in Sections 303 and 310:

Section 303
Light, Ventilation and Heating

303.1 Habitable rooms. All habitable rooms shall be provided with aggregate glazing area of not less than 8% of the floor area of such rooms. One-half of the required area of glazing shall be openable.[34]

Section 310
Exits

310.2 Emergency egress required. Every sleeping room shall have at least one openable window or exterior door approved for emergency egress or rescue. The units must be operable from the inside to a full clear opening without the use of a key or tool. Where windows are provided as a means of egress or rescue, they shall have a sill height of not more than 44 in. (1118 mm) above the floor.

310.2.1 Minimum size. All egress or rescue windows from sleeping rooms must have a net clear opening of 5.7 sq ft (0.530 m^2). The minimum net clear opening height shall be 22 in. (559 mm). The minimum net clear opening width shall be 20 in. (508 mm).

    Exception: Grade floor window may have a minimum net clear opening of 5 sq ft.[35]

The 8% requirement for the window area results in a minimum window size of 9.6 sq ft for a 10×12 ft bedroom. Thus, a 3×4 ft window would exceed the code requirements, but would barely provide an adequate amount of light for the bedroom. Most homebuyers will demand a much larger window area to create an atmosphere of illumination and vitality.

Kitchen

The **kitchen** is often considered to be the focal point of the household because it is where family members receive their nourishment, in the form of food as well as interpersonal contact and direction. This room, along with the bathrooms, are often considered to be the "hot buttons" that sell the house. As the primary room in the working/service zone, it should be easily accessible from other rooms that depend on it. Interference with the work role should be minimized by discouraging its use as a quasi-circulation zone to other parts of the house or outdoors. There also should be an efficient relationship between the work centers that define its role.

WORK CENTER ORIENTATION

The primary functions which the kitchen serves can be defined in terms of kitchen work centers and several associated centers. Jane Snow[36] defines these as follows:

1. Primary Work Centers
    a. Cleanup center
    b. Refrigerator center
    c. Cooking center
2. Secondary Work Centers
    a. Mixing center
    b. Serving center
3. Other Centers
    a. Eating center
    b. Planning center
    c. Entertainment center

THE WORK TRIANGLE

Kitchens often are designed around the sink as a focus because this primary component of the cleanup center is the most-used unit in the kitchen. The refrigeration center and the range/cook top center are typically positioned so that all three can be connected as a **work triangle**. The steps between these centers are kept to a reasonable level if the following guidelines are followed:

> In the most efficient layout, the sink forms the apex of the work triangle and is equidistant from the other two primary work centers, the refrigerator and the cook top. The arms of the triangle should not be less than 12 ft nor exceed 22 ft (the absolute maximum is 26 ft). Each arm may vary, of course, but in the 12-ft kitchen, the arms might be 4 ft × 4 ft × 4 ft, and in a 22-ft kitchen, they might be 6.5 ft × 7.5 ft × 8 ft. Ideally, the individual legs are between 4 and 7 ft long.[37]

In most kitchens, an attempt is made to place the sink under a window, probably for brightness and surveillance purposes; the refrigerator is placed closest to the door through which groceries are carried in; and the cooktop/range is placed closest to the dining area.

BASIC KITCHEN LAYOUTS

Based upon extensive studies using the work triangle as the base, it is generally agreed that the four most desirable layouts are: (1) the U-shape, (2) the L-shape, (3) the two-wall or corridor kitchen, and (4) the one-wall kitchen. These four often become the starting point for customizing the kitchen in a house.[38] A basic U-shaped kitchen with its related work triangle is shown in Fig. 5.15.

OTHER CONSIDERATIONS

The desired location of the kitchen on the floor plan is a direct function of the lifestyle of the members of the household. A strong argument for placing the kitchen in the back of the house, with the sink under a window, is that it allows the person preparing the meals to observe the children at play in the backyard. This location also places the kitchen in close proximity with the backyard patio or deck that might be used for entertaining. An argument also can be made for placing the kitchen at the front of the house because this allows the person in the kitchen to both observe people approaching the house and the children who are playing in the front yard. Regardless of the location of the kitchen, however, it should be in reasonable proximity to: (1) the garage, if groceries arrive from that location, and (2) the rooms to which food is transported for meals or entertaining.

Figure 5.15

A Basic U-Shaped Kitchen with the Work Triangle = 13 ft, 6 in.

Source: Jane Moss Snow, *Kitchens,* p. 38 (Washington, D.C.: The National Association of Homebuilders, 1987).

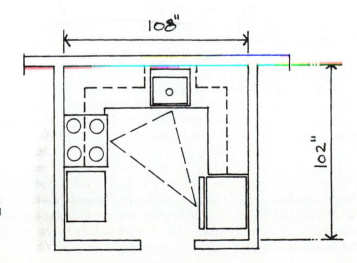

Bathrooms

The second most important selling "hot button" for a house, after the kitchen, usually is considered to be the number, style, and placement of the **bathrooms**. Although the typical house has only one kitchen, the time has long passed when one full bathroom (consisting of a bathtub/shower combination, lavatory, and toilet) is considered to be an acceptable standard.

LOCATION AND PURPOSE

In most houses with two or more baths, one of them is typically integrated into the master bedroom suite and has its own private entrance. In households where both adults work outside the home, the master bath typically is designed so that both adults can simultaneously use it.

The location and purpose of the second full bathroom in the house is dependent upon the consumer group being addressed. In households with children, the second bathroom should be placed in a hallway that is accessible to one or more bedrooms. In households without children (young professionals, retirees, empty-nesters), the second bathroom is for occasional use and for guests. Therefore, it should be easily accessible from the guest bedroom and the living/social zones of the house.

The third bathroom in a house is usually a half bath (containing a lavatory and toilet) or a powder room (containing only a lavatory). It is usually placed near the front entrance and/or primary living/social zone so that it is easily accessible by guests. It should not be too close to the living/social zone, however, because it presents acoustical problems and it should not be easily visible from that zone.

Ventilation is extremely important in bathrooms. That is why they were originally located on outside walls that had the benefit of natural ventilation through windows. Interior bathrooms without windows are now routinely used, provided there is sufficient mechanical ventilation that is directed to the outside. These fans are normally activated by the light switch in the bathroom.

Section 303.3 of the 1995 CABO Code addresses this issue as follows:

> **303.3 Bathrooms.** Bathrooms, water closet compartments, and other similar rooms shall be provided with aggregate glazing area in windows of not less than 3 sq ft (0.279 m^2), one-half of which must be openable.
>
> **Exception:** The glazed areas shall not be required where artificial light and an approved mechanical ventilation system capable of producing a change of air every 12 minutes are provided. Bathroom exhausts shall be vented directly to the outside.[39]

THE BATHROOM FLOOR PLAN

The minimum and average size bathrooms in Table 5.6 are 5×7 ft and 5.5 × 8 ft, respectively. The 5×7-ft size traditionally has been called the **basic bathroom**, with three fixtures consisting of the shower/tub, lavatory, and toilet. Figure 5.16 indicates both a *minimal* and a *desirable* basic bathroom floor for a one-wall configuration. Snow provides many other examples of such floor plans.[40]

Homebuyer groups that have needs or desires that go beyond the basic bathroom are usually focusing on the added features of **move-up bathrooms**, which might include, as a minimum, a whirlpool, a shower, and two vanities. Amenities such as integral dressing rooms, saunas, and exercise areas can also be added.[41]

Laundry Rooms

The average house has four general areas in which the **laundry facilities** may be installed. The location is determined by space, practicality, and preference, not necessarily in that order. The alternatives, all of which have both positive and negative attributes, include the kitchen, mud room, bedroom–bath area, or basement.

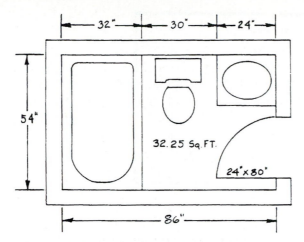

(a) A Minimal One-Wall Bathroom

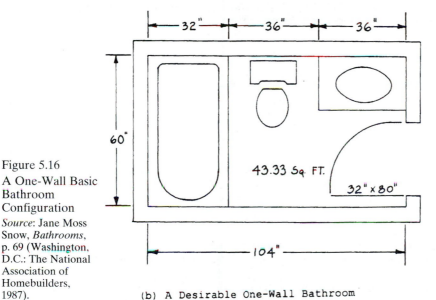

Figure 5.16
A One-Wall Basic
Bathroom
Configuration
Source: Jane Moss
Snow, *Bathrooms*,
p. 69 (Washington,
D.C.: The National
Association of
Homebuilders,
1987).

(b) A Desirable One-Wall Bathroom

Garages

Consideration must be given to the number of vehicles owned by the household when the size of the **garage** is determined. In addition, an estimate should be made of the other storage and/or workshop uses that will be delegated to the garage. Generally, a minimum 10×20-ft space is required for a one-car garage without allowing any space for other uses. This minimum size increases to 20×20 ft for a two-car garage. Figure 5.17 indicates the resultant dimensions for a one-car and two-car garage, respectively, when this minimum criterion is exceeded in order to provide adequate space for other uses.

When a garage is attached directly to the house, there is a danger associated with the exhaust fumes from the cars. Accordingly, Section 309 of Chapter 3 of the 1995 CABO Code provides the following requirements:

Section 309
Garages

309.1 Opening protection. Openings from a private garage directly into a room used for sleeping purposes shall not be permitted. Other openings between the garage and the residence shall be equipped with either solid wood doors not less than 1-3/8 in. (35 mm) in thickness or 20-minute fire-rated doors.

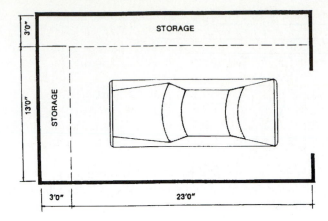

(a) A Recommended One-Car Garage with Storage

Figure 5.17 Recommended Dimensions for Garage with Storage

Source: Henry S. Harrison, *Houses: The Illustrated Guide to Construction, Design and Systems, 2nd Ed.* (Chicago, IL: Real Estate Education Company, 1992), pp. 100, 102.

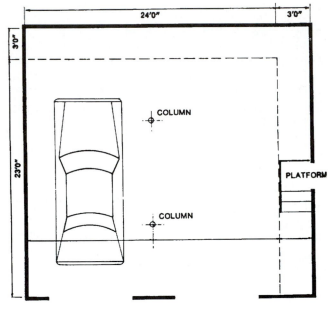

(b) A Recommended Two-Car Garage with Storage

309.2 Separation required. The garage shall be separated from the residence and its attic area by means of 1/2 in. (12.7 mm) gypsum board applied to the garage side.

309.3 Floor surface. Garage and carport floor surfaces shall be of approved noncombustible material. That area of floor used for parking of automobiles or other vehicles shall be sloped to facilitate the movement of liquids to a drain or toward the main vehicle entry doorway.[42]

Other Areas

The other areas of a house that should be mentioned since they must also be incorporated into the floor plan are: (1) patios and porches, (2) storage/closets, (3) hallways, and (4) stairs.

PATIOS AND PORCHES

Entertaining of guests, celebrations, and outdoor barbeques and picnics are a part of the social life of most households. The open space that is connected to the house, usually in the back to provide privacy, is often called a **patio**. Patios can range from a simple slab of concrete that provides a stable surface for a picnic table, benches, and a grill to an elaborate covered or uncovered terrace area or a wooden deck. When a patio area is incorporated into the floor plan of the house, there should be a direct passage from the working/service zone to the patio so that food, drinks, dishes, and other supplies can be transported outside easily.

Porches in the front of a house are an integral part of many traditional architectural housing styles. By the middle of the 20th Century, however, porches had virtually disappeared from the popular housing styles. Many feel that the rise of ranch/split-level communities and the popularity of outdoor living and entertaining in patio settings were the major causes of this decline.

Houses are still built with porches, however, for those people who either value the joy of sitting, rocking, and watching the neighbors or cars go by or who wish they had the time to pursue such leisurely activities.

STORAGE AND CLOSETS

A review of the stated consumer preferences in Table 2.4 indicates that one of the features which most homebuyers value is adequate storage and closet space. Three of the most popular places in a house for both short-term and long-term storage are: (1) the attic, (2) the garage, and (3) the basement. In addition, outdoor storage sheds are used to store yard tractors, lawn mowers, snow blowers, and garden tools.

Closets are required to store clothing and other items that cannot be conveniently placed in the previously noted locations. As a minimum, there usually is one closet in each bedroom and typically two in the master bedroom. These closets, which can be of either the wardrobe or walk-in type, are primarily devoted to clothing storage. In addition, books, camping equipment, etc., are also stored in them.

In addition to the clothes closets in the bedrooms, there usually is a combined family coat closet and guest closet near the front entrance to the house. This closet may be entirely for guests if a family coat closet is placed in the circulation zone between the kitchen and the garage.

At least one linen closet, typically located in the vicinity of the private/sleeping zone, is needed for bed linens, blankets, towels, and extra pillows. This closet usually will contain a number of shelves at various spacings. A general utility closet, which is probably placed in the vicinity of the working/service zone, is also required to store brooms, the vacuum cleaner, etc.

Storage space is also needed, of course, in the kitchen and bathrooms. In the kitchen, this usually takes the form of attractive and functional base cabinets, wall cabinets, and countertops that provide the style emphasis for that room. Walk-in pantries for the storage of groceries often are an added feature in move-up, custom, and luxury houses. In bathrooms, storage space is provided by medicine cabinets, wall and base cabinets, and, in move-up bathrooms, by separate walk-in dressing rooms.

The rooms in the living/social zone of the house (the dining room, the living room, and the family room) usually have their storage needs met either by the furniture in the room or by built-in wall and base cabinets.

HALLWAYS

In most houses, **hallways** are used as the primary circulation zone which connects the other zones of the house. Although they generally are viewed as a waste of space, they cannot be eliminated entirely unless a completely "open" floor plan without room definition is adopted. An attempt is usually made to minimize their usage as much as possible.

Hallways, however, must be wide enough to allow furniture and other bulky material to be moved in and out of rooms. They must also provide safe passage in case of an emergency. Accordingly, Section 311 of Chapter 3 of the 1995 CABO Code provides the following requirement:

Section 311
Doors and Hallways

311.1 General. The required exit door shall be a side-hinged door not less than 3 ft (914 mm) in width and 6 ft 8 in. (2032 mm) in height. Other exterior hinged or sliding doors shall not be required to comply with these minimum dimensions. **The minimum width of a hallway or exit access shall be not less than 3 ft** (914 mm).[43]

STAIRS

Stairs are an important part of the circulation zone in houses which are multi-storied. In addition to a functional role, stairs often are a major design component of the house. In times of emergency there are safety implications involved if stairs are too steep, too narrow, or poorly lighted. Excerpts from appropriate sections of Chapter 2 of the 1995 CABO Code which present the important information about issues related to stairs are provided. Reference should be made to the full text of these sections to obtain all of the design requirements that are specified.

Section 303
Light, Ventilation, and Heating

303.4 Stairway Illumination. All interior and exterior stairs shall be provided with a means to illuminate the stairs, including the landings and treads . . .

303.4.1 Light activation. The control for activation of the required interior stairway lighting shall be accessible at the top and bottom of each stair without traversing any step of the stair.[44]

Section 314
Stairways

314.1 Width. Stairways shall not be less than 36 in. (914 mm) in clear width at all points above the permitted handrail height and below the required headroom height. The minimum width at and below the handrail height shall not be less than 32 in. (813 mm) where a handrail is installed on one side and 28 in. (711 mm) where handrails are provided on both sides.

314.2 Treads and risers. The maximum riser height shall be 7-3/4 in. (197 mm) and the minimum tread depth shall be 10 in. (254 mm). . . .

314.3 Headroom. The minimum headroom in all parts of the stairway shall not be less than 6 ft 8 in. (2032 mm) measured vertically from the sloped plane adjoining the tread nosing or from the floor surface of the landing or platform.[45]

Section 315
Handrails and Guardrails

315.1 Handrails. Handrails having minimum and maximum heights of 30 in. and 38 in. (762 mm and 965 mm), respectively, measured vertically from the nosing of the treads, shall be provided on at least one side of stairways of three or more risers. Spiral stairways shall have the required handrail . . .[46]

ADDITIONAL BUILDING CODE REQUIREMENTS

Chapter 3 of the 1995 CABO Code, entitled "Building Planning,"[47] establishes certain **minimum** design requirements which affect the interior architectural design of a house. Some of these requirements have already been highlighted; others are found in the sections noted below. Although exterior *architectural* design requirements are not addressed directly in Chapter 3 of the 1995 CABO Code, the *technical* requirements related to exterior wall and roof coverings are presented in Chapters 7[48] and 9,[49] respectively, of that code.

Feature and Service Requirements

Minimum requirements for the basic features and services that must be provided in a house are covered in the following sections of the 1995 CABO Code:

Section 303   Light, Ventilation, and Heating
This section specifies the window requirements.

Section 304   Room Sizes

Section 305   Ceiling Height
Section 305 establishes the basic ceiling height as 7 ft, 6 in., with exceptions provided for kitchens and other areas.

Section 306   Sanitation

Section 307   Toilet, Bath, and Shower Spaces

Section 308   Glazing
Section 308 provides extensive coverage of the safety requirements associated with the use of glass in houses.

Section 309   Garages

Circulation Zone Requirements

Safety requirements related to the various parts of the circulation zone of the house are covered in the following sections of the 1995 CABO Code:

Section 310   Exits

Section 311   Doors and Hallways

Section 312   Landings

Section 313   Ramps

Section 314   Stairways

Section 315   Handrails and Guardrails

Fire Safety Requirements

The threat of fire in a house is always a concern to the homeowner. The following sections of the 1995 CABO Code establish requirements that guard against the danger of fire.

Section 316   Smoke Detectors

Section 317   Foam Plastic

Section 318   Flame Spread and Smoke Density

Section 319   Insulation

Section 320   Dwelling Unit Separation
Section 320 deals with the fire safety requirements in multifamily and townhouse situations where individual housing units must be separated from each other by walls and/or floor assemblies that have a minimum *fire resistance rating*.

Other Requirements

Additional building planning requirements specified in the 1995 CABO Code cover the following areas:

Section 321   Moisture Vapor Retarders

Section 322   Protection against Decay

Section 323   Protection against Termites

Section 324   Protection against Radon

CHAPTER SUMMARY

The first section of this chapter presented an architectural classification system which is based upon the attributes of **class**, **type**, and **style** as a means of organizing the decisions that must be made at the beginning of the design process. A major emphasis was placed on housing **style** since a decision about this attribute often determines the **class** and **type** of house that is designed. A number of traditional housing styles were presented within a regional framework in order to indicate the diversity of choice which is available to the architect, homebuilder, and homebuyer.

Closely associated with the exterior architectural design of the house is the organization and design of the living space within the house. The **use zone** analysis approach was used to indicate that a house can be divided into: (1) living/social, (2) private/sleeping, (3) working/service, (4) circulation, (5) guest entrance, and (6) family entrance zones. The approach was illustrated for one-story and two-story houses. Floor plan deficiencies that are commonly found in house designs were also presented.

The use-zone approach is the framework around which individual rooms and spaces in the house were analyzed. Particular attention was given to kitchens and bathrooms since consumer studies have shown that these two rooms are often the ones which "sell the house." The size, shape, and use guidelines provided should allow an individual to prepare a preliminary version of a floor plan which meets the particular criteria that are provided by a homebuyer.

The interior architectural design of a house must ensure that the **minimum** building code requirements are either met or exceeded. Accordingly, appropriate sections of "Chapter 3: Building Planning" of the 1995 CABO Code are referenced in various places in the chapter. The chapter concluded by providing an overview of what is contained in the remainder of Chapter 3 of the 1995 CABO Code.

NOTES

[1] *CABO One and Two Family Dwelling Code, 1995 Edition.* (Falls Church, VA: The Council of American Building Officials, 1995), pp. 7–24.

[2] Henry S. Harrison, *Houses: The Illustrated Guide to Construction, Design and Systems,* 2nd ed. (Chicago, IL: Real Estate Education Company, 1992), pp. 398–404.

[3] Ibid., p. 399.

[4] Ibid., p. 400.

[5] *Means 1995 Residential Cost Data,* 14th Annual Ed. (Kingston, MA: R.S. Means Co., Inc., 1994), p. 6.

[6] Ibid.

[7] Ibid.

[8] Ibid.

[9] Virginia McAlester and Lee McAlester, *A Field Guide to American Houses.* (New York, NY: Alfred A. Knopf, Inc., 1984), p. xi.

[10] Harrison, *Houses.*

[11] Lester Walker, *American Shelter: An Illustrated Encyclopedia of the American Home.* (Woodstock, NY: The Overlook Press, 1981).

[12] Mary Mix Foley, *The American Home.* (New York, NY: Harper and Row, 1980).

[13] Carol Rifkind, *A Field Guide to American Architecture.* (New York, NY: New American Library, 1980).

[14] John Milnes Baker, *American House Styles: A Concise Guide.* (New York, NY: W. W. Norton & Company, 1994).

[15] Jim Kemp, *American Vernacular: Regional Influences in Architecture and Interior Design.* (New York, NY: Viking Penguin Inc., 1987).

[16] Ibid., pp. 8 and 10.

[17] Harrison, *Houses.*

[18] McAlester and McAlester, *A Field Guide to American Houses,* p. 21.

[19] Harrison, *Houses,* p. 113.

[20] Ibid., p. 149.

[21]Ibid., p. 156.

[22]Kemp, *American Vernacular,* p. 140.

[23]Harrison, *Houses,* p. 142.

[24]Ernest R. Weidhaas, *Architectural Drafting and Construction,* 4th ed.   Boston, MA: Allyn and Bacon, 1989, pp. 117–118.

[25]Ibid, p. 143.

[26]Harrison, *Houses,* pp. 40–41.

[27]Ibid., p. 50.

[28]*1995 CABO Code,* p. 15.

[29]Ibid.

[30]Ibid., p. 16.

[31]Weidhaas, *Architectural Drafting,* p. 143.

[32]Harrison, *Houses,* p. 58.

[33]Ibid., *Houses,* p. 87.

[34]*1995 CABO Code,* p. 7.

[35]Ibid., p. 19.

[36]Jane Moss Snow, *Kitchens.*   Washington, D.C.: The National Association of Homebuilders, 1987, p. 7.

[37]Ibid., p. 8.

[38]Ibid., p. 13.

[39]*1995 CABO Code,* p. 15.

[40]Jane Moss Snow, *Bathrooms.*   Washington, D.C.: The National Association of Homebuilders, 1987, pp. 71–83.

[41]Ibid., p. 92.

[42]*1995 CABO Code,* p. 20.

[43]Ibid., p. 19.

[44]Ibid., p. 15.

[45]Ibid., p. 20.

[46]Ibid.

[47]Ibid., pp. 7–24.

[48]Ibid., pp. 81–88.

[49]Ibid., pp. 131–136.

BIBLIOGRAPHY

Baker, John Milnes, *American House Styles: A Concise Guide.*   New York, NY: W. W. Norton & Company, 1994.

CABO One and Two Family Dwelling Code, 1995 Edition.   Falls Church, VA: The Council of American Building Officials, 1995.

Foley, Mary Mix, *The American Home.*   New York, NY: Harper and Row, 1980.

Harrison, Henry S., *Houses: The Illustrated Guide to Construction, Design and Systems,* 2nd ed.   Chicago, IL: Real Estate Education Company, 1992.

Kemp, Jim, *American Vernacular: Regional Influences in Architecture and Interior Design.*   New York, NY: Viking Penguin Inc., 1987.

McAlester, Virginia and Lee McAlester, *A Field Guide to American Houses.*   New York, NY: Alfred A. Knopf, Inc., 1984.

Means 1995 Residential Cost Data, 14th Annual Ed.   Kingston, MA: R.S. Means Co., Inc., 1994.

Poppeliers, John C., et al., *What Style Is It?—A Guide to American Architecture.*   Washington, D.C.: The Preservation Press, 1983.

Rifkind, Carol, *A Field Guide to American Architecture.*   New York, NY: New American Library, 1980.

Snow, Jane Moss, *Bathrooms.*   Washington, D.C.: The National Association of Homebuilders, 1987.

Snow, Jane Moss, *Kitchens.*   Washington, D.C.: The National Association of Homebuilders, 1987.

Walker, Lester, *American Shelter: An Illustrated Encyclopedia of the American Home.*   Woodstock, NY: The Overlook Press, 1981.

Weidhaas, Ernest R., *Architectural Drafting and Construction,* 4th ed.   Boston, MA: Allyn and Bacon, 1989.

Williams, Gene B., *Be Your Own Architect.*   Blue Ridge Summit, PA: Tab Books, 1990.

6

Development of a Set of Contract Documents

INTRODUCTION

The basic decisions about the design of a house usually are made during the **planning phase** of the project in a series of meetings between the homebuyer and either an architect or a homebuilder. The **design phase**, which follows the planning phase, is the first step in transforming the dreams of the homebuyer into a reality. It typically includes the development of a set of **working drawings** and a statement of the **material and workmanship specifications** which establish the standard of performance for the construction phase.

The **construction phase** of the project begins when a homebuilding firm is selected to build the house. A construction contract, typically provided by the homebuilder and modified to meet the needs of the homebuyer, usually is signed before work begins at the building site. Two of the most important elements of the contract are: (1) the contract price, and (2) the duration of the construction phase. The homebuilder determines this information by developing a **cost estimate** and a **construction schedule** for the house.

Aspects of the design and construction phases that are associated with these topics are presented in this chapter. Additional information about the construction phase is provided in Chapter 7, where the major phases of the house construction process are presented.

THE DESIGN PHASE

Design Methodology

A number of different methods can be used to translate the program requirement decisions into a set of working drawings. Ernest R. Weidhaas identifies four of the methods that are widely used:

1. The Prototype House Method
2. The Template Method
3. The Interior Planning Method
4. The Overall Planning Method[1]

The Prototype House Method

The **prototype house** method of preliminary design is recommended for inexperienced persons because the design process starts by referring to: (1) the features of a particular house that has impressed the homebuyer, or (2) a set of house plans presented in a home magazine, a residential design book, or a newspaper which appeared interesting to the homebuyer. An example of such a source is shown in Fig. 6.1, which provides several possible architectural design options for the "Pauline" model that is part of the "Custom Series" of S&A Custom Built Homes, headquartered in State College, PA. Table 6.1 provides a listing of all of the standard features that are associated with this model.

Modifications and alterations can be made to the architectural design and the standard features of the prototype house if it meets the majority of the program requirements that were established during the planning phase. An example of the types of changes that were made to the basic "Pauline" model to meet the needs of a particular homebuyer are shown in the set of house plans in Appendix A.

It should be noted that this set of house plans is used as the **Case Study House** example, which illustrates the concepts and procedures presented in this chapter and throughout the remainder of the book.

The Template Method

Templates can be used during the preliminary design stage. The size and shape of each room can be determined by arranging furniture templates (i.e., scaled furniture replicas) into appropriate groupings in scaled drawings of the rooms in which the furniture will be placed. Once room decisions have been made, a set of room templates can be used to determine acceptable floor plan layouts for the house.

The Interior Planning Method

The stages of the design process in the **interior planning method** include: (1) thumbnail sketches, (2) preliminary sketches, and (3) finished sketches of the floor plan of the house. At each stage, there typically is interaction between the homebuyer and the designer in the form of a critical analysis of the current version of the design. The results of this analysis determine the design modifications that still must be made.

The Overall Planning Method

The interior planning method, which focuses on floor plans, essentially designs the house from the inside out, whereas the **overall planning method** considers both the interior and the exterior aspects of the house from the beginning of the design process. These aspects are integrated by developing preliminary elevation sketches in conjunction with preliminary floor plans and details.

Working Drawings

After final agreement between the homebuyer and the designer is reached, the finished sketches must be converted into a set of **working drawings**, which is similar to the set of nine drawings of the "Pauline" model which are shown in Appendix A. Some architects still rely on the traditional manual techniques and tools used by a draftsman. Most design offices, however, produce working drawings using Computer-Aided Design (CAD) software and equipment. Recent advancements in computer-based technology have greatly reduced the cost of CAD and computers in general. CAD systems are now in an affordable price range for most modest-sized architectural or homebuilder offices.

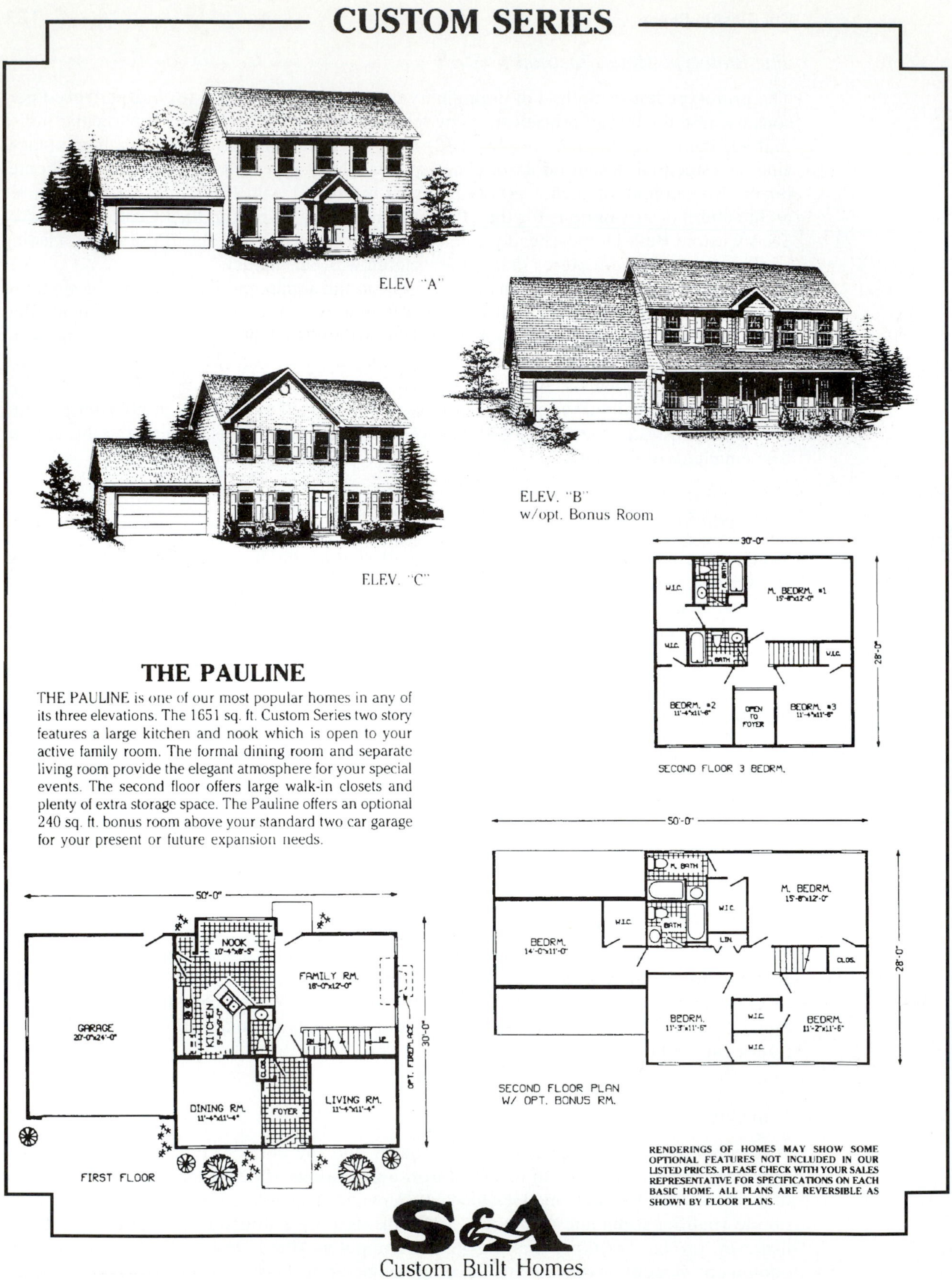
CUSTOM SERIES
ELEV "A"
ELEV. "B"
w/opt. Bonus Room
ELEV. "C"
THE PAULINE
THE PAULINE is one of our most popular homes in any of its three elevations. The 1651 sq. ft. Custom Series two story features a large kitchen and nook which is open to your active family room. The formal dining room and separate living room provide the elegant atmosphere for your special events. The second floor offers large walk-in closets and plenty of extra storage space. The Pauline offers an optional 240 sq. ft. bonus room above your standard two car garage for your present or future expansion needs.
SECOND FLOOR 3 BEDRM.
SECOND FLOOR PLAN
W/ OPT. BONUS RM.
FIRST FLOOR
GARAGE
NOOK
FAMILY RM.
KITCHEN
DINING RM.
FOYER
LIVING RM.
OPT. FIREPLACE
M. BEDRM.
BEDRM.
50'-0"
30'-0"
28'-0"
RENDERINGS OF HOMES MAY SHOW SOME OPTIONAL FEATURES NOT INCLUDED IN OUR LISTED PRICES. PLEASE CHECK WITH YOUR SALES REPRESENTATIVE FOR SPECIFICATIONS ON EACH BASIC HOME. ALL PLANS ARE REVERSIBLE AS SHOWN BY FLOOR PLANS.
S&A
Custom Built Homes

Figure 6.1
Architectural Design Options: The "Pauline" Model
Source: S&A Custom Built Homes (State College, PA).

TABLE 6.1   Standard Features of the Custom Series of S&A Custom Built Homes

DESIGNER INTERIORS
- Wall-to-wall plush carpeting
- Quality white plumbing fixtures
- Decorator quality lighting package
- No-wax kitchen and bath flooring
- Large selection of designer paint colors to choose from for one color throughout
- Flush lauan, birch, or six-panel masonite interior doors
- Ranch or colonial casing at doors, windows, and 3-1/4″ base at floor

ELABORATE KITCHEN AND BATH
- 30″ free-standing whirlpool range with hood
- Double bowl stainless steel sink
- Choice of formica countertops with 4″ backsplash
- Beautiful merillat kitchen and bath cabinetry
- Fluorescent light over sink
- Paneled soffit above cabinets
- Deluxe medicine cabinet
- Fan/light combination in full and three-quarter baths and exhaust fan in half bath
- Fiberglass tub–shower combination

ENERGY SAVING FEATURES
- Energy-efficient Comfort Home™ program
- R-38 ceiling insulation
- R-19 total wall cavity insulation
- R-19 floor insulation
- Electric baseboard heat with individual room thermostat
- Vapor barrier in exterior walls
- 52 gallon quick-recovery electric water heater
- Stanley steel insulated doors
- Ridge vent attic ventilation

MAINTENANCE FREE/SAFETY FEATURES
- Anderson High Performance vinyl clad insulated windows with screens
- Maintenance-free vinyl siding
- Aluminum soffit and fascia
- Aluminum gutters and downspouts
- Brick veneer as per model
- Shutters on front as per model

MAINTENANCE FREE/SAFETY FEATURES (continued)
- 25-year fiberglass shingles
- Frost-free front and rear hose bibs
- Two outside electrical outlets
- Outside light at all exterior doors
- Door chimes with two door pushbuttons
- Smoke detectors as required
- Vinyl window grilles on front as per model
- Two phone and two TV outlets

QUALITY CONSTRUCTION MATERIALS AND METHODS
- All framing materials, sizes, and procedures to meet local code
- 2″ × 4″ exterior wall with 1/2″ insulated sheathing
- 3/4″ T&G plywood floor
- Pressure-treated 2″ × 6″ sill plate
- 1/4″ underlayment in kitchen and bath (under vinyl)
- All construction-grade materials
- Drywall finish on walls and ceilings
- Foundation size and courses according to model
- Porch and stoops included according to model
- 3″ concrete basement floor
- Basement walls parged and asphalt coated
- 12 block courses on all bi-levels
- 3″ foundation drains at perimeter of house
- Sump pump or floor drains in basement
- 200 AMP service
- Type L copper water lines
- All plumbing and wiring to meet national building codes
- Washer and dryer facilities available
- One-year warranty on all materials and labor

OPTIONAL FEATURES
- Full range of customizing options are available to enhance your lifestyle

This is just a partial list of the many features included in our homes. The homes showcased in this book are only a sampling of the high-quality homes available from S&A Homes. Being a leader in the custom home industry has enabled us to have our own drafting and design department which can take your plans and ideas and make them into your very own dream home. Some homes shown in our brochures have optional features not included in our list price. Check plans and specifications on each basic home. All plans are reversible as shown by floor plan. Standard Features are subject to change without notice.

Source: *S&A Custom Built Homes.* (State College, PA, 1994).

Architectural Drafting Practices

Architects, engineers, homebuilders, estimators, building code officials, and homebuyers must be able to correctly interpret the information found on a set of working drawings if the house is to be constructed properly. Guidance, in terms of several basic architectural drafting concepts, is presented in order to assist in such interpretation activities.

Orthographic Projections

A residential structure is a three-dimensional object which can be described in a two-dimensional plane by means of a multiview or orthographic projection. The six orthographic sides of the "Pauline" model are shown in Fig. 6.2. As indicated in drawing A-3 (Appendix A), however, rarely are all six views used to depict architectural structures. Instead, only the four elevations (Front, Right Side, Rear, and Left Side) are typically shown. The roof view is usually replaced with a floor plan, and the bottom view may be inverted and shown as a foundation plan.

Horizontal Cutting Plane

Floor plans, such as those shown in drawings A-1 and A-2, and the foundation plan shown in drawing A-3 (Appendix A), convey the maximum amount of information about the interior architectural design features of a house. They are created by passing an imaginary **horizontal cutting plane** through the house (Fig. 6.3) at a typical distance of 3 ft, 3 ft 6 in., or 4 ft above each floor line (approximately midway between the floor and the ceiling). The portion above the horizontal cutting plane is assumed to be removed so that the location and size of interior partitions, doors, windows, and stairs can be determined easily.

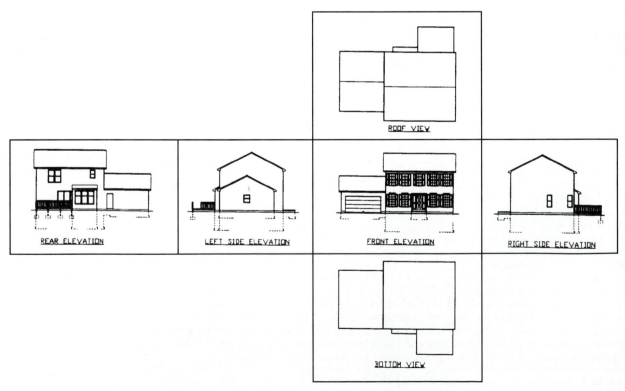

Figure 6.2
Orthographic Views of a Typical House

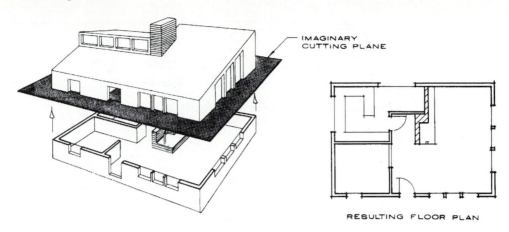

Figure 6.3

The Use of a Horizontal Cutting Plane to Obtain a Full Section

Source: Ernest R. Weidhaas, *Architectural Drafting and Construction*, 4th ed., p. 44 (Boston, MA: Allyn and Bacon, 1989).

Vertical Cutting Plane

Vertical cutting planes are used to highlight details which are needed during the construction phase. Section drawings are created when one of the sides that is cut is removed so that features on the side beyond the cutting plane can be observed. One of the most important full sections that is created by such a plane is the **structural section** (Fig. 6.4). Usually drawn to a 1/4 in. = 1 ft scale, it supplements the structural framing information provided on the floor plans.

A set of working drawings typically contains the following vertical sections:

WALL SECTIONS

Detailed information about the construction of an exterior wall is typically provided on a wall section (Fig. 6.5). It is usually drawn to a larger scale (1-1/2 in. = 1 ft) so that the detailed architectural features and vertical dimensions from the footing to the roof cornice area can be shown clearly.

Framing Section A-A on drawing A-5 (Appendix A), which serves a dual purpose as a full structural section and a wall section, is sufficient to show the size and nature of all exterior wall covering materials and the exact vertical location of foundation walls, floors, and ceilings, as well as the roof slope.

SPECIAL SECTIONS

Special sections, often called detail sections, are used to indicate any unusual construction techniques related to a particular wall section. They are also used to highlight the details related to other parts of the house, such as the stair detail which is shown in drawing A-5 (Appendix A).

Figure 6.4

A Vertical Cutting Plane Used to Produce a Full Structural Section

Source: Ernest R. Weidhaas, *Architectural Drafting and Construction*, 4th ed., p. 209 (Boston, MA: Allyn and Bacon, 1989).

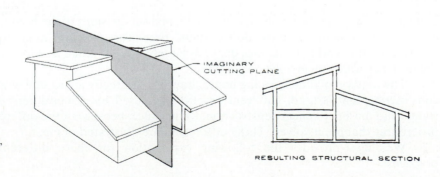

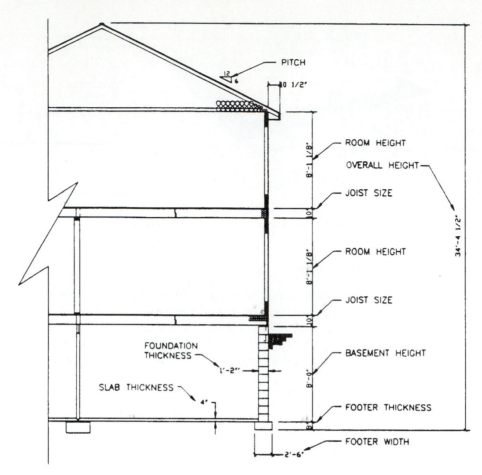

PITCH

10 1/2"

ROOM HEIGHT

OVERALL HEIGHT

JOIST SIZE

ROOM HEIGHT

JOIST SIZE

FOUNDATION
THICKNESS
1'-2"

BASEMENT HEIGHT

SLAB THICKNESS
4"

FOOTER THICKNESS

FOOTER WIDTH
2'-6"

34'-4 1/2"

8'-1 1/8"

8'-1 1/8"

8'-0"

Figure 6.5
An Exterior Wall
Section

Interior Elevations

The arrangement of partitions, closets, fixtures, appliances, and furniture can best be shown on floor plans which are created by using a horizontal cutting plane. The representation of the interior walls in the kitchen and the bathrooms, which often include particular arrangements of wall and base cabinets, soffits, countertops, and appliances, requires that interior elevation drawings be prepared. An example of such a drawing for a kitchen is provided in Fig. 6.6.

The sight arrows labeled 1/4, 2/4, 3/4, and 4/4 in Fig. 6.6 indicate the direction of the view and the drawing number on which that view appears. For example, label 1/4 indicates view #1, which is detailed on working drawing #4.

Dimensioning Guidelines

Dimensions indicate the width, height, length, and subdivisions of the house. They also indicate the location of windows, doors, stairs, fireplaces, etc. The number of dimensions included on the working drawings depends largely on how much freedom of interpretation the designer wants to give the homebuilder.

The method of dimensioning house drawings is directly associated with the method of construction that is used. Different practices are used for dimensioning masonry, wood frame, and masonry veneer construction, for instance, because of the different construction procedures that are involved. These differences are illustrated in Fig. 6.7.

Weidhaas provides the following explanation for these different dimensioning practices:

For example, in masonry construction, the widths of window and door openings are shown since these dimensions are needed to lay up the wall. Openings in a frame wall, however, are often dimensioned to their centerlines to simplify locating the window and door frames. In masonry construction, dimensions are given to the faces of the walls. In frame construction, overall dimensions are given to the outside faces of studs because these dimensions are needed first. Masonry partitions are dimensioned to their faces, whereas frame partitions are usually dimensioned to their centerlines. The thickness of masonry walls and partitions is indicated on the plan, but frame wall thicknesses are indicated on the detail drawings where construction details may be shown to a larger scale.

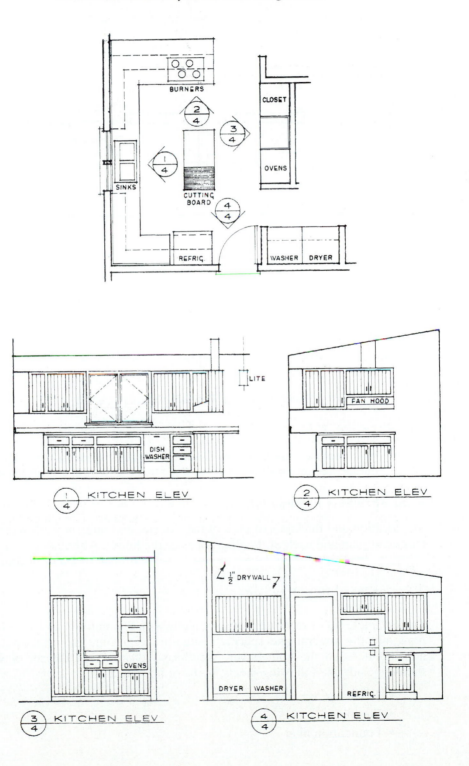

Figure 6.6
Vertical Interior
Elevations of the
Kitchen Walls
Source: Ernest R.
Weidhaas,
*Architectural
Drafting and
Construction*, 4th
ed., p. 197 (Boston,
MA: Allyn and
Bacon, 1989).

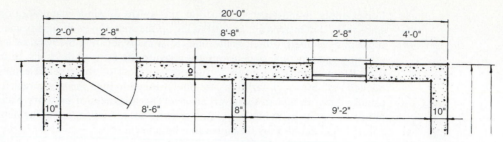

(a) Dimensioning Masonry Construction
(concrete, concrete block, solid brick, and cavity brick)

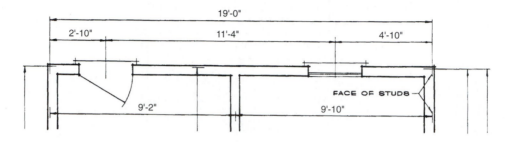

(b) Dimensioning Frame Construction

Figure 6.7
Differences in
Dimensioning
Practice
Source: Ernest R.
Weidhaas,
*Architectural
Drafting and
Construction*, 4th
ed., p. 61 (Boston,
MA: Allyn and
Bacon, 1989).

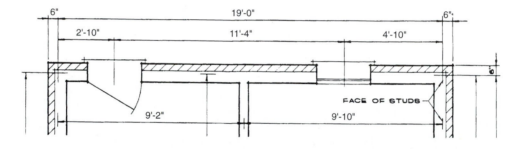

(c) Dimensioning Veneer Construction

Masonry veneer on a wood-frame wall is dimensioned as a frame wall would be (to the outside faces of the studs) since the wood frame is constructed before the veneer is laid up.[2]

Wall Thickness Representation

The location and thickness of each of the exterior and interior walls must be shown on the basement plan and each of the floor plans of the house. Although the finishes on the walls will vary, the thickness guidelines shown in Table 6.2 can be used for drafting purposes.

A Set of Working Drawings

The complete set of working drawings, along with the material and workmanship specifications and the general and specific conditions, represent the legal contract between the homebuilder and the homebuyer. Weidhaas indicates that the set of working drawings should include the following sheets, in the order indicated:

- Title page
- Plot plan
- Foundation plan

TABLE 6.2 Wall Thickness Drafting Guidelines

Foundation Walls	Thickness
Under wood-frame houses	
Concrete block	8″[a]
Poured concrete	10″
Under veneered construction	
Concrete block	10″
Poured concrete	10″
Under solid masonry construction	
Concrete block	12″
Poured concrete	12″

First and Second Floor Walls	Thickness
Wood-frame house	
Exterior walls	6″[b]
Interior partitions	5″
Brick-veneered walls	10″
Brick walls	
With two courses of brick	8″
With two courses of brick and air space	10″
With three courses of brick	12″
Concrete block walls	
Light	8″
Medium	10″
Heavy	12″

[a]With pilasters 16′ o.c.
[b]Assumes a 2″×4″ stud wall. Use 8″ thickness for a 2″×6″ stud wall

Source: Ernest R. Weidhaas, *Architectural Drafting and Construction*, 4th ed., pp. 175–176 (Boston, MA: Allyn and Bacon, 1989).

- First-floor plan
- Second-floor plan
- Elevations
- Sections
- Typical details
- Schedules (for doors, windows, finishes, etc.)
- Electrical requirements
- Heating and air conditioning
- Plumbing
- Ventilation
- Floor framing plan
- Roof framing plan
- Column schedule
- Structural details[3]

The information contained on some of these sheets will be discussed. It is important to note that a typical set of residential working drawings does not always contain each of these individual sheets. In some instances, information from several sheets is combined onto one sheet (i.e., heating and air conditioning and plumbing may be referred to as the mechanical phase).

The set of working drawings in Appendix A, for instance, represents an adequate representation of the "Pauline" model from the perspective of S&A Custom Built Homes. Another homebuilder or architect probably would use a different number of sheets and provide a different level of detail for a house of similar complexity.

Plot Plan

A **plot plan** is developed to provide all of the information necessary for locating the house on the site (Fig. 6.8). The plan will usually include the title of the sheet, plot and lot numbers of the house site, the scale of the drawing, and the north direction arrow. Also, a key is typically included to define the meaning of the various symbols.

The house size and location on the lot are necessary for staking out the house. It is extremely important to indicate the required setbacks on the drawing. With regard to topographical information, a reference elevation, either in relation to sea level or with reference to a benchmark elevation, is usually shown. Dotted lines typically indicate the contour information before construction, while solid lines indicate the contour information after construction. Drainage patterns and landscaping elements such as trees, planters, etc., may also be shown. Finished elevations of the basement, the first and second floors, and the garage are typically shown. The location of existing services which must be connected to the building, such as electric, water, gas, and sewer, should also be included on the plot plan. In addition, the following information is usually provided:

1. Lot area in square feet or acres.
2. Easements or rights-of-way.
3. Street location and widths, existing and proposed.
4. Width of sidewalk.
5. Curb locations and driveway entrances.
6. Building setbacks.

Foundation Plan

A **foundation plan**, such as Sheet A-3 for the "Pauline" model (Appendix A), indicates the overall dimensions of the house as measured between the outside faces of the concrete block foundation walls. The location and size of the footings for the foundation wall, basement columns, and wood deck are shown as dashed (hidden) lines. Appropriate elevation information is also provided. The thickness of the basement wall and the concrete floor, as well as the location of windows and offsets in the wall, are shown. Sheet A-3 also indicates that the garage (designated as unexcavated) is located on the left side of the house.

It is important to note that the structural framing system that **supports the first floor** is also shown on the Foundation Plan (i.e., size, spacing, and direction of the first-floor joists; size and location of the main girder and its column supports; structural framing for the stairs; and the LVL header under the eating nook in the kitchen).

First-Floor Plan

A **first-floor plan**, such as Sheet A-1 for the "Pauline" model (Appendix A), provides information about the guest and family entrances, living/social and working/service rooms, circulation zones, and stairwells as well as the closets, windows, and doors. Sheet A-1 also provides additional information about the garage and wood deck and uses architectural symbols to designate the various fixtures and appliances found in the kitchen and the powder room.

The structural framing system that **supports the second floor** is shown on the first-floor plan because the floor joists are supported by the bearing walls that are shown in this plan.

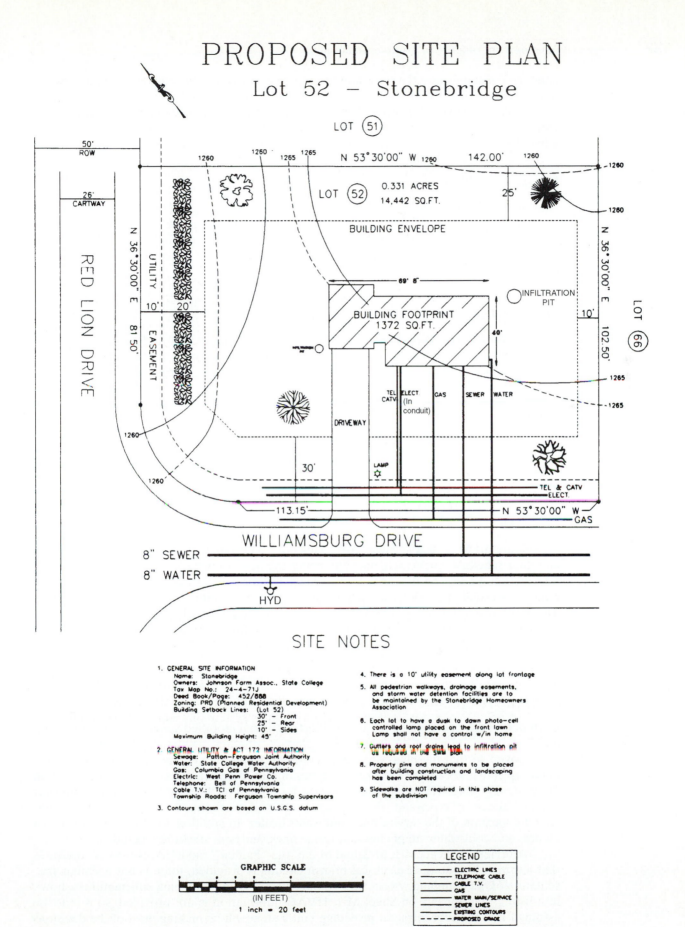

PROPOSED SITE PLAN
Lot 52 – Stonebridge

LOT (51)

N 53°30'00" W   142.00'

LOT (52)   0.331 ACRES
14,442 SQ.FT.

BUILDING ENVELOPE

BUILDING FOOTPRINT
1372 SQ.FT.

INFILTRATION PIT

RED LION DRIVE

UTILITY EASEMENT

N 36°30'00" E   81.50'

N 36°30'00" E   102.50'

LOT (66)

50' ROW

26' CARTWAY

10'   20'

69' 8"

40'

10'

DRIVEWAY

TEL CATV   ELECT (In conduit)   GAS   SEWER   WATER

30'

LAMP

TEL & CATV
ELECT.

113.15'   N 53°30'00" W

GAS

WILLIAMSBURG DRIVE

8" SEWER

8" WATER

HYD

SITE NOTES

1. GENERAL SITE INFORMATION
   Name: Stonebridge
   Owners: Johnson Farm Assoc., State College
   Tax Map No.: 24-4-71J
   Deed Book/Page: 452/888
   Zoning: PRD (Planned Residential Development)
   Building Setback Lines: (Lot 52)
       30' – Front
       25' – Rear
       10' – Sides
   Maximum Building Height: 45'

2. GENERAL UTILITY & ACT 172 INFORMATION
   Sewage: Patton-Ferguson Joint Authority
   Water: State College Water Authority
   Gas: Columbia Gas of Pennsylvania
   Electric: West Penn Power Co.
   Telephone: Bell of Pennsylvania
   Cable T.V.: TCI of Pennsylvania
   Township Roads: Ferguson Township Supervisors

3. Contours shown are based on U.S.G.S. datum

4. There is a 10' utility easement along lot frontage

5. All pedestrian walkways, drainage easements, and storm water detention facilities are to be maintained by the Stonebridge Homeowners Association

6. Each lot to have a dusk to dawn photo-cell controlled lamp placed on the front lawn Lamp shall not have a control w/in home

7. Gutters and roof drains lead to infiltration pit as required in the SWM plan

8. Property pins and monuments to be placed after building construction and landscaping has been completed

9. Sidewalks are NOT required in this phase of the subdivision

GRAPHIC SCALE

(IN FEET)

1 inch = 20 feet

LEGEND
- ELECTRIC LINES
- TELEPHONE CABLE
- CABLE T.V.
- GAS
- WATER MAIN/SERVICE
- SEWER LINES
- EXISTING CONTOURS
- PROPOSED GRADE

Figure 6.8
Information Contained on a Typical Plot Plan

Sheet A-1 also indicates the structural framing that is required to provide an entrance foyer that is open to the second floor as well as the pre-engineered roof trusses above the garage.

It should be noted that detailed information about the windows and doors is not provided on Sheet A-1; it appears on Sheet A-6, which presents a detailed window and door schedule.

Second-Floor Plan

A **second-floor plan**, such as Sheet A-2 for the "Pauline" model (Appendix A), typically provides information about the sleeping/private rooms as well as the master and secondary bathrooms. Sheet A-2 indicates that there is a roof over the garage at the second-floor level and that pre-engineered roof trusses define the roof structural system above the main section of the house.

Elevation Drawings

The **elevation drawings** provide information about the exterior architectural features on all four sides of the house (see Sheet A-4, Appendix A). Windows and doors are indicated, along with the roof and wall covering materials that are used. These drawings are usually shaded to reflect the final appearance of the house. Roof slope, overhang, and ventilation information is also provided.

Section Drawings and Typical Details

The information provided on these drawings (see Sheet A-5, Appendix A) was discussed earlier in this section when vertical cutting planes were addressed.

Schedules

A separate drawing such as Sheet A-6 (Appendix A) is typically used to present a **window and door schedule**. Such a schedule eliminates the need to include size and rough opening information directly on the already-overcrowded floor plans. Instead, a reference letter or number is usually provided for each window and door on either a plan or elevation view. This reference letter is keyed directly to the window and door schedule.

A review of Sheet A-6 indicates that detailed information about the type of cabinets and appliances in the kitchen and bathrooms is provided adjacent to the window and door schedule.

Electrical, HVAC, and Plumbing Systems

Each of these systems can be separately illustrated on a modified version of the foundation and floor plans. Much of the typical floor plan information, such as the dimensioning, can be removed so that the plans are not as cluttered when the systems are added. Electrical plans typically indicate the location and types of outlets and fixtures as well as the location of the service panel. HVAC plans provide the location of the heating and cooling equipment and any necessary types and sizes of ducts to the various rooms. The plumbing plans indicate the location of the service lines and water heater. In addition, the location and size of all hot- and cold-water supply lines, drainage lines, and vent stacks are included.

A review of sheets M-1, M-2, and M-3 of the "Pauline" model (Appendix A) indicates that although a routing of electrical branch circuits is provided, there is not a similar presentation of the plumbing system. The primary electrical and plumbing information is shown in a listing that appears on Sheet M-1. HVAC information is not provided, so it must be assumed that the subcontractor providing that system will be making most of the decisions about system design and placement.

Structural Framing

The primary source of information about the structural framing system for the "Pauline" model (Appendix A) is provided on the floor plans. Additional structural framing information is provided on the structural framing full-section view shown in Sheet A-5.

A set of working drawings for a larger house with a more complicated structural framing system may require the following additional sheets: (1) floor framing plan, (2) roof framing plan, (3) column schedule, and (4) structural details.

THE CONSTRUCTION PHASE

The design phase of the project is essentially complete when the final version of the working drawings is approved by the homebuyer. The construction phase begins after the homebuyer selects the homebuilder that will serve as either the **general contractor** or the **construction manager** on the construction site.

A number of milestones must be completed before construction activities can begin. One of the most important of these is the signing of the construction contract by the homebuyer and the homebuilder. Although practices vary, the essential **contract documents** which are usually considered to be part of the contract include:

1. An Agreement Form (i.e., Contract).
2. A Statement of General Conditions.
3. A Statement of Special Conditions associated with the project.
4. A Statement of the Material and Workmanship Specifications.
5. A Set of Working Drawings.

Examples of these documents for the Case Study House (i.e., the "Pauline" model) described in Appendix A are discussed in this section of the chapter.

Three of the essential elements of a legal contract that must be agreed upon by the parties to the contract are: (1) the **scope of the work**, (2) the **cost**, and (3) the **duration of the project**. The scope of work is defined by documents 4 and 5. The homebuilder must analyze the scope of work involved in order to determine a project **cost estimate** which will cover both the direct and indirect costs involved and also provide a reasonable profit for the firm. Closely associated with the cost-estimating process is the development of a **project schedule** which considers both the homebuyer's needs as well as the homebuilder's current workload.

Contract Documents

The actual Agreement Form, as well as the Statement of General Conditions and Special Conditions, are usually prepared by the homebuilder in consultation with an attorney to ensure that they are legally defensible. They are often modified versions of standard forms which have been developed by a professional society such as the American Institute of Architects.

Agreement and Conditions Documents

The **Agreement Form**, if used as a separate document, is typically brief because it simply states the essential elements of the agreement (i.e., scope of work, cost, duration, etc.) and includes a place for the signatures of the principal representatives of the parties involved in the contract. The scope of work is usually incorporated into the agreement by referencing the set of working drawings and the material and workmanship specifications.

The **Statement of General Conditions** defines the typical "set of ground rules" which the homebuilder has developed for all housing construction projects. These are usually written to protect the rights, and define the responsibilities, of all involved parties. The

Statement of Special Conditions is used to present the peculiar conditions that apply to the house being built (i.e., unusual geological conditions, portion of the work to be performed by the homebuyer, etc.). Negotiated agreements initiated by the homebuyer as modifications of the Statement of General Conditions are also often included.

Appendix B provides a copy of the residential construction contract form which is used by S&A Custom Built Homes of State College, PA.[4] A review of that document will indicate that both the Agreement Form and the Statement of General Conditions essentially have been incorporated into it. A separate Statement of Special Conditions was not prepared for the "Pauline" model shown in Appendix A.

Material and Workmanship Specifications

The set of working drawings is the first source of information about the material and workmanship specifications that must be met. For certain items, the type and manufacturer of the product that is to be used is directly stated on a drawing. In such cases, the manufacturer's installation specifications provide the standards that must be met. Notes on drawings are also often used to provide additional material and workmanship information. Examples of both of these situations can be found on the "Pauline" model drawings in Appendix A.

As indicated in Chapter 4, there are also minimum-level standards of performance provided in the model building codes. The 1995 CABO Code, for instance, devotes Chapter 4 to the foundation phase of house construction.[5] The information contained therein can be viewed as a supplement to the information contained on Sheet A-3 of the "Pauline" model (Appendix A) which specifies the foundation plan requirements.

Typically, it is necessary to provide additional information about the **material and workmanship specifications** that will be met in order to adequately inform the homebuyer, the lending agency, and others. These specifications can range from a simple listing of the standard features that will be provided (Table 6.1) to a formal document prepared by the architect who developed the set of working drawings. Many builders use a "Description of Materials" form that has been prepared by the USDA–Farmers Home Administration. An example of such a form describing the "Pauline" model is provided in Appendix C.

As an example of the level of detail provided on this form, it can be noted under the Foundations section that a 1–2–4 concrete mix (where 1–2–4 describes the proportion of components in the concrete mix) which has a compressive strength of 2500 psi will be used for the footings. In addition, two 1-3/4 in. × 9-1/4 in. LVL beams will be used as the main girder in the basement.

Cost Estimating

At some point in the interaction process between the homebuyer and either an architect or a homebuilder, a set of Working Drawings and a Statement of Material and Workmanship Specifications are complete enough so that a cost estimate can be prepared. A number of homebuilders may be asked to submit competitive lump-sum bids based upon these documents. Alternatively, the homebuilder who helped the homeowner reach this stage may be asked to prepare a sole-source lump-sum cost estimate for the project.

Types of Cost Estimates

A number of cost-estimating methods are used in the construction industry. Three methods that are probably most relevant and practical for the homebuilder are: (1) unit cost estimates, (2) systems cost estimates, and (3) detailed cost estimates.

UNIT COST ESTIMATES

Because they are the quickest and easiest to prepare, unit cost estimates are often used by the homebuilder to provide a "ball park" estimate for the homebuyer before the set of working drawings and specifications are totally complete. As a result, they usually also provide the lowest level of accuracy.

The cost per square foot version of this type of an estimate is used most often by homebuilders. A reasonable level of accuracy can be achieved if the homebuilder's historical unit cost data file has been categorized according to the types and sizes of houses that were built in the past. This historical data must be modified to account for cost changes over time, as well as the unique design features of the house being estimated.

A homebuilder can also use a cost-estimating reference manual to obtain reasonable cost per square foot estimating information. The *Means Residential Cost Data*[6] manual, which provides national average cost information, is widely used.

TOTAL COST PER SQUARE FOOT INFORMATION

At the macro-level, the estimated cost of a house can be determined using a single cost per square foot value. Figure 6.9 provides data in that format for an average two-story house

RESIDENTIAL | **Average** | **2 Story**

SQUARE FOOT COSTS

- **Simple design from standard plans**
- **Single family – 1 full bath, 1 kitchen**
- **No basement**
- **Asphalt shingles on roof**
- **Hot air heat**
- **Drywall interior partitions**
- **Materials and workmanship are average**

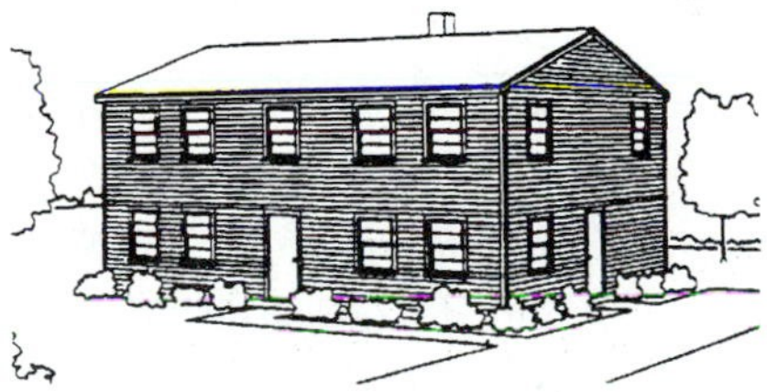

Base Cost Per Square Foot Of Living Area

	LIVING AREA										
EXTERIOR WALL	1000	1200	1400	1600	1800	2000	2200	2600	3000	3400	3800
Wood Siding - Wood Frame	76.35	68.90	65.90	63.85	61.40	59.00	57.50	54.20	51.00	49.70	48.35
Brick Veneer - Wood Frame	82.05	74.15	70.85	68.55	65.80	63.25	61.55	57.85	54.40	52.90	51.40
Stucco on Wood Frame	77.00	69.55	66.50	64.40	61.90	59.50	57.95	54.60	51.40	50.10	48.70
Solid Masonry	90.75	82.25	78.40	75.75	72.60	69.80	67.75	63.50	59.60	57.85	56.15
Finished Basement, Add	11.65	11.20	10.85	10.65	10.35	10.20	10.00	9.70	9.45	9.25	9.15
Unfinished Basement, Add	4.65	4.35	4.15	4.00	3.80	3.70	3.60	3.35	3.20	3.10	3.05

Modifications

ALTERNATES	Add to or deduct from costs per S.F. of living area
Cedar shake roof	$ + 1.00
Clay tile roof	+ 2.60
Slate roof	+ 3.60
Central air conditioning	
In heating ducts	+ 1.25
In separate ducts	+ 3.35
Heating systems	
Hot water	+ 1.60
Heat pump	+ 2.05
Electric heat	- .70
Not heated	- 2.55

BATHROOMS – TOTAL COST (includes plumbing and wall & floor finishes)	
Full bath - 3 fixtures	$ + 3361
Half bath - 2 fixtures	+ 2087

ADJUSTMENTS	
For two or more families (Add to total cost)	
Additional kitchen & entrance, each	+ 4058
Separate heating, each	+ 953
Townhouse/Rowhouse (Multiplier for costs per S.F. of living area)	
Inner unit	.90
End unit	.95

	Page
Wings & ells	44
Porches & breezeways	77
Finished attics	77
Appliances	77

	Page
Kitchen cabinets	78
Garages & carports	79
Site improvements	79

Use LOCATION FACTORS in the Reference Section

Figure 6.9
Base Square Foot Cost Information
Source: *Means Residential Cost Data 1995*, p. 30. Copyright R.S. Means Co., Inc., Kingston, MA, 617-585-7880, all rights reserved.

which is similar to the "Pauline" model (Appendix A). Information is provided for various size houses with four different exterior wall systems.

For instance, for a two-story house with 2,000 sq ft of living area, wood siding on a wood frame, and an unfinished basement:

1. Base Unit Cost   = $59.00/sq ft
2. Unfinished Basement Unit Cost = $3.70/sq ft
3. Total Base Unit Cost   = $62.70/sq ft
4. Total Base Cost = ($62.70)(2000) = $125,400

As indicated in Fig. 6.9, various design-related adjustments must be made. An adjustment for the location in which the house is built and an allowance for overhead and profit for the homebuilder must also be incorporated into the cost.

COMPONENT COST PER SQUARE FOOT INFORMATION

A division of the total unit cost ($62.70/sq ft) into unit costs which represent the various major components of the construction process provides additional useful information to the homebuilder, particularly when subcontracting decisions must be made. Means provides such a division (Fig. 6.10) when it divides the 2,000-sq-ft living area house with wood siding into the following components:

1. Site work   6. Interior
2. Foundation   7. Specialities
3. Framing   8. Mechanical
4. Exterior Walls   9. Electric
5. Roofing   10. Overhead

Additional specification information for each component is provided, along with man-hour, material, labor, and total cost data. Thus, the following information related to the **electrical subcontract component** can be obtained:

Manhours   =   .039 (2,000) = 78 manhours
Material cost   =   .61 (2,000) = $1,220
Labor cost   =   .90 (2,000) = $1800
Total electrical cost = 1.51(2,000)   = $3,020

It is important to note that even though the electrical work consists of wires, receptacles, lighting fixtures, etc., the unit costs are provided in a cost per square foot of living area basis. The total living area is the multiplier that is always used in the calculations related to each component.

SYSTEMS COST ESTIMATES

Homebuilders who do not want to place their entire reliance on the cost per square foot approach to estimating, but who do not have the resources to develop detailed cost estimates, often adopt the intermediate-level **systems cost estimating** method. This method, which Means designates as the **assemblies method,**[7] evaluates the costs of individual built-up systems in a unit of measure (cost per lineal foot, cost per square foot) that is appropriate to the system being analyzed.

The **foundation system** for the "Pauline" model (Appendix A), for example, consists of a **concrete footing subsystem** which supports a **concrete block wall subsystem**. Systems cost information, of the form shown in Fig. 6.11, can be used to quickly estimate the cost of the concrete footing subsystem.

The total systems cost of the 8 in. thick by 18 in. wide footing, for instance, is $7.75 per lineal foot of foundation wall (Fig. 6.11). This is a composite cost which includes a specified volume of concrete (usually priced on a cubic yard basis), the cost of reinforcing steel (usu-

Average 2 Story			Living Area 2000 S.F.		
			Perimeter	135 L.F.	
		MAN-HOURS	COST PER SQUARE FOOT OF LIVING AREA		
			MAT.	LABOR	TOTAL
1 Site Work	Site preparation for slab; trench 4' deep for foundation wall.	.034		.49	.49
2 Foundation	Continuous concrete footing 8" deep x 18" wide; cast-in-place concrete wall, 8" thick, 4' deep, 4" concrete slab on 4" crushed stone base, trowel finish.	.066	1.72	2.19	3.91
3 Framing	2" x 4" wood studs, 16" O.C.; 1/2" plywood sheathing; 2" x 6" rafters 16" O.C. with 1/2" plywood sheathing, 4 in 12 pitch; 2" x 6" ceiling joists 16" O.C.; 2" x 8" floor joists 16" O.C. with bridging and 5/8" plywood subfloor; 1/2" waferboard subfloor on 1" x 2" wood sleepers 16" O.C.	.131	5.50	4.54	10.04
4 Exterior Walls	Horizontal beveled wood siding; 15# felt building paper; 3-1/2" batt insulation; wood double hung windows; 3 flush solid core wood exterior doors; storms and screens.	.111	8.51	4.38	12.89
5 Roofing	240# asphalt shingles; #15 felt building paper; aluminum flashing; 6" attic insulation. Aluminum gutters and downspouts.	.024	.36	.67	1.03
6 Interiors	1/2" drywall, taped and finished, painted with primer and 1 coat; softwood baseboard and trim, painted with primer and 1 coat; finished hardwood floor 40%, carpet with underlayment 40%, vinyl tile with underlayment 15%, ceramic tile with underlayment 5%; hollow core doors.	.232	11.02	9.24	20.26
7 Specialties	Kitchen cabinets - 14 L.F. wall and base cabinets with laminated plastic counter top; medicine cabinet, stairs.	.021	1.04	.38	1.42
8 Mechanical	1 lavatory, white, wall hung; 1 water closet, white; 1 bathtub with shower; porcelain enamel steel, white; 1 kitchen sink, stainless steel, single; 1 water heater, gas fired, 30 gal.; gas fired warm air heat.	.060	2.07	1.52	3.59
9 Electrical	200 Amp. service; romex wiring; incandescent lighting fixtures, switches, receptacles.	.039	.61	.90	1.51
10 Overhead	Contractor's overhead and profit and plans.		2.15	1.71	3.86
Total			32.98	26.02	59.00

SQUARE FOOT COSTS

Figure 6.10
Base Component Square Foot Cost Information

Source: *Means Residential Cost Data 1995*, p. 31. Copyright R.S. Means Co., Inc., Kingston, MA, 617-585-7880, all rights reserved.

ally priced on a pound basis), etc. The total cost of a house can be estimated by combining all of the individual system costs that have been determined in a similar fashion.

Components of a Detailed Cost Estimate

The **detailed cost estimating** approach attempts to incorporate all of the components of cost at the individual resource level into the total contract price of the house. In its simplest terms, as shown in Fig. 6.12, job-level costs (which are shown below the line), when accumulated, result in the **total project cost**.

Adjustments to that cost, in the form of load multiplier costs, a general overhead assessment to cover home office costs, and an allowance for profit and contingency are made in order to determine the **total contract price**.

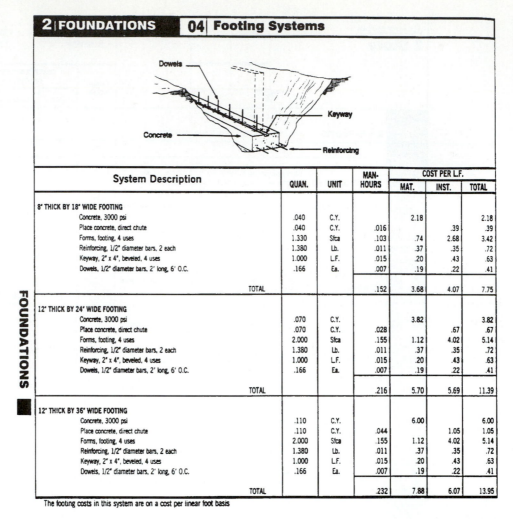

2 | FOUNDATIONS   04 | Footing Systems

System Description	QUAN.	UNIT	MAN-HOURS	COST PER L.F.		
				MAT.	INST.	TOTAL
8" THICK BY 18" WIDE FOOTING						
Concrete, 3000 psi	.040	C.Y.		2.18		2.18
Place concrete, direct chute	.040	C.Y.	.016		.39	.39
Forms, footing, 4 uses	1.330	Sfca	.103	.74	2.68	3.42
Reinforcing, 1/2" diameter bars, 2 each	1.380	Lb.	.011	.37	.35	.72
Keyway, 2" x 4", beveled, 4 uses	1.000	L.F.	.015	.20	.43	.63
Dowels, 1/2" diameter bars, 2' long, 6' O.C.	.166	Ea.	.007	.19	.22	.41
TOTAL			.152	3.68	4.07	7.75
12" THICK BY 24" WIDE FOOTING						
Concrete, 3000 psi	.070	C.Y.		3.82		3.82
Place concrete, direct chute	.070	C.Y.	.028		.67	.67
Forms, footing, 4 uses	2.000	Sfca	.155	1.12	4.02	5.14
Reinforcing, 1/2" diameter bars, 2 each	1.380	Lb.	.011	.37	.35	.72
Keyway, 2" x 4", beveled, 4 uses	1.000	L.F.	.015	.20	.43	.63
Dowels, 1/2" diameter bars, 2' long, 6' O.C.	.166	Ea.	.007	.19	.22	.41
TOTAL			.216	5.70	5.69	11.39
12" THICK BY 36" WIDE FOOTING						
Concrete, 3000 psi	.110	C.Y.		6.00		6.00
Place concrete, direct chute	.110	C.Y.	.044		1.05	1.05
Forms, footing, 4 uses	2.000	Sfca	.155	1.12	4.02	5.14
Reinforcing, 1/2" diameter bars, 2 each	1.380	Lb.	.011	.37	.35	.72
Keyway, 2" x 4", beveled, 4 uses	1.000	L.F.	.015	.20	.43	.63
Dowels, 1/2" diameter bars, 2' long, 6' O.C.	.166	Ea.	.007	.19	.22	.41
TOTAL			.232	7.88	6.07	13.95

The footing costs in this system are on a cost per linear foot basis

Figure 6.11 Systems Cost-Estimating Information for a Concrete Foundation Footing
Source: *Means Residential Cost Data 1995*, p. 96. Copyright R.S. Means Co., Inc., Kingston, MA, 617-585-7880, all rights reserved.

HOMEBUILDER DIRECT COSTS

Direct costs are those which contribute directly to the construction of the house. These costs account for the homebuilder's contributions of labor, material, and equipment resources to the project to complete the work that is performed by the homebuilder's own forces. "Other" direct costs include the costs of permits, etc.

SUBCONTRACTOR DIRECT COSTS

As noted in Chapter 3, many homebuilders have become **construction managers** who subcontract a large percentage of the work on a house. During the cost-estimating stage, therefore, the homebuilder must establish reasonable **subcontractor direct costs**, usually in the form of lump-sum bids, for the portions of the work that will be subcontracted.

JOB INDIRECT COSTS

Indirect costs are those project costs that do not contribute directly to the finished project, but which still are necessary to perform the work properly. These costs, which sometimes are referred to as **job overhead** costs, include: (1) administrative costs for field supervision and other management expenses, (2) support service costs for maintaining the project field office and site facility, and (3) subcontract costs for surveying, materials testing, and other job-related services provided by others. Some homebuilders simply calculate job overhead costs as a percentage of the total direct job cost, with common values for this allowance

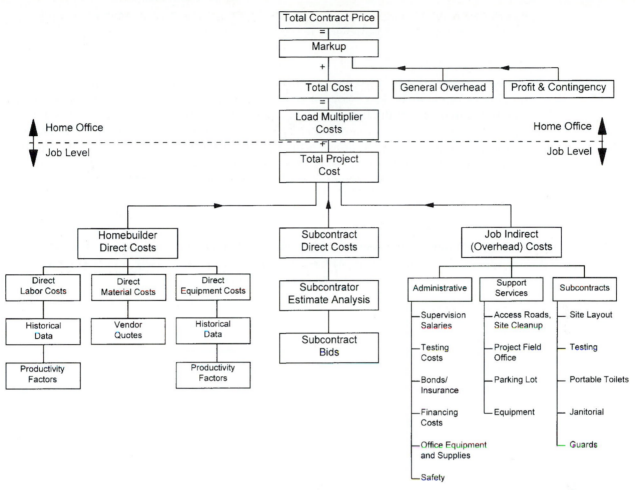

Figure 6.12
Cost Components of the Total Contract Price

ranging from 5% to 15%. A more preferable approach is to identify all of the job overhead costs and determine an estimated cost for each of them.

TOTAL PROJECT COST

The **total project cost,** as shown in Fig. 6.12, is the sum of the: (1) homebuilder direct costs, (2) subcontractor direct costs, and (3) job indirect costs. The final summation represents the estimator's best guess about the actual costs that will be incurred **at the site of construction**.

The total project cost must be increased before the total contract price is determined to allow for such items as insurances, taxes, labor, overhead, general overhead, profit, and contingencies.

LOAD MULTIPLIER COSTS

Load multipliers can be used to account for some of the additional costs associated with the major components of the total project cost. These include: (1) a Labor Load Multiplier, which is applied by the homebuilder to recover costs due to Workmen's Compensation Insurance, Social Security tax, federal and state unemployment tax, and similar mandated expenses; (2) a Material Load Multiplier, which is typically added to the material costs to cover the sales tax that applies in the state in which the house is being built and any appropriate material handling and storage costs; (3) a Subcontract Load Multiplier, which covers the construction management costs associated with scheduling, coordinating, and controlling the work of the subcontractors; and (4) an "Other Cost" Load Multiplier, which may have to be applied on a case-by-case basis.

GENERAL OVERHEAD

General overhead costs (also called operating overhead costs) are those that are incurred in support of the homebuilder's overall construction program. These costs must be divided among the houses that the firm is working on during a particular time period. Some of the types of costs that may fall into this category are: (1) home office management and staff salaries; (2) home office rent, office insurance, office utilities, office supplies, and furniture; (3) legal and accounting fees; (4) advertising and travel costs; etc. A typical method of accounting for these costs is as a **percentage of the total project cost** for the house.

PROFIT AND CONTINGENCY

The allowance for profit and contingency can also be included as a percentage of Total Costs. **Profit and contingency** are either combined and labeled only as profit, or are identified separately as a part of the **Mark-Up** of the project. Depending on market conditions, profit typically may vary from 0% to 20% of the Total Cost.

TOTAL CONTRACT PRICE

As indicated in Fig. 6.12, the **total contract price** is determined by summing the following values:

Total Contract Price = Total Job Direct and Indirect Costs + Load Multiplier Costs + General Overhead + Profit and Contingency

Cost Account Systems

The detailed cost-estimating approach provides the highest level of estimating accuracy. Many Level I and Level II homebuilders cannot use this approach, however, because they have not maintained an adequate historical database that is structured so that specific information can be obtained easily from it.

Such a structured database is necessary for Level III and Level IV homebuilders, however, because the preparation of the **cost estimate** for a particular house is only one of the key elements of the **cost management systems** used by the homebuilders. Another key element is the **cost control** procedure that is used while the house is being built.

In order for the cost estimate and cost control procedures to be part of an **integrated cost management system**, the homebuilder must adopt a standardized cost account (i.e., cost code) system which is initiated for each project when the cost-estimating process starts.

The National Association of Home Builders has developed a cost code system which is directly applicable to the homebuilding industry.[8] The best-known example, however, is the MASTERFORMAT system of classification and numbering developed by the Construction Specifications Institute (CSI).[9] This system, widely accepted in the industry, consists of the basic work breakdown structure of the CSI system that is shown in Appendix D. In addition to an introductory division which includes "Bidding Requirements" and "Contracting Requirements," there are 16 major divisions which cover all of the phases of a building project. Figure 6.13 illustrates how the 16-division system applies to a typical single-family detached house.

Division 1, entitled "General Requirements," is typically used to account for the Job Indirect (Overhead) Costs shown in Fig. 6.12. Each of the 16 divisions can be subdivided further into many more specific cost subcategories.

Development of a Detailed Cost Estimate

Table 6.3 presents an **estimate summary** report for the "Pauline" model (Appendix A) which was prepared using the detailed cost-estimating methodology shown in Figure 6.12. It should be noted that the homebuilder direct labor, material and equipment costs, as well as

BUILDING COMPONENTS BY DIVISION

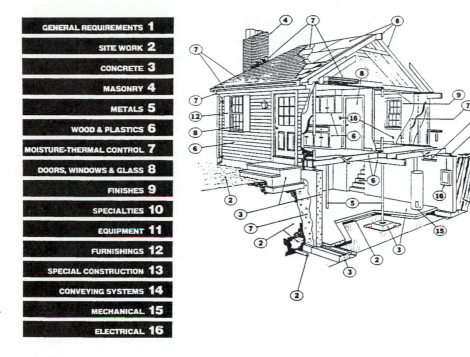

Figure 6.13
Residential
Building
Components by
CSI Division

*Source: Means
Repair and
Remodeling
Estimating Seminar
Workbook*, p. 28.
Copyright R.S.
Means Co., Inc.,
Kingston, MA, 617-
585-7880, all rights
reserved.

GENERAL REQUIREMENTS **1**
SITE WORK **2**
CONCRETE **3**
MASONRY **4**
METALS **5**
WOOD & PLASTICS **6**
MOISTURE-THERMAL CONTROL **7**
DOORS, WINDOWS & GLASS **8**
FINISHES **9**
SPECIALTIES **10**
EQUIPMENT **11**
FURNISHINGS **12**
SPECIAL CONSTRUCTION **13**
CONVEYING SYSTEMS **14**
MECHANICAL **15**
ELECTRICAL **16**

TABLE 6.3   Estimate Summary Report—"Pauline" Model

SUBDIVISION	Worley Acres				ESTIMATE NUMBER		4056
LOCATION	416 Julo Way				SHEET NUMBER		39/40
CSI	Estimate Summary Report—Total Contract Price				DATE		9/26/95

		Labor	Material	Subcontract	Equipment	Other	Total
1.00	General Conditions	4,525	433	—	825	4,516	10,299
2.00	Site Work	1,209	364	—	1,408	—	2,981
3.00	Concrete	—	—	3,611	—	—	3,611
4.00	Masonry	—	—	7,141	—	—	7,141
5.00	Steel	—	160	—	—	—	160
6.00	Wood and Plastics	6,596	14,474	3,500	1,440	—	26,011
7.00	Thermal/Moisture Protection	2,883	4,040	—	—	—	6,923
8.00	Doors and Windows	1,324	9,597	—	—	—	10,921
9.00	Finishes	4,479	4,966	—	—	—	9,445
10.00	Special Conditions	314	797	—	—	—	1,111
11.00	Equipment	43	956	—	—	—	998
15.00	Mechanical—Plumbing Sub.	—	—	4,600	—	—	4,600
15.00	Mechanical—HVAC Sub.	—	—	6,500	—	—	6,500
16.00	Electrical Subcontract	—	—	4,800	—	—	4,800
	Total Project Cost	21,373	35,787	30,152	3,673	4,516	95,501
	Load Multipliers	× .348	× .06	× .05	× .0	× .04	—
	Direct Costs Load	7,438	2,147	1,508	—	181	11,274
	TOTAL COST	$28,811	$37,934	$31,660	$3,673	$4,697	$106,775

Total Markup:	7,474	General Overhead = .07 × 106,775	7,474
	12,567		114,249
	$20,041	Profit = .11 × 114,249	12,567
		TOTAL CONTRACT PRICE	$126,816

the subcontractor and "other" costs, are indicated for only 13 of the 16 CSI divisions because it was assumed that the other three divisions (12: Furnishings, 13: Special Construction, and 14: Conversion Systems) did not apply to the "Pauline" model. It was also assumed that the homebuilder's cost estimator, in a meeting with the Vice President of Construction, made the decision to subcontract the following phases of the project:

1. Division 3.00—The Concrete Phase
2. Division 4.00—The Masonry Phase
3. Division 6.00—The Kitchen Cabinet System
4. Division 14.00—The Plumbing Phase
5. Division 15.00—The HVAC Phase
6. Division 16.00—The Electrical Phase

The estimate summary report does not list each of the **Job Indirect** costs shown in Fig. 6.12 separately because they all are absorbed into Division 1.00—General Conditions. A total for each of the cost columns is presented at the bottom of the report just above the calculations which determine the **Load Multiplier** costs and the **Mark-up** value.

DETAILED ESTIMATE SHEET

The basic source document for the information that appears in Table 6.3 for each CSI division is the **detailed estimate sheet.** Pricing the estimate involves developing a complete and detailed enumeration of all direct and indirect costs for a project in a format which identifies the appropriate cost codes for the phase of work being estimated. Table 6.4, for instance, presents the detailed cost estimate sheet which determines the cost of the interior drywall (i.e., gypsum wallboard) that is required for the "Pauline" model. As noted, the appropriate major CSI division is 090-000—Finishes, and the appropriate subcategory for plaster and gypsum board is 092-000.

The five different uses of drywall in the "Pauline" model are indicated and each is assigned its own cost code number. The uses are evaluated separately because the crew productivity and material costs are not the same for all use categories. The column headings of interest are quantity/units, labor, and material since it is assumed that the work will be performed by the homebuilder's own forces.

The estimated labor and material costs for the Drywall subcategory are $1,998 and $1,522, respectively. These costs, when added to the estimated costs for all of the other subcategories included in Division 090-000—Finishes, result in the Labor and Material entries of $4,479 and $4,966, respectively, in Division 9.00—Finishes in Table 6.3.

GENERAL CONDITIONS COSTS

One of the three key cost components of Total Project Cost is **Job Indirect Cost** (Fig. 6.12). All of the estimated costs of this component are included in CSI Division 1.00—General Conditions (Table 6.3). A detailed estimate sheet, such as the one shown in Table 6.5, provides the support documentation for all of these costs and indicates the type of expense involved (i.e., labor, material, equipment, subcontract, and other).

In terms of labor indirect cost, the largest expense shown in Table 6.5 is the salary of the field manager. Additional labor, as well as material expense, is incurred in building the siltation fence, clean-up of the job site, and installing various safety measures (i.e., guardrails, hole covers, etc.). Several other labor and material expenses that are typically included under the General Conditions division are shown in Table 6.5. The only equipment expenses are for the rental of a refuse dumpster and a portable toilet.

Included in the "Other" cost category are the costs of each of the permits which the homebuilder must obtain prior to starting the project. There are also "other" costs related to insurance (typically builder's risk, completed operations, and general liability), the interest expense on the construction loan, and closing costs. It is assumed that the sale of the house will be handled by the homebuilder's sales staff at the model home in the subdivision

TABLE 6.4 Detailed Estimate Sheet for CSI Divisions 090000—Finishes and 092608—Drywall

SUBDIVISION	Worley Acres
LOCATION	416 Julo Way
CSI	Division 090000—Finishes and 092608—Drywall

ESTIMATE NUMBER	4056
SHEET NUMBER	31/40
DATE	9/22/95

	Quantity/Units	Labor[a]	Material[a]	Subcontract	Equipment	Other	Total
092-608-0350 1/2 in. interior drywall walls—taped and finished	3,914 SF	$.27/SF × 3,914 = $1,056	$.21/SF × 3,914 = $822	—	—	—	$1,878
092-608-0550 1/2 in. interior greenboard walls—taped and finished	512 SF	$.27/SF × 512 = $138	$.29/SF × 512 = $149	—	—	—	$287
092-608-1050 1/2 in. interior drywall ceilings—taped and finished	2,000 SF	$.35/SF × 2,000 = $700	$.23/SF × 2,000 = $460	—	—	—	$1,160
092-608-1250 1/2 in. interior greenboard ceilings—taped and finished	110 SF	$.35/SF × 110 = $39	$.28/SF × 110 = $31	—	—	—	$70
092-608-2050 5/8 in. interior drywall walls—taped and finished	240 SF	$.27/SF × 240 = $65	$.25/SF × 240 = $60	—	—	—	$125
SHEET TOTALS		$1,998	$1,522				$3,520

[a]All costs are "bare" costs, without overhead and profit.

TABLE 6.5   Detailed Estimate Sheet for the "General Conditions" Costs of the "Pauline" Model

SUBDIVISION	Worley Acres			ESTIMATE NUMBER		4056	
LOCATION	416 Julo Way			SHEET NUMBER		4/40	
CSI	Division 010000—General Conditions			DATE		9/18/95	

		Labor	Material	Subcontract	Equipment	Other	Total
1	Field Manager $35,000/10 houses/year	3,500	—	—	—	—	3,500
2	Temporary Water/Electric	—	—	—	—	125	125
3	Temporary Siltation Fence	32	108	—	—	—	140
4	Surveying	211	—	—	—	—	211
5	Dumpster Rental (3 mo.)	—	—	—	825	—	825
6	Portable Toilet Rental (3 mo.)	—	—	—	—	75	75
7	Safety Measures	150	100	—	—	—	250
8	Blue Prints	170	125	—	—	—	295
9	Building Permit	—	—	—	—	225	225
10	Water Permit	—	—	—	—	748	748
11	Sewer Permit	—	—	—	—	150	150
12	Power Company Permit	3	—	—	—	8	11
13	Zoning Permit	—	—	—	—	75	75
14	Insurance	—	—	—	—	210	210
15	Interest Cost—Const. Loan	—	—	—	—	1,130	1,130
16	Closing Costs	—	—	—	—	1,270	1,270
17	Punch List Items	—	—	—	—	500	500
18	Intermediate/Final Cleanup	459	100	—	—	—	559
	SHEET TOTALS	4,525	433	—	825	4,516	10,299

in which the house is built. Since sales expense is assumed to be a corporate general overhead expense for the firm, no direct sales commission charges are indicated. Punch list items are also included in the "other" cost category because it is difficult to estimate an appropriate labor/material breakdown for these expenses when the estimate is being prepared.

TOTAL CONTRACT PRICE

As noted in Table 6.3, the **Total Contract Price** that the homebuilder would present to the homebuyer for the "Pauline" model, exclusive of the cost of the lot, is $126,816. This represents a total mark-up above the **Total Project Cost** of $126,816/$95,501 = 1.33 or 33%.

Project Scheduling

The homebuyer places a great deal of importance on the total contract price developed by the homebuilder. Knowledge about the duration of the construction phase is also important to homebuyers who must sell their present houses in order to finance the new one. The requirement to develop a schedule for each house also forces the homebuilder to give some up-front consideration to the potential problems that may develop.

Cost/Time Relationship

The homebuilder can usually develop a general plan and a time schedule at the same time that the cost estimate is being prepared. For the portion of the work which the homebuilder completes with the firm's own forces, the relationship between time and cost in Fig. 6.14 indicates that the key factor which links the two together for a particular construction activity is **crew productivity rate.** This factor determines the **labor unit cost** when it is combined with the **crew cost** for the activity, and it determines the **duration** of the activity when it is combined with the **quantity take-off** information about the activity.

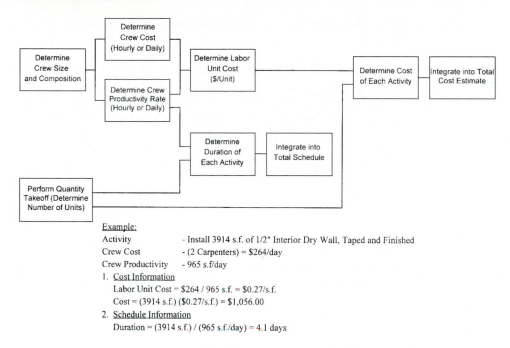

Example:
Activity   - Install 3914 s.f. of 1/2" Interior Dry Wall, Taped and Finished
Crew Cost   - (2 Carpenters) = $264/day
Crew Productivity - 965 s.f/day
1. Cost Information
   Labor Unit Cost = $264 / 965 s.f. = $0.27/s.f.
   Cost = (3914 s.f.) ($0.27/s.f.) = $1,056.00
2. Schedule Information
   Duration = (3914 s.f.) / (965 s.f./day) = 4.1 days

Figure 6.14
The Time/Cost
Relationship of
Construction
Activities

This relationship also applies to subcontractors as they plan and schedule their own work. The homebuilder's schedule will probably directly incorporate each subcontractor's time duration information—however, without direct knowledge of their crew productivity rates.

Overview of Project Scheduling

Project scheduling actually consists of the three distinct phases of: (1) planning, (2) scheduling, and (3) controlling. The **planning phase** defines the sequence in which the construction activities will be performed. Decisions related to a particular construction activity (i.e., installation of siding) include the identification of: (1) the activities which precede it, (2) the activities which follow it, and (3) the activities that can be performed at the same time as it.

Scheduling provides a **time framework** for the project plan. It begins when the time duration of each construction activity is estimated, perhaps by using the relationships established in Fig. 6.14. The most widely accepted method of presenting the activity sequence and time duration information is the **bar-chart** format which represents each activity as a bar that is plotted on a horizontal time line.

Scheduling can also be performed by using a **network model** to represent the activity sequence and time duration information. Network modeling, which was first introduced to the construction industry in the late 1950s as the Critical Path Method (CPM), has evolved into the following three formats: (1) Activity-on-Arrow (AOA), (2) Activity-on-Node (AON), and (3) Precedence Diagramming Method (PDM).[10]

The final phase of a project scheduling system consists of **controlling** the construction schedule to ensure that the house can be turned over to the homebuyer on the date indicated. The original schedule serves as a **performance standard** to which the actual schedule which occurs at the job site is compared. A frequency of updating is usually established in order to evaluate the actual progress which has taken place on the project.

Comparison of the Three Network Modeling Techniques

The **Activity-on-Arrow** (AOA) network model, which was the version originally applied in the construction industry, represents each activity as an arrow which connects two nodes (called events). It requires restraints, called *dummy activities,* to represent various logic relationships. Although still used, the AOA has been widely replaced by the **Activity-on-Node (AON)** network model, which is easier to construct because each activity is completely described by its own node.

The **critical activities** on the project, those that form the **critical path** that determines the duration of the project, can be found on either AOA or AON networks by performing a *forward pass* and a *backward pass* set of calculations. An assumption made with both of these network models is that all activities that precede a particular activity must be completely finished before the activity can start. This is commonly called a Finish-to-Start relationship.

The **Precedence Diagraming Method** (PDM) is essentially an AON network model that has been enhanced so that activities can overlap each other, a condition which often exists on a construction project.

Development of an Activity-on-Node Network Schedule

A number of references, including ones published by the National Association of Home Builders[11] and the Associated General Contractors,[12] provide a detailed explanation of the calculation procedures associated with the three network modeling techniques. Therefore, the steps involved in the development of an Activity-on-Node network schedule that are presented in relation to the "Pauline" model focus on the concepts rather than the mathematical relationships underlying the procedure.

STEP 1: DEVELOPMENT OF AN ACTIVITY LIST AND PRECEDENCE RELATIONSHIPS

Two types of activities are typically defined on a network: **production activities,** which require labor, material, and equipment resources and contribute directly to the project, and **nonproduction activities,** such as ordering materials, delivery time of materials, wait for building inspections, etc.

Table 6.6, which resulted from a detailed analysis of the working drawings for the "Pauline" Model (Appendix A), provides a description of the macro-level activities on the project and the sequencing relationships between them. It is assumed that consultation between the planner/scheduler and the cost estimator resulted in a common agreement about the duration of each of the activities.

STEP 2: PREPARATION OF THE PROJECT PLAN IN AN ACTIVITY-ON-NODE NETWORK FORMAT

The commonly accepted elements of an AON network are shown in Fig. 6.15, which is an excerpt of the complete AON network for the "Pauline" model. Each node, represented by a box, indicates a different activity that is identified by name and number at particular locations in the box. The relationship between activities that is defined in Table 6.6 is indicated by directional links in Fig. 6.15. Activity 130: Insulation, for instance, cannot begin until Activity 110: Siding, Activity 90: Plumbing Rough, and Activity 100: HVAC Rough have been completed. The duration of each activity is indicated at a specific location in the box because it will be used directly in the computations.

STEP 3: DETERMINATION OF THE EARLY START/EARLY FINISH INFORMATION FOR EACH ACTIVITY AND THE DURATION OF THE PROJECT

Four different activity times for each activity are indicated on Fig. 6.15. The **early start date (ES)** for an activity is the earliest point in time that the activity can begin. The **early finish date (EF)** for an activity is obtained when its duration is added to its ES. A **forward pass** computation, which proceeds from the initial activity to the last activity in the network (i.e., from left to right), is used to determine both the ES and EF values for each activity in the network as well as the **duration** of the project.

The Construction Phase

147

TABLE 6.6   Activity List for the "Pauline" Model

Activity Number	Activity Description	Depends Upon	Followed By	Duration
10	SITE LAYOUT	—	20	1 DAY
20	BASEMENT EXCAVATION	10	30	2 DAYS
30	BLOCK FOUNDATION	20	40	12 DAYS
40	BACKFILL AND SETTLEMENT	30	50	11 DAYS
50	FRAMING	40	60, 70	10 DAYS
60	EXT. DOORS & WINDOWS	50	80, 100	2 DAYS
80	ELECTRICAL ROUGH	60	110	2 DAYS
110	SIDING	80	130, 140	10 DAYS
70	ROOF	50	90	1 DAY
90	PLUMBING ROUGH	70	130	2 DAYS
100	BRICK	60	120, 130	5 DAYS
120	CONCRETE SLABS	100	140	3 DAYS
130	INSULATION	110, 90, 100	150	2 DAYS
140	EXTERIOR TRIM	110, 120	180	1 DAY
150	DRYWALL	130	160	10 DAYS
160	INTERIOR TRIM & DOORS	150	170	14 DAYS
170	PAINT	160	200, 190	14 DAYS
180	FINAL GRADE	140	210	1 DAY
190	TILE & VINYL FLOORING	170	230	2 DAYS
200	KITCHEN & VANITIES	170	220, 240	5 DAYS
210	DECK	180	250	3 DAYS
220	PLUMBING FINAL	200	270	2 DAYS
230	CARPET FLOORING	190	260	2 DAYS
240	ELECTRICAL FINAL	200	270	3 DAYS
250	DRIVEWAY	210	280	2 DAYS
260	ATTIC INSULATION	230	310	1 DAY
270	BASEMENT CEILING INSULATION	220, 240	290	1 DAY
280	GRASS	250	300	2 DAYS
290	PUNCH LIST	270	310	2 DAYS
300	CONTINGENCY	280	310	10 DAYS
310	CLEAN UP	260, 290, 300	—	3 DAYS

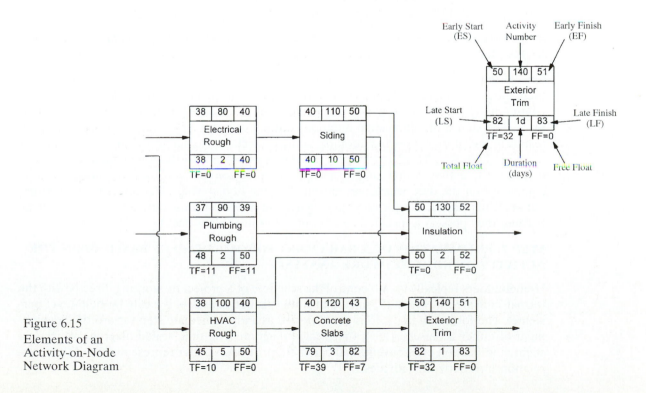

Figure 6.15
Elements of an
Activity-on-Node
Network Diagram

The forward pass computation results in ES and EF values for Activity 140: Exterior Trim of 50 and 51, respectively (Fig. 6.15). This indicates that the earliest date that Activity 140 can start is Day 50 and the earliest date that it can be completed is Day 51.

STEP 4: DETERMINATION OF THE LATE START/LATE FINISH INFORMATION FOR EACH ACTIVITY

Two other activity times indicated on Fig. 6.15 are the latest start date (LS) and the latest finish date (LF). The **latest finish date (LF)** of an activity is the latest point in time that the activity can be completed without extending the completion time of the entire project. The **latest start time (LS)** for an activity is determined when its duration is subtracted from its LF.

A **backward pass** computation, which proceeds from the last activity to the initial activity in the network(i.e., from right to left), is used to determine the LS and LF values for each activity on the network.

The backward pass computation results in LS and LF values for Activity 140: Exterior Trim of 82 and 83, respectively (Fig. 6.15). This indicates that the latest date that Activity 140 can start is Day 82 and the latest date that it can be completed is Day 83. Thus, there is a maximum period of time of 32 days (LS − ES = 82 − 50 = 32 or LF − EF = 83 − 51 = 32) during which the one-day long Activity 140 can be performed without affecting the duration of the project that was determined by the forward pass.

STEP 5: DETERMINATION OF THE TOTAL FLOAT OF EACH ACTIVITY

The 32-day period determined for Activity 140: Exterior Trim in Fig. 6.15 is called its **Total Float (TF)**. Since Activity 140 can be performed at any time during the 32-day period, it is designated as a **noncritical activity** because there is flexibility involved with its scheduling. Other activities in the network, by virtue of the forward and backward passes, are found to have a total float = 0; these are designated as **Critical Activities**. Activity 110: Siding is such an activity because its earliest start and latest start dates (i.e., Day 40), as well as its earliest finish and latest finish dates (i.e., Day 50), are the same.

The scheduling of critical activities does not have any flexibility. If Activity 110: Siding, for instance, does not start on Day 40 and does not end on Day 50, the entire project will be delayed the same number of days that the completion of Activity 140 is extended. As noted in Fig. 6.15, another float value, called **Free Float (FF)** which relates each activity to the activities which follow it, also has a value of 0 for critical activities.

STEP 6: IDENTIFICATION OF THE CRITICAL PATH ON THE ACTIVITY-ON-NODE NETWORK

One of the major contributions of all three network modeling techniques is that each of them identifies the critical path through the entire project. This critical path consists of the continuous chain of critical activities (i.e., those with TF and FF values of 0) through the network. The sum of the individual duration values of all of the activities on the critical path results in the shortest possible duration of the project for the network relationship of activities that is shown. This knowledge is extremely important to the homebuilder because it indicates the activities that must receive direct attention and control if the house is to be completed on the date which was indicated to the homeowner. The noncritical activities, although important, do not need as much close attention until they also become critical because they possess a specified amount of float (also called slack).

STEP 7: PREPARATION OF A BAR CHART FOR THE PROJECT BASED UPON THE ACTIVITY-ON-NODE NETWORK ANALYSIS

Homebuilders typically do not control the schedule of a project by working directly with the format indicated in Fig. 6.15, since that format is not understood easily by field-level personnel. The format in Fig. 6.15 is often viewed as an intermediate step toward the development of either a bar-chart or a time-scaled diagram. The time-scaled diagram, which is a slight modification of a conventional bar chart, provides a better representation of the interrelationships between activities.

Activity-on-Node Network Schedule for the "Pauline" Model

Figure 6.16 presents an Activity-on-Node network for the "Pauline" Model which was developed using Steps 1 through 6 of the procedure described previously. The 31 activities in the network define the construction process at the macro-level. This level is appropriate when the homebuilder is determining the duration of the project for the homebuyer, providing preliminary information to subcontractors, and identifying the critical activities that require attention. The network should be expanded to include activities which recognize the ordering of material, the delivery of material, and a more detailed breakdown of the macro-level activities in order to identify other important activities. For instance, Activity 50: Framing could be divided into:

> Activity 52: Basement Girder and Posts
> Activity 54: First Floor Platform
> Activity 56: First Floor Walls, etc.

Such an expansion of the network would provide the homebuilder with a greater level of schedule control in the field.

As noted in Fig. 6.16, the estimated duration of the construction phase is 104 working days, which translates into 21 weeks, or approximately five months. A desirable alternative is to present the schedule information in terms of calendar days. This can be done easily if a computer-based scheduling program is used. Once the appropriate calendar day corresponding to the first day of the project is defined in the program, the scheduling information will be translated automatically into calendar days. All intervening holidays and weekends will also be considered.

The activities which determine the critical path, and hence the duration of 104 days, are indicated with a check mark (✓). All of these activities have total float (TF) and free float (FF) values equal to zero. These are the activities that do not have any flexibility in terms of their schedule. Thus, if Activity 150: Drywall takes 14 days (instead of 10 days) to complete, and nothing is done later in the project to change the plans indicated, the project duration will be increased by 4 days to 108 working days.

Activity 270: Attic Insulation, which is noncritical, has both a total float and a free float of 6 days. Thus, if the shipment of the insulation material is delayed from the 94th day to the 97th day, it should not greatly concern the homebuilder because the overall project duration of 104 days will not be affected by this delay.

Bar-Chart Schedule for the "Pauline" Model

The bar chart that was developed using the network-generated information from Fig. 6.16 is shown in Fig. 6.17. The horizontal time scale is shown in working days for consistency purposes; it should be converted into calendar days before it is issued to the field manager. The activities on the vertical axis have been grouped into logical categories to represent major phases of the work. This basic bar-chart format can be enhanced, particularly if a computer-based scheduling system is used, to indicate some of the sequencing relationships between activities and to allow for planned and actual progress to be indicated efficiently.

Computer-Based Cost Estimating and Construction Scheduling

The discussion of cost estimating and construction scheduling has emphasized the principles involved and indicated the important information that can be generated. The guidance provided is equally applicable whether these functions are performed manually or are computer based.

As a homebuilder evolves from a Level I homebuilder to the higher levels (Fig. 3.4), a point is reached, in terms of the construction output, where computer-based cost estimating and construction scheduling must be adopted. As a homebuilder begins to consider that step, it will become apparent that there are many software programs on the market that will perform the various required functions.

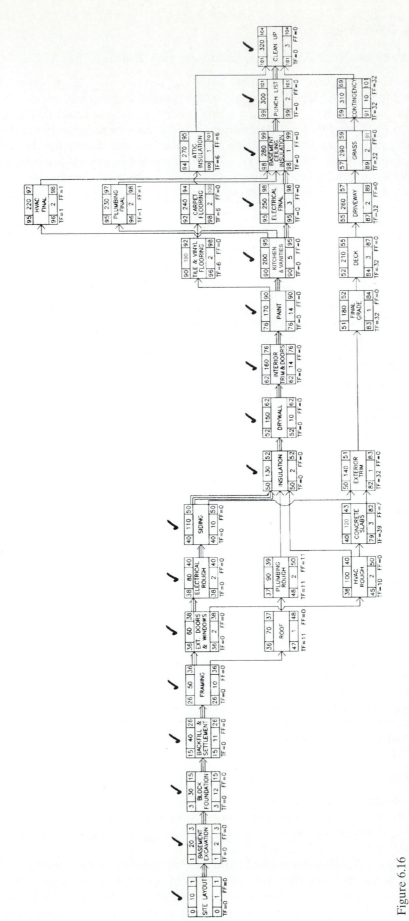

Figure 6.16
Macro-level Activity-on-Node Network for the "Pauline" Model

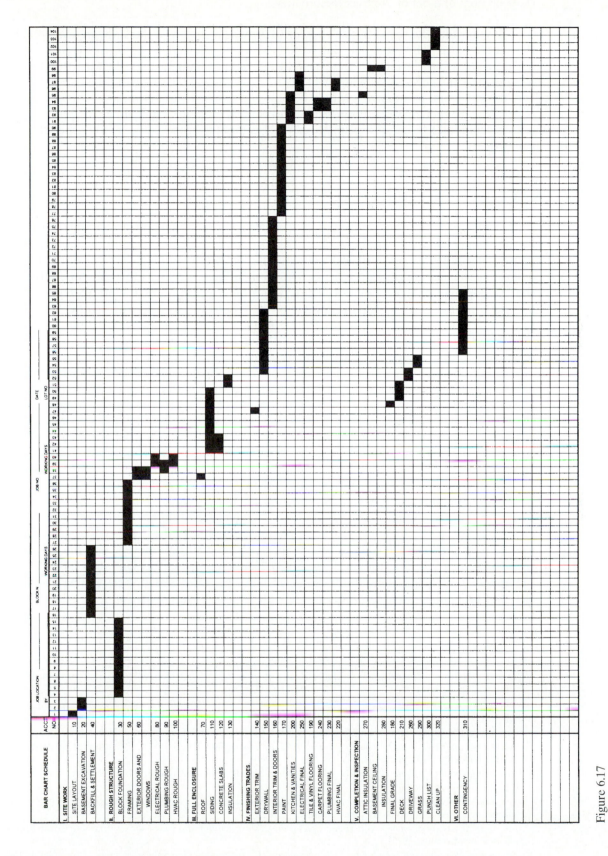

Figure 6.17

An Activity-on-Node Based Bar-Chart Schedule for the "Pauline" Model

151

It is beyond the scope of this chapter to provide a review of the software available in the marketplace. Therefore, the interested reader is referred to *Software Review*[13] and *Software Directory,*[14] which are published by the National Association of Home Builders to meet the specific need for such information. It is also beyond the scope of this chapter to discuss the important features that should be present in the software programs. The interested reader is referred to two reports of the Housing Research Center at Penn State which address computer-based cost estimating[15] and computer-based construction scheduling,[16] respectively.

CHAPTER SUMMARY

Three of the essential elements of the legal contract between the homebuyer and the homebuilder are: (1) the scope of the work, (2) the cost, and (3) the duration of the construction phase. The scope of work is typically described in the set of working drawings of the house and the supportive contract documents. The **Design Phase** section of this chapter addressed the development of the set of working drawings by reviewing design methodology, architectural drafting practices, and the content of the set of working drawings for the "Pauline" model that is used as the **Case Study House** for the remaining chapters in the book.

The **Construction Phase** section of this chapter began by describing the typical content of a set of contract documents and providing an example for the Case Study House. The cost element of the legal contract was addressed by discussing the types of cost estimates which a homebuilder could develop. Primary emphasis is placed on the components of a **Detailed Cost Estimate** since this is the most accurate one that can be prepared by the homebuilder. Each of the components of such an estimate was discussed. An introduction to a code account system, which is based upon the CSI MASTERFORMAT breakdown, was provided. An analysis of some of these cost estimating components for the Case Study House was also provided.

The Construction Phase section of the chapter concluded with a presentation of the scheduling methods that can be used to determine the duration of the construction phase. The relationship between a cost estimate and a **Project Schedule** was discussed. The typical types of schedules that can be used were then presented, with the primary focus placed on the Activity-On-Node network schedule approach. The steps involved with the development of such a schedule for the Case Study House were then presented to indicate the methodology involved. A brief discussion of the computer-based cost estimating and construction scheduling programs that are available was then presented.

NOTES

[1]Ernest R. Weidhaas, *Architectural Drafting and Construction*, 4th ed.   (Boston, MA: Allyn and Bacon, 1989), p. 162.

[2]Ibid., p. 60.

[3]Ibid., p. 172.

[4]"Residential Construction Contract," provided by S&A Custom Built Homes, State College, PA.

[5]*CABO One and Two Family Dwelling Code, 1995 Edition* (Falls Church, VA: The Council of American Building Officials, 1995), pp. 25–34.

[6]*Means Residential Cost Data 1995*, 14th ed.   (Kingston, MA: R.S. Means Company, Inc., 1994).

[7]Ibid., p. v.

[8]*Accounting and Financial Management for Builders, Remodelers and Developers.* (Washington, D.C.: The National Association of Home Builders, 1994).

[9]*Masterformat—Master List of Numbers and Titles for the Construction Industry.* (Alexandria, VA: Construction Specifications Institute, 1995).

[10]*Construction Planning and Scheduling.* (Washington, D.C.: Associated General Contractors of America, 1994), p. 55.

[11]Jerry Householder, *Scheduling for Home Builders,* 2nd ed. (Washington, D.C.: Home Builder Press, The National Association of Home Builders, 1990).

[12]*Construction Planning and Scheduling.*

[13]*Software Review: Approved Product Summaries for Builders.* (Washington, D.C.: The National Association of Home Builders, 1993).

[14]*Software Directory for Home Builders and Remodelers.* (Washington, D.C.: The National Association of Home Builders, 1993).

[15]Jack Julo and Jack Willenbrock, *HRC Research Series: Report No. 26, Technology Transfer Manual for Computer-Based Cost Estimating.* (University Park, PA: The Housing Research Center at Penn State, 1994).

[16]Nicole Fleming, *Computerized Scheduling Software for Residential Builders: An Evaluation and Selection System.* (unpublished Master of Science thesis, Department of Architectural Engineering, The Pennsylvania State University, 1995).

BIBLIOGRAPHY

Accounting and Financial Management for Builders, Remodelers and Developers. Washington, D.C.: The National Association of Home Builders, 1994.

CABO One and Two Family Dwelling Code, 1995 Edition. Falls Church, VA: The Council of American Building Officials, 1995.

Construction Planning and Scheduling. Washington, D.C.: Associated General Contractors of America, 1994.

Fleming, Nicole, *Computerized Scheduling Software for Residential Builders: An Evaluation and Selection System.* University Park, PA: unpublished Master of Science thesis, Department of Architectural Engineering, The Pennsylvania State University, 1995.

Householder, Jerry, *Scheduling for Home Builders,* 2nd ed. Washington, D.C.: Home Builder Press, The National Association of Home Builders, 1990.

Julo, Jack and Jack Willenbrock, *HRC Research Series: Report No. 26, Technology Transfer Manual for Computer-Based Cost Estimating.* University Park, PA: The Housing Research Center at Penn State, 1994.

Masterformat—Master List of Numbers and Titles for the Construction Industry. Alexandria, VA: Construction Specifications Institute, 1995.

Means 1995 Residential Cost Data, 14th ed. Kingston, MA: R.S. Means Company, Inc., 1994.

"Residential Construction Contract," provided by S&A Custom Built Homes, State College, PA.

Software Directory for Home Builders and Remodelers. Washington, D.C.: The National Association of Home Builders, 1993).

Software Review: Approved Product Summaries for Builders. Washington, D.C.: The National Association of Home Builders, 1993.

Weidhaas, Ernest R., *Architectural Drafting and Construction,* 4th ed. Boston, MA: Allyn and Bacon, 1989.

7

The Residential
Construction Process

INTRODUCTION

This chapter provides a transition between the qualitative coverage of the housing industry in the previous chapters and the technical details associated with each phase of house design and construction which is found in Chapters 8–14. The major phases of the residential construction process are illustrated with a series of progress photos that were taken during the construction of two "Pauline" models that are part of the "Custom Series" of S&A Custom Built Homes of State College, PA. These models, which are designated as House A and House B in this text when such differentiation is necessary, were built in State College, PA, in 1992/1993 and 1995, respectively. House A, the Case Study House whose plans appear in Appendix A, serves as the example for most of the design illustrations shown in Chapters 8–14.

SITE PREPARATION

Existing subsoil conditions must be investigated before a new house can be constructed on a particular lot. This step can involve a professional soil investigation, which includes test borings, or it can be accomplished by checking the performance of existing houses near the site. The type of soil present on the lot (i.e., does it possess an adequate bearing capacity? is it unstable or expansive in the presence of water? etc.) is extremely important since it governs the corresponding design of the foundation. In certain situations, the existing soil conditions may be so poor that they must be removed and replaced before construction can begin.

After the subsoil conditions have been determined and the foundation design has been adjusted accordingly, the site can be prepared for construction. The trees and other surface material, such as organic matter, vegetation, humus, etc., first must be removed from the foundation and garage area so that work can begin on virgin undisturbed soil.

It is important to remove trees for a sufficient distance around the foundation so that equipment can operate easily when excavating or backfilling. Appropriately located trees should be left on the site, however: (1) for landscaping purposes, (2) to provide shade during the summer months, and (3) to act as part of a wind barrier system.

154

Stake-Out of the Site

Stake-out refers to the process of transferring the information contained on the plot plan to the ground. It is one of the first and most critical operations of the entire building process. Often there is very little room for error between the location of the house and the required setbacks. Staking out is usually performed by a surveyor. The homebuilder also may perform the task.

Traditional stake-out begins by establishing the outside corners of the foundation with small stakes. Then, since these stakes will be disturbed when excavating, larger stakes are typically driven 4 ft beyond the foundation lines, three at each corner (Fig. 7.1). Batter boards are nailed to these stakes so that their tops are at the same elevation. By using a plumb bob and stretching twine across the batter boards at the correct locations, foundation lines can be established when necessary. It is extremely important when staking out that the corners of the house are squared. This may be assured by using surveying instruments, measuring diagonals, or utilizing the principle of the 3–4–5 triangle.

Excavation

Before **excavation** is begun, the topsoil should be stripped to a depth of about 6 in. and stockpiled for future use. This topsoil is usually suitable for landscaping purposes and is expensive to purchase. Topsoil removal and foundation excavation are completed using a bulldozer, backhoe, or similar piece of equipment.

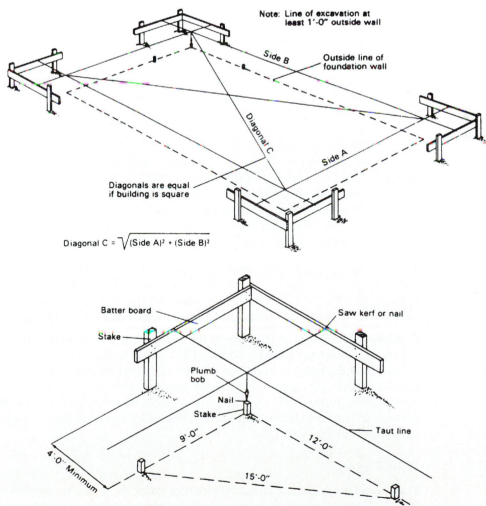

Figure 7.1
Staking and Laying Out the House

Source: Gerald E. Sherwood and Robert C. Stroh, *Wood Frame House Construction*, p. 14 (Mineola, NY: Dover Publications, Inc., 1989).

Figure 7.2
Foundation
Excavation
Completed
Source: This photo,
and all other photos
appearing in this
chapter, are from
the construction of
two "Pauline" mod-
els that were built in
State College, PA,
in 1992/93 and 1995,
respectively.

Excavation usually proceeds to the level of the top of the footings (Fig. 7.2). Continuous wall footings and individual spread footings then are excavated using a backhoe. The wall footing excavation must be wide enough to provide working space when constructing the foundation wall and installing the necessary draintile. During the excavation process, adequate safety precautions must be followed to ensure that failure of the created embankment does not occur.

FOUNDATION SYSTEM

From a structural and basement moisture standpoint, the foundation is one of the most critical components of the house. The weight of the entire structure must be transmitted through the foundation wall and footings to the ground below. Geographic location usually determines the type of foundation that is required. Climate, subsurface conditions, and frost depth are the principal determinants.

In the eastern part of the United States, full basements are commonly provided. Slab on grade and crawl space alternatives are often used when the excavation of a basement becomes prohibitively expensive or the existing subsurface conditions provide an unstable situation. Whatever type of foundation is used, however, it is important that the footings extend below the frost grade (the depth below the surface that the ground freezes in the winter). Local building codes usually define this required depth since it varies in different geographical areas.

The focus of this section of the chapter is on the full basement alternative.

Footings

Footings, which serve as the base of the foundation wall and the basement columns, transmit the superimposed load to the soil. The size of the footing is dependent upon the magnitude of the load being transmitted and the allowable bearing capacity of the soil at the site. Poured concrete is commonly used for footings, although advances in treated wood foundations have proven that they are also a viable alternative. If the footing is inadvertently excavated to a depth below that which is necessary, the excavation should be filled with concrete, not backfilled with soil.

Wall Footings

Wall footings are placed around the perimeter of the house to support the foundation wall (Fig. 7.3). One commonly used method of determining the size of the wall footings for normal soil conditions is based on the governing foundation wall thickness. As a general rule,

Figure 7.3
Poured Concrete
Continuous Wall
Footing

the footing depth is made equal to the foundation wall thickness and the footing width is established as twice the foundation wall thickness. More appropriate methods of determining the wall footing size include reference to the local building code or the completion of a structural analysis based on the soil-bearing pressure that has been determined.

Column Footings

Concrete spread footings are placed beneath the basement columns that support the main basement girder, which usually is located near the centerline of the house (see Fig. 7.9). The size of the concrete spread footing is dependent upon the superimposed load on the columns from the main girder, the allowable soil-bearing capacity, and the spacing of the columns. The size of the footings can be determined by reference to the local building code or as a result of a structural analysis.

Steel posts are usually placed directly on these footings and the concrete floor is often poured around the posts. This practice, however, does not allow the column to be replaced easily at a later date. To avoid this problem, the top of the footings can be poured to the top of the concrete slab height.

Foundation Wall

The purpose of the **foundation wall** is to provide a moisture-free enclosure for the basement or crawl space and to carry some of the superimposed loads from the wood frame system above and transmit them to the footings. Foundation walls are typically constructed of concrete block or cast-in-place concrete. Pressure-treated wood foundations and frost-protected shallow foundations offer alternatives that are accepted by some codes. This section focuses on the concrete block foundation wall (Figs. 7.4 and 7.5).

When constructing the foundation wall, concrete blocks are laid with mortar joints on all sides. The block courses start at the footing and usually are laid in a common bond pattern (staggered vertical joints). Once the courses are completed, a cement mortar parging and a waterproofing coating such as asphalt are applied to the wall wherever it will come into contact with backfilled earth (Fig. 7.5). These coatings repel water and, in conjunction with proper site drainage, usually ensure a dry basement.

Basement door and window frames should be keyed in the foundation wall for rigidity and to prevent air leakage. Concrete lintel beams are used to span over the window and door openings when the openings do not correspond to the top of the foundation wall. Anchor bolts or straps which fasten the sill plate of the wood framing system to the foundation wall (Figs. 7.5 and 7.10) usually extend through the top two rows of blocks.

Footing drains are often used around foundations enclosing basements in order to drain away subsurface water. These drains and a satisfactory system for preventing rain

Figure 7.4
Footings
Completed and
Foundation Wall
Started

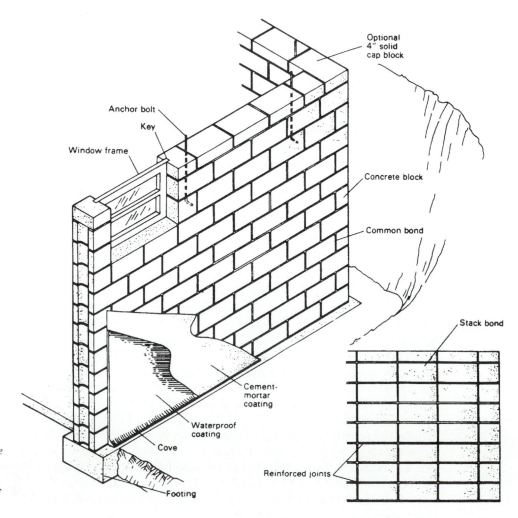

Figure 7.5
Concrete Block
Foundation Wall
and Continuous
Wall Footing
Source: Gerald E.
Sherwood and
Robert C. Stroh,
*Wood Frame House
Construction*, p. 22
(Mineola, NY:
Dover Publications,
Inc., 1989).

Figure 7.6
Plastic Drainage
Pipe Footing
Drain

runoff from entering the soil immediately adjacent to the foundation wall can help to prevent wet and damp basements. Perforated plastic pipe is normally placed at the bottom of the footing on a 2-in. gravel bed (Fig. 7.6). The joints between sections of the pipe are usually covered with asphalt felt or some other similar material. Another 6–8 in. of gravel are placed over the drain pipe. The drain is typically sloped away from the building into a storm sewer or ditch. Local building codes usually provide the details for the acceptable alternatives that may be used.

Basement and Garage Floors

Basement and garage floors are normally finished with a concrete floor even though the area may not contain habitable rooms. Structurally, the concrete floor in the basement helps to resist the horizontal soil pressure which tends to push in the bottom of the foundation wall. These floors are cast in place after all under-slab connections to the sewer and water lines have been made. At least one floor drain should be installed in both the basement floor and the garage floor.

A base for the concrete floor usually consists of 4 in. of compacted gravel which is installed on top of the compacted fill that has been used in the basement and in the garage (Fig. 7.7). The gravel base breaks the capillary action between the soil and the concrete, and helps to maintain a drier floor by storing and then removing any groundwater that may seep

Figure 7.7
Completed
Foundation Wall
for House B -
Garage Area
Backfilled with
Compacted Fill
(view from the
front of the
garage looking
toward the future
family room at
the upper right)

below the concrete slab. For each of the foundation alternatives, it is important to use a polyethylene vapor barrier above the gravel. The vapor barrier protects untreated wood framing members from ground moisture.

WOOD STRUCTURAL FRAMING SYSTEM

Although the high price fluctuations and increases in the price of lumber in the period from 1992 to 1995 have created interest among homebuilders in alternative structural systems (i.e., light-gage structural steel, masonry, poured concrete, and others), wood structural framing systems are still the most widely used in the United States.

Most homebuilders in the United States prefer to use the **platform framing** system shown in Fig. 7.8. Load-bearing and non-load-bearing walls for each story are built on separate platforms of joists, band joists, and subfloor.

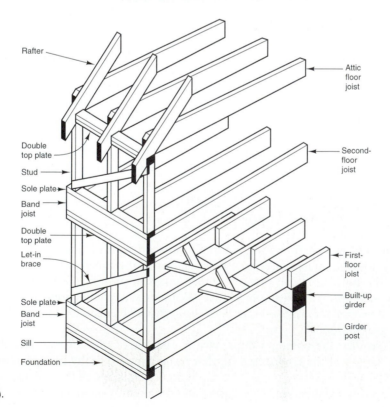

Figure 7.8
Basic Platform
Framing System
Source: Charlie
Wing, *The Visual
Handbook of
Building and
Remodeling*, p. 141
(Emmaus, PA:
Rodale Press, 1990).

Two of the lesser-used systems are **balloon framing** and **post and beam**. In the balloon framing system (see Fig. 9.3), the vertical studs are continuous from the sill plate on the foundation to the top plate on the wall of the highest story. Because of this design detail, firestopping of the vertical wall cavities is required.

The post and beam system, also called the timber framing system, has a long history in both the United States and Europe. It has regained a market niche in the 1980s and 1990s in the United States in house designs which incorporate exposed, widely spaced timber frames and use exterior wall cladding made of insulated stress skin panels. Because of the joint detailing that is required, the post and beam system demands a higher degree of carpentry skills than the other two systems.

Because of its wide acceptance in the United States, the focus of the following discussion will be on the **platform framing system**.

Basement Columns

The columns in the basement, acting in conjunction with the foundation walls, support the first-floor wood framing system. The columns transmit the load from the main basement girder to the concrete spread footings (Fig. 7.9). The columns commonly used in houses are made of wood or steel. The spacing and size of the columns is determined by the load transmitted from the girder that they must support.

Main Basement Girder

The primary purpose of the main girder in the basement is to support one end of each of the first-floor joists that it supports and to transmit the load from these joists to the basement columns (Fig. 7.9). A main girder is needed to provide center support for two sets of floor joists because the width of a typical house exceeds the allowable floor joist span for a single joist when conventionally sized lumber is used. The main girder usually bears on the foundation wall at each of its ends.

Floor Framing System

Before the actual framing of the first-floor platform can begin, a sill plate must be attached to the top of the foundation wall. The sill plate, which is usually attached to the foundation with anchor bolts or straps 8 ft on center (o.c.), acts as the nailing ledger for the floor joists which will be placed on it (Fig. 7.10). The sill plate is commonly a 2 in. × 6 in. member made of pressure-treated wood. It is also advisable to install a sill sealer between the sill plate and the foundation wall. The sill sealer helps to prevent cold air from entering the building due

Figure 7.9
Steel Pipe Columns Supporting the Main Basement Girder and Being Supported by a Poured Concrete Spread Footings (note the end bearing support for the main basement girder is provided by the notch in the foundation wall)

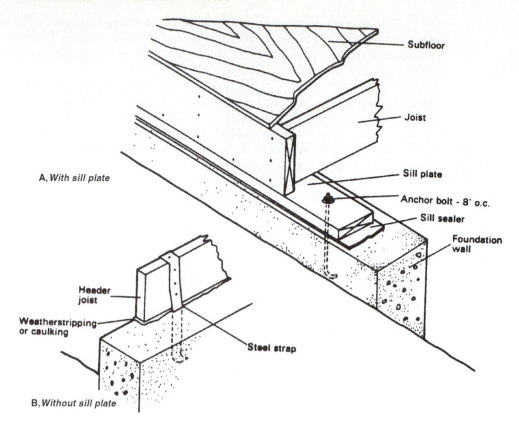

Figure 7.10
Attachment of
Sill Plate to the
Foundation Wall
Source: Gerald E.
Sherwood and
Robert C. Stroh,
*Wood Frame House
Construction*, p. 39
(Mineola, NY:
Dover Publications,
Inc., 1989).

to the uneven nature of the foundation wall. In certain areas of the United States, termite shields are also required.

An example of the role of a sill plate in the garage area is shown in Fig. 7.11. In this case, the stud wall rests directly on the sill plate.

Floor Joists

Each floor is structurally supported by **floor joists**, which commonly are 2 in. × 8 in., 2 in. × 10 in., or 2 in. × 12 in. members. The required size and spacing of the floor joists are determined by the floor loads and the span distance. The first-floor joists span from the girder to the foundation wall. They bear directly on the sill plate on one end and either on top of, or into the side of, the girder on the other end.

Figure 7.11
Sill Plate, Bottom
(Sole Plate), and
Wall Studs in the
Garage Area

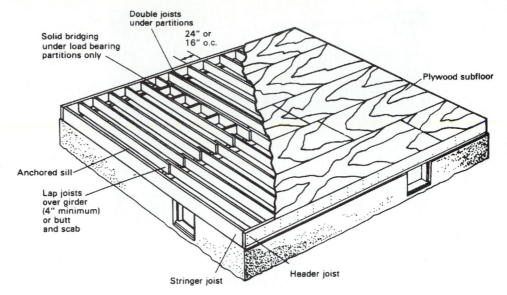

Double joists
under partitions
24" or
16" o.c.

Solid bridging
under load bearing
partitions only

Plywood subfloor

Anchored sill

Lap joists
over girder
(4" minimum)
or butt
and scab

Stringer joist

Header joist

Figure 7.12
Typical Platform
Framed
Construction

Source: Gerald E.
Sherwood and
Robert C. Stroh,
*Wood Frame House
Construction*, p. 46
(Mineola, NY:
Dover Publications,
Inc., 1989).

The complete floor joist system is generally of the platform framed type (Fig. 7.12). Header joists are used across the ends of the floor joists. The header joist temporarily braces the floor joists prior to the installation of the subfloor and helps to support the wall stud loads. The end joists that run parallel to the floor joists are typically called stringer joists.

Floor joists supporting the second floor are typically supported at one end by the exterior bearing wall of the house and at the other end by an interior bearing wall. These floor joists typically overlap the interior bearing wall and rest on the top plate of the exterior bearing wall. When it becomes necessary, floor joists can be attached to the side of another member by using a joist hanger (Fig.7.13).

Bridging between the floor joists is required by some codes. Bridging provides the floor system with additional stability and strength by distributing the loads over a larger floor area than the area on which the loads occur. Some homebuilders, however, feel that floor bridging is not necessary when the floor joists are 12 in. in depth or less; in fact, they feel that such bridging may contribute to squeaking floors.

When bridging is used, it is commonly of the metal cross type. Other types of bridging include wood cross bridging (usually 1 in. × 3 in. members) and solid wood bridging (using members that are the same size as the floor joists).

Joists should at least be doubled under load-bearing partition walls that are parallel to the joists. Solid blocking should be used in place of doubled joists when access from below

Figure 7.13
Joist Hanger
Support System

Figure 7.14
Plywood Floor
Sheathing on the
Second Floor and
a Stair Opening
that is Perpen-
dicular to the
Floor Joists

is required due to the installation of heating ducts in the load-bearing partition. However, it is not necessary to double the joists under non-load-bearing parallel partitions.

Floor Sheathing

Floor sheathing or subflooring is used over the floor joists to provide a working platform and a base for an underlayment layer and the finish flooring material. Floor sheathing materials include boards of up to 8 in. in width, plywood (Fig. 7.14), and reconstituted wood panels (i.e. particleboard, oriented strand board, flakeboard). The attachment requirements for each of these sheathing materials can be determined from the local building code or the manufacturer.

Floor Openings

Openings in the floor for stairs, fireplaces, etc. should be planned, if possible, so that their long dimension is parallel to the joists in order to minimize the number of joists that are disturbed. In such a situation, the openings can be framed with trimmers running in the direction of the joists and with headers which support the cut-off tail joists (Fig. 7.15). Typically, when the opening is less than 4 ft wide and within 4 ft of the end of the span, single headers and trimmers can be used. For header lengths greater than 4 ft, most codes require that double headers and trimmers be used.

Figure 7.16 indicates the more complicated framing system that is required for the floor opening situation which corresponds to the situation in Fig. 7.14.

Figure 7.15
Floor Framing for
Stairwell Opening
Parallel to the
Floor Joists
Source: Gerald E.
Sherwood and
Robert C. Stroh,
*Wood Frame House
Construction*, p. 57
(Mineola, NY:
Dover Publications,
Inc., 1989).

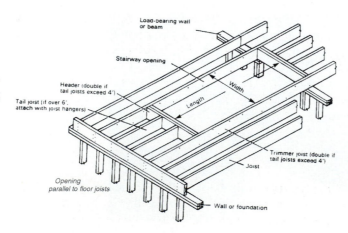

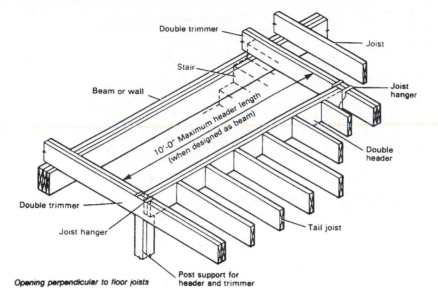

Figure 7.16
Floor Framing for
Stairwell Opening
Perpendicular to
the Floor Joists
Source: Gerald E.
Sherwood and
Robert C. Stroh,
*Wood Frame House
Construction*, p. 57
(Mineola, NY:
Dover Publications,
Inc., 1989).

Stairs

Stairway design should consider safety and the need for adequate headroom and the movement of furniture. Most stairways are either of the straight-run type or include various alternatives with mid-height landings. Basic dimension considerations for stairway design include headroom clearance, stairway width, stair tread run, and stair rise (Fig. 7.17). The generally accepted dimensions are:

- Minimum headroom clearance—6 ft, 8 in.
- Minimum stairway width—36 in.
- Minimum stair tread run—9 in.
- Maximum stair rise—8-1/4 in.

A rule of thumb that gives a comfortable proportion between the riser height (R) and the tread run (T) is 2R + T = 25. For example, risers 7-1/2 in. high and treads 10 in. wide result in a set of stairs which is comfortable for humans to traverse. A homebuilder should consult the local governing building code to verify these dimensions.

Wall Framing System

The **wall framing** stage in platform framed construction begins after the first-floor platform has been completed. After the first-floor walls are finished, the second floor is constructed and so on. The most common method of constructing the walls is known as **tilt-up construction**, where the walls are constructed on the ground or on a previous platform and then tilted up into place and anchored to the platform. The wall framing consists of all the framing required to construct the exterior and interior walls of the house (Fig. 7.18). Most walls consist of studs, top and bottom plates, bracing, and window framing members.

The first-floor and second-floor wall framing stages of the platform framing system are shown in Fig. 7.19.

Wall Studs

Wall studs are the primary components of the wall framing system that transmit the loads from above. Wall studs are typically 2 in. × 4 in. or 2 in. × 6 in. members and they usually are spaced 12, 16, or 24 in. o.c. Metal studs are available but currently are not as popular as

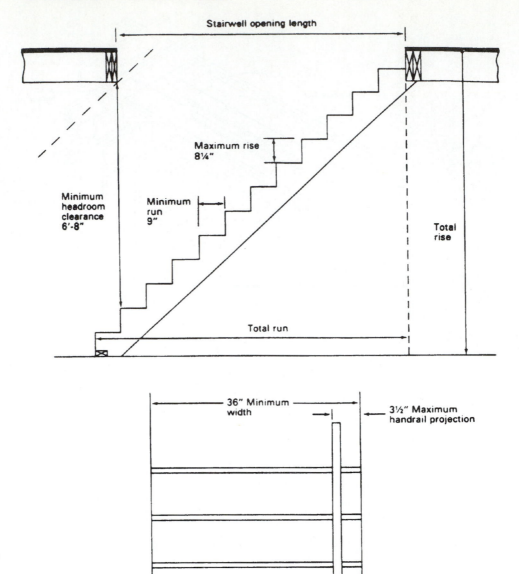

Figure 7.17
Stairway Design
Requirements
and Terminology
Source: Gerald E.
Sherwood and
Robert C. Stroh,
*Wood Frame House
Construction*, p. 53
(Mineola, NY:
Dover Publications,
Inc., 1989).

wood studs. When 2 in. × 6 in. members are used for the exterior walls, greater loads can be supported and more insulation can be installed in the wall cavities, thereby increasing the thermal resistance of the wall.

Wall studs are used in both bearing and non-load-bearing partitions (Fig. 7.20). **Bearing partitions** are those which support a floor, ceiling, or roof above. **Non-load-bearing partitions** are those which do not support a ceiling, floor, or roof but mainly divide the living space in the house. Non-load-bearing partitions can generally be constructed of a smaller-sized member than bearing partitions. Reference to the local building code that governs construction is required to verify the specific size of studs that can be used.

Several different arrangements of the studs at the outside corners of the house can be used. Blocking between two corner studs is the traditional method used to provide a nailing surface for the interior wall system [Fig. 7.21(a)]. A variation of this traditional method results in a three-stud corner [Fig. 7.21(b)]. A two-stud corner, which uses less lumber and provides more space for insulation, uses wallboard backup clips [Fig. 7.21(c)]. At first glance, this variation does not appear to result in significant cost savings, but if a builder is constructing 1,000 houses per year, a reduction of approximately 8,000 studs can be realized.

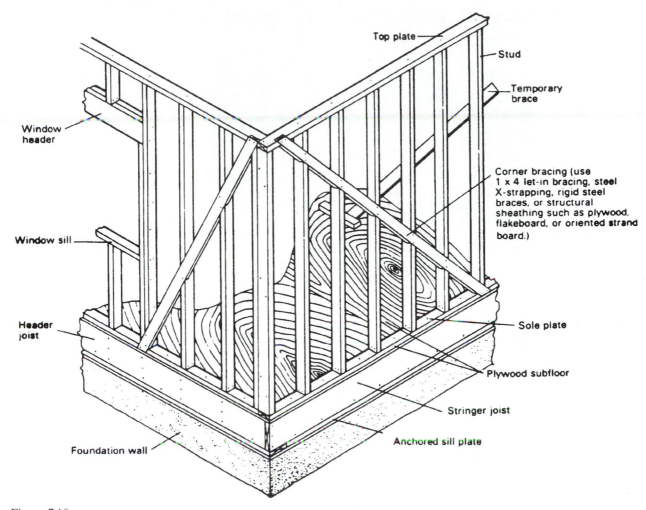

Figure 7.18
Wall Framing with Platform Construction
Source: Gerald E. Sherwood and Robert C. Stroh, *Wood Frame House Construction*, p. 63 (Mineola, NY: Dover Publications, Inc., 1989).

Figure 7.19
First- and Second-Floor Wall Framing Stage—Platform Framing System

Figure 7.20
Wall Studs
between the
Second Floor and
the Roof before
the Roof Trusses
are Installed

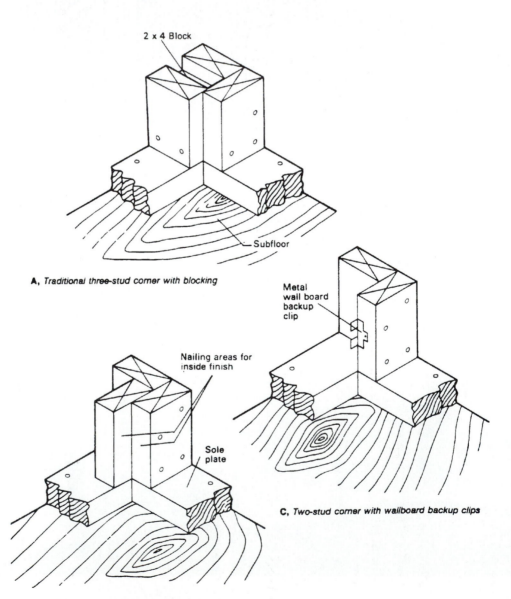

Figure 7.21
Corner Stud
Assembly
Alternatives
Source: Gerald E.
Sherwood and
Robert C. Stroh,
*Wood Frame House
Construction*, p. 66
(Mineola, NY:
Dover Publications,
Inc., 1989).

2 x 4 Block

Subfloor

A, *Traditional three-stud corner with blocking*

Metal
wall board
backup
clip

Nailing areas for
inside finish

Sole
plate

C, *Two-stud corner with wallboard backup clips*

B, *Three-stud corner without blocking*

Interior walls must be fastened to all exterior walls where they intersect. A nailing surface is once again required for the interior wall system and is commonly achieved as shown in Fig. 7.22(a). An alternative to this arrangement also uses wallboard backup clips [Fig. 7.22(b)]. Savings of the same magnitude associated with the corner arrangements can also be achieved in these situations.

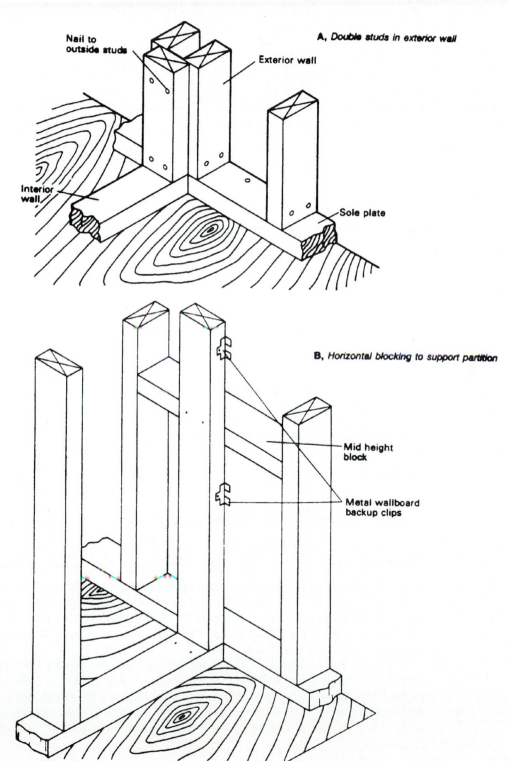

Figure 7.22
Intersection of Interior Partition and Exterior Wall
Source: Gerald E. Sherwood and Robert C. Stroh, *Wood Frame House Construction*, p. 67 (Mineola, NY: Dover Publications, Inc., 1989).

Top and Bottom Plates

Since wall studs are generally precut to a length of 93 in., a single sole plate and double top plates are used to provide the common 8-ft ceiling height (Fig. 7.20). When the house is designed with the roof or floor framing members located directly above the wall studs, a single top plate can be used (Fig. 7.18). However, when the roof or floor framing members bear on the top plate between studs, the top plate must be doubled. Savings of the same magnitude mentioned earlier can also be achieved by using single top plates whenever possible.

Bracing

All exterior corners of the house must be braced in order to prevent racking, or leaning of the wall, due to high winds or other horizontal forces. The appropriate **bracing** can be achieved by any of the following alternatives: 1 in. × 4 in. let-in bracing (Fig. 7.18), steel X-bracing (Fig. 7.20), rigid steel braces, or a structural sheathing such as plywood or oriented strand board.

Window and Door Framing

The members used to span over window and door openings are called headers (Fig. 7.18). As the width of the opening increases, and the magnitude of the load being transmitted through the wall becomes larger, it is necessary to increase the depth of these members. A header in a 2 in. × 4 in. wall commonly consists of two 2-in. thick members separated with 3/8-in. lath or plywood strips. The strips allow the faces of the header to be flush with the studs. When a 2 in. × 6 in. wall stud is being used, the header once again consists of two 2-in. members, but now blocking is used to bring the header flush with the studs. In this situation, it is possible to insert insulation between the blocking, thus increasing the thermal resistance of the wall. Headers are supported at the ends by the jack studs (Fig. 7.23), in what is commonly known as a double stud assembly. The short members above the header are called cripple studs.

Roof Framing System

The **roof framing system** is also an important contributor to the structural integrity of a house. The roof must be able to support snow loads in colder climates and to resist uplift forces in areas which are prone to high wind loads. In high-wind locations, hurricane strapping is often required. In addition, the roof must be capable of supporting the workers that are on the roof during the construction phase. The roof framing system usually consists of either: (1) roof rafters and ceiling joists or (2) manufactured roof trusses.

Ceiling Joists

In a roof rafter/ceiling joist system, the **ceiling joists** are typically 2 in. × 6 in., 2 in. × 8 in., or 2 in. × 10 in. members which are spaced 12 in., 16 in., or 24 in. o.c., depending upon the width of the house and the dead loads and live loads from the attic that they must carry. They are usually supported by the exterior walls and a bearing wall located somewhere near the centerline of the house (Fig. 7.24).

During the construction phase, the ceiling joists are usually installed before the roof rafters so that they are available to resist the tendency for the walls of the house to be pushed out due to the horizontal thrust imposed by the roof rafters. Because they are thrust-resisting members, they must be nailed securely to the top plate of both the outer and inner walls. They also should be lapped and nailed together or butted with wood or have metal cleats attached.

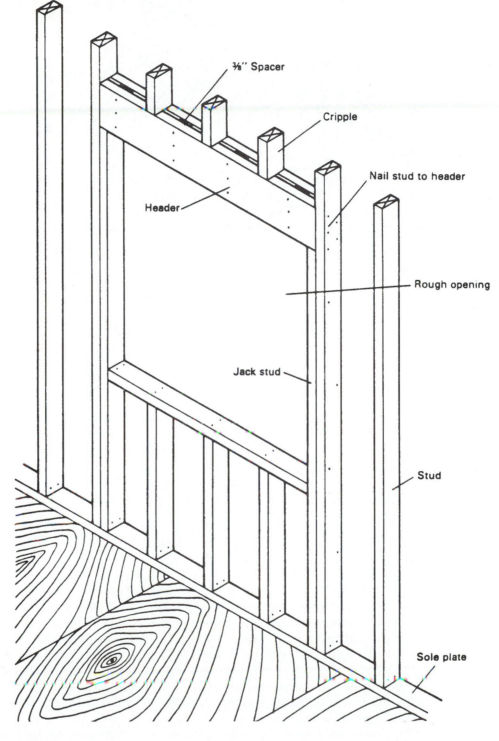

¹⁄₈" Spacer

Cripple

Nail stud to header

Header

Rough opening

Jack stud

Stud

Sole plate

Figure 7.23
Traditional
Header Assembly
over Window and
Door Openings
Source: Gerald E.
Sherwood and
Robert C. Stroh,
*Wood Frame House
Construction,* p. 69
(Mineola, NY:
Dover Publications,
Inc., 1989).

Roof Rafters

Once the ceiling joists are nailed in place, the **roof rafters** can be installed. Common sizes of roof rafters include 2 in. × 6 in., 2 in. × 8 in., or 2 in. × 10 in. members spaced 12 in., 16 in., or 24 in. o.c., depending upon the width of the house and the magnitude of the dead and live loads imposed on the roof.

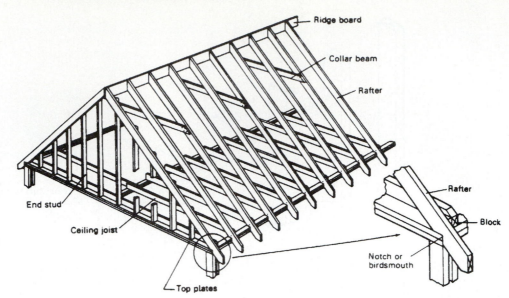

Figure 7.24
Roof
Rafter/Ceiling
Joist Roof
Framing System
Source: Gerald E.
Sherwood and
Robert C. Stroh,
*Wood Frame House
Construction*, p. 81
(Mineola, NY:
Dover Publications,
Inc., 1989).

The rafters from both sides are usually connected at the top to a ridge board (Fig. 7.24). Due to the angle cut in the rafters, the ridge board must be a larger size than the rafter. The ridge board runs the entire length of the roof and helps to distribute the roof load among several rafters. The location where the rafter bears on the exterior wall is notched to form a snug fit. The notch is commonly known as a "bird's mouth." The end of the roof is known as the gable and is the location where the roof and wall join.

Collar beams (i.e., collar ties) are often required when roof spans are long and the slopes are flat. Steeper slopes and shorter spans also may require collar beams, but only between every third rafter pair. Collar beams are usually 1 in. × 6 in. or 2 in. × 4 in. members.

Manufactured Roof Trusses

Roof trusses that have been designed and manufactured in a factory to meet the architectural requirements of both simple and complicated roof lines have become a widely accepted alternative to the conventional roof rafter/ceiling joist system (Fig. 7.25). A truss is a rigid framework of triangular shapes which is placed across the width of the house and nailed directly to the top plates. Trusses require no interior bearing partition near the middle of the span since they are supported entirely by the exterior walls. This allows freedom of interior partition location without affecting the structural integrity of the house. Metal truss plates (Fig. 7.26) are used to attach the various members of the truss together when manufactured.

Trusses have become increasingly popular over the years for the following reasons: (1) trusses can span greater distances than the roof rafter and ceiling joist system, (2) the erection of trusses requires less field labor since they are usually built in a factory and cranes can be used for their installation (Fig. 7.27), and (3) trusses typically use less material than the equivalent roof rafter and ceiling joist systems. In addition, roof trusses are "engineered" by the firm that supplies them for the specific set of house plans submitted by the homebuilder. Thus, the homebuilder has the "protection" of a professional engineer's stamp on the set of roof plans. All of these factors typically result in roof trusses being cost effective and appealing to the homebuilder.

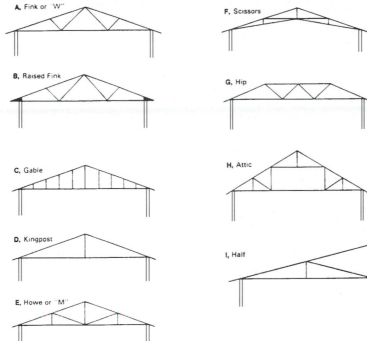

Figure 7.25
Common Roof
Trusses
Source: Gerald E.
Sherwood and
Robert C. Stroh,
*Wood Frame House
Construction*, p. 74
(Mineola, NY:
Dover Publications,
Inc., 1989).

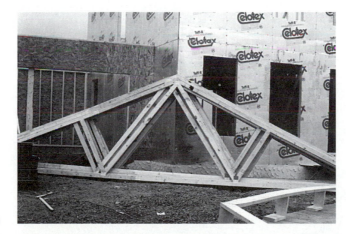

Figure 7.26
Conventional
Fink Roof
Trusses with
Metal Truss Plate
Joint Connections

Figure 7.27
Installation of the
Roof Trusses on
House B

Roof Sheathing

The **roof sheathing** is the covering applied over the roof rafters and trusses to provide racking resistance to the roof framing and to provide a surface for attaching the roof covering materials. Plywood and oriented strand board are commonly used sheathing materials. Roof sheathing edges that run perpendicular to the roof framing must be supported by wood blocking or fastened together with metal fasteners.

These metal fasteners, which are called "H" clips (Fig. 7.28), are the least costly and most common method of providing the necessary edge support. The "H" clips also provide space between the plywood, oriented strand board, and other roof sheathing materials to allow for expansion and contraction of the materials caused by thermal and moisture changes. Without such an allowance, the sheathing has a propensity to buckle or curl at the edges.

Roof Coverings

The selection of a roof covering is governed by the initial cost, the expected life of the roof covering, local building code requirements, house design, and the homebuilder's or homeowner's preferences. Wood, fiberglass, and organic asphalt (Fig. 7.29) shingles, shakes, tile, and slate are the most common roof coverings currently being used for houses.

Flashing

All roof penetrations—such as those required for skylights, pipes, dormers, and chimneys—and the locations where different angled roofs meet must be sealed to avoid leaks. Roof surfaces that butt into vertical walls also must have protection against leaks. This protective

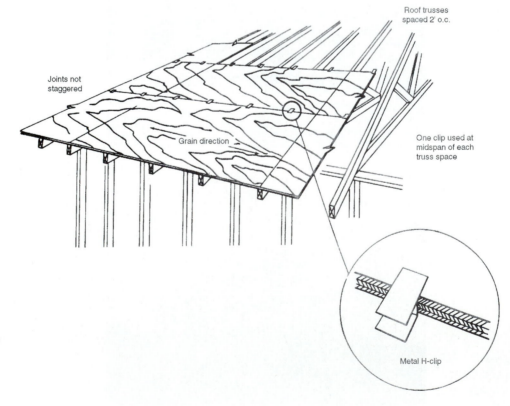

Figure 7.28
Plywood Roof Sheathing with "H" Clips for Edge Support

Source: Gerald E. Sherwood and Robert C. Stroh, *Wood Frame House Construction*, p. 86 (Mineola, NY: Dover Publications, Inc., 1989).

Figure 7.29
Asphalt Shingles—
Front of House A

seal is known as **flashing**. The flashing should be a rust- and corrosion-resistive metal such as copper, galvanized steel, lead, or aluminum or it can be a heavy, reinforced fabric. It is important to note that almost all metals eventually rust or corrode. Problems due to galvanic action also occur if dissimilar metals are not kept apart properly.

Ventilation

One of the most important aspects of the roof framing stage is ensuring that adequate **ventilation** of the attic is achieved. Since airtight construction has become more common, water vapor tends to find its way into the attic space and accumulate under the roofing. Condensation occurs when this warm, moist air comes into contact with the cold underside of the roofing during the winter. This can lead to the formation of ice dams on the roof, roof failure, or the buckling of roofing materials.

Building codes currently require that: (1) enclosed attics and (2) enclosed roof rafter spaces formed when ceilings are applied directly to the underside of roof rafters (i.e., cathedral ceilings) must have cross-ventilation created by ventilated openings which are protected against the entrance of rain or snow. With proper ventilation, air can circulate freely in the attic and remove the water vapor before it can condense.

One of the most common methods of ventilation used is a combination of soffit vents and ridge vents (Fig. 7.30), which provides a natural draft ventilation from the bottom to the

Figure 7.30
Ridge Vent and
Soffit Vents

Figure 7.31
Plastic Vents
which Allow Air
Circulation in the
Attic

top of the attic. Building codes usually specify the minimum area of openings for a certain square footage of attic surface. If batt insulation is placed on the underside of the roof covering, plastic vents should be used to allow air circulation into the attic (Fig. 7.31). If this is not done, the insulation may be pushed into the soffit and then air may not be able to pass into the attic.

MECHANICAL SYSTEMS

The **mechanical systems** of a house usually include the (1) heating, ventilating, and air conditioning (HVAC) system, (2) plumbing system, and (3) electrical system. Individual chapters presented later in this book discuss each of these systems in detail. The purpose of this section is to highlight some of the key details associated with these systems. The progress photos that are shown were recorded during the rough-in stage of construction of the two "Pauline" models (i.e., before the interior finish stage).

HVAC System

The **HVAC system** currently installed in most houses is the forced air type, meaning that air is forced through a ductwork system (Figs. 7.32–7.34) and delivered to the various rooms of the house. The unit that forces the air through the ductwork is commonly the fan or blower in a gas-fired furnace (Fig. 7.34) or a heat pump.

A detailed explanation of all of the components and the design of the HVAC system for the Case Study House is presented in Chapter 12 of this book.

Plumbing System

The **plumbing system** is divided into two main categories: water supply and distribution, and sanitary drainage. The house must have a hot and cold water supply, and it must be connected to a sewer system so that all of the waste and used water can be drained from the house. Cold water is usually supplied to the house through the basement wall. A branch of

Figure 7.32
Ductwork
Located in an
Interior Partition

Figure 7.33
Ductwork
Directly
Connected to the
High-Efficiency
Gas Furnance

the cold-water line then enters a domestic water heater to create a hot-water line (Fig. 7.34). Hot- and cold-water lines (Figs. 7.35 and 7.36) are then distributed throughout the house to the necessary locations.

Parts of the sanitary drainage system are shown in Figs. 7.35 and 7.36. The components of the sanitary drainage system that are in the basement are shown in Fig. 7.37.

A detailed explanation of all of the components and the design of the plumbing system for the Case Study House is presented in Chapter 13 of this book.

Electrical System

All electrical energy supplied to power-consuming devices and appliances within a house must pass through the electrical service equipment, where it is metered and protected. Electrical current, after passing through the meter, enters an electrical distribution panel

Figure 7.34
High-Efficiency
Gas-Fired
Furnace (left) and
Domestic Water
Heater (right)
Located in the
Basement

Figure 7.35
Hot- and Cold-
Water Lines and
Sanitary Drainage
under the First
Floor

Figure 7.36
Hot- and Cold-
Water Supply
and Sanitary
Drainage: The
Combined Bath
and Shower Unit
on the Second
Floor

Figure 7.37
Above-Slab and
Below-Slab
Sanitary
Drainage System
in the Basement
(before the con-
crete slab is
placed)

Figure 7.38
An Electrical
Distribution
Panel in the
Basement

(also called a panelboard, circuit breaker panel, or fuse box). The electrical distribution panel (Fig. 7.38) acts as a terminal where branch circuits that run to various loads throughout the house begin and end.

The branch circuits provide electricity to all fixtures, switches, and plugs. Code requirements stipulate that all outlets must be installed in an approved electric box (Fig. 7.39). A typical example of the electrical wiring and components involved is shown in Fig. 7.40, which indicates the service to the kitchen on the first floor of the Case Study House (see Sheet A-2 in Appendix A).

A detailed explanation of all of the components and the design of the electrical system for the Case Study House is presented in Chapter 14 of this book.

Figure 7.39
Electrical Service
to a Typical
Electrical Box

Figure 7.40
Electrical Wiring
and Electric Box
Providing Service
to the Kitchen
Area on the First
Floor

EXTERIOR FINISHING STAGE

Siding

A number of coverings can be used to finish the exterior of the house. Wood siding can be obtained in several different patterns and can be finished naturally, stained, or painted. Vinyl siding, which essentially has replaced aluminum siding in many areas of the United States, requires little maintenance beyond periodic cleaning (Fig. 7.41). A masonry veneer (Fig. 7.42) constructed of brick or stone can also be used for all or part of the exterior wall finish. It is normal practice to install the masonry veneer with a 3/4-in. space between the veneer and the wall sheathing to provide room for the bricklayer's fingers when setting the brick. Corrugated metal brick ties are used to attach the brick to the sheathing. Weep holes should be located 4 ft o.c. to couple the air space to the outdoors for moisture control purposes.

Figure 7.41
Vinyl Siding
Installation
Completed—
Front of House A

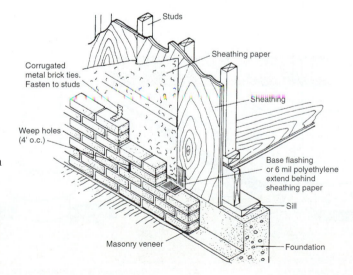

Figure 7.42
Masonry Veneer
Siding Installation
Source: Gerald E.
Sherwood and
Robert C. Stroh,
*Wood Frame House
Construction*, p. 128
(Mineola, NY:
Dover Publications,
Inc., 1989).

Gutters and Downspouts

Gutters and **downspouts**, which are not required in all parts of the United States, carry water off the roof and away from the foundation (Fig. 7.43). The most commonly used gutter is the type which is hung from the edge of the roof. Gutters may be of the half-round or formed type and typically are constructed of aluminum, galvanized steel, or vinyl. Gutters obviously should be installed with a slight pitch (such as 1/4 in. in 10 ft). Downspouts are round or rectangular; the round type is used with the half-round gutters. They are usually corrugated to provide extra stiffness and strength.

Landscaping

The final phase of the exterior finishing stage consists of the various **landscaping** tasks, which typically include: (1) planting grass seed or placing sod, (2) planting shrubs, and (3) completing the driveway and all walkways to the house.

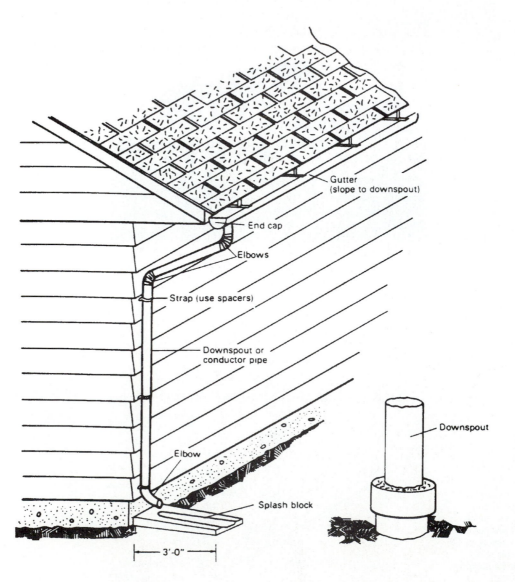

Figure 7.43
Gutter and
Downspout
Installation
Source: Gerald E.
Sherwood and
Robert C. Stroh,
*Wood Frame House
Construction*, p. 109
(Mineola, NY:
Dover Publications,
Inc., 1989).

A, *Downspout with splash block*   **B, *Downspout connected directly to storm sewer***

INTERIOR FINISHING STAGE

After the mechanical system rough-in stage has been completed, the interior of the house can be finished. Some of the key components of the interior finishing stage include the installation of the insulation and vapor retarder in the walls, wall and ceiling finishes, and the trimwork.

Insulation and Vapor Retarder

The most common type of **insulation** currently being used in houses is fiberglass (Fig. 7.44). It is composed of very fine inorganic fibers made from rock, slag, or glass, with other materials added to enhance service properties. Available forms include flexible batts and blankets (with and without facings), semirigid and rigid boards (with and without facings), and a loose form which is blown or poured. In the United States, batt and blanket insulations usually have a kraft vapor retarder paper facing. They are also available with aluminum-foil facings and an unfaced form held in place by pressure.

   Vapor retarders are used in walls and ceilings, and sometimes in floors, to prevent water vapor from migrating to a cold surface where it may condense or freeze. Common vapor retarders include polyethylene (Fig. 7.44) and the facings on insulation batts and blankets. Vapor retarders are not recommended when wood-framed houses are built in hot, humid southern climates.

Figure 7.44
Polyethylene
Vapor Retarder
Installed over
Fiberglass Batt
Insulation in the
Wall

Figure 7.45
Completed
Gypsum Board
Installation in the
Kitchen Area on
the First Floor

Wall and Ceiling Finish

The most commonly used interior **wall and ceiling finish** is gypsum board. It has the advantages of being economical, noncombustible, and easy to install and repair. Standard gypsum board is available in 3/8-in., 1/2-in., and 5/8-in. thicknesses. It also is available as a fire-resistive Type X gypsum board in 1/2-in. and 5/8-in. thicknesses. Type X gypsum board is used to build separation walls between independent living units and between the garage and the general living area in single-family houses. Standard gypsum board can be used throughout the house, while 1/2-in.-thick water-resistive gypsum board (which has a green paper face) should be used in bathrooms and kitchens.

Gypsum board is usually installed on walls in a horizontal manner (Fig. 7.45). It can be attached to a wood or steel framing system with screws that should be driven with the heads slightly below the surface (Fig. 7.46). Nails can also be used to fasten the gypsum board to a wood framing system. Metal corner beads are used on all exterior corners to ensure that all walls are straight (Fig. 7.47).

Joint compound is then used to fill in over the screws or nails and tape is applied over the joints. Once the joint compound has dried, it should be sanded smooth. The joint compound/sanding process is usually repeated two or three times in order to achieve the best results.

Figure 7.46
Installation of
Gypsum Board
on the Ceiling of
the First Floor
Using Screws

Figure 7.47
Metal Corner
Bead Installation
on the Low
Diagonal Wall in
the Kitchen on
the First Floor

Once the entire process is completed, the walls can be painted. Another popular wall covering is paneling, which may be made of plywood, hardboard, or particleboard.

Floor Coverings

A wide variety of **floor coverings** is available for application over the subfloor. The usual considerations in making a selection are maintenance, durability, comfort, aesthetics, and initial cost. Commonly used floor coverings include carpet, sheet vinyl, vinyl tile, hardwood, and ceramic tile.

Trimwork and Cabinetry

The interior trimwork can be finished once the interior wall, ceiling, and floor coverings have been completed. The **trimwork** includes the wood that is used around doorway openings (Fig. 7.48), as well as baseboard and crown molding that are used in each of the rooms. The trimwork can either be stained or painted.

The primary **cabinetry** is installed in the bathrooms and in the kitchen. Figure 7.49 indicates an intermediate stage of completion of the kitchen cabinet system.

Figure 7.48
Doorway
Opening
Trimwork

Figure 7.49
Installation of the
Cabinetry System
in the Kitchen

CHAPTER SUMMARY

This chapter provided a summary of the various phases involved in the construction of a house. The chapter presented an overview of the complete construction process; it did not provide an overwhelming amount of details and specifications. As a result, in some sections of the chapter important details related to a particular construction phase may have been omitted. In such cases, the reader is encouraged to reference the local building code, a detailed house construction text, the manufacturer of a particular product, or the appropriate chapter in the remainder of this book for the necessary guidelines.

NOTE

[1]Gerald E. Sherwood and Robert C. Stroh, *Wood Frame House Construction* (Mineola, NY: Dover Publications, Inc., 1989), p. 57.

BIBLIOGRAPHY

Sherwood, Gerald E. and Robert C. Stroh, *Wood-Frame House Construction.*   Mineola, NY: Dover Publications, Inc., 1989.

Wing, Charlie, *The Visual Handbook of Building and Remodeling.*   Emmaus, PA: Rodale Press, 1990.

8

Wood: Its Properties and Strength

INTRODUCTION

The overview of the residential construction process in Chapter 7 indicated that most houses rely on a wood structural framing system. For such structural applications, **wood** is commonly available as either sawn timbers, dimension lumber, glue laminated members (glulams), or other engineered products such as laminated veneer lumber, strand board, and plywood. Wood is the hard, fibrous material that makes up the tree under the bark. **Lumber** is the wood that has been sawn or planed. **Timber** is lumber with a smallest dimension of at least five inches. The principal advantages of wood over the other common structural materials are (1) its economy, (2) its appearance, and (3) its ease of working and reworking. Other advantages include better durability (for some applications), a high strength/weight ratio, and excellent thermal insulating properties. This chapter provides an introduction to the physical and mechanical properties and behavior of structural lumber.

GENERAL CONSIDERATIONS

Wood is unique as a structural material because it is a biological material. Consequently, one must cope with the natural variability of its strength and stiffness properties through quality control strategies. The structural properties of wood vary between species (pine, maple, oak, etc.), between trees of the same species (high or low density), and even within a single tree of the same species.

Wood is anisotropic; i.e., its structural properties are direction dependent. This behavior is easily understood by considering the structure of the wood section shown in Fig. 8.1. The longitudinal (L) axis represents a line parallel to the tree trunk, the radial (R) axis is directed outward from the center of the trunk, and the tangential (T) axis is normal to both the R and L axes. In the T–R plane, the annual growth rings are visible. The grain, which is visible in the T–L and L–R planes, is the side view of the annual rings. The wood structure varies with direction. Likewise, the wood strength varies with direction. Specifically, the tensile or compressive strength parallel to the grain is significantly greater than in directions oblique to the grain, the weakest direction being perpendicular to the grain. On the other hand, the resistance to shear-type failure is greatest in the direction perpendicular to the

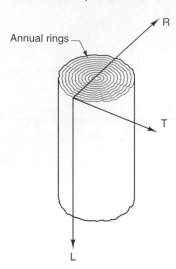

Figure 8.1
Wood Section

grain and least in the direction parallel to the grain. It is easier to shear the bonds between annual rings (fibers) than to shear the fibers themselves.

Wood is not homogeneous. Features such as knots, changes in grain direction, and separations between and within annual rings occur to some degree in nearly all commercially available lumber. As will be discussed later, these features, which are collectively termed **wood characteristics**, influence wood strength.

The specifications for structural design in wood are presented in the *National Design Specification for Wood Construction*.[1] This is the primary specification for wood design in the United States. Some of the material presented in this chapter will be based upon the provisions of this specification, hereinafter referred to as NDS®. Design values for wood construction are summarized in the NDS® Supplement.[2]

Classification

The two classes of trees commonly used for construction purposes are **softwoods** and **hardwoods**. Softwoods actually include some species that are quite hard, while hardwoods include some species whose wood is very soft. Softwoods mainly consist of the needle-leaved conifers (cone bearing) and of the evergreen type (i.e., they do not drop their leaves for the winter season). All hardwood trees have broad leaves and are of the deciduous type (i.e., they shed their leaves in winter). Some of the typical species within each classification are indicated in Table 8.1.

TABLE 8.1   Typical Tree Classifications

Hardwoods	Softwoods
Ash	Cedar
Aspen	Fir
Basswood	Hemlock
Beech	Pine
Birch	Redwood
Cottonwood	Spruce
Maple	
Oak	
Poplar	

Moisture Content

The **moisture content** (MC) is a measure of the amount of moisture contained in the wood. It is the weight of the water the wood contains expressed as a percentage of the weight of the wood when it is oven dry.

$$MC = 100 \times \left(\frac{\text{original weight} - \text{oven-dry weight}}{\text{oven-dry weight}} \right)$$

Moisture in wood may be bound or free. **Bound moisture** is the water that has been absorbed by the wood fibers. **Free moisture** is found in the cavities (spaces) between the fibers.

Once a tree is cut, the wood begins to dry and the free moisture begins to evaporate. The free moisture will continue to evaporate until it is completely gone. It is only after all the free moisture has evaporated that any of the absorbed moisture in the wood fibers will evaporate. The point at which all of the free moisture in the cavities has evaporated, but the wood fibers are still fully saturated, is referred to as the **fiber saturation point** (FSP). In most wood species, the fiber saturation point occurs at a moisture content of approximately 25–30%. Over time, some of the moisture in the fibers will diffuse into the surrounding air until the wood achieves its **equilibrium moisture content** (EMC). The EMC of wood is dependent upon the temperature and humidity of the air. For example, EMC = 8% at 50°F and 40% r.h., 12% at 70°F and 65% r.h., and 20% at 80°F and 90% r.h.[3] EMCs are plotted versus relative humidity for three dry-bulb temperatures in Fig. 8.2. Recommended moisture contents for several end uses are listed in Table 8.2.

After the tree is cut into lumber, it must usually be dried below the FSP before it can be used for construction purposes. The lumber can be air dried, but should not be used until reaching a moisture content of 19% or less. This process usually takes several months. Above 19% moisture content, wood is classified as **green lumber**, while lumber with a moisture content less than 19% is classified as **seasoned (dry) wood**. In order to speed up the drying process, the lumber may be placed in a kiln, which uses heat to evaporate the moisture. Most structural lumber is dried to below 19% moisture content.

Wood Grain

The process of sawing lumber from logs results in a visible pattern showing the annual rings on the end of the lumber or the pattern of longitudinal fibers on the surface (Fig. 8.3). Generally, all lumber is cut from a log either tangential or radial to the growth rings. Not all

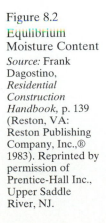

Figure 8.2
Equilibrium
Moisture Content
Source: Frank
Dagostino,
*Residential
Construction
Handbook*, p. 139
(Reston, VA:
Reston Publishing
Company, Inc.,®
1983). Reprinted by
permission of
Prentice-Hall Inc.,
Upper Saddle
River, NJ.

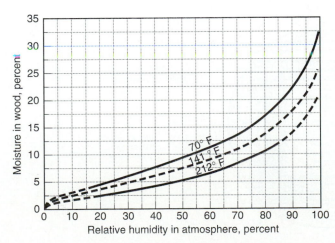

TABLE 8.2   Recommended Approximate Moisture Contents

Use of Lumber	Dry Western Areas		Damp and Coastal Southern Areas		Remainder of United States	
	Average	Ind. Pieces	Average	Ind. Pieces	Average	Ind. Pieces
Exterior trim, siding, sheathing, and framing[a]	8–10	7–12	11–13	9–14	11–13	9–14
Interior trim woodwork, and softwood flooring	5–7	4–9	10–12	8–13	7–9	5–10
Hardwood flooring	6–7	5–8	10–11	9–12	7–8	6–9

[a]Framing lumber of higher moisture content is commonly used in construction because lumber with the moisture content recommended may not be available except on special order.

Source: Frank Dagostino, *Residential Construction Handbook*, p. 140 (Reston, VA: Reston Publishing Company, Inc.,® 1983). Reprinted by permission of Prentice-Hall, Inc., Upper Saddle River, NJ.

of the pieces, however, will conform to these two definitions. Several of the possible wood grain directions are identified in Fig. 8.4. These grain patterns include:

Edge Grain: Lumber is cut from logs radially to the growth rings and will have the annual rings run approximately at a right angle to the face.

Angle Grain: Lumber is cut from the logs radially to the growth rings, with some being at less than a 45° angle from the face.

Flat Grain: Lumber is cut from the logs tangentially to the growth rings; the growth rings run approximately parallel to the surface.

Cross Grain: Lumber is cut such that wood fibers do not run parallel to the length of the board. Somewhere along the surface the ends of the fibers have been cut off, leaving a rough area that will be difficult to finish neatly.

Figure 8.3 Lumber Cut from Logs

Source: Frank Dagostino, *Residential Construction Handbook*, p. 141 (Reston, VA: Reston Publishing Company, Inc.,® 1983). Reprinted by permission of Prentice-Hall Inc., Upper Saddle River, NJ.

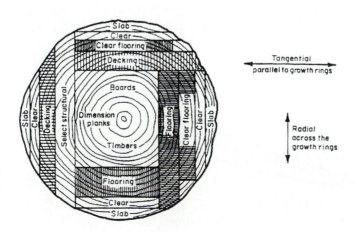

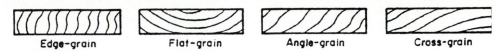

Edge-grain   Flat-grain   Angle-grain   Cross-grain

Figure 8.4
Grain Patterns
Source: Frank Dagostino, *Residential Construction Handbook*, p. 142 (Reston, VA: Reston Publishing Company, Inc.,® 1983). Reprinted by permission of Prentice-Hall Inc., Upper Saddle River, NJ.

STRUCTURAL BEHAVIOR OF WOOD MEMBERS

The wood design professional must be familiar with the basic structural behavior of wood when loaded in compression, tension, bending, or shear. Wood used in the structural framing of a house will be affected by these load cases. The fundamental behaviors, as well as some other important terms used in wood design, are discussed herein.

Compression

When a wood member is used as a column, the loads are applied in the axial direction (parallel to the grain) and through the centroidal axis of the cross section (Fig. 8.5). The stress developed in the column is defined as:

$$f_c = \frac{P(\text{lbs})}{A(\text{in}^2)}$$

When axial loads are applied at some eccentricity, or when centric axial loads are applied in combination with transverse loads, the member is a beam-column. Then, the stresses are defined as:

$$f = \frac{P}{A} + \frac{M}{S}$$

where M = bending moment (lb–in.)
S = section modulus (in^3)

Column lengths are usually many times greater than the least lateral dimension. One example of a wood structural member designed as a column is the vertical member (stud) in load-bearing walls (Fig. 8.6). It should be noted that the load-carrying capacity of the studs is greatly increased over that of individual free-standing studs by attached structural sheathing, such as plywood or particleboard.

The load-carrying capacity of a wood column is inversely proportional to the square of the column length, all other factors being constant. A "short" column (Fig. 8.7) may be able to carry a load P_1; however, if the length is increased from L_1 to L_2, the column will buckle at a smaller load, $P_2 < P_1$. Columns are often laterally supported (e.g., braced with blocking

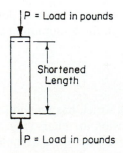

P = Load in pounds

Shortened
Length

Figure 8.5
Compression

P = Load in pounds

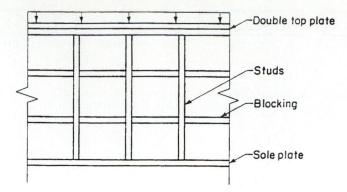

Figure 8.6
Studs as Columns

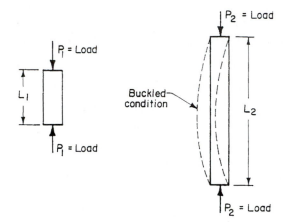

Figure 8.7
"Short" vs.
"Long" Columns

in a stud wall) at certain intervals to reduce their "unsupported length" and thereby increase their load capacity. Indeed, if the column is braced continuously with respect to both cross-sectional axes, $P_2 = P_1$ in Fig. 8.7.

Compressive loads can also be applied perpendicular to the grain of the wood. Such a situation is the case of the first-floor joist which spans from the foundation wall to the spliced connection over the built-up girder in a basement (Fig. 8.8). If the compressive force is too great, the wood will be crushed at the floor joist/built-up girder interface, causing an unacceptable permanent displacement of the supported member. Wood sills and supports must be designed to prevent excessive bearing-compressive stresses. The compressive strength of wood perpendicular to the grain is much lower than the compressive strength parallel to the grain. Potential crushing situations are avoided by adjusting the bearing length between the load-transmitting member and the girder, or sill, accordingly.

Tension

When a force tends to stretch or elongate a member, tensile stresses develop (Fig. 8.9). If the tensile force, P, is axial and if it is directed through the longitudinal centroidal axis, the tensile stress is:

$$f_t = \frac{P(\text{lbs})}{A(\text{in}^2)}$$

Simple structural analysis illustrates that a roof truss which has pinned joints and is loaded only at the joints contains both compression and tension members (Fig. 8.10). Under gravity loads, the lower chords are loaded in tension as are two of the web members.

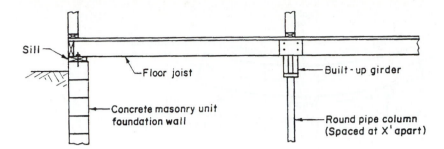

Figure 8.8
First-Floor Joist
over Girder

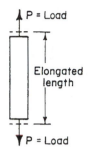

Figure 8.9
Tension

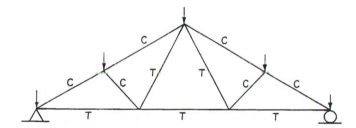

Figure 8.10
Roof Truss
Stresses

Bending

Bending members are characterized as those having one dimension (the longitudinal) significantly larger than the other two, and having loads applied normal to the longitudinal axis. When a transverse load is applied to a beam, the beam begins to deflect (bend), thus creating bending stresses (Fig. 8.11). **Bending** produces compression on the loaded side of the member and tension on the opposite side. In the case of pure bending, the neutral axis (the axis of zero strain and stress) coincides with the centroidal axis of the cross section. The maximum bending stresses occur in the fibers furthest from the central axis, e.g., at either the top or bottom of the beam. The distribution of compressive and tensile stresses due to bending at section "x" through a beam is approximately linear (Fig. 8.12).

The maximum flexural stress in the beam at station x occurs at the top (compression) and bottom (tension) fibers and is expressed as:

$$f_b = \frac{M}{S}$$

where f_b = the maximum flexure stress at the section being considered (lbs/in^2)
   M = the calculated bending moment at the section being considered (in.–lb)
   S = the section modulus of the member ($S = I/C = bd^2/6$ for solid rectangular members) (in^3)
   I = area moment of inertia ($I = bd^3/12$ for solid rectangular members) (in^4)
   C = distance from the neutral axis to the outer fiber (in.)

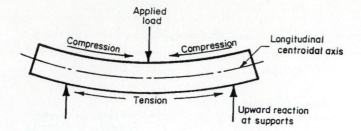

Figure 8.11
Bending

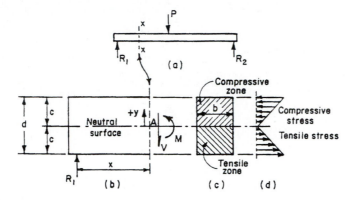

Figure 8.12
Bending Stress
Distribution

Horizontal Shear

A **shear stress** results from the tendency of two equal and parallel forces, acting in opposite directions, to cause adjoining surfaces of a member to slide on one or the other (Fig. 8.13). Flexural shear stresses in wood beams are termed **horizontal shear**.

As load is applied to a beam, the wood fibers above the neutral axis tend to move horizontally with respect to the fibers below the neutral axis (Fig. 8.14). This movement results in fiber stresses which are referred to as horizontal shear. The distribution of horizontal shearing stresses due to bending on a section through a beam is illustrated in Fig. 8.15. It should be noted that the maximum horizontal shear stress occurs at the centroidal axis of the cross section. The shear force (Figs. 8.12 and 8.15) is the resultant of the horizontal shear stress distribution (Fig. 8.15) acting at a given cross section.

The maximum horizontal shearing stress in beams of a solid rectangular cross section is defined as:

$$(f_v)_{max} = \frac{3V}{(2bd)}$$

where: $(f_v)_{max}$ = the maximum horizontal shearing stress at the neutral axis of the beam cross section (lb/in^2)
V = total vertical shear at the cross section selected (lb)
b = width of beam (in.)
d = depth of beam (in.)

Vertical Shear

Vertical shear develops at the interface between the beam and the reactive support (Fig. 8.16). Vertical shear develops in the beam just to the inside of the beam supports. Wood and steel both behave well in tension and compression and are able to resist shear forces proportional to their strengths. In wood, vertical shear is seldom a problem because

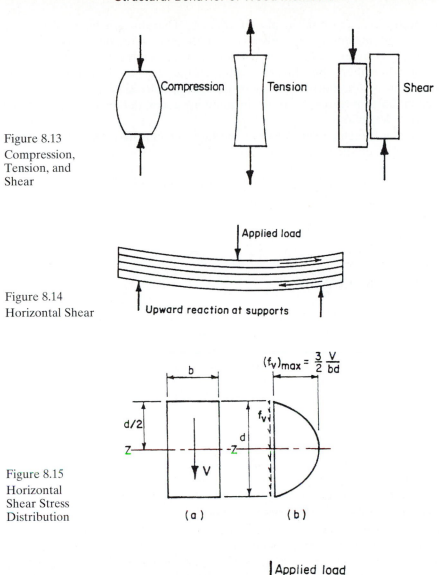

Figure 8.13
Compression,
Tension, and
Shear

Figure 8.14
Horizontal Shear

Figure 8.15
Horizontal
Shear Stress
Distribution

Figure 8.16
Vertical Shear

the shear force acts perpendicular to the wood fibers. The beam will fail in horizontal shear (parallel to the fibers) long before failing in vertical shear.

Deflection

In addition to the flexural and shearing stresses, consideration must be given to the **deflection** of members subjected to transverse loads. Some deflection exists in all beams, and the designer must ensure that the deflection does not exceed certain prescribed limits. A beam may be adequate to support the imposed loading without exceeding stress limitations, but at the same time the deflection may be so great that damage to secondary non-load-bearing components may occur, e.g., cracking in plaster ceilings. Therefore, a beam should be designed for both strength in bending and stiffness, even though a large deflection does not necessarily constitute fracture of the beam. Deflection limitations are generally taken to be 1/360 of the span for beams that support plaster ceilings. Of course, the local building code should always be consulted for specific provisions governing deflection. For example, for some cases deflection limits may only be 1/180. In some instances, deflection requirements are limited to live load deflections.

Beam deflection is directly related to the **modulus of elasticity (E)** of the beam material. The modulus of elasticity, the ratio between the unit stress and the unit strain below the elastic limit of the material, is a measure of its stiffness. For structural steel, $E = 29,000,000$ psi, and for wood, depending on the species and grade, E varies from something less than 1,000,000 psi to about 2,400,000 psi.

DESIGN METHODS

Allowable Strength Design (ASD)

Current practice in the United States for designing wood structural members is based on **allowable strength design** (ASD). That is, the required load capacity of a member is determined from service loads. The load capacity is based on allowable stresses, which for the most part are below the elastic limit of the wood, and therefore can be calculated using familiar strength-of-materials equations for linearly elastic materials. These equations were discussed in the immediately preceding section. ASD will be used in this text.

Load Resistance Factor Design (LRFD)

Specifications for **Load Resistance Factor Design** (LRFD) for wood have been developed by AFPA and, indeed, Canadian practice already uses limit state design for wood. Factored service loads and ultimate load capacities are used in LRFD. The LRFD specification was adopted as an alternative in the U.S. for wood design in 1996.

Applications

The primary structural members of a house frame and the associated stress states are listed in Table 8.3. It is important to note that the sheathing which is used on the roof to span between the roof rafters (or trusses) and the sheathing which is used on each floor span

TABLE 8.3   Structural Applications

Structural Member	Stress States
1. Roof Rafter, Ceiling Joist, Floor Joist, Built-Up Girder	a. Bending Stress (psi)—f_b b. Horizontal Shear Stress (psi)—f_v c. Deflection (in.)—Δ d. Bearing Stress (psi)—$f_{c\perp}$ e. Combined Bending and Axial Stress—$f_b + (f_c \text{ or } f_t)$
2. Columns, Stud in Bearing Wall, Certain Members of a Truss	a. Axial Compressive Stress (psi)—f_c b. Combined Compressive and Bending Stresses—$f_c + f_b$
3. Certain Members of a Truss	a. Axial Tensile Stress (psi)—f_t b. Combined Tensile and Bending Stresses—$f_t + f_b$

between the floor joists are analyzed as a beam since they are loaded transversely to the panel. Also, many members in the building frame are subjected to combined stress states, e.g., compression plus bending.

The stresses induced by the service loads are designated by the lowercase f_i and are often called **actual**, or **calculated**, **stresses**. The structural member is satisfactory provided the calculated stress, f_i, does not exceed the wood design strength, F_i.

WOOD STRENGTH

The structural properties of wood are influenced by a number of factors, e.g., species, density, moisture content, slope of grain, temperature, load duration, and defects. Some of the most important factors that must be considered are discussed next.

As noted earlier, there can be a significant difference in the strength and stiffness of wood between species. However, the strengths of the two most common species used in the United States, Douglas Fir (DF) and Southern Pine (SP), are nearly equal. Spruce–pine–fir, another common species used in residential construction, is somewhat weaker and less stiff than DF and SP.

Wood density influences strength within the same species. In dense wood, the annual rings are more closely spaced than in lighter samples. Generally, the greater the wood density, the greater its strength.

Moisture content (MC) also influences the strength of wood. Generally, the strength properties of wood decrease as the moisture content increases, but only up to the FSP. Above FSP, the additional moisture adds weight to the wood but does not affect the strength. Wood also shrinks and swells as the moisture content varies. With increasing moisture content (again, only up to the fiber saturation point), the fibers swell and all dimensions of the lumber become larger. With reduction in the moisture content, the fibers shrink and all dimensions decrease. With these changes in dimensions, other section properties, such as the moment of inertia, are also affected. This is of particular importance if a member is sawn to standard size when green and then used at 15% moisture. The final member size will be smaller than standard, resulting in a change in the section's load-carrying capacity. Generally, the decrease in section properties as MC decreases from 19% to 15% is offset by increases in wood strength. In addition, if the moisture content of the wood changes greatly after installation, warpage can occur.

The direction of loading, particularly the direction of the primary stress with respect to the direction of the wood grain, influences the strength of wood. In Figs. 8.17(a)–(c), the direction of loading and the primary stress (compressive) are collinear. Of these configurations, the strength of the wood in Fig. 8.17(a) is greatest, whereas it is least in Fig. 8.17(c). In Figs. 8.17(d)–(f), the load is vertical, but the primary stress (flexural) is horizontal. Of these three configurations, the descending order of flexural strength is (d)–(e)–(f). If the slope of the grain to the direction of primary stress is less than 1:12, no significant strength reduction is observed. However, if the slope of grain increases to 1:6, there is a 50–60% reduction in strength.

Member size also significantly influences the load-carrying capacity of wood, because deviations between ideal elastic behavior increase with member size. In flexural members, the design approach assumes $f_b = M/S$ and a triangular stress distribution across the beam section. In reality, wood is not elastic and the stress distribution is not linear. The significance of this discrepancy becomes more pronounced as the beam depth increases; the deeper the beam, the lower its bending strength (stress).

Temperature does not influence wood strength unless temperatures exceed 100°F and are sustained. Members subject only to periodic temperatures between 100° and 150°F experience no permanent loss of strength.

Defects, or wood characteristics such as knots, checks, shakes, and decay, also influence the strength of wood. Strength is reduced by the introduction of stress concentrations and reduction of net sections. The amount of strength reduction depends upon the defect

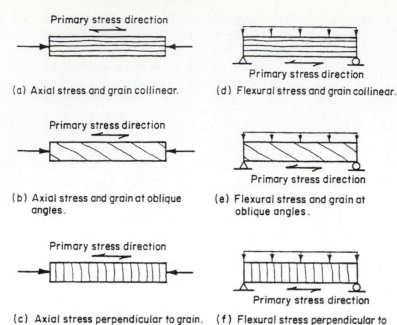

Figure 8.17
Angles between
the Wood Grain
and Primary
Stress

(a) Axial stress and grain collinear.

(b) Axial stress and grain at oblique angles.

(c) Axial stress perpendicular to grain.

(d) Flexural stress and grain collinear.

(e) Flexural stress and grain at oblique angles.

(f) Flexural stress perpendicular to grain.

size and location and upon the type of load. For example, in a simply supported timber beam, knots near the neutral axis would be less critical than those near the outer fiber. Similarly, knots located near the end supports, where the bending moment equals zero, are less critical than those located in the middle half of the span. The wood grading and classification scheme accounts for the influence of wood characteristics on strength properties.

The structural strength of wood is also greatly influenced by the type of stress, i.e., tensile, compressive, flexural, or shear. For example, a No. 1 grade Spruce–Pine–Fir 2×4 at 15% moisture content has the following allowable stresses:

Flexure—F_b	= 1,310 psi
Tension Parallel to Grain—F_t	= 640 psi
Horizontal Shear—F_v	= 70 psi
Compression Parallel to Grain—F_c	= 1,270 psi

The elastic limit and the ultimate strength of wood both increase as the duration of loading decreases. Similarly, deflections and elongations for long-term loads are greater than for short-duration loads of the same magnitude. The design values listed in the NDS®, and most other wood engineering references, are for normal duration of loading. **Normal load duration** is defined as fully stressing a member to its allowable design stress for a period of ten years, either continuously or cumulatively over its expected life.

Figure 8.18 illustrates the influence of load duration upon the relative strength of wood. The design values for normal loading, except modulus of elasticity and $F_{c\perp}$, are adjusted for other durations of loads. When a member is fully stressed to the design value by the application of the full maximum load permanently, or for a total of more than ten years either continuously or for cumulative periods, 90% of the design value for normal loading conditions should be used. Likewise, when the cumulative duration of the full maximum load does not exceed the following durations, all the design values, except E and $F_{c\perp}$, may be increased by:

- 15% for two months duration, as for snow
- 25% for seven days duration (such as for construction loading)
- 60% for wind or earthquake
- 100% for impact

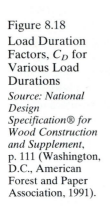

Figure 8.18
Load Duration
Factors, C_D for
Various Load
Durations
*Source: National
Design
Specification® for
Wood Construction
and Supplement,*
p. 111 (Washington,
D.C., American
Forest and Paper
Association, 1991).

GRADING AND ALLOWABLE STRESSES

Structural lumber is either **visually stress rated** (VSR) or **machine stress rated** (MSR). Visually stress rated lumber is assigned allowable stresses based upon the member size and visual observation of quality and defects by an inspector. Machine rated lumber is assigned allowable stress values based upon observed performance of the pieces in a standard deflection-type test.

Visually Stress Rated Lumber

Wood is classified by its size and strength. A simple flow diagram of the sizing, grading, and adjustment scheme used to assign strength values to visually stress rated lumber is shown in Fig. 8.19. The general size classifications are boards, dimension lumber, and timbers. **Boards** include any lumber less than two inches thick, e.g., 1×4's, 1×8's. **Dimension lumber** includes any lumber between two and four inches thick and greater than or equal to two inches wide.

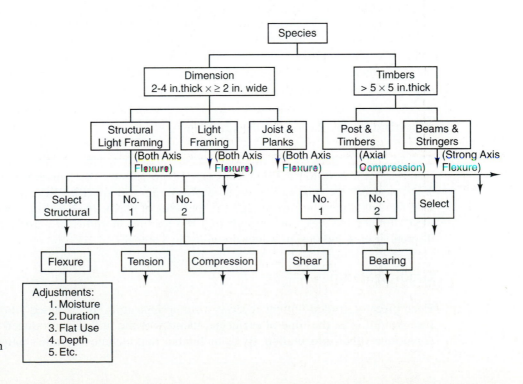

Figure 8.19
Lumber Grading
and Classification
System

Timbers, which include lumber nearly square in cross section and greater than five inches on a side, are graded primarily for use as axial compression members. Boards are generally not stress rated. They are, however, graded for appearance and for sheathing purposes.

Dimension lumber is further classified as structural light framing, light framing, joists and planks, and beams and stringers. **Structural light framing** is relatively high grade lumber which is limited in size to members two to four inches thick and two to four inches wide. **Light framing** is similar except that it is limited to lower quality grades. Both light framing classes are graded with respect to bending about both major axes. **Joists and plank** classes are also graded with respect to bending about both major axes. **Beams and stringers** include members greater than five inches thick and at least two inches wider than they are thick. These members are primarily graded with respect to bending about the strong axis. Within each size classification, structural lumber is graded with respect to lumber density and the number, size, and location of defects. Typical grades within a size include, from highest to lowest quality: (1) dense select structural, (2) select structural, (3) No. 1 dense, (4) No. 1, (5) No. 2 dense, (6) No. 2, (7) No. 3, (8) stud, (9) construction, (10) standard, and (11) utility. The dense grades are only found in Southern Pine lumber; Southern Pine also has special grades: nondense select structural, nondense No. 1, and nondense No. 2. **Nondense** is a grade wherein all dense material has been removed. The size and grade schemes are illustrated schematically in Fig. 8.19.

NDS Allowable Strength Tables

Allowable tensile, flexural, compressive, and shear strengths for design with visually graded lumber are published for each species and grade in the NDS® supplement. The allowable stresses published by NDS® for hardwood, cedar, and red pine dimension lumber, as well as all hardwood and softwood timbers, are equal to the ultimate strength of a green, defect-free specimen times a strength reduction factor (safety factor). The strength reduction factor varies with the grade of lumber, the size of lumber, and the mode of loading. Published allowable stresses are on the order of 1/6 to 1/12, the ultimate strength of a "perfect" straight-grained, defect-free piece of lumber. The allowable stresses for all other dimension softwoods are based on in-grade testing of individual pieces of lumber. Published in-grade stresses are approximately 1/6 the ultimate stress values.

The allowable stresses for visually stress rated spruce–pine–fir, hem–fir, and Southern Pine dimension lumber and timbers are listed in Tables 8.4 and 8.5. The values in Table 8.4 for spruce–pine–fir and hem–fir are for 12-inch-deep members and moisture content less than 19%. The values for Southern Pine in Table 8.5 are for MC < 19% and for the size listed for dimension lumber and for members 12 inches deep for timbers. Adjustments for depths and moisture contents different from those listed are summarized in Table 8.6.

Machine Stress Rated (MSR) Lumber

The allowable stresses for **machine stress rated** dimension lumber range from 900 to 3,300 psi with E values ranging from 1.0×10^6 to 2.6×10^6 psi. The grade designation xxxf–YYE replaces the No. 1, No. 2, etc. designations and specifies the allowable fiber stress, xxx, for single member usage, and the modulus of elasticity, YY. For example, 1800f-1.6E MSR lumber has flexural strength of 1,800 psi and modulus of elasticity of 1.6×10^6 psi. Typical grades and design values are given in Table 8.7.

Grading Stamp

Each piece of graded lumber receives a stamp indicating the assigned grade, species, moisture condition at the time of surfacing, identity of the mill that produced the lumber, and sometimes other information. By using lumber that includes a **grade stamp**, the builder can

TABLE 8.4 Design Values for Visually Stress Rated (VSR) Spruce-Pine-Fir and Hem-Fir Dimension Lumber and Timber

Species and Grade	Size Classification	Design values in pounds per square inch (psi)						Grading Rules Rules Agency
		Bending F_b	Tension parallel to grain F_t	Shear parallel to grain F_v	Compression perpendicular to grain $F_{c\perp}$	Compression parallel to grain F_c	Modulus of Elasticity E	
Spruce–Pine–Fir		Dimension Lumber[a,b,c,d,e]						
Select Structural	2″–4″ thick	1,250	675	70	425	1,400	1,500,000	NLGA
No. 1/No. 2		875	425	70	425	1,100	1,400,000	
No. 3	2″ & wider	500	250	70	425	625	1,200,000	
Stud	2″–4″ thick	675	325	70	425	675	1,200,000	
Construction		975	475	70	425	1,350	1,300,000	
Standard	2″–4″ wide	550	275	70	425	1,100	1,200,000	
Utility		250	125	70	425	725	1,100,000	
Hem–Fir		Dimension Lumber[a,b,c,d,e]						
Select Structural	2″–4″ thick	1,400	900	75	405	1,500	1,600,000	WCLIB WWPA
No. 1 and Better		1,050	700	75	405	1,350	1,500,000	
No. 1		950	600	75	405	1,300	1,500,000	
No. 2	2″ & wider	850	500	75	405	1,250	1,300,000	
No. 3		500	300	75	405	725	1,200,000	
Stud	2″–4″ thick	675	400	75	405	800	1,200,000	
Construction		975	575	75	405	1500	1,300,000	
Standard	2″–4″ wide	550	325	75	405	1300	1,200,000	
Utility		250	150	75	405	850	1,100,000	

Design values in pounds per square inch (psi)

Species and Grade	Size Classification	Bending F_b	Tension parallel to grain F_t	Shear parallel to grain F_v	Compression perpendicular to grain $F_{c\perp}$	Compression parallel to grain F_c	Modulus of Elasticity E	Grading Rules Rules Agency
Spruce–Pine–Fir				Timbers[a,c,f]				
Select Structural	Beams and Stringers	1,100	650	65	425	775	1,300,000	NLGA
No. 1	5″×5″ and larger	900	450	65	425	625	1,300,000	
No. 2		600	300	65	425	425	1,000,000	
Select Structural	Post and Timbers	1,050	700	65	425	800	1,300,000	
No. 1	5″×5″ and larger	850	550	65	425	700	1,300,000	
No. 2		500	325	65	425	500	1,000,000	
Hem–Fir				Timbers[a,c,f]				
Select Structural	Beams and Stringers	1,250	725	70	405	925	1,300,000	WWPA
No. 1	5″×5″ and larger	1,050	525	70	405	775	1,300,000	
No. 2		675	325	70	405	475	1,100,000	
Select Structural	Posts and Timbers	1,200	800	70	405	975	1,300,000	
No. 1	5″×5″ and larger	950	650	70	405	850	1,300,000	
No. 2		525	350	70	405	375	1,100,000	

[a]Normal load duration, MC ≤ 19%; d = 12 in.
[b]Size factor adjustments per Table 8.6a.
[c]Wet service factor adjustment per Table 8.6c (when MC ≥ 19).
[d]Repetitive use factor (C_r = 1.15) applicable to F_b when dimension lumber used as joists, truss chords, rafters, studs, or planks which are in contact or spaced not more than 24 in. apart, are not less than three in number, and are joined by floor, roof or other load-distributing elements adequate to support the design load.
[e]Flat use factor adjustment per Table 8.6b.
[f]When d > 12 in., multiply F_b by C_f = $(12/d)^{1/9}$.

Source: *National Design Specification® for Wood Construction and Supplement.* (American Forest & Paper Association, 1991.)

TABLE 8.5 Design Values for Visually Stress Rated (VSR) Southern Pine Dimension Lumber and Timber

Design values in pounds per square inch (psi)

Dimension Lumber[a,b,c,d,e]

Species and Grade	Size Classification	Bending F_b	Tension parallel to grain F_t	Shear parallel to grain F_v	Compression perpendicular to grain $F_{c\perp}$	Compression parallel to grain F_c	Modulus of Elasticity E	Grading Rules Agency
Southern Pine								
Dense Slct. Struct.	2"–4" thick	3,050	1,650	100	660	2,250	1,900,000	
Select Structural		2,850	1,600	100	565	2,100	1,800,000	
Non-D Slct. Struct.		2,650	1,350	100	480	1,950	1,700,000	
No. 1 Dense		2,000	1,100	100	660	2,000	1,800,000	
No. 1	2"–4" wide	1,850	1,050	100	565	1,850	1,700,000	
No. 1 Non-Dense		1,700	900	100	480	1,700	1,600,000	
No. 2 Dense		1,700	875	90	660	1,850	1,700,000	
No. 2		1,500	825	90	565	1,650	1,600,000	
No. 2 Non-Dense		1,350	775	90	480	1,600	1,400,000	
No. 3		850	475	90	565	975	1,400,000	
Stud		875	500	90	565	975	1,400,000	
Construction	2"–4" thick	1,100	625	100	565	1,800	1,500,000	
Standard		625	350	90	565	1,500	1,300,000	
Utility	4" wide	300	175	90	565	975	1,300,000	
Dense Slct. Struct.		2,700	1,500	90	660	2,150	1,900,000	SPIB
Select Structural		2,550	1,400	90	565	2,000	1,800,000	
Non-D Slct. Struct.		2,350	1,200	90	480	1,850	1,700,000	
No. 1 Dense	2"–4" thick	1,750	950	90	660	1,900	1,800,000	
No. 1		1,650	900	90	565	1,750	1,700,000	
No. 1 Non-Dense		1,500	800	90	480	1,600	1,600,000	
No. 2 Dense	5"–6" wide	1,450	775	90	660	1,750	1,700,000	
No. 2		1,250	725	90	565	1,600	1,600,000	
No. 2 Non-Dense		1,150	675	90	480	1,500	1,400,000	
No. 3		750	425	90	565	925	1,400,000	
Stud		775	425	90	565	925	1,400,000	
Dense Slct. Struct.		2,450	1,350	90	660	2,050	1,900,000	
Select Structural		2,300	1,300	90	565	1,900	1,800,000	
Non-D Slct. Struct.		2,100	1,100	90	480	1,750	1,700,000	
No. 1 Dense	2"–4" thick	1,650	875	90	660	1,800	1,800,000	
No. 1		1,500	825	90	565	1,650	1,700,000	
No. 1 Non-Dense	8" wide	1,350	725	90	480	1,550	1,600,000	
No. 2 Dense		1,400	675	90	660	1,700	1,700,000	
No. 2		1,200	650	90	565	1,550	1,600,000	
No. 2 Non-Dense		1,100	600	90	480	1,450	1,400,000	
No. 3		700	400	90	565	875	1,400,000	

Design values in pounds per square inch (psi)

Species and Grade	Size Classification	Bending F_b	Tension parallel to grain F_t	Shear parallel to grain F_v	Compression perpendicular to grain $F_{c\perp}$	Compression parallel to grain F_c	Modulus of Elasticity E	Grading Rules Agency
Dense Slct. Struct.		2,150	1,200	90	660	2,000	1,900,000	
Select Structural		2,050	1,100	90	565	1,850	1,800,000	
Non-D Slct. Struct.		1,850	950	90	480	1,750	1,700,000	
No. 1 Dense	2″–4″ thick	1,450	775	90	660	1,750	1,800,000	
No. 1		1,300	725	90	565	1,600	1,700,000	
No. 1 Non-Dense	10″ wide	1,200	650	90	480	1,500	1,600,000	
No. 2 Dense		1,200	625	90	660	1,650	1,700,000	
No. 2		1,050	575	90	565	1,500	1,600,000	
No. 2 Non-Dense		950	550	90	480	1,400	1,400,000	
No. 3		600	325	90	565	850	1,400,000	SPIB
Dense Slct. Struct.		2,050	1,100	90	660	1,950	1,900,000	
Select Structural		1,900	1,050	90	565	1,800	1,800,000	
Non-D Slct. Struct.		1,750	900	90	480	1,700	1,700,000	
No. 1 Dense	2″–4″ thick	1,350	725	90	660	1,700	1,800,000	
No. 1		1,250	675	90	565	1,600	1,700,000	
No. 1 Non-Dense	12″ wide	1,150	600	90	480	1,500	1,600,000	
No. 2 Dense		1,150	575	90	660	1,600	1,700,000	
No. 2		975	550	90	565	1,450	1,600,000	
No. 2 Non-Dense		900	525	90	480	1,350	1,400,000	
No. 3		575	325	90	565	825	1,400,000	

Southern Pine — (Wet Service Conditions) Timbers[a,f,g]

Species and Grade	Size Classification	Bending F_b	Tension parallel to grain F_t	Shear parallel to grain F_v	Compression perpendicular to grain $F_{c\perp}$	Compression parallel to grain F_c	Modulus of Elasticity E	Grading Rules Agency
D Slct Struct		1,750	1,200	110	440	1,100	1,600,000	
Select Struct		1,500	1,000	110	375	950	1,500,000	
No. 1 Dense		1,550	1,050	110	440	975	1,600,000	
No. 1	5′×5″ and	1,350	900	110	375	825	1,500,000	
No. 2 Dense	larger	975	650	100	440	625	1,300,000	
No. 2		850	550	100	375	525	1,200,000	
Dense Struct. 86		2,100	1,400	145	440	1,300	1,600,000	
Dense Struct. 72		1,750	1,200	120	440	1,100	1,600,000	
Dense Struct. 65		1,600	1,050	110	440	1,000	1,600,000	SPIB

[a] Normal load duration.
[b] For depth (width) listed.
[c] For MC ≤ 19%; multiply by C_m values in Table 8.6c when MC > 19%.
[d] Use C_r = 1.15 if footnote 4, Table 8.4, conditions are satisfied.
[e] Use flat use factor per Table 8.6b.
[f] For depth ≤ 12 in.; for $d \geq$ 12 in., multiply F_b by $C_F = (12/d)^{1/9}$.
[g] No further adjustment for C_m required (MC > 19%).

Source: *National Design Specification® for Wood Construction and Supplement* (American Forest & Paper Association, 1991.)

TABLE 8.6   Adjustment Factors for Base Design Values in Tables 8.4 and 8.5

Table 8.6a Size Factor—C_F: Tabulated bending, tension, and compression parallel to grain design values in Table 8.4 and dimension lumber 2″ to 4″ thick shall be multiplied by the following size factors:

Grades	Width	F_b Thickness 2″ & 3″	F_b Thickness 4″	F_t	F_c
	2″, 3″, & 4″	1.5	1.5	1.5	1.15
	5″	1.4	1.4	1.4	1.1
Select Structural	6″	1.3	1.3	1.3	1.1
No. 1 & Btr.	8″	1.2	1.3	1.2	1.05
No. 1, No. 2, No. 3	10″	1.1	1.2	1.1	1.0
	12″	1.0	1.1	1.0	1.0
	14″ & wider	0.9	1.0	0.9	0.9
Stud	2″, 3″, & 4″	1.1	1.1	1.1	1.05
	5″ & 6″	1.0	1.0	1.0	1.0
Construction & Standard	2″, 3″, & 4″	1.0	1.0	1.0	1.0
Utility	4″	1.0	1.0	1.0	1.0
	2″ & 3″	0.4	—	0.4	0.6

Table 8.6b Flat Use Factor—C_{fu}: Bending design values adjusted by size factors are based on edgewise use (load applied to narrow face). When dimension lumber is used flatwise (load applied to wide face), the bending design value, F_b, shall also be multiplied by the following flat use factors:

Width	Thickness 2″ & 3″	Thickness 4″
2″ & 3″	1.0	—
4″	1.1	1.0
5″	1.1	1.05
6″	1.15	1.05
8″	1.15	1.05
10″ & wider	1.2	1.1

Table 8.6c Wet Service Factor, C_M: When dimension lumber is used where moisture content will exceed 19% for an extended time period, design values shall be multiplied by the appropriate wet service factors from the following table:

F_b	F_t	F_v	$F_{c\perp}$	F_c	E
0.85*	1.0	0.97	0.67	0.8**	0.9

*when $(F_b)(C_F) \leq 1150$ psi, $C_M = 1.0$ **when $(F_b)(C_F) \leq 750$ psi, $C_M = 1.0$

Source: *National Design Specification® for Wood Construction and Supplement.* (American Forest & Paper Association, 1991).

be assured of getting the specified quality of lumber, and the owner or building inspector can be assured that the material actually being furnished conforms to the construction specifications.

An example of a stamp for visually stress rated lumber is illustrated in Fig. 8.20. The numeral 12 identifies the mill that produced the lumber. The logo in the circle certifies that the grading was done under WWPA (Western Wood Products Association) rules and supervision. The species group (i.e., Douglas Fir) is displayed in the triangle on the lower right. The label S-DRY on the upper right indicates that the wood was dry (seasoned) when it was surfaced. The large number in the center indicates that the piece bearing this stamp is of Grade No. 2.

TABLE 8.7 Design Values for Selected Machine Stress Rated (MSR) Lumber[a,b,c,d]

Species and Commercial Grade	Size Classification	Design values in pounds per square inch (psi)				Grading Rules Agency
		Bending F_b	Tension parallel to grain F_t	Compression parallel to grain F_c	Modulus of Elasticity E	
Machine Stress Rated (MSR) Lumber[a,b,c,d]						
900f-1.0E		900	350	1,050	1,000,000	WCLIB
1,200f-1.2E		1,200	600	1,400	1,200,000	NLGA, SPIB, WCLIB, WWPA
1,350f-1.3E		1,350	750	1,600	1,300,000	SPIB, WCLIB, WWPA, NLGA
1,450f-1.3E		1,450	800	1,625	1,300,000	NLGA, WCLIB, WWPA
1,500f-1.3E		1,500	900	1,650	1,300,000	SPIB
1,500f-1.4E		1,500	900	1,650	1,400,000	NLGA, SPIB, WCLIB, WWPA
1,650f-1.4E		1,650	1,020	1,700	1,400,000	SPIB
1,650f-1.5E		1,650	1,020	1,700	1,500,000	NLGA, SPIB, WCLIB, WWPA
1,800f-1.6E		1,800	1,175	1,750	1,600,000	NLGA, SPIB, SCLIB, WWPA
1,950f-1.5E	2″ & less in thickness	1,950	1,375	1,800	1,500,000	SPIB
1,950f-1.7E		1,950	1,375	1,800	1,700,000	NLGA, SPIB, WWPA
2,100f-1.8E		2,100	1,575	1,875	1,800,000	NLGA, SPIB, WCLIB, WWPA
2,250f-1.6E	2″ & wider	2,250	1,750	1,925	1,600,000	SPIB
2,250f-1.9E		2,250	1,750	1,925	1,900,000	NLGA, SPIB, WWPA
2,400f-1.7E		2,400	1,925	1,975	1,700,000	SPIB
2,400f-2.0E		2,400	1,925	1,975	2,000,000	NLGA, SPIB, WCLIB, WWPA
2,550f-2.1E		2,550	2,050	2,025	2,100,000	NLGA, SPIB, WWPA
2,700f-2.2E		2,700	2,150	2,100	2,200,000	NLGA, SPIB, WCLIB, WWPA
2,850f-2.3E		2,850	2,300	2,150	2,300,000	SPIB, WWPA
3,000f-2.4E		3,000	2,400	2,200	2,400,000	NLGA, SPIB
3,150f-2.5E		3,150	2,500	2,250	2,500,000	SPIB
3,300f-2.6E		3,300	2,650	2,325	2,600,000	SPIB
900f-1.2E	2″ & less in thickness	900	350	1,050	1,200,000	NLGA, WCLIB
1,200f-1.5E		1,200	600	1,400	1,500,000	NLGA, WCLIB
1,500f-1.8E		1,500	900	1,650	1,800,000	WCLIB
1,800f-2.1E	6″ & wider	1,800	1,175	1,750	2,100,000	NLGA, WCLIB

[a]For normal load duration and MC < 19%.
[b]Use flatwise use adjustment factors for 2″ and 3″ thick in Table 8.6b.
[c]Use wet service factor in Table 8.6c.
[d]Use C_r = 1.15 for F_b if conditions in footnote 4, Table 8.4, are satisfied.

Source: *National Design Specification® for Wood Construction and Supplement*, p. 29 (American Forest & Paper Association, 1991).

Figure 8.20
Grading Stamp for Visually
Graded Lumber
*Source: Western Lumber Grading
Rules*, p. 25 (Portland, OR: Western
Wood Products Association, 1988).

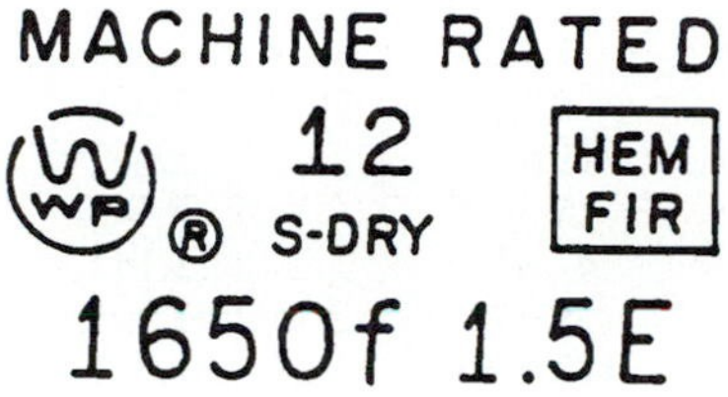

Figure 8.21
Grading Stamp for Machine
Stress Rated Lumber
*Source: Western Lumber
Grading Rules*, p. 28 (Portland,
OR: Western Wood Products
Association, 1988).

Fig. 8.21 shows an example grade stamp for machine stress rated lumber. This stamp indicates which mill produced and graded the lumber, under what agency rules and supervision it was graded, and its moisture content when it was surfaced. The species, allowable stress, and modulus of elasticity are also shown. In this example, E is 1.5 million psi, and F_b is 1,650 psi.

STANDARD SIZES AND SECTION PROPERTIES

Wood member cross sections may be either "full" or "dressed" size. For example, a full-size 2×4 has cross-sectional dimensions of two and four inches, whereas a dressed 2×4 has dimensions of $1\frac{1}{2}$ inches and $3\frac{1}{2}$ inches. In either case, the nominal size is 2×4. To identify a member accurately, both the nominal size and type of section must be identified. The majority of structural lumber is dressed.

The most common nominal sizes of lumber are 2, 4, 6, 8, 10, and 12 inches. The dimensions, section properties, and weights of the more common dressed lumber sizes are listed in Table 8.8. The section moduli and moments of inertia are given with respect to the cross-sectional axis parallel to the b-dimension in Column 1. That is, the section modulus of a dressed 2×4 equals 3.063 in^3, whereas the section modulus of a 4×2 equals 1.313 in^3. When a flexural load is applied to the normal face, bending takes place about the strong axis and the appropriate section modulus is that of a 2×4. This section is termed "loaded on edge." On the other hand, when the load is applied to the wide face, bending occurs about the weak axis and the section modulus for a 4×2 is used. Such an application is termed "loaded flatwise."

TABLE 8.8   Section Properties of Dressed Lumber

Nominal Size b (in.) d	Standard Dressed Size (S4S) b (in.) d	Area of Section A (in²)	Moment of Inertia I (in⁴)ᵃ	Section Modulus S (in³)ᵃ	Mass in Pounds Per Linear Foot of Piece when Density of Wood per Cubic Foot Equals:	
					30 lbs	35 lbs
1 × 3	.75 × 2.5	1.875	0.977	0.781	0.391	0.456
1 × 4	.75 × 3.5	2.625	2.680	1.531	0.547	0.638
1 × 6	.75 × 5.5	4.125	10.398	3.781	0.859	1.003
1 × 8	.75 × 7.25	5.438	23.817	6.570	1.133	1.322
1 × 10	.75 × 9.25	6.938	49.466	1.695	1.445	1.686
1 × 12	.75 × 11.25	8.438	88.989	15.820	1.758	2.051
2 × 3	1.5 × 2.5	3.750	1.953	1.563	0.781	0.911
2 × 4	1.5 × 3.5	5.250	5.359	3.063	1.094	1.276
2 × 5	1.5 × 4.5	6.750	11.391	5.063	1.406	1.641
2 × 6	1.5 × 5.5	8.250	20.797	7.563	1.719	2.005
2 × 8	1.5 × 7.25	10.875	47.635	13.141	2.266	2.643
2 × 10	1.5 × 9.25	13.875	98.932	21.391	2.891	3.372
2 × 12	1.5 × 11.25	16.875	177.979	31.641	3.516	4.102
2 × 14	1.5 × 13.25	19.875	290.775	43.891	4.141	4.831
3 × 1	2.5 × .75	1.875	0.088	0.234	0.391	0.456
3 × 2	2.5 × 1.5	3.750	0.703	0.938	0.781	0.911
3 × 4	2.5 × 3.5	8.750	8.932	5.104	1.823	2.127
3 × 5	2.5 × 4.5	11.250	18.984	8.438	2.344	2.734
3 × 6	2.5 × 5.5	13.750	34.661	12.604	2.865	3.342
3 × 8	2.5 × 7.25	18.125	79.391	21.901	3.776	4.405
3 × 10	2.5 × 9.25	23.125	164.886	35.651	4.818	5.621
3 × 12	2.5 × 11.25	28.125	296.631	52.734	5.859	6.836
4 × 1	3.5 × .75	2.625	0.123	0.328	0.547	0.638
4 × 2	3.5 × 1.5	5.250	0.984	1.313	1.094	1.276
4 × 3	3.5 × 2.5	8.750	4.557	3.646	1.823	2.127
4 × 4	3.5 × 3.5	12.250	12.505	7.146	2.552	2.977
4 × 5	3.5 × 4.5	15.750	26.578	11.813	3.281	3.838
4 × 6	3.5 × 5.5	19.250	48.526	17.646	4.010	4.679
4 × 8	3.5 × 7.25	25.375	111.148	30.661	5.286	6.168
4 × 10	3.5 × 9.25	32.375	230.840	49.911	6.745	7.869
4 × 12	3.5 × 11.25	39.375	415.283	73.828	8.203	9.570
6 × 1	5.5 × .75	4.125	0.193	0.516	0.859	1.003
6 × 2	5.5 × 1.5	8.250	1.547	2.063	1.719	2.005
6 × 3	5.5 × 2.5	12.750	7.161	5.729	2.865	3.342
6 × 4	5.5 × 3.5	19.250	19.651	11.229	4.010	4.679
6 × 6	5.5 × 5.5	30.250	76.255	27.729	6.302	7.352
6 × 8	5.5 × 7.5	41.250	193.359	51.563	8.594	10.026
6 × 10	5.5 × 9.5	52.250	392.963	82.729	10.885	12.700
6 × 12	5.5 × 11.5	63.250	697.068	121.229ᵃ	13.177	15.373
8 × 1	7.25 × .75	5.438	0.255	0.680	1.133	1.322
8 × 2	7.25 × 1.5	10.875	2.039	2.719	2.266	2.643
8 × 3	7.25 × 2.5	18.125	9.440	7.552	3.776	4.405
8 × 4	7.25 × 3.5	25.375	25.904	14.803	5.286	6.168
8 × 6	7.5 × 5.5	41.250	103.984	37.813	8.594	10.026
8 × 8	7.5 × 7.5	56.250	263.672	70.313	11.719	13.672
8 × 10	7.5 × 9.5	71.250	535.859	112.813	14.844	17.318
8 × 12	7.5 × 11.5	86.250	950.547	165.313	17.969	20.964

(continued)

TABLE 8.8   *(continued)*

Nominal Size b (in.) d	Standard Dressed Size (S4S) b (in.) d	Area of Section A (in²)	Moment of Inertia I (in⁴)[a]	Section Modulus S (in³)[a]	Mass in Pounds Per Linear Foot of Piece when Density of Wood per Cubic Foot Equals:	
					30 lbs	35 lbs
10 × 1	9.25 × .75	6.938	0.325	0.867	1.445	1.686
10 × 2	9.25 × 1.5	13.875	2.602	3.469	2.891	3.372
10 × 3	9.25 × 2.5	23.125	12.044	9.635	4.818	5.621
10 × 4	9.25 × 3.5	32.275	33.049	18.885	6.745	7.869
10 × 6	9.5 × 5.5	52.250	131.714	47.896	10.885	12.700
10 × 8	9.5 × 7.5	71.250	333.984	89.063	14.844	17.318
10 × 10	9.5 × 9.5	90.250	678.755	142.896	18.802	21.936
10 × 12	9.5 × 11.5	109.250	1,204.026	209.396	22.760	26.554
12 × 1	11.25 × .75	8.438	0.396	1.055	1.758	2.051
12 × 2	11.25 × 1.5	16.875	3.164	4.219	3.516	4.102
12 × 3	11.25 × 2.5	28.125	14.648	11.719	5.859	6.836
12 × 4	11.25 × 3.5	39.375	40.195	22.969	8.203	9.670
12 × 6	11.5 × 5.5	63.250	159.443	57.979	13.177	15.373
12 × 8	11.5 × 7.5	86.250	404.297	107.813	17.969	20.964
12 × 10	11.5 × 9.5	109.250	821.651	172.979	22.760	26.554
12 × 12	11.5 × 11.5	132.250	1457.505	253.479	27.552	32.144

[a]Section property for axis parallel to the b-dimension.

Source: *National Design Specification*® *for Wood Construction and Supplement*, pp. 10–11 (American Forest & Paper Association, 1991).

CHAPTER SUMMARY

This chapter has provided an overview of wood with regard to its use in residential construction. The attention was mainly focused on the structural properties of wood that are appropriate for design and construction purposes.

NOTES

[1]*National Design Specification for Wood Construction* (Washington, D.C.: American Forest and Paper Products Association [AF&PA], 1991).

[2]*Supplement to the National Design Specification for Wood Construction* (Washington, D.C.: AF&PA, 1991).

[3]*Wood Handbook: Wood as an Engineering Material.* Agricultural Handbook No. 72, p. 3.11 (Washington, D.C.: United States Department of Agriculture, Forest Service, 1987).

BIBLIOGRAPHY

Dagostino, Frank R. *Residential Costruction Handbook.* Reston, VA: Reston Publishing Company, Inc., 1983.

National Design Specification for Wood Construction. Washington, D.C.: American Forest and Paper Products Association (AF&PA), 1991.

Supplement to the National Design Specification for Wood Construction, Supplement. Washington, D.C.: AF&PA, 1991.

Western Lumber Grading Rules. Portland, OR: Western Wood Products Association, 1988.

Wood Handbook: Wood as an Engineering Material. Agricultural Handbook No. 72. Washington, D.C.: United States Department of Agriculture, Forest Service, 1987.

9

Foundation and Floor System Requirements

INTRODUCTION

This chapter is part of a four-chapter sequence which starts with an introduction to the basic structural characteristics of wood in Chapter 8 and ends with an engineering-level analysis of the design of wood structural systems in Chapter 11. The types of wood structural framing systems that are commonly used in the United States and the design process that governs such systems are introduced first. The primary emphasis in this chapter is placed on the design of the foundation and floor systems. Chapter 10 provides an understanding of the remaining structural systems in a house.

The primary building code context for the design criteria is the 1995 edition of the *CABO One and Two Family Dwelling Code*,[1*] hereinafter called the 1995 CABO Code. The **Building Section** of that code[2] includes:

Chapter 2: Building Definitions
Chapter 3: Building Planning
Chapter 4: Foundations
Chapter 5: Floors
Chapter 6: Wall Construction
Chapter 7: Wall Covering
Chapter 8: Roof Ceiling Construction
Chapter 9: Roof Coverings
Chapter 10: Chimneys and Fireplaces

Chapter 2, "Building Definitions",[3] provides a brief, two-page listing of the general building definitions that are commonly used. The parts of Chapter 3, "Building Planning",[4] that deal with requirements that affect the architectural design of the house were covered

---

[*]Sections, figures, and tables from this Code are cited for instructional purposes only, and the treatment of the Code provisions contained in the design solutions illustrated in this chapter are not intended as a substitute for a complete and thorough understanding of all building code provisions to which any particular project may be subject.

in Chapter 5 of this book. The remainder of Chapter 3, Chapter 4, "Foundations", and Chapter 5, "Floors", are covered in this chapter.

STRUCTURAL FRAMING SYSTEMS

Based upon the broad range of housing types and styles presented in Chapter 5 of this book, it initially appears that each house design requires its own unique structural framing system. Although this may be true for a small percentage of houses, usually in the custom or luxury classes, most can be designed using a **conventional** wood structural framing system. If it can be determined that the proposed framing system for a particular house is conventional, then the adequacy of the system can be directly determined by the homebuilder or the building code official by referring to the specific requirements and structural loading tables provided in the 1995 CABO Code.

In the situation where the proposed framing system is judged to be **nonconventional**, perhaps because of some innovation that is not directly covered in the Code, the 1995 CABO Code requires the homebuilder or designer to provide supportive design and testing documentation before a structural adequacy decision can be made by the building code official.[5]

Conventional Wood Structural Framing Systems

The 1995 CABO Code recognizes the **platform framing** system and the **balloon framing** system as **conventional framing systems** that do not require additional design and testing documentation (Fig. 9.1). A more detailed presentation of the two conventional framing systems (Figs. 9.2 and 9.3) indicates that both systems use the same types of primary structural elements. One of the major differences in the two systems is associated with the type of wall studs that are used.

As mentioned briefly in Chapter 7, the exterior wall studs in the platform framing system (Fig. 9.2) only extend between the floor platforms. As a result, the load-bearing stud walls between the first- and second-floor platforms can be installed after the first-floor platform has been placed on the foundation walls and the main basement girder. The completed first-floor platform serves as a convenient and safe working surface for the fabrication and erection of the stud walls between the first and second floors. These stud walls then provide support for the floor joist/sheathing system which constitutes the second-floor platform. The process is then repeated for the load-bearing stud walls between the second-floor platform and the roof.

The balloon framing system (Fig. 9.3) is designed with wall studs that extend from the sill plate on the foundation wall all the way to the double plate at the top of the wall that supports the roof. A 1×4 in. ribbon strip (Fig. 9.3), or some other alternative, is required to support the second-floor joist system. As a result, the balloon framing system is not as easy to install as the platform framing system. In addition, the platform framing system eliminates the need for the firestopping that is required in the balloon framing system to inhibit the spread of fire between floors through the wall system.

Platform Frame Options

A number of different options are available when the platform framing system is adopted. Several of these are presented in order to indicate the framing flexibility which is possible.

Roof Rafter/Ceiling Joist Option

An elevation view of the cross section of a house that is designed in accordance with the platform framing approach is shown in Fig. 9.4. The primary roof structural system which

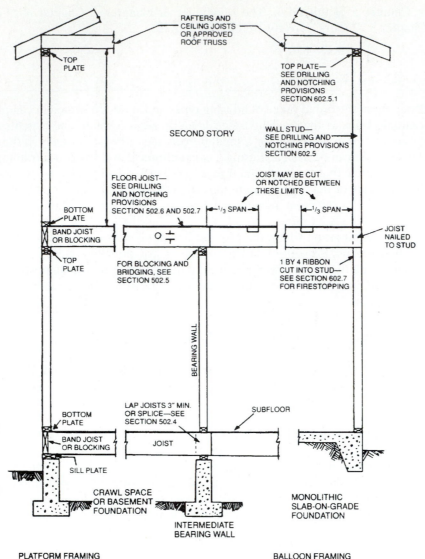

RAFTERS AND
CEILING JOISTS
OR APPROVED
ROOF TRUSS

TOP
PLATE

TOP PLATE—
SEE DRILLING
AND NOTCHING
PROVISIONS
SECTION 602.5.1

SECOND STORY

WALL STUD—
SEE DRILLING AND
NOTCHING PROVISIONS
SECTION 602.5

FLOOR JOIST—
SEE DRILLING
AND NOTCHING
PROVISIONS
SECTION 502.6 AND 502.7

JOIST MAY BE CUT
OR NOTCHED BETWEEN
THESE LIMITS

BOTTOM
PLATE

1/3 SPAN

1/3 SPAN

BAND JOIST
OR BLOCKING

JOIST
NAILED
TO STUD

TOP
PLATE

FOR BLOCKING AND
BRIDGING, SEE
SECTION 502.5

1 BY 4 RIBBON
CUT INTO STUD—
SEE SECTION 602.7
FOR FIRESTOPPING

BEARING WALL

Figure 9.1
"Conventional"
Wood Structural
Framing Systems
Recognized by
the 1995 CABO
Code
Source: Reproduced
from the 1995 edi-
tion of the *CABO
One and Two
Family Dwelling
Code,*™ copyright ©
1995, with the per-
mission of the pub-
lisher, The
International Codes
Council, Whittier,
CA, Fig. 602.3a,
p. 67.

BOTTOM
PLATE

LAP JOISTS 3" MIN.
OR SPLICE—SEE
SECTION 502.4

SUBFLOOR

BAND JOIST
OR BLOCKING

JOIST

SILL PLATE

CRAWL SPACE
OR BASEMENT
FOUNDATION

MONOLITHIC
SLAB-ON-GRADE
FOUNDATION

INTERMEDIATE
BEARING WALL

PLATFORM FRAMING

BALLOON FRAMING

NOTE: See Figure 403.1a for other foundation types

carries the external combination of snow load, live load, and wind load, in conjunction with the roof dead load, consists of: (1) roof rafters, (2) a ridge board, and (3) roof sheathing. The roof rafters are typically braced with collar ties that are installed at the top one-third of the attic space and spaced four feet apart.[6]

A structural system, such as the one shown in Fig. 9.5(a), would fail if only the roof rafters carried the external roof loads to the exterior bearing walls. This failure would occur because the exterior bearing walls would not, by themselves, be able to resist the horizontal thrust applied to them due to the load carried by the roof rafters.

When the ceiling joists and the collar ties are included as a part of the roof structural system, the load imposed on the roof can be satisfactorily carried. In such a system the ceiling joists serve the same purpose as the bottom chord of a truss. In areas where roof assemblies are subject to wind uplift pressures of 20 pounds per square foot or greater, the roof rafters must be tied down to the exterior walls with either metal straps or anchors.[7]

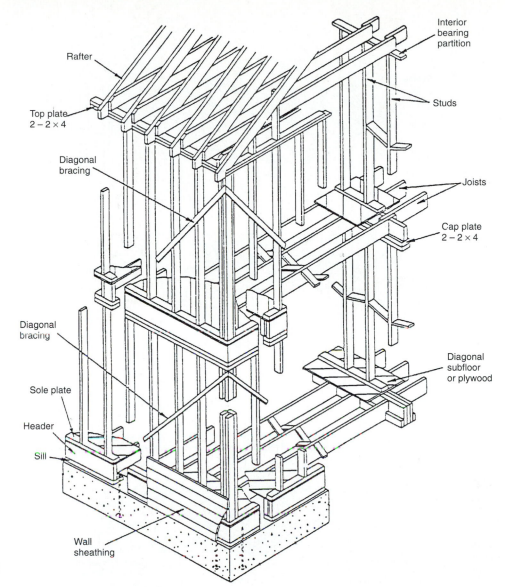

Figure 9.2
The Platform
Framing System
Source: Reiner, J.,
Methods and
Materials of
Residential
Construction,
©1981, p. 131.
Reprinted by per-
mission of Prentice
Hall, Upper Saddle
River, New Jersey.

Another important aspect of the platform framing system is that both the exterior as well as the interior walls shown in Fig. 9.4 are considered to be load bearing. The exterior bearing walls carry their portion of the total vertical load down through the foundation wall to the undisturbed soil beneath the house. Part of that load, as shown in Fig. 9.6, is due to the weight of the wall itself; the remainder is due to the portions of the roof and floor loads which are transmitted to the exterior bearing walls by the roof rafters, ceiling joists, and floor joists.

The interior bearing wall plays a similar role in that it carries the portion of the floor load that is transmitted to it by the combination of ceiling and floor joists feeding in from both the left and right. These loads, as well as the weight of the interior bearing walls on each floor, are ultimately transmitted to a main house girder which spans over a series of columns (Fig. 9.7). The ends of the girder are supported by the foundation wall at the ends of the house.

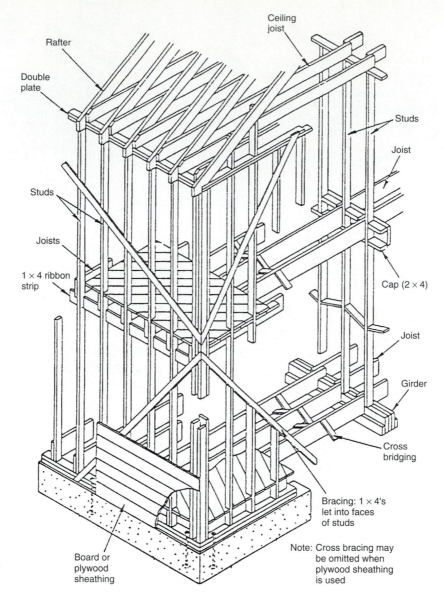

Figure 9.3
The Balloon
Framing System
Source: Reiner, J.,
*Methods and
Materials of
Residential
Construction,*
©1981, p. 134.
Reprinted by per-
mission of Prentice
Hall, Upper Saddle
River, New Jersey.

Ceiling
joist

Rafter

Double
plate

Studs

Joist

Studs

Joists

Cap (2 × 4)

1 × 4 ribbon
strip

Joist

Girder

Cross
bridging

Bracing: 1 × 4's
let into faces
of studs

Board or
plywood
sheathing

Note: Cross bracing may
be omitted when
plywood sheathing
is used

Engineered Roof Truss Options

One of the technological innovations which has been widely accepted by homebuilders is the pre-engineered roof truss alternative to the roof rafter/ceiling joist system. Figure 9.8 indicates the platform framing system which incorporates roof trusses. In addition to reducing the erection time of the roof, one of the other advantages achieved is that the interior bearing wall between the second floor and the truss can be replaced by a non-bearing wall. This is possible because the roof truss can be designed to span between the exterior bearing walls without exceeding allowable deflection requirements.

The 1995 CABO Code permits the use of roof trusses.[8] However, structural tables which can be used to design or verify the structural adequacy of roof trusses are not included in the 1995 CABO Code. The design is usually verified by a licensed professional engineer

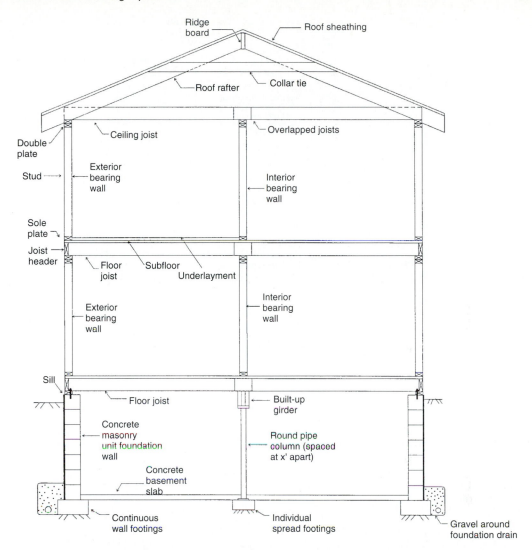

Figure 9.4
Primary
Structural
Elements of a
Basic Platform
Framing System:
Roof Rafter/
Ceiling Joist
Option

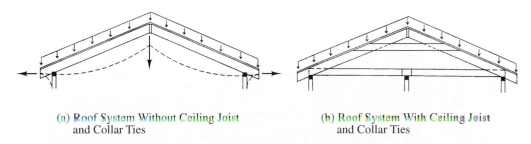

Figure 9.5
Structural
Behavior of Roof
Systems

(a) Roof System Without Ceiling Joist
and Collar Ties

(b) Roof System With Ceiling Joist
and Collar Ties

representing the truss manufacturer stamping the truss drawings. Guidance about the design of engineered roof trusses is published by the Wood Truss Council of America.[9]

Engineered Roof Truss/Floor Truss Option

The 1995 CABO Code also permits the use of pre-engineered floor trusses in place of conventional floor joists.[10] Once again, a licensed professional engineer must verify the adequacy of such systems by placing a seal on the truss drawings.

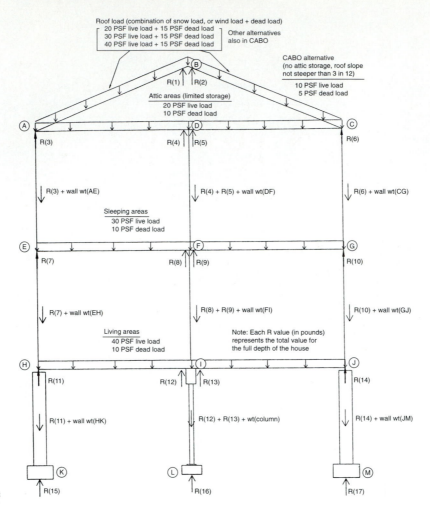

Figure 9.6
Distribution of
Forces through
the Structural
Frame of a House

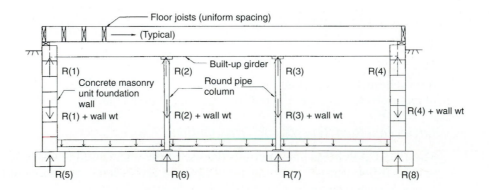

Figure 9.7
Main House
Girder Support
System

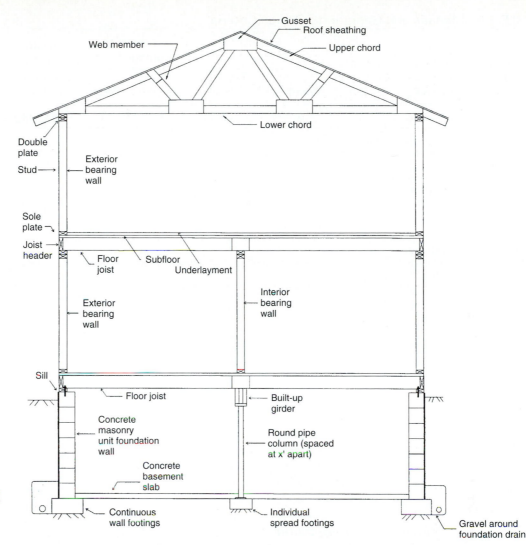

Figure 9.8
Primary
Structural
Elements of a
Basic Platform
Framing System:
Engineered Wood
Roof Truss
Option

DESIGN CRITERIA

When the working drawings for a house are developed, consideration must be given to: (1) designing the house to resist climatic and natural forces as well as the loads incurred in the house, and (2) ensuring the safety and habitation of the homeowner through proper interior architectural design. The 1995 CABO Code provides minimum design and review guidelines in both of these areas for architects, homebuilders, building code officials, and homeowners.[11] With regard to the first area of consideration, requirements related to: (1) climatic and geographic design criteria, (2) dead load, (3) live load, (4) roof load, and (5) deflection are provided.

Climatic and Geographic Design Criteria

The climatic and geographic design criteria for a particular house depend upon where the house is located in the United States. Table 9.1 identifies the climatic and design criteria that must be established by the local governmental jurisdiction.

TABLE 9.1 Climatic and Geographic Design Criteria

Roof Snow Load (pounds per S.F.)	Wind Pressure[a] (pounds per S.F.)	Seismic Condition by Zone	Subject to Damage from[b]		Subject to Damage from[c]		Winter Design[d] Temp. for Htg. Facilities	Radon-Resistant Construction Required[e]
			Weathering	Frost Line Depth	Termite	Decay		
30 psf[f]	70 mph	1	severe	36 in.	Yes	Yes	7°F (−14°C)	No

For SI: 1 pound per S.F. (psf) = 0.0479 kN/m².

[a]The jurisdiction shall fill in this part of the table with wind design loads determined from the Wind Probability Map . . . .

[b]Weathering may require a higher strength concrete or grade of masonry than necessary to satisfy the structural requirements of this code. The weathering column shall be filled in with the weathering index (i.e., "negligible," "moderate," or "severe") for concrete as determined from the Weathering Probability Map . . . . The grade of masonry units shall be determined from ASTM C 34, C 55, C 62, C 73, C 90, C 129, C 145, C 216, or C 652. The frost line depth may require deeper footings than indicated in Figure 403.1a. The jurisdiction shall fill in the frost line depth column with "yes" or "no" including the minimum depth of footing below finish grade.

[c]The jurisdiction shall fill in this part of the table with "yes" or "no" depending on whether there has been a history of local damage.

[d]If heating facilities are not required in this climate, enter "None Required."

[e]The jurisdiction in areas of high radon potential, as indicated by Zone 1 on the U.S. EPA Map of Radon Zones . . . or as determined from other locally available data, shall fill in this part of the table with "yes."

[f]Indicated requirements are for Centre County, PA.

Eight maps which indicate representative values for each of these criteria throughout the United States are provided in the 1995 CABO Code. The representative values shown in Table 9.1 are applicable in the region under the jurisdiction of the Centre Region Code Enforcement Office, which is located in Centre County, Pennsylvania.

Roof Load Determination

The roof of the house, according to the 1995 CABO Code,[12] must be designed to carry the **larger** of the following two loads: (1) a specified minimum live load (Table 9.2) which accounts for the live loads imposed on the roof during the construction period or at a later time, and (2) the snow load specified in Table 9.1.

The design roof load for the Case Study House is determined next to illustrate the procedure.

LIVE LOAD

The minimum live load on the roof, in areas of the United States where snow load is not a problem, is specified in Table 9.2. Both the slope of the roof and the tributary loaded area (which depends on the spacing of either the roof rafters or the roof trusses) must be considered when this value is determined.

For the "Pauline" Model Case Study House (Appendix A),* according to Sheet A-5— Framing Section A-A:

Roof slope = 6 in 12 (26.6°)
Rafter length = 15/cos 26.6° = 16.8 ft
Spacing of trusses = 2.0 ft
Tributary area = 16.8(2.0) = 33.6 S.F.
Minimum live load = 16 psf (Table 9.2)

This uniform live load of 16 psf is assumed to act on **the horizontal projection** of the roof.

TABLE 9.2   Minimum Roof Live Loads in
Pounds–Force per Square Foot of Horizontal
Projection

| Roof Slope | Tributary Loaded Area in Square Foot for Any Structural Member | | |
	0 to 200	201 to 600	Over 600
Flat or rise less than 4 ins. per ft (1:3)	20	16	12
Rise 4 ins. per foot (1:3) to less than 12 ins. per foot (1:1)	16	14	12
Rise 12 ins. per foot (1:1) and greater	12	12	12

For SI: 1 square foot = 0.0929 m², 1 psf = 0.0479 kN/m².

Source: Reproduced from the 1995 edition of the *CABO One and Two Family Dwelling Code*,™ copyright © 1995, with the permission of the publisher, The International Codes Council, Whittier, CA, Table 301.5, p. 14.

SNOW LOAD

In addition to the roof snow load information provided by the local building code official in Table 9.1, the 1995 CABO Code also provides a snow load map[13] which can be used to determine the average **ground snow load** in a particular location. According to the CABO map, the estimated ground snow load value for central Pennsylvania, which is the assumed location of the Case Study House, is approximately 20 psf.

The snow load acting on the roof is different than the ground snow load because it depends on the surroundings in which the house is situated as well as the design of the roof. A flat roof, for instance, will tend to retain a greater amount of the snow that has fallen than a high-pitched roof. Interconnecting roofs also provide an opportunity for drifting snow to accumulate. Unfortunately, the 1995 CABO Code does not provide guidance about how to convert the ground snow load into the equivalent roof snow load.

The *1996 BOCA National Building Code*[14] does provide such guidance. Roof snow load formulas are provided for **flat-roof and low-slope snow loads** and for **sloped-roof snow loads**.

For the flat-roof and low-slope snow load condition, for instance, it is noted:

The snow load on unobstructed flat roofs and roofs having a slope of 30 degrees (0.2 rad) or less (P_f) shall be calculated in pounds per square foot using the following formula:

$$P_f = C_e \times I \times P_g$$

where:

C_e = snow exposure factor . . .
I = snow load importance factor . . .
P_g = ground snow load expressed in pounds per square foot . . .[15]

Tables 9.3 and 9.4 provide information about C_e and I, respectively. The use of these tables is illustrated for the Case Study House example. Since the Case Study House is assumed to be located in central Pennsylvania, the detailed ground snow load map of the Eastern United States found in the 1996 BOCA National Building Code[16] indicates an estimated P_g = 30 psf for that location.

For the Case Study House:

$$\text{Roof slope} = 26.6° \text{ (less than } 30°)$$
$$P_f = C_e \times I \times P_g$$

TABLE 9.3   Snow Exposure Factor (C_e)

Roofs located in generally open terrain extending one-half mile or more from the structure	0.6
Structures located in densely forested or sheltered areas	0.9
All other structures	0.7

where

P_f = roof snow load expressed in psf of horizontal projection
C_e = snow exposure factor
I = importance factor
P_g = ground snow load

$$C_e = 0.7 \text{ (Table 9.3, ``All other structures'')}$$
$$I = 1.0 \text{ (Table 9.4)}$$
$$P_g = 30 \text{ psf (for Central Pennsylvania)}$$
$$\text{Thus, } P_f = (0.7)(1.0)(30) = 21 \text{ psf}$$

It is important to note that: (1) the estimated **ground snow load** of 20 psf obtained from the map in the 1995 CABO Code is considerably lower than the estimated **ground snow load** of 30 psf obtained from the map in the 1996 BOCA Code and (2) a conservative value of 30 psf of **roof snow load** has been designated in Table 9.1 for Centre County, PA. These discrepancies indicate the importance of basing the calculations on the procedures specified in the actual code that has been adopted by the local governmental body.

DESIGN ROOF LOAD

The **design roof load** choice for the Case Study House, based upon: (1) the previously noted live load and snow load calculations, (2) the assumption that a design roof load had not been specified by the local building code official in Table 9.1, and (3) that the 1996 BOCA Code had been adopted by the local governmental body, would have been:

$$\left.\begin{array}{l}\text{Minimum live load} = 16 \text{ psf}\\\text{Roof snow load} = 21 \text{ psf}\end{array}\right\} \text{Design Roof Load} = 21 \text{ psf}$$

Wind Load Determination

Wind pressure results from the force applied to all exterior surfaces of the house that are exposed to wind. Depending upon the design of the house, and the orientation of the wind in relation to the house, some exterior surfaces will experience **positive pressure** (i.e., pressure which pushes the surface inward against the structural frame of the house). Other exterior surfaces will experience **negative pressure** or **uplift** (i.e., pressure which pulls the surface away from the structural frame of the house). These pressures must be considered in the structural design of the house if they are of a sufficient magnitude.

For instance, the 1995 CABO Code indicates:

Exterior walls subject to wind pressures of 30 pounds per square foot (1.44 kN/m²) or greater . . . shall be designed in accordance with accepted engineering practice.[17]

In addition, with regard to roof tie-down requirements, the 1995 CABO Code indicates:

Roof assemblies subject to wind uplift pressures of 20 pounds per square foot (0.958 kN/m² or greater . . . shall have rafter or truss ties provided. . . . The resulting uplift forces from the rafter or truss ties shall be transmitted to the foundation.[18]

TABLE 9.4   Importance Factor (I)

Nature of Occupancy	Wind Load Importance Factor (I)[a]		Snow Load Importance Factor (I)
	100 Miles[b] from Hurricane Oceanline, and in Other Areas	At Hurricane Oceanline[c]	
All buildings and structures except those listed below	1.00	1.10	1.0
Occupancies in Use Group A in which more than 300 people congregate in one area	1.15	1.23	1.1
Buildings and structures having essential facilities, including buildings containing any one or more of the indicated occupancies: 1. Fire, rescue, and police stations 2. Use Group 1-2 having surgery or emergency treatment facilities 3. Emergency preparedness centers 4. Designated shelters for hurricanes 5. Power-generating stations and other utilities required as emergency backup facilities 6. Primary communication facilities	1.15	1.23	1.2
Buildings and structures that represent a low hazard to human life in the event of failure, such as agricultural buildings, production greenhouses, certain temporary facilities, and minor storage facilities	0.90	1.00	0.8

[a]For regions between the hurricane oceanline and 100 miles inland, the importance factor (I) shall be determined by linear interpolation.
[b]1 mile = 1.61 km.
[c]Hurricane oceanlines are the Atlantic and Gulf of Mexico coastal areas.

The process of determining whether wind pressure must be considered in a particular geographical location begins by using Fig. 9.9 to determine the **basic wind speed** for the location of the house. Examination of Fig. 9.9 indicates that one of the areas of the United States that is most prone to high wind velocities is the coastal region extending from Texas along the Gulf of Mexico to the tip of Florida and from the tip of Florida to Maine along the Atlantic Ocean. This is a region that frequently experiences hurricane-related devastation. There is also a broad region of the United States which parallels this coastline, but which is further inland, that is exposed to a basic wind speed of 70 mph. The entire state of Pennsylvania lies, for instance, in that region.

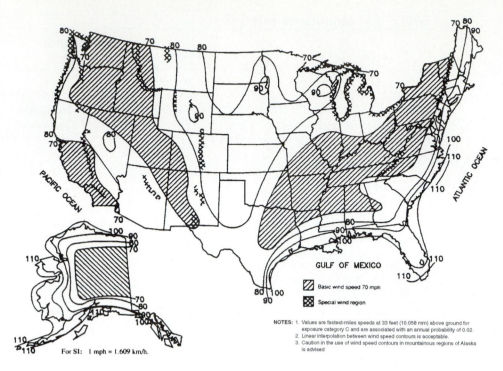

Figure 9.9
Basic Wind Speed
Map
Source: Reproduced
from the 1995 edi-
tion of the *CABO
One and Two
Family Dwelling
Code,*™ copyright ©
1995, with the per-
mission of the pub-
lisher, The
International Codes
Council, Whittier,
CA, Fig. 301.2d,
p. 12.

Wind speeds can be converted into wind pressures by using Table 9.5. A determination of whether wind pressure must be incorporated into the design process must consider
the exposure classification of the house.

According to Table 9.5, it appears that wind pressure on the walls does not have to be
considered if the two-story Case Study House is exposed to a basic wind speed of 70 mph
and is located in an Exposure A/B, which is fairly typical for central Pennsylvania. In addition, **roof uplift pressure** does not have to be considered in an Exposure A/B unless the basic
wind speed is assumed to be 100 mph.

Although it appears that wind loads for the Case Study House are not significant if it
is built in central Pennsylvania, that certainly would not be the case, for instance, if it were
built in Cape Hatteras, North Carolina. In such a situation, Table 9.5 would be used to identify a wind situation that needs an engineering analysis. One of the sources of information
for a wind analysis is the *1996 BOCA National Building Code.*[19] The procedures presented
permit the designer to estimate both the magnitude and the direction (i.e., positive or negative) of the pressure on all of the walls and roof surfaces of the house.

Seismic Risk

Seismic risk can best be described as a building's resistance to ground motion. In areas of
high seismic risk, the designer must evaluate the building's diaphragm, load path, ductility,
period and resonance, shear wall, torsion, strength, and stiffness. The 1995 CABO Code[20]
provides a seismic risk map which indicates that central Pennsylvania, for example, is
located in a seismic risk zone of 0.

Seismic risk is severe for parts of California and Alaska, which are in seismic risk zones
of 4. These regions of the United States require their own structural design standards to
guard against this risk.

Weathering

Weathering is a term applied to climatic conditions such as exposure to rainfall and freezing
temperatures. The exterior components of the building, especially concrete products, must
be designed to withstand prolonged adverse weather conditions or extreme changes in these

TABLE 9.5   Design Wind Loads (pounds per S.F.)

Exposure Classification	Basic Wind Speed (mph)	One Story		Two Story		Three Story	
		Walls	Roof Uplift	Walls	Roof Uplift	Walls	Roof Uplift
A/B	80	N/A	N/A	N/A	N/A	N/A	N/A
	90	N/A	N/A	N/A	N/A	N/A	20
	100	N/A	N/A	N/A	21	N/A	25
C	70	N/A	N/A	N/A	N/A	N/A	N/A
	80	N/A	20	N/A	22	N/A	25
	90	N/A	26	N/A	28	N/A	31
	100	N/A	32	32	35	35	39
	110	35	38	38	42	43	47
D	70	N/A	20	N/A	22	N/A	24
	80	N/A	27	N/A	28	N/A	31
	90	32	37	36	40	39	43
	100	42	46	44	49	49	53
	110	50	55	54	59	59	64

For SI: 1 in. = 25.4 mm, 1 ft = 304.8 mm, 1 psf = 0.0479 kN/m^2, 1 mph = 1.609 km/h.

NOTES:

1. Select exposure classification from the descriptions below:

   Exposure A/B—Urban and suburban areas, or other terrain with numerous closely spaced obstructions having the size of single-family dwellings or larger in the upwind direction for a distance of at least 1,500 ft.

   Exposure C—Open terrain with scattered obstructions having heights generally less than 30 ft.

   Exposure D—Flat, unobstructed areas exposed to wind flowing over large bodies of water extending inland from the shoreline a distance of 1,500 ft.

2. Wind speeds may be obtained from the Basic Wind Speed Map (Figure 301.2d). Wind speeds represent the following: 70 = 0 to 70 mph, 80 = 71 to 80 mph, 90 = 81 to 90 mph, 100 = 91 to 100 mph, 110 = 101 to 110 mph. Hawaii has a basic wind speed of 80 mph and Puerto Rico has a basic wind speed of 95 mph.

3. Building heights used to determine design wind loads are: One story = 20 ft, two story = 30 ft, three story = 50 ft. Interpolation between building height values and between map contours is acceptable.

4. Uplift loads act normal to the roof or overhang.

5. N/A = no design is required in accordance with Sections 602.3 and 802.11 of the 1995 CABO Code.

6. Buildings over 50 ft in height, or with unusual constructions or geometric shapes, with overhanging eave projections greater than 24 ins. or located in special wind regions or localities shall be designed in accordance with the provisions of ASCE 7-88.

Source: Reproduced from the 1995 edition of the *CABO One and Two Family Dwelling Code,*™ copyright © 1995, with the permission of the publisher, The International Codes Council, Whittier, CA, Table 301.2b,

conditions. A weathering probability map for concrete is presented in the 1995 CABO Code.[21] The weathering probability in central Pennsylvania, for example, is defined as **severe**.

Frost Depth

Frost depth, or frost penetration, is expressed as the number of inches below the surface that the ground is frozen during the winter months. Although frost depths are provided throughout the United States, frost is dependent upon soil conditions and elevations, which can deviate drastically from the published guidelines of the U.S. Weather Bureau. The local jurisdiction, therefore, should determine the frost depth. The frost penetration depth map which is presented in the Application and Commentary to the 1992 CABO Code[22] indicates a frost penetration depth of approximately 40 in., for example, for central Pennsylvania.

Termite Infestation and Decay

The 1995 CABO Code also provides guidelines about the probability for **termite infestation**[23] and **decay**[24] of ordinary wood products. An examination of the maps presented indicates that the probability of termite infestation is determined to be moderate to heavy in central Pennsylvania, for example, while the decay probability is determined to be slight to moderate. These results affect design parameters in other sections of the 1995 CABO Code.

For instance, the 1995 CABO Code identifies the types of termite protection methods that can be used in areas which are prone to termite damage. These include chemical soil treatment, pressure-treated wood, naturally termite-resistant wood, and physical barriers.[25] It also is noted that pressure preservatively treated lumber, of a particular species and grade, can be used in those locations that are particularly prone to decay damage.[26]

Winter Design Temperature

One of the factors that must be considered when an energy analysis of a house is performed (see Chapter 12) is the conductivity energy losses of the house. In that determination, it is necessary to evaluate the difference between the indoor air temperature and an outside ambient temperature during the months of December through February. The outside temperature that is typically used is the 97.5% design condition. It is expected that temperatures lower than this will occur only 2.5% of the time during those months.

The 1995 CABO Code provides a map which presents the "Isolines of the 97.5 Percent Winter . . . Design Temperature (°F)" information for the United States.[27] It appears that in central Pennsylvania, for instance, this value varies between 10°F and about 5°F. Values for specific cities in Pennsylvania are provided in Chapter 12.

Radon

The 1995 CABO Code[28] requires that all new houses shall be constructed using radon control methods where they are required. An "EPA Map of Radon Areas" is also provided.[29] As noted on that map, most of central Pennsylvania, for instance, is in Zone 1, which is the most severe zone. Specific requirements for the radon control methods that are required in that zone are provided in the 1995 CABO Code.[30]

Other Design Criteria

In addition to climatic and geographic design criteria, a homebuilder or architect must also consider the dead loads and live loads imposed upon the structural framework of the house as well as the deflections caused by these loads. The 1995 CABO Code[31] provides the following guidelines about these structural aspects.

Dead Loads

The **dead load** category includes the weight of all permanent features of the house such as the roof, floors, walls, interior partitions, fixed service equipment, etc.

The minimum values for some of the design dead loads for the typical components used in a house are provided in Table 9.6.

Live Loads

The **live loads** used in the structural analysis and design of a house represent the actual loads that are applied to the house because of its use and occupancy. Wind, earthquakes, and dead loads are not included in this category.[32] The person performing the structural analysis does not have the freedom to select a unique set of live loads for each house. Instead, the 1995 CABO Code specifies that the **uniformly distributed live loads** shown in Table 9.7 shall be used to **represent** the actual live loads that will be experienced in the house.[33]

TABLE 9.6 Minimum Design Dead Loads for Typical Residential Components

Component	Load (psf)
CEILINGS	
Acoustical fiber tile	1
Gypsum board (per 1/8" thickness)	0.55
Mechanical duct allowance	4
Plaster on tile or concrete	5
Plaster on wood lath	8
Suspended steel channel system	2
Suspended metal lath & cement plaster	15
Suspended metal lath & gypsum plaster	10
Wood-furring suspension system	2.5
COVERINGS, ROOF AND WALL	
Asbestos–cement shingles	4
Asphalt shingles	2
Cement tile	16
Clay tile (for mortar, add 10 lb.):	
Book tile, 2"	12
Book tile, 3"	20
Ludowici	10
Roman	12
Spanish	19
Composition:	
Three-ply ready roofing	1
Four-ply felt and gravel	5.5
Five-ply felt and gravel	6
Copper or tin	1
Corrugated asbestos–cement roofing	4
Deck, metal, 20 gauge	2.5
Deck, metal, 18 gauge	3
Decking, 2" wood (Douglas fir)	5
Decking, 3" wood (Douglas fir)	8
Fiberboard, 1/2"	0.75
Gypsum sheathing, 1/2"	2
Insulation, roof boards (per in. thick)	
Cellular glass	0.7
Fibrous glass	1.1
Fiberboard	1.5
Perlite	0.8
Polystyrene foam	0.2
Urethane foam with skin	0.5
Plywood (per 1/8" thickness)	0.4
Rigid insulation, 1/2"	0.75
Skylight, metal frame,	
3/8" wired glass	8
Slate, 3/16"	7
Slate, 1/4"	10
Waterproofing membranes:	
Bituminous, gravel-covered	5.5
Bituminous, smooth surface	1.5
Liquid applied	1.0
Single-ply, sheet	0.7
Wood sheating (per 1" thickness)	3
Wood shingles	3
FLOORS AND FLOOR FINISHES	
Asphalt block (2"), 1/2" mortar	30
Cement finish (1") on	
concrete–stone fill	32
Ceramic or quarry tile (3/4") on	
1/2" mortar bed	16

Component	Load (psf)
Ceramic or quarry tile (3/4")	
on 1" mortar bed	23
Concrete fill finish (per in. thick)	12
Hardwood flooring, 7/8"	4
Linoleum or asphalt tile, 1/4"	1
Marble and mortar on	
stone–concrete fill	33
Slate (per in. thickness)	15
Solid flat tile on 1" mortar base	23
Subflooring, 3/4"	3
Terrazzo (1-1/2") directly on slab	19
Terrazzo (1") on stone–concrete fill	32
Terrazzo (1"), 2" stone concrete	32
Wood block (3") on mastic, no fill	10
Wood block (3") on 1/2" mortar base	16

FLOORS, WOOD-JOIST (NO PLASTER)—DOUBLE WOOD FLOOR

Joist Sizes (in.)	12" Spacing (lb/ft²)	16" Spacing (lb/ft²)	24" Spacing (lb/ft²)
2 × 6	6	5	5
2 × 8	6	6	5
2 × 10	7	6	6
2 × 12	8	7	6

Component	Load (psf)
FRAME PARTITIONS	
Movable steel partitions	4
Wood or steel studs, 1/2" gypsum	
board each side	8
Wood studs, 2 × 4, unplastered	4
Wood studs, 2 × 4, plastered 1 side	12
Wood studs, 2 × 4, plastered 2 sides	20
FRAME WALLS	
Exterior stud walls:	
2 × 4 @ 16", 5/8" gypsum	
insulated, 3/8" siding	11
2 × 6 @ 16", 5/8" gypsum	
insulated, 3/8" siding	12
Exterior stud walls with brick veneer	48
Windows, glass, frame and sash	8
MASONRY PARTITIONS AND WALLS	
Clay tile:	
4"	18
6"	24
8"	34
Concrete block, heavy aggregate:	
4"	30
6"	42
8"	55
12"	85
Concrete block, light aggregate:	
4"	20
6"	28
8"	38
12"	55

Source: Reproduced from the 1992 edition of the *Application and Commentary, CABO One and Two Family Dwelling Code*™, copyright © 1992, with the permission of the publisher, the International Codes Council, Table 201.3, p. 9.

TABLE 9.7   Minimum Uniformly Distributed Live
Loads (in pounds per square foot—lb/ft^2)

Use	Live Load
Balconies (exterior)	60
Decks	40
Fire escapes	40
Garages (passenger cars only)	50
Attics (no storage with roof slope not steeper than 3 in 12)	10
Attics (limited attic storage)	20
Dwelling units (except sleeping rooms)	40
Sleeping rooms	30
Stairs	40[a]
Guardrails and handrails (a single concentrated load applied in any direction at any point along the top)	200

For SI: 1 psf = 0.0479 kN/m^2, 1 sq in. = 645 mm^2.
[a]Individual stair treads shall be designed for the uniformly distributed live load or a 300-pound concentrated load acting over an area of 4 sq. ins., whichever produces the greater stresses.

Source: Reproduced from the 1995 edition of the *CABO One and Two Family Dwelling Code*,™ copyright © 1995, with the permission of the publisher, The International Codes Council, Whittier, CA, Table 301.4, p. 8.

Live Load Deflection

The structural members used in a house experience initial deflections as the dead loads are naturally distributed throughout the structural framework. This **dead load deflection**, for the case of the simply supported beam shown in Fig. 9.10, is assumed to have its maximum value at the center of the beam's span. As live loads are applied to the house, such a beam typically will experience an additional **live load deflection**, with the maximum value again assumed to be at the center of the beam's span.

The 1995 CABO Code[34] places limits on the maximum allowable value for the live load deflection for various structural components. This is to ensure that the finish material applied to these components is not damaged due to excessive deflection. The limits are expressed as a specified fraction of either the length of the component or the height of the wall or partition (see Table 9.8).

Figure 9.10
Comparison of Dead Load Deflection and Live Load Deflection for a Simply Supported Beam

Source: Reproduced from the *Application and Commentary*, 1992 edition of the CABO One and Two Family Dwelling Code,™ copyright © 1992, with the permission of the publisher, The International Codes Council, Whittier, CA, Fig. 201.6 p. 8.

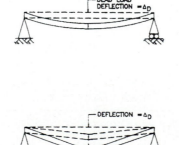

TABLE 9.8   Allowable Deflection of Structural
Members

Structural Member	Allowable Deflection
Rafters having slopes greater than 3/12 with no finished ceiling attached to rafters	$L/180$
Interior walls and partitions	$H/180$
Floors and plastered ceilings	$L/360$
All other structural members	$L/240$

NOTES: L = span length
              H = span height

Source: Reproduced from the 1995 edition of the *CABO One and Two Family Dwelling Code*,™ copyright © 1995, with the permission of the publisher, The International Codes Council, Whittier, CA, Table 301.6, p. 14.

FOUNDATIONS

The foundations of houses can be constructed with: (1) full or partial basements, (2) crawl spaces, or (3) the first-floor slab placed directly on grade (Fig. 9.11).

The foundation walls for the options shown in Fig. 9.11(a), (b), and (d) are typically constructed of masonry units (i.e., concrete block), unreinforced concrete, or reinforced concrete. These foundation walls are typically supported by a poured concrete footing. Stemwall foundations, which consist of a poured concrete foundation wall bearing directly on the soil without a traditional concrete footing, are also being used.[35] In addition, wood foundation walls that are part of a **permanent wood foundation system** are another alternative for the homebuilder and the homebuyer to consider.

Slab-on-grade foundations typically consist of either a monolithic concrete system [Fig. 9.11(c)] or an independent concrete slab supported by an independent foundation wall or footing [Fig. 9.11(d)]. Local climatic conditions will determine if these foundation systems must be insulated.

In those jurisdictions that adopt the 1995 CABO Code, Chapter 4, "Foundations",[36] controls the design and construction of the foundation and foundation spaces for all houses. Although both concrete/masonry unit (concrete block) foundations and wood foundations are permissible alternatives in Chapter 4, only the former is discussed in detail since it is the one that is used most frequently by homebuilders.

Important Considerations

A complete foundation system consists of one of the alternatives shown in Fig. 9.11 and the soil upon which it is placed. Both parts must be capable of carrying the loads of the house. As noted in the 1995 CABO Code: "The foundation and its structural elements shall be capable of accommodating all superimposed live, dead and other loads according to Section 301 and all lateral loads in accordance with the provisions of this code."[37]

A homebuilder can determine the load that the soil can carry (its load-bearing capacity) either from a geotechnical evaluation or by referring to Table 9.9. In those situations where expansive, compressible, shifting, or unusual soil characteristics are suspected, soil tests are necessary. If such soils are found, they must be removed and replaced.

Failure of a foundation also can occur due to water-related problems. As a result, adequate drainage away from the house is required to prevent such damage. This can be accomplished by providing either a minimum fall of grade of 6 ins. in the first 10 ft from the house, or by using drains or swails.[38]

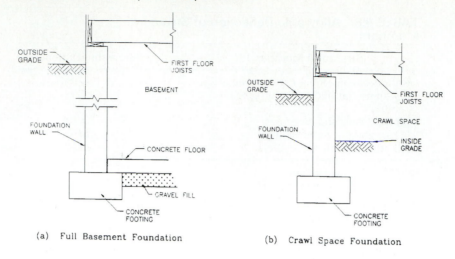

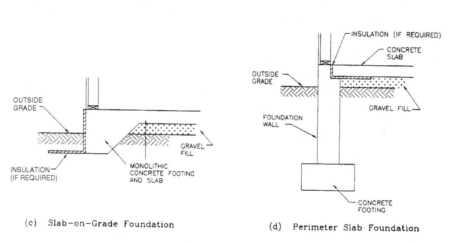

Figure 9.11

Types of Foundations for Houses (not all details shown)

Source: Ernest R. Weidhaas, *Architectural Drafting and Construction*, 4th ed, p. 242 (Boston, MA: Allyn and Bacon, 1989) (modified).

TABLE 9.9   Presumptive Load-Bearing Values of Foundation Materials

Class of Material	Load-Bearing Pressure (pounds per square foot)
Crystalline bedrock	12,000
Sedimentary rock	6,000
Sandy gravel or gravel	5,000
Sand, silty sand, clayey sand, silty gravel, and clayey gravel	3,000
Clay, sandy clay, silty clay, and clayey silt	2,000

For SI: 1 psf = 0.0479 kN/m².

Source: Reproduced from the 1995 edition of the *CABO One and Two Family Dwelling Code,*™ copyright © 1995, with the permission of the publisher, The International Codes Council, Whittier, CA, Table 401.4.1, p. 25.

Materials

Information about both foundations and concrete is provided in the 1995 CABO Code.[39] The minimum specified compressive strength of concrete for various uses is shown in Table 9.10. Air-entrained concrete is required in areas where concrete is subjected to weathering (Table 9.1). The minimum cement contents for concrete mixtures for exterior porches, carport slabs, and steps that will be exposed to freezing and thawing in the presence of deicing chemicals is also specified.

Footings

Requirements for the design of **footings** for both wood and concrete/concrete block foundations are provided in the 1995 CABO Code.[40] Specific requirements include:

> All exterior walls shall be supported on continuous solid masonry or concrete footings, wood foundations, or other approved structural systems which shall be of sufficient design to support safely the loads imposed as determined from the character of the soil and, except when erected on solid rock or otherwise protected from frost, shall extend below the frost line . . .[41]

The depth of the frost line can be determined from Table 9.1. The minimum sizes for concrete and masonry footings can be determined from Table 9.11 and Fig. 9.12 (which indicates the various types of pre-approved footing/footing wall assemblies).

TABLE 9.10 Minimum Specified Compressive Strength of Concrete

Type or Locations of Concrete Construction	Minimum Specified Compressive Strength[a] (f'_c)		
	Weathering Potential[b]		
	Negligible	Moderate	Severe
Basement walls and foundations not exposed to the weather	2,500	2,500	2,500[c]
Basement slabs and interior slabs on grade, except garage floor slabs	2,500	2,500	2,500[c]
Basement walls, foundation walls, exterior walls, and other vertical concrete work exposed to the weather	2,500	3,000[d]	3,000[d]
Porches, carport slabs, and steps exposed to the weather, and garage floor slabs	2,500	3,000[d,e]	3,000[d,e]

For SI: 1 psi = 6.895 kPa.
[a]At 28 days psi.
[b]See Table 301.2a of the Code for weathering potential.
[c]Concrete in these locations which may be subject to freezing and thawing during construction shall be air-entrained concrete in accordance with Footnote d.
[d]Concrete shall be air entrained. Total air content (percent by volume of concrete) shall not be less than 5% or more than 7%.
[e]See Section 402.2 of the Code for minimum cement content.

Source: Reproduced from the 1995 edition of the *CABO One and Two Family Dwelling Code*,™ copyright © 1995, with the permission of the publisher, The International Codes Council, Whittier, CA, Table 402.2, p. 26.

TABLE 9.11   Minimum Width of Concrete or Masonry
Footings (inches)

	Load-Bearing Value of Soil (psf)					
	1,500	2,000	2,500	3,000	3,500	4,000
Conventional Wood Frame Construction						
1-story	16	12	10	8	7	6
2-story	19	15	12	10	8	7
3-story	22	17	14	11	10	9
4-in. Brick Veneer over Wood Frame or 8-in. Hollow Concrete Masonry						
1-story	19	15	12	10	8	7
2-story	25	19	15	13	11	10
3-story	31	23	19	16	13	12
8-in. Solid or Fully Grouted Masonry						
1-story	22	17	13	11	10	9
2-story	31	23	19	16	13	12
3-story	40	30	24	20	17	15

For SI: 1 in. = 25.4 mm, 1 psf = 0.0479 kN/m^2.

Source: Reproduced from the 1995 edition of the *CABO One and Two Family Dwelling Code,*™ copyright © 1995, with the permission of the publisher, The International Codes Council, Whittier, CA, Table 403.1a, p. 28.

Requirements for **stepped footings** and for **frost-protected footings** (i.e., slab-on-grade footings that are insulated so they do not have to extend below the frost line) are also provided.

Foundation Walls

Foundation walls, as noted earlier, may be constructed of concrete, masonry units (i.e., concrete block), and wood. The 1995 CABO Code[42] specifies that foundation walls must extend at least 6 ins. above the finished grade (to prevent termite and moisture damage). In addition, it is noted that:

> Backfill adjacent to the wall shall not be placed until the wall has sufficient strength and has been anchored to the floor, or has been sufficiently braced to prevent damage by the backfill. **Exception**: Such bracing is not required for walls having less than 3 ft (914 mm) of unbalanced backfill.[43]

The **unbalanced fill** (H) is the height of the finished grade above the top of the basement floor slab (Fig. 9.13).

Design requirements for foundation walls for houses are indicated as follows:

> Foundation walls shall be constructed in accordance with the provisions of this section or in accordance with ACI 318, ACI 318.1, NCMA TR68-A or ACI 530/ASCE 5/TMS 402 or other approved structural systems. **Exception**: When ACI 530/ASCE 5/TMS 402 is used to design masonry foundation walls, project drawings, typical details and specifications are not required to bear the seal of the architect or engineer responsible for design.[44]

If unstable soil and ground conditions do not exist and the house is located in Seismic Zone 0, 1, or 2 (Table 9.1), then Table 9.12 provides the basic foundation wall design information. Table 9.13 must be used when: (1) unstable soil conditions exist or the foundation

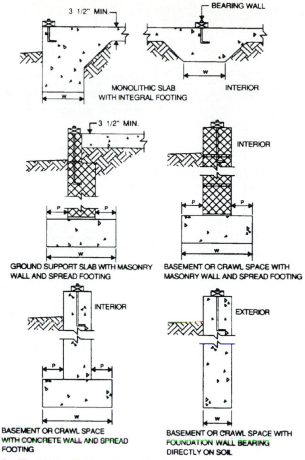

Figure 9.12

Concrete and Masonry Foundation Details

Source: Reproduced from the 1995 edition of the *CABO One and Two Family Dwelling Code*,™ copyright © 1995, with the permission of the publisher, The International Codes Council, Whittier, CA, Fig. 403.1a, p. 26.

For **SI:** 1 inch = 25.4 mm, 1 foot = 304.8 mm.

NOTES:
1. Exterior footings shall extend to below the frost line unless otherwise protected against frost heave. In no case shall exterior footings be less than 12 in. below grade.
2. Footing widths (W) shall be based on the load-bearing value of the soil in accordance with Table 401.4.1 or shall be designed in accordance with accepted engineering practice.
3. Spread footings shall be a minimum of 6 in. thick, and footing projections (P) shall be a minimum of 2 in. and shall not exceed the footing thickness.
4. Footings shall be supported on undisturbed natural soil or engineering fill.
5. The sill plate or floor system shall be anchored to the foundation with 1/2-in. -diameter bolts placed 6 ft. on center and not more that 12 in. from corners. Bolts shall extend a minimum of 15 in. into masonry or 7 in. into concrete. Sill plates shall be protected against decay where required by Section 322.
6. Pier and column footing sizes shall be based on the tributary load and allowable soil pressure in accordance with Table 502.3.3b.

extends to or below the seasonal high groundwater table or (2) the house is located in Seismic Zone 3 or 4.[45] As noted in Table 9.13, certain situations require an engineering design.

In addition, **accepted engineering practice**, rather than Tables 9.12 and 9.13, must be used if: (1) foundation walls are subject to more pressure than would be exerted by backfill having an equivalent fluid weight of 30 pounds per cubic foot[46] or (2) foundation walls are "subject to more lateral pressure than would be exerted by backfill consisting of freely draining sands and gravel classified as Group I according to the United States Soil Classification System or soils having an equivalent fluid weight of greater than 30 pounds per cubic foot."[47]

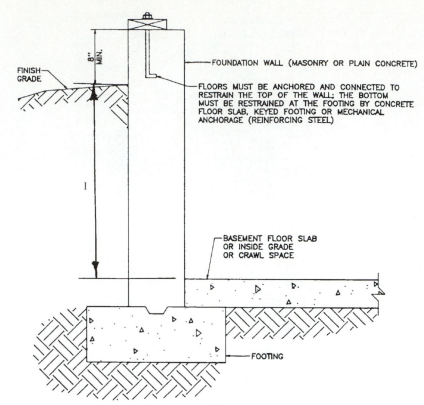

Figure 9.13
Bracing of
Foundation Walls
Against Lateral
Earth Pressure
Source: Reproduced
from the
Application and
Commentary, 1992
edition of the
*CABO One and
Two Family
Dwelling Code,*™
copyright © 1992,
with the permission
of the publisher,
The International
Codes Council,
Whittier, CA, Fig.
304.2 (modified),
p. 39.

Case Study House

The adequacy of the foundation design of the Case Study House (it is assumed to be located in central Pennsylvania) can be verified in terms of the requirements in the 1995 CABO Code, as follows:

Actual Design (Appendix A)

- Concrete strength: 2500 psi: material specifications, Appendix C
- Exterior wall footings: 18 in. × 8 in.: Sheet A-5
- Foundation wall: 10 in. concrete block, assume fully grouted: Sheet A-5
- Column footings in basement: 2 ft × 2 ft × 1 ft: Sheets A-3, A-5
- Height of unbalanced fill: 8 ft − (8 in. + 3 in.) = 7 ft. 1 in.: Sheet A-5
- Two-story house
- Soil-bearing capacity: assume 2,500 psf

Verification of Design

- Concrete Strength—Table 9.10: Use 2,500-psi concrete in severe weathering zone if elements are not exposed to the weather. Use air-entrained concrete if placed in winter. *Therefore, 2,500-psi concrete strength is satisfactory.*
- Exterior Wall Footings—Figure 9.12: Minimum thickness = 6 in.; minimum width = 10 in. + 2 in. + 2 in. = 14 in. Table 9.11: Minimum width = 12 in. for a two-story conventional wood frame house, soil-load-bearing capacity = 2,500 psi. *Therefore, a 18 in. × 8 in. wall footing is satisfactory.*

TABLE 9.12   Minimum Thickness and Allowable Depth
of Unbalanced Fill for Unreinforced Masonry and
Concrete Foundation Walls[a,b] where Unstable Soil
or Groundwater Conditions do not Exist in Seismic
Zones, 0, 1, or 2

Foundation Wall Construction	Nominal Thickness[c] (ins.)	Maximum Depth of Unbalanced Fill[a] (ft)
Masonry of hollow units, ungrouted	8	4
	10	5
	12	6
Masonry of solid units	6	3
	8	5
	10	6
	12	7
Masonry of hollow or solid units, fully grouted	8	7
	10	8
	12	8
Plain concrete	6[d]	6
	8	7
	10	8
	12	8
Rubble stone masonry	16	8
Masonry of hollow units reinforced vertically with No. 4 bars and grout at 24 ins. on center. Bars located not less than 4-1/2 ins. from pressure side of wall	8	7

For SI: 1 in. = 25.4 mm, 1 ft = 304.8 mm.
[a]Unbalanced fill is the difference in height of the exterior and interior finish ground levels. Where an interior concrete slab is provided, the unbalanced fill shall be measured from the exterior finish ground level to the top of the interior concrete slab.
[b]The height between lateral supports shall not exceed 8 ft.
[c]The actual thickness shall not be more than 1/2 in. less than the required nominal thickness specified in the table.
[d]6-in. plain concrete walls shall be formed on both sides.

Source: Reproduced from the 1995 edition of the *CABO One and Two Family Dwelling Code*,™ copyright © 1995, with the permission of the publisher, The International Codes Council, Whittier, CA, Table 404.1.1a, p. 31.

- Foundation Wall—Table 9.12 can be used for a Seismic Risk Zone of 0 or 1 if unstable soil or groundwater conditions do not exist. For fully grouted masonry units, a 10-in. foundation wall can sustain an unbalanced fill height of 8 ft, 0 in. *Therefore, a 10-in. concrete block wall is satisfactory* (actual unbalanced fill = 7 ft, 1 in.).
- Column footings in basement—Adequacy of design cannot be determined by using the information provided previously from Chapter 4, "Foundations", of the 1995 CABO Code. Chapter 5, "Floors" must be consulted.[48]

Foundation Drainage, Waterproofing, and Dampproofing

Foundation systems must be protected against the infiltration of water in order to avoid damage to the foundation elements and to prevent the "wet or damp basement" from becoming a problem for the homeowner. The 1995 CABO Code[49] provides the require-

TABLE 9.13   Requirements for Masonry or Concrete Foundation Walls Subjected to No More Pressure than Would be Exerted by Backfill Having an Equivalent Fluid Weight of 30 Pounds per Cubic Foot Located in Seismic Zone 3 or 4 or Subjected to Unstable Soil Conditions

Material Type	Height of Unbalanced Fill in Feet[a]	Length of Wall between Supporting Masonry or Concrete Walls in Feet	Minimum[b] Wall Thickness in Inches[c]	Required Reinforcing	
				Horizontal Bar in Upper 12 in. of Wall	Size and Spacing of Vertical Bars
Hollow Masonry	4 or less	unlimited	8	not required	not required
	more than 4	design required	design required	design required	design required
Concrete or Solid Masonry[d]	4 or less	unlimited	8	not required	not required
	more than 4	less than 8	8	2-No. 3	No. 3 @ 18″ o.c.
	8 or less	8 to 10	8	2-No. 4	No. 3 @ 18″ o.c.
	8 or less	10 to 12	8	2-No. 5	No. 3 @ 18″ o.c.
	more than 8	design required	design required	design required	design required

For SI: 1 in. = 25.4 mm, 1 ft = 304.8 mm, 1 pound per cubic foot (pcf) = 0.1572 kN/m^3.
[a]Backfilling shall not be commenced until after the wall is anchored to the floor.
[b]Thickness of concrete walls may be 6 in., provided reinforcing is placed not less than 1 in. or more than 2 in. from the face of the wall not against the earth.
[c]The actual thickness shall not be more than 1/2 in. less than the required thickness specified in the table.
[d]Solid masonry shall include solid brick or concrete units and hollow masonry units with all cells grouted.

Source: Reproduced from the 1995 edition of the CABO One and Two Family Dwelling Code,™ copyright © 1995, with the permission of the publisher, The International Codes Council, Whittier, CA, Table 404.1.1b, p. 31.

ments which address those objectives. The **first line of defense** for concrete and masonry foundations (unless the soil is well drained) consists of a drain system which must be installed around all concrete or masonry foundations enclosing habitable or usable spaces located below grade. Specific details for such a system for concrete and masonry foundations are provided.[50] Similar information is provided for wood foundations.

The requirements for a **second line of defense** which involves waterproofing and dampproofing are also provided. For concrete and masonry foundations, for instance, **waterproofing** consists of installing a waterproofing membrane on the exterior face of all foundation walls enclosing habitable or storage space. It is required in situations where a high water table or other severe soil–water conditions exist.[51] **Dampproofing**, which is used whenever waterproofing is not required, consists of 3/8-in. Portland cement parging covered with a coating layer of specified material for masonry walls. Concrete walls require only one layer of dampproofing material.[52]

Figure 9.14 provides an example of a water-protective system for a concrete foundation wall which combines some of the elements discussed.

Additional Topics

The additional foundation-related topics that are presented in Chapter 4 of the 1995 CABO Code include:

- Foundation insulation
- Columns
- Crawl space

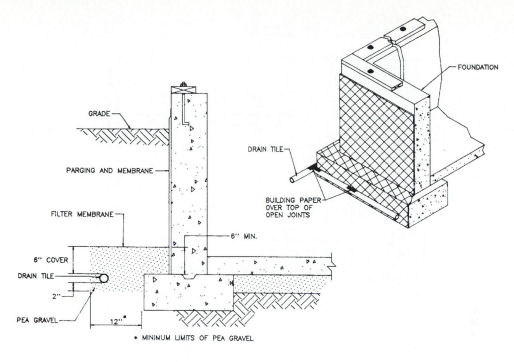

Figure 9.14
A Water
Protection
System for
Habitable Space
Below Grade
Source: Reproduced
from the
Application and
Commentary, 1992
edition of the
*CABO One and
Two Family
Dwelling Code,*™
copyright © 1992,
with the permission
of the publisher,
The International
Codes Council,
Whittier, CA, Fig.
305.1a, p. 43.

FLOOR SYSTEMS

Chapter 5, "Floors", of the 1995 CABO Code[53] presents the requirements which ensure that the loads imposed on the floor component of either a platform framing or a balloon framing structural system are supported adequately and also properly transmitted to the rest of the structural frame. The primary structural elements of the floor framing system are shown in Fig. 9.15.

Identification of Wood Properties

The 1995 CABO Code requires that all load-bearing dimension lumber used for joists, beams, and girders must be identified by a grade mark or certificate of inspection issued by an approved agency (see Figs. 8.20 and 8.21). The information contained in a grade mark should be sufficient to determine values for F_b, the allowable stress in bending, and E, the modulus of elasticity of the lumber being used.[54] These are the only values needed when the adequacy of bending members is determined using the procedure specified in Chapter 5 of the 1995 CABO Code.

Sources of Information

The 1995 CABO Code provides information, in a visual stress rated (VSR) format, about F_b and E values for the most commonly used species and grades of lumber.[55] Table 9.14 presents an excerpt of that table for Hem–Fir lumber. Similar information is also provided in the 1995 CABO Code in a machine stress rated (MSR) format.

Case Study House Example

An example related to the first-floor joists in the Case Study House will indicate how appropriate F_b and E values can be determined from both of these sources of information.

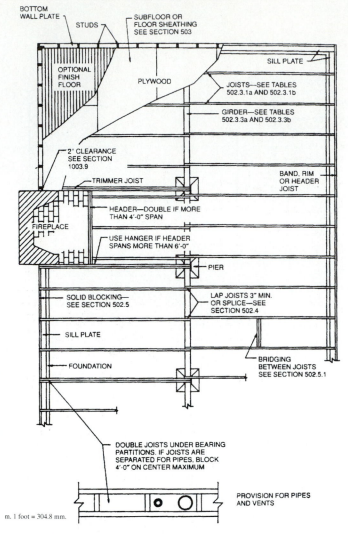

Figure 9.15
Primary
Structural
Elements of the
Floor Framing
System

Source: Reproduced
from the 1995 edition of the *CABO
One and Two
Family Dwelling
Code,*™ copyright ©
1995, with the permission of the publisher, The
International Codes
Council, Whittier,
CA, Fig. 502.2,
p. 36.

m. 1 foot = 304.8 mm.

Actual Design (Appendix A)

- Type of lumber—No. 2 Hem–Fir, surface dry, used at less than 19% moisture content: Sheet A-3 and Appendix C, Material Specifications
- First-Floor Joists—2 in. × 10 in. floor joists, spaced at 16 in. o.c.: Sheet A-3 and Sheet A-5

Determination of F_b and E

- Table 9.14—1995 CABO Code

$$F_b = 1075 \text{ psi (normal duration loading)}$$
$$E = 1.3 \times 10^6 \text{ psi}$$

- Table 8.4—1991 National Design Specification for Wood Construction and Supplement (N.D.S.)[56]

Base $F_b = 850$ psi (dimension lumber, 2–4 in. thick, 2 in. and wider)
$E = 1.3 \times 10^6$ psi
$C_f =$ Size factor (Table 8.6a) for No. 2 grade, 10 in. wide, 2 in. thick
$C_f = 1.10$
$C_r =$ Repetitive Use Factor (Footnote d of Table 8.4)
$C_r = 1.15$
$F_b = 850(1.10)(1.15) = 1075$ psi

TABLE 9.14   Design Values for Hem–Fir Dimension Lumber—Visual Grading

These "F_b" values are for use where three or more repetitive members are spaced not more than 24 in. apart. For wider spacing or for single- or double-member headers or beams, the "F_b" values should be reduced 13%. Values for surfaced dry or surfaced green lumber apply at 19% maximum moisture content in use.

| Species and Grade | Size | Normal Duration | Design Value in Bending "F_b" | | Modulus of Elasticity "E" | Grading Rules Agency |
			Snow Loading	7-day Loading		
			Hem–Fir			
Select Structural		2,415	2,775	3,020	1,600,000	
No. 1 & Btr		1,810	2,085	2,265	1,500,000	
No. 1		1,640	1,885	2,050	1,500,000	
No. 2		1,465	1,685	1,835	1,300,000	
No. 3	2″ × 4″	865	990	1,080	1,200,000	West Coast
Stud		855	980	1,065	1,200,000	Lumber
Construction		1,120	1,290	1,400	1,300,000	Inspection
Standard		635	725	790	1,200,000	Bureau
Utility		290	330	360	1,100,000	Western Wood
Select Structural		2,095	2,405	2,615	1,600,000	Products Association
No. 1 & Btr		1,570	1,805	1,960	1,500,000	
No. 1	2″ × 6″	1,420	1,635	1,775	1,500,000	
No. 2		1,270	1,460	1,590	1,300,000	
No. 3		750	860	935	1,200,000	
Stud		775	895	970	1,200,000	
Select Structural		1,930	2,220	2,415	1,600,000	
No. 1 & Btr		1,450	1,665	1,810	1,500,000	
No. 1	2″ × 8″	1,310	1,510	1,640	1,500,000	
No. 2		1,175	1,350	1,465	1,300,000	
No. 3		690	795	865	1,200,000	
Select Structural		1,770	2,035	2,215	1,600,000	
No. 1 & Btr.		1,330	1,525	1,660	1,500,000	
No. 1	2″ × 10″	1,200	1,380	1,500	1,500,000	
No. 2		1,075	1,235	1,345	1,300,000	
No. 3		635	725	790	1,200,000	
Select Structural		1,610	1,850	2,015	1,600,000	
No. 1 & Btr		1,210	1,390	1,510	1,500,000	
No. 1	2″ × 12″	1,095	1,255	1,365	1,500,000	
No. 2		980	1,125	1,220	1,300,000	
No. 3		575	660	720	1,200,000	

Source: Reproduced from the 1995 edition of the *CABO One and Two Family Dwelling Code*,™ copyright © 1995, with the permission of the publisher, The International Codes Council, Whittier, CA, excerpt from Table 502.3.1c, pp. 49–50.

Comment on Results

The values of F_b and E obtained using the two approaches agree after the appropriate adjustments are made to the tabulated values in Table 8.4. Table 9.14 is really a simplified version of the full N.D.S.-based Table 8.4. It provides **direct information about F_b** (for the normal duration, snow loading, and seven-day loading conditions) **and about E** for joists and rafters. These members are normally used as repetitive members. It does not, however, provide information about: (1) beams and stringers, (2) posts and timbers, and (3) decking, or about the other structural properties of wood members.

It also should be noted that the original structural member selection for the Case Study House was made according to the requirements in the 1989 CABO Code. According to that version of the CABO Code, the following values for No. 2 Hem–Fir could be used: $F_b =$ 1150 psi, $E = 1.4 \times 10^6$ psi.[57]

The downward adjustment of lumber property values between the 1989 CABO Code and the 1995 CABO Code illustrates a very important point: *The homebuilder should indicate on the working drawings the particular version of the CABO Code that was specified by the building code official at the time that the house was designed.*

Allowable Spans for Floor Joists

Figure 9.4 indicates that the floor joists supporting the second floor typically span from the exterior bearing wall to the interior bearing wall, while the floor joists supporting the first floor typically span from the foundation wall to the main girder in the basement of the house. Table 9.7 and Fig. 9.6 indicate that the uniform live load for sleeping areas is assumed to be 30 psf, and the uniform live load for living areas is assumed to be 40 psf. In addition, a uniform dead load of 10 psf is used to represent the weight of the floor framing and floor covering components.

Description of Span Tables

Chapter 5 of the 1995 CABO Code provides tables which can be used for both the 30- and 40-psf live loading conditions to determine the size and spacing of various species and grades of the floor joists that are required to span the distances shown in Fig. 9.4. Excerpts from these tables, which correspond to the 2 in. × 10 in. joist size used in the Case Study House, are provided in Tables 9.15 and 9.16.

It is important to note that the use of tables such as Tables 9.15 and 9.16 is restricted to a very particular loading condition. *Tables 9.15 and 9.16 can be used only for the case of*

TABLE 9.15   Allowable Spans for Floor Joists: Living Area

40 lbs per sq ft live load (All rooms except those used for sleeping areas and attic floors).
Strength—Live load of 40 lbs per sq ft plus dead load of 10 lbs per sq foot determines the fiber stress value shown.
DESIGN CRITERIA:
Deflection—For 40 lbs per sq ft live load
Limited to span in inches divided by 360.

Joist Size and Spacing		Modulus of Elasticity, "E," in 1,000,000 psi								
(in.)	(in.)	0.4	0.6	0.8	1.0	1.1	1.2	1.3	1.4	1.5
	12.0	11-4 450	13-0 590	14-4 720	15-5 830	15-11 890	16-5 940	16-10 990	17-3 1,040	17-8 1,090
	13.7	10-10 470	12-5 620	13-8 750	14-9 870	15-3 930	15-8 980	16-1 1,040	16-6 1,090	16-11 1,140
	16.0	10-4 500	11-10 650	13-0 790	14-0 920	14-6 980	14-11 1,040	15-3 1,090	15-8 1,150	16-0 1,200
2×10	19.2	9-9 530	11-1 690	12-3 840	13-2 970	13-7 1,040	14-0 1,100	14-5 1,160	14-9 1,220	15-1 1,280
	24.0	9-0 570	10-4 750	11-4 900	12-3 1,050	12-8 1,120	13-0 1,190	13-4 1,250	13-8 1,310	14-0 1,380
	32.0			10-4 1,000	11-1 1,150	11-6 1,240	11-10 1,310	12-2 1,380	12-5 1,440	12-9 1,520

For SI: 1 inch = 25.4 mm, 1 pound per square inch = 6.895 kPa, 1 pound per square foot = 0.0479 kN/m^2.

NOTE: The extreme fiber stress in bending, "F_{br}," in pounds per square inch is shown below each span.

Source: Modified from the 1995 edition of the *CABO One and Two Family Dwelling Code*,™ copyright © 1995, with the permission of the publisher, The International Codes Council, Whittier, CA, excerpt from Table 502.3.1a, p. 37.

TABLE 9.16   Allowable Spans for Floor Joists: Sleeping Area

30 lbs per sq ft live load
(All rooms used for sleeping areas and attic floors).
Strength—Live load of 30 lbs per sq ft plus dead load of 10 lbs per sq foot determines the fiber stress value shown.
DESIGN CRITERIA:
Deflection—For 30 lbs per sq ft live load
Limited to span in inches divided by 360.

Joist Size and Spacing		Modulus of Elasticity "E," in 1,000,000 psi								
(in.)	(in.)	0.4	0.6	0.8	1.0	1.1	1.2	1.3	1.4	1.5
	12.0	12-6	14-4	15-9	17-0	17-6	18-0	18-6	19-0	19-5
		440	570	700	810	860	910	960	1,010	1,060
	13.7	11-11	13-8	15-1	16-3	16-9	17-3	17-9	18-2	18-7
		460	600	730	840	900	950	1,010	1,060	1,110
	16.0	11-4	13-0	14-4	15-5	15-11	16-5	16-10	17-3	17-8
		480	630	770	890	950	1,000	1,060	1,110	1,160
2×10	19.2	10-8	12-3	13-6	14-6	15-0	15-5	15-10	16-3	16-7
		510	670	810	940	1,010	1,070	1,130	1,180	1,240
	24.0	9-11	11-4	12-6	13-6	13-11	14-4	14-8	15-1	15-5
		550	720	880	1,020	1,080	1,150	1,210	1,270	1,330
	32.0			11-4	12-3	12-8	13-0	13-4	13-8	14-0
				960	1,120	1,200	1,260	1,330	1,400	1,470

For SI: 1 inch = 25.4 mm, 1 pound per square inch = 6.895 kPa, 1 pound per square foot = 0.0479 kN/m².

NOTE: The extreme fiber stress in bending, "F_b" in pounds per square inch is shown below each span.

Source: Modified from the 1995 edition of the *CABO One and Two Family Dwelling Code*,™ copyright © 1995, with the permission of the publisher, The International Codes Council, Whittier, CA, excerpt from Table 502.3.1b, p. 39.

simply supported members which carry a uniformly distributed load. In addition, the "repetitive member use" F_b value can be used only if the floor joists are spaced not more than 24 in. on center.

The intersection of a particular *joist spacing* row and a particular E column (which corresponds to the *modulus of elasticity*, E, of a particular species and grade of lumber) results in a cell which contains two numbers. The upper number in the cell indicates the maximum allowable *clear span* of the specified joist and the lower number indicates the *maximum bending stress value*, f_b, experienced by the joist. The size, spacing, species, and grade of joist selected for a particular clear span must be such that the indicated *actual f_b value does exceed the allowable F_b value.*

Two additional points about the span tables should be noted. The first is that the clear span value, as shown in Fig. 9.16, is taken as the distance between the faces of the supports at either end of the joist. The second point is that the clear span distance in a particular cell is sometimes determined by the allowable live load deflection criteria of the $L/360$ rather than the maximum bending stress criterion.

Use of Span Tables

The span tables provided in Chapter 5 of the 1995 CABO Code can be used by a building code official who is checking a particular house design or inspecting the installation of the structural framing system in a house. The tables can also be used by an architect or homebuilder who is designing the flooring systems for a new house. The procedures that are used in these two types of applications are indicated in the form of checklists:

Figure 9.16
Definition of
Clear Span
Distance
Source: Reproduced
from the Applica-
tion and Commen-
tary, 1992 edition of
the *CABO One and
Two Family
Dwelling Code,*™
copyright © 1992,
with the permission
of the publisher,
The International
Codes Council,
Whittier, CA, Fig.
602.2.1a, p. 115.

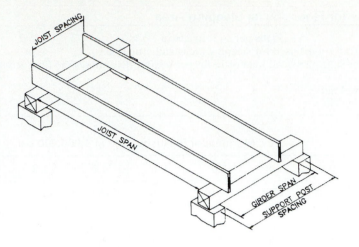

I. Checklist for Inspection or Checking
   A. Determine:
      1. Grade, species, and moisture condition
      2. Size of joist
      3. Span (clear)
      4. Joist spacing
      5. Loading condition (sleeping area or other)
   B. Enter Table 9.14 and determine:
      1. F_b (allowable fiber stress in bending)
      2. E (modulus of elasticity)
   C. Enter appropriate standard span table (Table 9.15 or 9.16) with:
      1. F_b
      2. E
      3. Size of joist
      4. Joist spacing
   D. Read allowable span and compare with the actual span
II. Checklist for Design
   A. Determine:
      1. Span
      2. Joist spacing desired
   B. Enter standard span table (Table 9.15 or 9.16) and determine:
      1. Size of joist
      2. F_b
      3. E
   C. Enter Table 9.14 with:
      1. F_b
      2. E
   D. Read grade and species of various sizes of lumber permitted[58]

In the design process, it is possible to select any joist spacing desired. From a modularity viewpoint, however, the six joist spacings indicated in Tables 9.15 and 9.16 are probably used most often because the overall 8-ft distance that results accommodates the 4 ft × 8 ft size of a sheet of floor sheathing material. This feature is indicated in Fig. 9.17.

Case Study House Example

A review of Sheet A-3 of the working drawings for the Case Study House (Appendix A) indicates that there are two types of members supporting the *first floor*. The first type consists of the primary floor joist system which includes:

1. 12" SPACING – 8 SPACES, 9 JOISTS

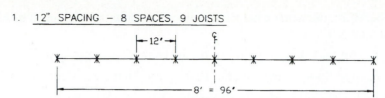

2. 13.7" SPACING – 7 SPACES, 8 JOISTS

13.7" X 7 = 96"

3. 16" SPACING – 6 SPACES, 7 JOISTS

16.0" X 6 = 96"

4. 19.2" SPACING – 5 SPACES, 6 JOISTS

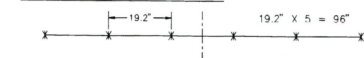

19.2" X 5 = 96"

5. 24" SPACING – 4 SPACES, 5 JOISTS

24.0" X 4 = 96"

6. 32" SPACING – 3 SPACES, 4 JOISTS

32.0" X 3 = 96"

Figure 9.17
Typical Floor
Joist Spacings

1. 2 in. × 10 in., 12 ft long @ 16 in. o.c.
2. 2 in. × 10 in., 14 ft long @ 16 in. o.c. (not shown: these members span between the stairway and the rear foundation wall)
3. 2 in. × 10 in., 16 ft long @ 16 in. o.c.

The primary floor joist system can be evaluated using Table 9.15.

The second type includes the double 2 in. × 10 in. members which are used to frame the stairway opening. It also includes the 2 in. × 10 in., 18-ft-long floor joists that span between the center girder and the rear foundation wall and are partially supported by the two 1-3/4 in. × 9-1/4 in. LVL (laminated veneer lumber) members. *Both of these situations can be evaluated only by using the structural analysis procedures discussed in Chapter 11 of this book.*

The primary joist system supporting the *second floor* (Sheet A-1) consists of:

1. 2 in. × 10 in., 12 ft long @ 16 in. o.c.
2. 2 in. × 10 in., 14 ft long @ 16 in. o.c.
3. 2 in. × 10 in., 16 ft long @ 16 in. o.c.

These members can be evaluated using Table 9.16. *The two 2 in. x 10 in. floor joists which frame the open foyer must, however, be evaluated using the structural analysis procedures discussed in Chapter 11 of this book.*

Therefore, the worst-case primary floor joist situation that must be examined for structural adequacy is

- First floor—2 in. × 10 in., 16 ft long @ 16 in. o.c., living area
- Second floor—2 in. × 10 in., 16 ft long @ 16 in. o.c., sleeping area

The **Checklist for Inspection or Checking** can be used to determine if these members are adequately sized for the loading conditions indicated.

FIRST FLOOR JOISTS

I. Checklist for Inspection and Checking
   A. Determine:
      1. Grade, species, and moisture condition
            No. 2 Hem–Fir, Surfaced Dry
      2. Size of joist
            2 in. x 10 in.
      3. Span (clear)

The 1995 CABO Code allows the clear span to be used when the structural adequacy of a floor joist is being determined. Unfortunately, clear span information is not directly provided on the working drawings. Figure 9.18 indicates that a tedious calculation is required to determine that the clear span of a 16-ft-long, 2 in. × 10 in. first-floor joist is 15-ft., 4-3/4 in. Therefore, it is recommended that the actual length of 16 ft be used initially to determine structural adequacy. If it appears that the final answer is "close," then a further refinement of the floor joist span length may be required.

Therefore, Span = 16 ft.

II. Enter Table 9.14 and determine:
   A. F_b (allowable fiber stress in bending)
         For No. 2 Hem–Fir, Surfaced Dry, this was found in the earlier example
         to be F_b = 1,075 psi
   B. E (modulus of elasticity)
         For No. 2 Hem–Fir, Surfaced Dry, this also was found in the earlier example to be E = 1,300,000 psi
   C. Enter Table 9.15 with:
      1. F_b ·   1,075 psi
      2. E 1,300,000 psi
      3. Size of joist   2 in. × 10 in.
      4. Joist spacing   16 in. o.c.
   D. Read the allowable span and compare with the actual span

The allowable span of 2 in. × 10 in. No. 2 Hem–Fir joists with a spacing of 16 in. o.c. is 15 ft., 3 in. The actual maximum bending stress, f_b, at that span distance is 1,090 psi, which exceeds the allowable bending stress, F_b, of 1,075 psi. Thus, these joists are inadequate,

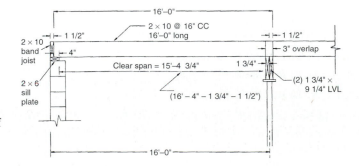

Figure 9.18
Determination of the Clear Span Distance

according to the requirements of the 1995 CABO Code, even if the clear span distance of 15 ft., 4-3/4 in. (Fig. 9.18) is considered.

The reason for the inadequacy should again be noted, because it raises an important point. As mentioned earlier, the Case Study House was originally designed under the requirements of the **1989 CABO Code** which indicated that $F_b = 1,150$ psi and $E = 1.4 \times 10^6$ psi for No. 2 Hem–Fir. Using those criteria, the 2 in. $\times$ 10 in. joists with a spacing of 16 in. o.c. are clearly adequate. *Thus, the adequacy of the structural design for a particular house is sometimes associated with the version of the building code under which it was designed.*

An examination of Table 9.15 indicates that several alternatives are available to ensure that the first-floor joists that span 16 ft are in compliance with the 1995 CABO Code.

Alternative #1—The simplest alternative is probably to upgrade the species and grade of 2 in. $\times$ 10 in. lumber that is used. Although a large number of choices of wood offering different combinations of F_b and E are available, it is logical to start with No. 1 Hem–Fir. If this upgrade is used, the appropriate structural values are:

$$\text{No. 1 Hem–Fir: } F_b = 1,200 \text{ psi}, E = 1.5 \times 10^6 \text{ psi (Table 9.14)}$$

Then, according to Table 9.15:

$$\text{Allowable span (16 in. o.c.)} = 16 \text{ ft, actual } f_b = 1,200 \text{ psi}$$

Thus, No. 1 Hem–Fir floor joists can be used.

Alternative #2—The spacing of the joists that span 16 ft can be reduced if the home-builder wants to use 2 in. $\times$ 10 in. No. 2 Hem–Fir joists throughout the first floor. If the spacing is reduced to either 13.7 in. o.c. or 12 in. o.c., for instance,

$$\text{No. 2 Hem–Fir: } F_b = 1,075 \text{ psi}, E = 1.3 \times 10^6 \text{ psi}$$

Then, according to Table 9.15:

Spacing = 13.7 in. o.c. − allowable span = 16 ft, 1 in.
Actual f_b = 1,040 psi
Spacing = 12 in. o.c. − allowable span = 16 ft, 10 in.
Actual f_b = 990 psi

Thus, either spacing satisfactorily meets the requirements.

Additional Alternatives—Another alternative which the homebuilder could consider would be to use 2 in. $\times$ 12 in. floor joists at an appropriate spacing. Table 502.3.1a in the 1995 CABO Code (from which Table 9.15 was obtained) can be consulted for such information. In essence, an optimum design decision which considers size, spacing, and species/grade of lumber must also consider an economic variable in the analysis.

A slightly different approach is necessary when Table 9.15 is used to determine the structural adequacy of the other No. 2 Hem–Fir 2 in. $\times$ 10 in. first-floor joists, which are also spaced at 16 in. o.c.

- For the 14-ft span:
  Table 9.15 indicates that a floor joist with an E value of 1.0×10^6 psi can span 14 ft while experiencing a maximum bending stress of $f_b = 920$ psi. Thus, No. 2 Hem–Fir floor joists with $F_b = 1,075$ psi and $E = 1.3 \times 10^6$ psi are adequate.
- For the 12-ft span:
  Table 9.15 indicates that a floor joist with an E value of 0.8×10^6 psi can span 13 ft while experiencing a maximum bending stress of $f_b = 790$ psi. Thus, No. 2 Hem–Fir floor joists with $F_b = 1,075$ psi and $E = 1.3 \times 10^6$ psi are adequate.

The 14-ft and 12-ft span situations indicate that the use of No. 2 Hem–Fir actually represents an **overdesign** for the use intended since a lower-level species/grade combination, which might be less expensive, could have been used. Local market conditions for different types of lumber and the problems involved with mixing various types of lumber on the same project often make the overdesign choice the correct alternative.

SECOND-FLOOR JOISTS

Essentially, the same procedure as just described is followed when determining the adequacy of the second-floor joists. The major difference is that Table 9.16, which corresponds to the sleeping area situation and a specified live load of 30 psf, is used instead of Table 9.15. A comparison of the two tables should be made to understand the basic differences/similarities involved.

The pertinent information consists of:

- No. 2 Hem–Fir − F_b = 1,075 psi, $E = 1.3 \times 10^6$ psi
- 2 in. × 10 in. floor joists at 16 in. o.c.
- For the 16-ft. span
  According to Table 9.16:
      Allowable span = 16 ft, 10 in., f_b = 1,060 psi

Thus, No. 2 Hem–Fir floor joists can be used.

- For the 14-ft span
  According to Table 9.16:
      A floor joist with an $E = 0.8 \times 10^6$ psi can span 14 ft, 4 in. while sustaining a maximum f_b = 770 psi

Thus, No. 2 Hem–Fir floor joists can be used:

- For the 12-ft span
  According to Table 9.16:
      A floor joist with an $E = 0.6 \times 10^6$ psi can span 13 ft while sustaining a maximum f_b = 630 psi

Thus, No. 2 Hem–Fir floor joists can be used.

Several observations should be made about the second-floor joist results. The first is that the downward adjustment of the structural properties of No. 2 Hem–Fir lumber between the 1989 CABO Code and the 1995 CABO Code did not result in a design change for the 16-ft span situation. This is probably due to the lower value of the live load that is used for the sleeping floor. The second observation relates to the overdesign for the 14-ft and 12-ft spans that is similar to the first-floor joist situation.

Additional Structural Design Considerations

In Chapter 8 of this book, Table 8.3 presented the primary structural members of a house frame and their associated stress states. The portion of Table 8.3 that applies to members that are experiencing primarily bending stress is shown in Table 9.17. It is clear that bending stresses are considered in Tables 9.15 and 9.16. Deflection is also accounted for by considering the modulus of elasticity, E, as well as the deflection limitation of span length/360 in the designation of the allowable span length. As noted in Chapter 8, bearing stress is

TABLE 9.17   Stress States of Structural Members Exposed to Bending-Induced Loads

Structural Member	Situation	Symbol Describing the Situation
1. Floor Joist, Ceiling Joist, Built-Up Girder	a. Bending Stress (psi)	f_b
	b. Horizontal Shearing Stress (psi)	f_v
	c. Deflection (in.)	Δ
	d. Bearing Stress (psi)	$f_{c\perp}$

accounted for by assuring that there is a satisfactory bearing length between the load-transmitting member and its support point.

It appears, therefore, that the only stress situation that has not been directly considered in the 1995 CABO Code-based procedure is the horizontal shearing stress. In that regard, it can be shown by structural analysis that the horizontal shearing stress is not as critical in the design of a joist, beam, etc., as bending moment and deflection. Horizontal shearing stress usually governs the design of short, very heavily loaded members. It can be assumed that as Tables 9.15 and 9.16 were developed, if such a shear governing situation applied, the allowable span for those situations was properly adjusted.

Allowable Main Girder Spans

Figure 9.6 indicates that the entire load transmitted by the middle of the house is typically carried by the main girder in the basement. Figure 9.7 indicates that this load is transmitted to a series of columns (usually either wood or steel pipe) which transmit the load to individual spread footings under the columns. In addition, the two end spans of the main girder are also supported by the foundation wall.

The individual point loads which are transmitted to the main girder by the floor joists are usually represented as a uniform load since the joists are closely spaced. The main girder cannot, however, be treated as if it were a floor joist because it is not a repetitive member.

Two additional complications must be considered. The first relates to the total uniform load which the girder is assumed to carry. The magnitude of the load depends upon the number of floors which the girder must support and whether or not a portion of the roof/attic loads are transmitted to the girder (i.e., is the house designed according to Fig. 9.4 or Fig. 9.8?). The second complication relates to whether the main girder is structurally represented as a series of short, independent, uniformly loaded, simple beams or as a uniformly loaded continuous beam over a number of supports.

As a result, a different set of tables is used in the 1995 CABO Code[59] to analyze or design the main girder. The information presented in these tables is shown in Girder Table #1 (see Table 9.18) and Girder Table #2 (see Table 9.20).

TABLE 9.18   Girder Table #1: Allowable Span for Girders Supporting Only One Floor[a]

Size of Wood Girder[b]		Floor Live Load (psf)	Spacing between Girders or between Girders and Load-Bearing Walls[c]				
			4 ft	6 ft	8 ft	10 ft	16 ft
4×4	—	30	5'-6"	4'-6"	3'-6"	3'-0"	2'-6"
		40	5'-0"	4'-0"	3'-6"	3'-0"	2'-6"
4×6	—	30	8'-0"	6'-6"	5'-6"	5'-0"	4'-6"
		40	7'-6"	6'-0"	5'-6"	4'-6"	4'-0"
4×8	6×6	30	11'-0"	9'-0"	8'-0"	7'-0"	5'-6"
		40	10'-0"	8'-6"	7'-6"	6'-6"	5'-0"
4×10	6×8	30	14'-0"	11'-6"	10'-0"	8'-6"	6'-0"
		40	13'-0"	10'-6"	9'-6"	8'-6"	5'-6"
4×12	6×10	30	16'-6"	14'-0"	12'-0"	11'-0"	9'-0"
		40	16'-0"	12'-6"	11'-0"	11'-0"	8'-0"

For SI: 1 in. = 25.4 mm, 1 ft = 304.8 mm, 1 psf = 0.0479 kN/m^2.
[a]Allowable spans may be interpolated between tributary loads shown in the table. Span and girder sizes may be computed independently of the table in accordance with accepted engineering practice.
[b]Spans are based on No. 2 lumber.
[c]The spacing is the tributary load to the girder. It is found by adding the unsupported spans of the floor structure on each side which are supported by the girder and dividing by 2.

Source: Reproduced from the 1995 edition of the *CABO One and Two Family Dwelling Code,*™ copyright © 1995, with the permission of the publisher, The International Codes Council, Whittier, CA, Table 502.3.3a, p. 41.

Girder Table #1

The girder table that is the simpler one to use, but which is more restrictive, is Table 9.18. It can be used when the main girder supports only one floor of a house. Two hypothetical houses are analyzed in order to illustrate the use of Table 9.18. Hypothetical House #1, as shown in Fig. 9.19, contains one main girder with a different spacing distance on each side to indicate how Table 9.18 is used in such a situation. Figure 9.20 presents Hypothetical House #2, which illustrates the case of two parallel main girders which are spaced 10 ft apart. Note that both of these houses meet the requirement of the *girder supporting only one floor* since the roof truss in each case carries the loads through the exterior bearing walls directly to the foundation wall.

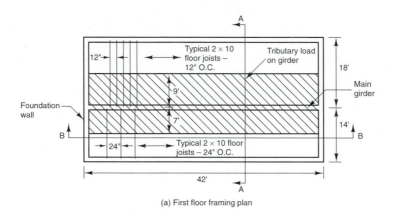

(a) First floor framing plan

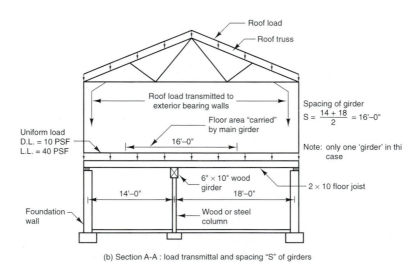

(b) Section A-A : load transmittal and spacing "S" of girders

Figure 9.19
Hypothetical
House #1

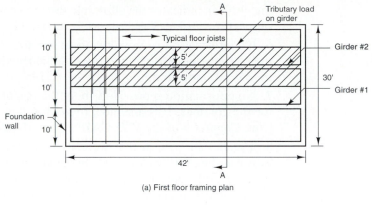

(a) First floor framing plan

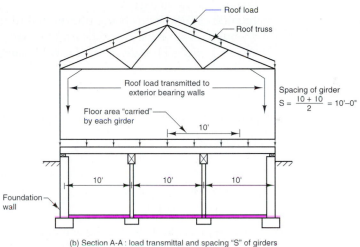

Figure 9.20
Hypothetical
House #2

(b) Section A-A : load transmittal and spacing "S" of girders

Several important points should be noted about Table 9.18:

1. Only wood girders of No. 2 grade lumber (with no species information provided) are covered in the table. In certain instances, two girder sizes are presented as acceptable alternatives for the same loading condition.

2. If the sizes of girders presented in the table are considered too restrictive by the builder, footnote (a) provides the alternative of an engineering design that would require approval by the building code official/inspector.

3. Two live load choices, one for bedrooms (i.e., 30 psf) and the other for living areas (i.e., 40 psf), are provided. Although not directly stated, it probably can be assumed that a dead load of 10 psf is considered in the table because this would be consistent with the assumptions made in the floor joist tables.

4. The right-hand side of the table is divided into a number of **"Spacing of Girder— S"** columns. Footnote (c) indicates the procedure for determining "S" for a given house design. As noted in Figs. 9.19(a) and (b) for the one-girder situation, the value of S (which measures the floor area that is carried by the girder) is:

$$S = (14 + 18)/2 = 16 \text{ ft}$$

(Note: The distance to the foundation wall, not to an adjacent girder, is considered.)

For the two-girder situation, as shown in Fig. 9.20, the value of S is:

$$S = (10 + 10)/2 = 10 \text{ ft}$$

(simply the average spacing of the girders)

5. The distance indicated in each individual "cell" of the table, as shown in Fig. 9.19(c), indicates the allowable distance between the wood or steel columns which support the girder. This distance is referred to as the **span of the girder**.

6. No information is provided about the size, species, or grade of the lumber used in the columns supporting the girder. Similarly, the size of the steel columns that can be used is not indicated.

7. No information is provided about the size of the spread footing that should be used for the columns. It is assumed, therefore, that the information presented in Fig. 9.12 from Chapter 4 of the 1995 CABO Code should be used for this purpose.

8. The design criteria that were used to develop Table 9.18 are not indicated. As shown in Fig. 9.19(c), one possible assumption could be that the girder is continuous over several supports, which would result in a smaller-sized girder being used. Alternatively, it can be assumed that the girder consists of a series of short, simple beams. This is an extremely conservative assumption which is not always cost effective. In addition, the loads being transmitted to the girder from the floor joists can be represented as a series of concentrated loads instead of the equivalent uniform load. Again, there is no indication of which assumption was made.

EXAMPLE: HYPOTHETICAL HOUSE #1

The adequacy of the girder design for Hypothetical House #1 can be determined as follows:

Given: Size of girder = 6 in. $\times$ 10 in.
             Spacing of girders = 16 ft
             The girder supports a living area; therefore, the live load = 40 psf
Assumption: The grading stamp indicates that No. 2 grade lumber (or better) is being used.

As indicated in Table 9.18, the allowable span of the girder = 8 ft. Since the actual span = 7 ft [Fig. 9.19(c)], the girder design is considered to be satisfactory.

It should be noted that if the average spacing "S" had been 14 ft, for instance, instead of 16 ft, the allowable span of the girder could have been determined by interpolation, as follows:

When S = 10 ft, then the girder span = 10 ft.
When S = 16 ft, then the girder span = 8 ft.
Therefore, when S = 14 ft, the girder span = 10 ft $-$ 4/6(10 ft $-$ 8 ft) = 8 ft, 8 in.
Thus, when S = 14 ft, the allowable distance between columns is 8 ft, 8 in.
NOTES: **a.** It is not possible to determine the appropriate size column which should be used for this application.
       b. Given the limitations of Table 9.18, it may be worthwhile for the homebuilder to retain the services of an architect or an engineer to structurally design the main girder (see Chapter 11) if an optimum design is desired.

EXAMPLE: HYPOTHETICAL HOUSE #2

Less information about the structural framing is provided in Fig. 9.20 for Hypothetical House #2. Neither the size of the two girders nor their span lengths (i.e., the distance between the columns) is specified.

Given: Size of girders = ?
             Spacing of girders = 10 ft
             The girders support a living area; therefore, the live load = 40 psf
Assumption: The grading stamp indicates that No. 2 grade lumber (or better) is being used.

TABLE 9.19 Possible Girder Size/Girder Span
Combinations according to Table 9.18 when
the Girder Spacing = 10', 0"

Size of Wood Girder	Allowable Span of Girder
4"×4"	3'-0"
4"×6"	4'-6"
4"×8" or 6"×6"	6'-6"
4"×10" or 6"×8"	8'-6"
4"×12" or 6"×10"	10'-0"

The possible girder size/girder span combinations that can be used are shown in Table 9.19. The homebuilder should perform an economic analysis to determine which option is to be selected.

Girder Table #2

The other girder table which is provided in the 1995 CABO Code is shown in an abbreviated format in Table 9.20. Several important points should be noted about Table 9.20:

1. The load imposed on the **main center girder** of the house varies directly with the width of the house; as the house becomes wider, the load increases. As a result, information related to four different commonly encountered housing widths (24, 26, 28, and 32 ft) are provided. Two of these are shown in Table 9.20 so that the important concepts can be highlighted.

2. The load imposed on the **main center girder** of the house also varies directly with the number of floors that ultimately are being supported by the girder. As noted in footnote (a), the values in Table 9.20 are based on the assumption that a **clear-span trussed roof** is used, which implies that the roof and attic loads are supported by the exterior walls (Fig. 9.8). As a result:

- One-story construction implies that there is not a load-bearing center wall between the first floor and the ceiling.
- Two-story construction implies that there is only a center bearing wall between the first floor and the second floor. The space between the second floor and the second-floor ceiling does not require a center bearing wall (Fig. 9.8).
- Three-story construction implies that there is a center bearing wall between the first floor and the second floor and between the second floor and the third floor. The space between the third floor and the third-floor ceiling does not require a center bearing wall.

3. The increased load carried by the main center girder for a given house width as the number of stories increases can be illustrated for the 28-ft-wide house case if a girder composed of four 2 in. × 12 in. members is used. The decrease in allowable column spacing and the increase in the size of the footing is:

Number of Stories	Maximum Span (ft-in.)	Footing Size (in.)
1	11 ft	22 in. × 22 in.
2	8 ft, 2 in.	28 in. × 28 in.
3	6 ft, 10 in.	30 in. × 30 in.

4. The increased load carried by the main center girder for a given number of stories as the house width increases can be illustrated for the two-story house case if a girder

TABLE 9.20　Girder Table #2: Allowable Spans for Built-Up Wood Center Girders and Footing Sizes for Girder Support Columns

Width of Structure (ft)	Girder Size (in.)	One Story		Two Story		Three Story	
		Max. Span (ft-in.)	Footing Size (in.)	Max. Span (ft-in.)	Footing Size (in.)	Max. Span (ft-in.)	Footing Size (in.)
24	3-2×8	6-7	17×17*	4-11	20×20	4-1	22×22
	4-2×8	7-8	19×19*	5-8	21×21	4-9	24×24
	3-2×10	8-5	20×20*	6-3	23×23	5-3	25×25
	4-2×10	9-9	21×21	7-3	24×24	6-1	27×27
	3-2×12	10-3	22×22	7-8	25×25	6-4	27×27
26	Not shown: See 1995 CABO Code, Table 502.3.3b						
28	3-2×8	6-2	17×17*	4-7	21×21	3-10	23×23
	4-2×8	7-1	18×18*	5-3	22×22	4-5	24×24
	3-2×10	7-10	19×19	5-10	23×23	4-10	26×26
	4-2×10	9-0	20×20	6-9	25×25	5-7	28×28
	3-2×12	9-6	21×21	7-1	26×26	5-11	28×28
	4-2×12	11-0	22×22	8-2	28×28	6-10	30×30
32	Not shown: See 1995 CABO Code, Table 502.3.3b						

For SI: 1 in. = 25.4 mm, 1 ft = 304.8 mm, 1 psf = 0.0479 kN/m², 1 psi = 6.895 kPa.

NOTES:
1. Values shown are for a clear-span trussed roof, a load-bearing center wall on the first floor in two-story construction, and a load-bearing center wall on the first and second floor in three-story construction.
2. Spans based on allowable bending moment F_b = 1,000 pounds per square inch (psi) for repetitive members.
3. Footing size based on 2,000 psf soil-bearing capacity; footing thickness shall be one-half (minimum) the width of the footing.
4. 4×4 posts may be used at these (*) locations; 6×6 posts, or 4×4 posts or 3-inch diameter steel columns with bearing plates or equivalent area are acceptable in all locations.

Source: Modified from the 1995 edition of the *CABO One and Two Family Dwelling Code,*™ copyright © 1995, with the permission of the publisher, The International Codes Council, Whittier, CA, excerpt from Table 502.3.3b, p. 41.

composed of four 2 in. × 12 in. members is used. The decrease in allowable column spacing and the increase in the size of the footing is:

Width of House (ft)	Maximum Span (ft-in.)	Footing Size (in.)
24 ft	8 ft, 10 in	27 in. × 27 in.
28 ft	8 ft, 2 in.	28 in. × 28 in.

5. In order to develop Table 9.20, certain additional assumptions had to be made. Footnote (b), for instance, indicates that it was assumed that the species/grade of lumber used had an allowable bending stress of F_b = 1,000 psi. Footnote (c) indicates that the soil-bearing capacity was assumed to be 2,000 psf, which according to Table 9.9 is a fairly conservative assumption. It was also assumed that the footing thickness was to be a minimum of one-half of the footing width.

The combinations which allow 4 in. × 4 in. wood columns to be used are highlighted. In all other cases, either 6 in. × 6 in. wood columns or 4 in. × 4 in. wood posts/3 in. diameter pipe columns with bearing plates of an area equivalent to 6 in. × 6 in. are acceptable.

EXAMPLE: CASE STUDY HOUSE

Initially, it appears that Table 9.20 provides the girder size, girder span, column size, and footing size information which a homebuilder can use for the commonly encountered main girder situations. A careful analysis of Sheet A-5 of the Case Study House (Appendix A), however, indicates that Table 9.20 cannot be used for the "Pauline" model. Even though the requirement of a clear-span truss roof is met, the bearing wall between the first and second floors and the main girder are not centered (the distances from those load-bearing members to the side walls are 16 ft and 12 ft). As a result, the structural analysis procedures discussed in Chapter 11 of this text must be used to design the main girder system for the Case Study House.

These structural analysis procedures would also have to be used if the design of the "Pauline" model included a conventional roof-rafter/ceiling-joist system (with the associated bearing wall between the first floor and the ceiling of the second level), such as the one shown in Fig. 9.4.

EXAMPLE: MODIFIED CASE STUDY HOSE

The use of Table 9.20 can be illustrated by considering the modified version of the Case Study House which is shown in Fig. 9.21. A center girder and a center bearing wall between the first and second floors, as well as a clear span trussed roof, have been assumed. In addition, the 3-in. pipe columns are spaced so that the span of the main girder is 7 ft, 6 in. (which corresponds to the spacing shown on Sheet A-3).

For the 28-ft-wide row and the two-story house column shown in Table 9.20, the following design information can be obtained:

- Required size of girder = four 2 in. × 12 in. members
- Maximum allowable girder span = 8 ft, 2 in.
- Size of spread footing = 28 in. × 28 in. × 14 in.
- Size of pipe column = 3 in. with a 6 in. × 6 in. bearing plate

Table 9.20 assumes that the species/grade of lumber will have an F_b = 1,000 psi. Reference to Table 9.14 indicates that F_b = 980 psi for a No. 2 Hem–Fir 2 in. × 12 in. member and F_b = 1,095 psi for a No. 1 Hem–Fir 2 in. × 12 in. member. In order to comply with the requirements of Table 9.20, therefore, No. 1 Hem–Fir lumber must be used for the main girder.

This information can be compared with the following information about the actual design used:

- Actual size of girder = two 1-3/4 in. × 9-1/4 in. LVL members (Sheets A-3 and A-5)
- Actual girder span = 7 ft, 6 in. (Sheet A-3)
- Size of spread footing = 24 in. × 24 in. × 12 in. (Sheets A-3 and A-5)
- Size of pipe column = 3 in.

Since LVL headers were used, a direct comparison with the results from Table 9.20 cannot be made. Since the actual girder span is less than the allowable span for the four 2 in. × 12 in. girders, it could replace the LVL members without the need to adjust the distance between columns. The actual size of the spread footing is less than the required value. This is probably due to the fact that a soil-bearing capacity greater than 2,000 psf was assumed when the Case Study House was designed.

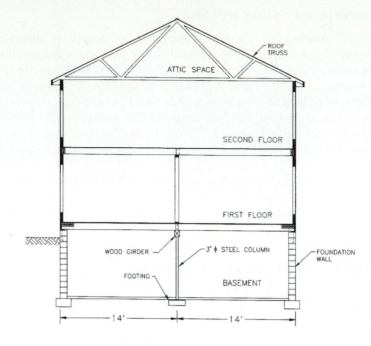

(a) Structural Section

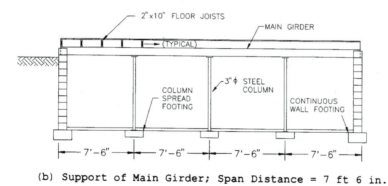

Figure 9.21
Modified Case
Study House (a)
Structural Section
(b) Support of
Main Girder;
Span Distance—
7 ft, 6 in.

(b) Support of Main Girder; Span Distance = 7 ft 6 in.

Additional Floor Framing Details

In addition to establishing the requirements related to wood properties, allowable spans for floor joists, and allowable main girder spans, the 1995 CABO Code also provides detailed information about specific aspects of floor framing. Included are requirements related to:

- Blocking and subflooring
- End-jointed lumber
- Joists under bearing partitions
- Bearing
- Floor systems
- Joist framing
- Lateral restraint at supports
- Bridging
- Drilling and notching
- Holes

- Fastening
- Framing of openings
- Floor trusses
- Draftstopping required[60]

These detailed requirements, although they seldom appear directly on a set of working drawings, should become part of a homebuilder's quality management checklist for the floor framing phase of the construction process.

Floor Sheathing

Floor sheathing can be applied in either a double-floor or a single-layer format. The traditional double-floor system consists of: (1) a **subfloor** which is applied directly to the structural framing members, and (b) a **finish floor** or **underlayment**. Alternatively, a single layer of underlayment grade plywood or wood structural panel, with tongue-and-groove edges, may be applied directly to the structural framing members. The single-layer sheathing, which can either be nailed or glued to the floor joists, provides savings in labor and material and is appropriate, for instance, in areas where wall-to-wall carpet or resilient flooring will be used.

Span Rating Tables

The 1995 CABO Code[61] presents requirements for: (1) lumber sheathing, (2) plywood sheathing, and (3) particleboard. Allowable spans and loads for various thicknesses of plywood and wood structural-use panels (which include performance-rated plywood, oriented strandboard, and composite panels) are shown in Table 9.21. The applications for these products in Table 9.21 include roofing, subfloor, and combination subfloor underlayment. The span limitations for each of these applications are associated with the specific grade of plywood that is used. Since plywood is performing a load-carrying function when it is used as part of a roof or floor system, it must conform to a known quality-control standard. Proof of conformance is provided by a grade mark or a certificate of inspection from an approved agency.

Span ratings for C–D, C–C, sheathing are provided in Table 9.21 in the format 24/0, 24/16, etc. In this format, the numerator and denominator represent the allowable span (in inches) for roof and floor applications, respectively. Thus, a 24/0 rating indicates that the sheathing can be used on a roof when the rafters or roof trusses are spaced at 24 in. o.c., but it cannot be used on a floor.

Case Study House Example

The floor sheathing material that is used in the Case Study House consists of 4 ft × 8 ft sheets of 3/4-in. tongue-and-groove plywood that is glued to the floor joists (Sheet A-5 in Appendix A and Item 7, "Subflooring" in Appendix C). Although this appears to be sufficient for the carpeted areas of the house, Item 21, "Special Floors and Wainscot" of Appendix C indicates that an additional layer of 1/4-in. underlayment is used under the vinyl flooring in the kitchen and bathrooms.

Concrete Floors

Concrete floors are typically installed in the basement and garage of the house. The 1995 CABO Code[62] provides a brief set of requirements for the slab-on-ground situation in which the floor loads are transmitted by the slab and directly distributed to the supporting soil.

TABLE 9.21 Allowable Spans and Loads for Plywood and Wood Structural Panels for Roof and Subfloor Sheathing and Combination Subfloor Underlayment[a,b,c]

Span Rating	Nominal Panel Thickness (in.)	Maximum Span (in.)* With Edge Support	Maximum Span (in.)* Without Edge Support	Load (psf at max. span) Total Load	Load (psf at max. span) Live Load	Max. Span (in.)
C–D, C–C, Sheathing*		Roof*				Subfloor*
12/0	5/16	12	12	40	30	0
16/0	5/16, 3/8	16	16	40	30	0
20/0	5/16, 3/8	20	20	40	30	0
24/0	3/8, 7/16, 1/2	24	20*	40	30	0
24/16	7/16, 1/2	24	24	50	40	16
32/16	15/32, 1/2, 5/8	32	28	40	30	16*
40/20	19/32, 5/8, 3/4, 7/8	40	32	40	30	20*
48/24	23/32, 3/4, 7/8	48	36	45	35	24
Underlayment, C–C Plugged, Single Floor*		Roof*				Combination Subfloor Underlayment*
16 o.c.	19/32, 5/8	24	24	50	40	16*
20 o.c.	19/32, 5/8, 3/4	32	32	40	30	20*
24 o.c.	23/32, 3/4	48	36	35	25	24
32 o.c.	7/8, 1	48	40	50	40	32
48 o.c.	1-3/32, 1-1/8	60	48	50	40	48

For SI: 1 in. = 25.4 mm, 1 psf = 0.0479 kN/m².

[a]The allowable loads were determined using a dead load of 10 psf. If the dead load exceeds 10 psf, then the live load shall be reduced accordingly.

[b]Panels continuous over two or more spans with long dimension perpendicular to supports. Spans shall be limited to values shown because of the possible effect of concentrated loads.

[c]Applies to panels 24 in. or wider.

*See Table 503.2.1.1a in the Code for other important footnotes.

Source: Reproduced from the 1995 edition of the *CABO One and Two Family Dwelling Code,™* copyright © 1995, with the permission of the publisher, The International Codes Council, Whittier, CA, Table 503.2.1.1a, p. 43.

Structural slabs (which are required in some areas of the United States because of existing soil conditions or seismic activity) are not addressed since they must be designed properly by a licensed professional engineer.

Specific Requirements

The appropriate design details for the slab-on-ground situation are provided in Fig. 9.12. As indicated, the minimum slab thickness that is specified is 3-1/2 in. Some of the other specific requirements which could become a part of the homebuilder's quality management program include:

- The minimum concrete compressive strength is specified as 2,500 psi unless weather exposure requires greater strength and air-entrained concrete (see Table 9.10).
- All vegetation, top soil, and foreign material must be removed from the slab areas and the fill that is placed should be compacted to assure uniform support of the slab.
- Slabs below grade require that a 4-in.-thick base course of specified material be placed on the prepared subgrade unless the existing soil is well drained or is a sand–gravel mixture.

- An approved **vapor barrier** must be placed between the concrete floor slab and the base course or the prepared subgrade where no base course exists. The vapor barrier may be omitted in certain specified areas.[63]

Specific requirements for crack control joints or steel reinforcement for the concrete slabs are not provided in the 1995 CABO Code.

Case Study House Example

Sheet A-1 of the set of working drawings for the Case Study House (Appendix A) indicates that a 4-in. concrete slab is used in the garage, and Sheet A-3 indicates that a 3-in. concrete slab is specified for the basement. Similar information about the thickness of the basement slab as well as the use of a "poly vapor barrier" under the slab is shown on Sheet A-5. The requirement for 2 in. to 4 in. of crushed stone fill under the basement slab and the thickness specifications for the vapor barrier are provided in Item 6, "Floor Framing" in Appendix C. There does not appear to be any additional specifications for the floor slab in the garage in Appendix C, nor does there appear to be a specific reference to the compressive strength of the concrete for either the basement or the garage slab.

In order to ensure compliance with the 1995 CABO Code, it appears that the thickness of the basement slab should be increased to 3-1/2 in. and there must be a specific reference added about the compressive strength of the concrete that is used.

Other Floors

Requirements for two additional flooring systems are presented in Chapter 5 of the 1995 CABO Code. These are **Treated-Wood Floors** (On Ground),[64] and **Metal Floors**.[65] Only a one-sentence requirement about metal floors related to straightness and freedom from defects is included. As a result, neither of these floor systems is discussed in this chapter of the book.

CHAPTER SUMMARY

The objective of this chapter was to present the structural systems of a house within a context that provides a consistent theme and framework. The context that was chosen is the 1995 CABO Code because it was developed as a universal building code for residential construction that could be adopted anywhere in the United States.

The first part of this chapter introduces a number of conventional wood structural framing systems and indicates that the platform framing system is commonly used in the United States. The design of the various structural systems in a house cannot begin until the design criteria have been established. Therefore, the second major part of the chapter discussed how the climatic and geographical design criteria, as well as the dead and live loads, are established within the 1995 CABO Code context.

Foundation design, both in terms of the loads that must be accommodated as well as the problems associated with foundation drainage, waterproofing, and dampproofing, is then addressed. As a part of the presentation on Foundations, the Case Study House (Appendix A) is introduced and the foundation design of that house is evaluated.

Floor systems are addressed next. At the beginning of the floor system section, the wood properties first introduced in Chapter 8 of this book are related to the selection of floor joists. An example related to the Case Study house explains how the span tables for floor joists found in the 1995 CABO Code are used. One of the major structural members of the structural frame of a house is the main girder in the basement. A number of examples are provided to indicate how the girder tables found in the 1995 CABO Code are used. Additional floor-related topics are also presented briefly.

The fairly comprehensive treatment of foundation and wood structural floor system requirements in this chapter should provide students, homebuilders, and building code officials with a more organized understanding of this important phase of house construction.

NOTES

[1] *CABO One and Two Family Dwelling Code, 1995 Edition* (Falls Church, VA: The Council of American Building Officials, 1995).

[2] Ibid., pp. 5–142.

[3] Ibid., pp. 5–6.

[4] Ibid., pp. 7–24.

[5] Ibid., Section 108, p. 2.

[6] *Cost-Effective Home Building: A Design and Construction Handbook* (Washington, DC: The Home Builder Press, The National Association of Home Builders, 1994), p. 60.

[7] *1995 CABO Code*, Section 802.12, p. 91.

[8] Ibid., Section 802.11, p. 91.

[9] *Metal Plate Connected Wood Truss Handbook* (Madison, WI: Wood Truss Council of America, 1993).

[10] *1995 CABO Code*, Section 502.10, p. 42.

[11] Ibid., Chapter 3, pp. 7–24.

[12] Ibid., Section 301.5, p. 7.

[13] Ibid., Fig. 301.2e, p. 12.

[14] *The BOCA National Building Code, 1996*, Section 1608.0 (Country Club Hills, IL: Building Code Officials and Code Administrators International, Inc., 1996), pp. 162–168.

[15] Ibid., Section 1608.4, p. 162.

[16] Ibid., Fig. 1608.3(1), p. 163.

[17] *1995 CABO Code*, Section 602.3.1, p. 63.

[18] Ibid., Section 802.12, p. 91.

[19] *1996 BOCA Code*, Section 1609.0, pp. 168–177.

[20] *1995 CABO Code*, Fig. 301.2b, p. 10.

[21] Ibid., Fig. 301.2c, p. 11.

[22] *Application and Commentary, CABO One and Two Family Dwelling Code, 1992 Edition* (Falls Church, VA: The Council of American Building Officials, 1992), Fig. 201.2, p. 7.

[23] *1995 CABO Code*, Fig. 301.2f, p. 13.

[24] Ibid., Fig. 301.29, p. 13.

[25] Ibid., Section 323,1, p. 24.

[26] Ibid., Section 322.1, p. 23.

[27] Ibid., Fig. 301.2a, p. 9.

[28] Ibid., Section 324, p. 24.

[29] Ibid., Fig. 301.2b, p. 14.

[30] Ibid., Appendix F, pp. 329–330.

[31] Ibid., Section 301, p. 7.

[32] *Application and Commentary*, p. 8.

[33] *1995 CABO Code*, Section 301.4, p. 7.

[34] Ibid., Section 301.6, p. 7.

[35] *Cost Effective Home Building*, pp. 15–16.

[36] *1995 CABO Code*, Chapter 4, pp. 25–34.

[37] Ibid., Section 401.2, p. 25.

[38] Ibid., Section 401.3, p. 25.

[39] Ibid., Section 402, pp. 25–26.

[40] Ibid., Section 403, pp. 26–29.

[41] Ibid., Section 403.1, p. 26.

[42] Ibid., Section 404.1.3, p. 30.

[43] Ibid., Section 404.1.3.1, p. 31.

[44] Ibid., Section 404.1, p. 30.

[45] Ibid., Section 404.1.1, p. 30.

[46] Ibid., Section 404.1.2, p. 30.

[47] Ibid., Section 404.2, p. 31.

48Ibid., Chapter 5, pp. 35–62.

49Ibid., Section 405, pp. 32–33.

50Ibid., Section 405.1, p. 32.

51Ibid., Section 406.2, p. 33.

52Ibid., Section 406.1, p. 33.

53Ibid., Chapter 5, p. 35–62.

54Ibid., Section 502.1, p. 35.

55Ibid., Table 502.3.1.c, pp. 45–61.

56National Design Specification for Wood Construction (Washington, D.C.: American Forest and Paper Products Association [AF&PA], 1991); and Supplement to the National Design Specification for Wood Construction (Washington, D.C.: American Forest and Paper Products Association [AF&PA], 1991).

57*CABO One and Two Family Dwelling Code, 1989 Edition*, p. 334 (Fall Church, VA: The Council of American Building Officials, 1989).

58*1992 CABO Application and Commentary*, pp. 112–113.

59*1995 CABO Code*, Section 502.3.3, p. 35.

60Ibid., Section 502, pp. 35, 42.

61Ibid., Section 503, p. 42.

62Ibid., Section 505, p. 44.

63Ibid.

64Ibid., Section 504, p. 44.

65Ibid., Section 506, p. 44.

BIBLIOGRAPHY

Application and Commentary, CABO One and Two Family Dwelling Code, 1992 Edition. Falls Church, VA: The Council of American Building Officials, 1992.

The BOCA National Building Code, 1996. Country Club Hills, IL: Building Code Officials and Code Administrators International, Inc., 1996.

Cost-Effective Home Building: A Design and Construction Handbook. Washington, D.C.: The Home Builder Press, The National Association of Home Builders, 1994.

CABO One and Two Family Dwelling Code, 1989 Edition. Falls Church, VA: The Council of American Building Officials, 1989.

CABO One and Two Family Dwelling Code, 1995 Edition. Falls Church, VA: The Council of American Building Officials, 1995.

Metal Plate Connected Wood Truss Handbook. Madison, WI: Wood Truss Council of America, 1993.

National Design Specification for Wood Construction. Washington, D.C.: American Forest and Paper Products Association (AF&PA), 1991.

Olin, Harold B., J. L. Schmidt, and Walter H. Lewis. *Construction Principles, Materials and Methods*, 4th Ed. Chicago, IL: The Institute of Financial Education, 1980.

Reiner, L. E. *Methods and Materials of Residential Construction.* Englewood Cliffs, NJ: Prentice-Hall, Inc., 1981.

Weidhaas, E. R. *Architectural Drafting and Construction*, 4th Ed. Boston, MA: Allyn and Bacon, 1989.

10

Wall and Roof System Requirements

INTRODUCTION

Homebuilders and building code officials naturally devote a great deal of attention to the design and construction of the foundation and the floor systems of a house. The environmental and structural adequacy of a house, however, is also a direct function of the phases of work, which include: (1) wall construction, (2) wall covering, (3) roof and ceiling construction, (4) roof covering, and (5) chimneys and fireplaces. These phases, which are addressed in Chapters 6–10 of the 1995 CABO Code,[1]* are the focus of Chapter 10 of this book.

WALL SYSTEMS

The wall systems of a house serve a number of purposes. **Exterior walls**, for instance, provide protection against the outside environment for the occupants of the house. They also permit the maintenance of a controlled indoor air and temperature environment which resists the hourly and daily fluctuations experienced outside the house. **Foundation walls** provide resistance to the lateral pressures created by the soil surrounding the basement or crawl space. **Interior walls** serve a different purpose because they allow the architect or homebuilder to divide the interior of the house into the various use zones desired by the homebuyer.

As noted in Chapter 9 of this book, some of these walls also serve an important structural function because they transmit the loads imposed upon them by the roof, attic, and floors to the supporting soil. It is important to study the set of working drawings for each house in order to: (1) determine how the loads are transmitted through the structural frame and (2) identify the walls that are **load bearing** and the ones that are **non-load bearing**.

Chapter 6 of the 1995 CABO Code[2] controls the design and construction of all walls and partitions for houses that are built in jurisdictions that have adopted that code. The two

*Sections, figures, and tables from this Code are cited only for instructional purposes and the treatment of the Code provisions contained in the design solutions illustrated in this chapter are not intended as a substitute for a complete and thorough understanding of all building provisions to which any particular project may be subject.

primary types of wall systems that are addressed are: (1) wood framed walls and (2) masonry walls. Light-gage steel wall systems are covered only by a one-sentence reference. Home-builders desiring to use steel framing systems, stress–skin panel wall systems, or other structural wall alternatives must, therefore, follow the provided procedures related to the approval of alternative materials and systems.[3]

A limited set of requirements for windows, sliding glass doors, plywood and structural panels, and particleboard is also provided. Wall coverings are typically installed on the structural framing which defines the interior and exterior walls of a house. Design and construction requirements for these wall coverings are provided in Chapter 7 of the 1995 CABO Code.[4]

Wood Wall Systems

Floor system design starts with an evaluation of the loads imposed on a floor joist or girder and includes the selection of member sizes so that the allowable structural properties of a particular species and grade of lumber are not exceeded. **Wall system design**, based upon the requirements of Chapter 6 of the 1995 CABO Code,[5] does not follow the same approach. Instead, it is assumed that if the specified wall framing requirements are followed, the resulting wall system will satisfactorily carry the structural loading existing in one- and two-family houses. Determination of loads and stresses on particular components of the wall system and size selection are not part of the wall design procedure.

Essential Aspects of Wood Wall System Design

The role of exterior and interior bearing walls in the preapproved platform and balloon framing systems is shown in Fig. 9.1. Additional detail about the primary load-bearing members in an exterior bearing wall is provided in Fig. 10.1.

The load-bearing members—which include (1) a single or double top plate, (2) wall studs, and (3) a bottom plate—must be identified by a grade mark which provides sufficient information to determine F_b and E for the lumber that is used. Studs must be a minimum No. 3, standard, or stud grade lumber.[6] Utility grade lumber may, however, be used for bearing studs that do not support floors and for non-load-bearing studs if the spacing specified for such an application is met.

The primary loading system on a wall is in the vertical direction. This load is resisted by the interaction of the wall studs and the wall sheathing. An exterior wall is also exposed to lateral loads due to wind. Design which considers these lateral loads, based upon accepted engineering practice, must be performed if the wind pressure exceeds 30 psf.[7]

Exterior Wood Walls

One of the key aspects of exterior bearing wall design is the spacing restrictions that are placed on the wall studs.[8] Spacing of 16 in. and 24 in. is allowed for studs not more than 10 ft in length subject to the criteria indicated in Table 10.1. The various stud loading conditions identified in Table 10.1 are illustrated in Fig. 10.2. (Note: The stud E condition in Fig. 10.2 is not directly addressed in Table 10.1.)

Additional requirements about the spacing of wall studs and the use of single top plates instead of double top plates are provided.[9]

Interior Wood Partitions

The framing requirements for interior load-bearing partitions are the same as those for exterior bearing walls. Non-load-bearing partitions, however, only require a single top plate and can be constructed with 2 in. × 3 in. studs spaced at 24 ins. o.c. or 2 in. × 4 in. flat studs spaced 16 in. o.c.[10]

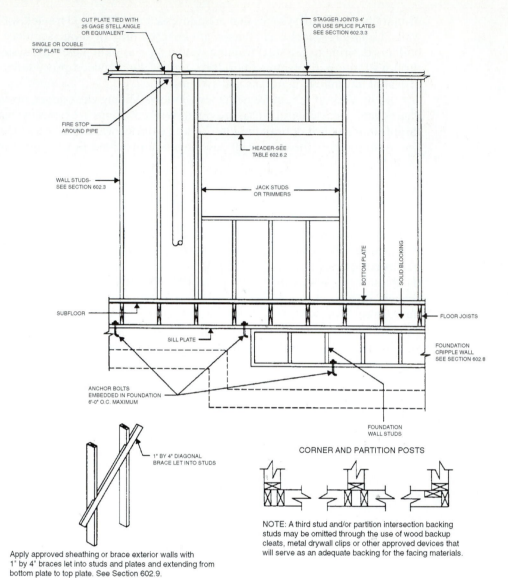

Figure 10.1
Framing Details
for Exterior Load-
Bearing Walls
Source: Reproduced
from the 1995 edi-
tion of the *CABO
One and Two
Family Dwelling
Code,*™ copyright
© 1995, with the
permission of the
publisher, The
International Codes
Council, Whittier,
CA, Fig. 602.3b,
p. 68.

NOTE: A third stud and/or partition intersection backing
studs may be omitted through the use of wood backup
cleats, metal drywall clips or other approved devices that
will serve as an adequate backing for the facing materials.

Apply approved sheathing or brace exterior walls with
1" by 4" braces let into studs and plates and extending from
bottom plate to top plate. See Section 602.9.

TABLE 10.1   Maximum Stud Spacing

Stud Size	Supporting Roof and Ceiling Only	Supporting One Floor, Roof, and Ceiling	Supporting Two Floors, Roof, and Ceiling	Supporting One Floor Only
2×4	24[a]	16	—	24[a]
3×4	24[a]	24	16	24
2×5	24	24	—	24
2×6	24	24	16	24

For SI: 1 in. = 25.4 mm.
[a]Shall be reduced to 16 in. if Utility grade studs are used.

Source: Reproduced from the 1995 edition of the *CABO One and Two Family Dwelling Code,*™ copyright
© 1995, with the permission of the publisher, The International Codes Council, Whittier, CA, Table 602.3d,
p. 66.

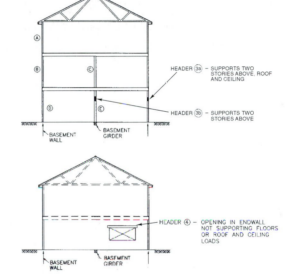

Figure 10.2
Bearing Well Loading Conditions for Wall Studs and Wall Opening Headers
Source: Cost Effective Homebuilding: A Design and Construction Handbook (Washington, DC: The Home Builder Press, The National Association of Home Builders, 1994), Fig. 5.5 (modified), p. 52.

Exterior Wood Wall Bracing

The structural integrity of a house against lateral load and racking can be ensured only if the exterior walls are properly braced. As shown in Fig. 10.1, such bracing at the exterior corners of the house can consist of 1 in. × 4 in. let-in bracing or approved sheathing material. Specific requirements for the following alternative types of wall bracing are provided:[11]

- Approved metal strap devices.
- Wood structural panels subject to the stud spacing criteria indicated in Table 10.2.
- Particleboard subject to the stud spacing criteria indicated in the 1995 CABO Code.[12]
- Gypsum sheathing, wallboard, veneer base, fiberboard sheathing, and Portland cement plaster applied over metal lath also can be used, subject to the restrictions specified in the 1995 CABO Code.

The general installation requirements for wall bracing as a function of the **seismic zone** in which the house is built is shown in Table 10.3. Exceptions to these requirements are indicated.[13]

TABLE 10.2 Allowable Stud Spacing for
Wood Structural Panel Sheathing

| Panel Span Rating | Panel Nominal Thickness (in.) | Maximum Stud Spacing (in.) | |
| | | Siding nailed to:[a] | |
		Stud	Sheathing
12/0, 16/0, 20/0, or Wall—16 o.c.	5/16, 3/8	16	16[b]
24/0, 24/16, 32/16, or Wall— 24 o.c.	3/8, 7/16, 15/32, 1/2	24	24[c]

For SI: 1 in. = 25.4 mm.
[a]Blocking of horizontal joints shall not be required.
[b]Plywood sheathing 3/8 in. thick or less shall be applied with long dimension across studs.
[c]Three-ply plywood panels shall be applied with long dimension across studs.

Source: Reproduced from the 1995 edition of the *CABO One and Two Family Dwelling Code*™, copyright © 1995, with the permission of the publisher, The International Codes Council, Whittier, CA, Table 602.3b, p. 66.

TABLE 10.3 Wall Bracing Requirements

Seismic Zone	Condition[a]	Type of Brace	Amount of Bracing[b,c]
0, 1, and 2	One story; top of two or three story; first story of two story; second story of three story	1″ × 4″ let-in bracing or structural sheathing	Located at each end and at least every 25′ of wall length
	First story of three story	Structural sheathing	Minimum 48″-wide panels. Located as required for let-in bracing
3 and 4	One story; top of two or three story	1″ × 4″ let-in bracing or structural sheathing	Located at each end and at least every 25′ of wall length
	First story of two story; second story of three story	Structural sheathing	25% of wall length to be sheathed
	First story of three story	Structural sheathing	40% of wall length to be sheathed

For SI: 1 in. = 25.4 mm, 1 ft = 304.8 mm.
[a]Foundation wall panels braced same as story above.
[b]Where structural sheathing is used, each braced panel must be at least 48 in. wide.
[c]Structural sheathing and let-in bracing shall be located at each end or as near thereto as possible.

Source: Reproduced from the 1995 edition of the *CABO One and Two Family Dwelling Code*™, copyright © 1995, with the permission of the publisher, The International Codes Council, Whittier, CA, Table 602.9, p. 71.

Fastening Requirements

Information about acceptable safe practice regarding the fastening together of the various structural components is important for the framing subcontractor and the building code inspector because it directly affects the structural integrity of the house. Such fastener schedule and alternate attachment information is provided in the form shown in Table 10.4.[14]

Wood Wall Opening Headers

It would be structurally ideal, from a load-carrying perspective, if there were no openings for windows, doors, archways, etc. in either the exterior or interior load-bearing walls of a house. Unfortunately, no one would enjoy living in such a house. Therefore, a provision must be made to allow openings to be placed in load-bearing walls, while at the same time ensuring that the vertical loads above the openings are properly transmitted around the opening (Fig. 10.3).

The provision most commonly used, as indicated in Fig. 7.23, consists of double 2× lumber headers which are installed directly above the wall opening with short cripple studs installed between the header and the wall top plate. The ends of the header are supported by extra **jack studs** that are positioned immediately adjacent to the conventional wood studs. Alternatives allowed in the 1995 CABO Code include: (1) a single nominal 4-in. header which directly replaces the double 2× headers,[15] (2) a single nominal 2-in. header provided that it is of adequate size to support all of the imposed loads[16], and (3) a plywood box header.[17]

FACTORS INFLUENCING HEADER DESIGN

The size of the header affected by the load distribution situation shown in Fig. 9.6 and the load transmission pattern shown in Fig. 10.3 are influenced by the following factors:

- The size of the opening.
- The grade and species of lumber used for the header.

TABLE 10.4   Fastener Schedule for Structural Members

Description of Building Elements	Number & Type of Fastener	Spacing of Fastener
Joist to sill or girder, toe nail	3-8d	—
1″ × 6″ subfloor or less to each joist, face nail	2-8d 2 staples, 1-3/4″	—
2″ subfloor to joist or girder, blind and face nail	2-16d	—
Sole plate to joist or blocking, face nail	16d	16″ o.c.
Top or sole plate to stud, end nail	2-16d	—
Stud to sole plate, toe nail	3-8d or 2-16d	—
Double studs, face nail	10d	24″ o.c.
Double top plates, face nail	10d	24″ o.c.
Double top plates, minimum 48″ offset of end joints, face nail in lapped area	4-10d	—
Top plates, laps at corners and intersections, face nail	2-10d	—

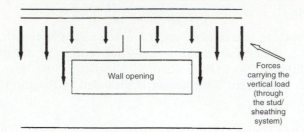

Figure 10.3
Effect of Wall
Opening on Load
Transmission in a
Load-Bearing
Wall

- The magnitude of the roof loads and floor loads (in psf) carried by the load-bearing wall containing the header.
- The width (or depth) of the house. It should be obvious that for a given length of house, as its width is increased there will be a greater total roof load and total floor loads because of the larger areas of the roof and the floors. Since this total load is carried by the same-length exterior load-bearing walls, the lbs/lineal ft imposed on the wall will increase as the house becomes wider. As a result, the size of the header must increase.
- The location of the opening in the wall. A window opening between the first and second floor is exposed to a higher load (lbs/linear ft) than a window opening between the second floor and the roof.
- The existence of interior load-bearing partitions which "share" the load with the exterior bearing walls. If there are no interior bearing walls (i.e., because roof trusses and floor trusses are used), then a much higher load is imposed on a header which spans an exterior wall opening.
- The type of exterior sheathing which is used. This factor may not be as obvious as the ones noted earlier, but it also is important and must be considered. The primary load-carrying members in a load-bearing wall are, of course, the studs. The total strength of the load-bearing wall is a function, however, of the stud/sheathing system because the continuity provided by the sheathing "ties" the individual studs into a composite unit. Thus, a structural sheathing such as plywood will provide greater structural integrity to the wall than an insulation board type of sheathing. Credit should, therefore, be given for the effect of the sheathing when the allowable span for a given size of header is determined.
- The interaction between various openings in the same load-bearing wall. As shown in Fig. 10.4(a), for instance, when a window or door opening is directly beneath an opening in the story above, the header spanning the upper opening carries the roof load around the lower openings. As a result, the headers over the first-story windows and door only have to carry the second-floor loads. In Fig. 10.4(b), however, the first-story window and door headers must carry both the roof and second-story loads.

HEADER SIZE SELECTION

A theoretical structural analysis of how loads are transmitted through a stud/sheathing bearing wall system is fairly complex. Fortunately, the practical implementation of the analysis has been simplified greatly in the 1995 CABO Code, as shown in Table 10.5. Fig. 10.2 illustrated some of the conditions specified in Table 10.5.

The factors which influence header design that were described earlier are not directly considered when the size of the header is selected. This represents a major change from the 1992 CABO Code, which included a set of tables which used these factors as decision criteria.[18]

Although Table 10.5 does not specifically state that reference is being made to **double 2×** members, their use is implied by the following statement:

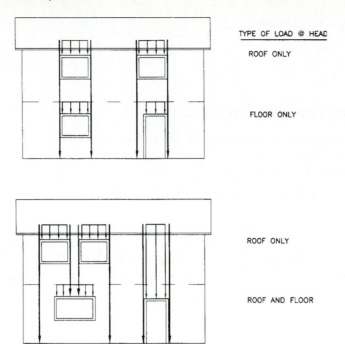

Figure 10.4
Possible
Combinations of
Roof and Floor
Loads Acting on
Different Headers
in a Two-Story
House

*Source: Cost
Effective
Homebuilding: A
Design and
Construction
Handbook*
(Washington, DC:
The Home Builder
Press, The National
Association of
Home Builders,
1994), Fig. 5.5
(modified), p. 50.

The allowable spans for nominal **4-in. single headers** and **2-in. double headers** in bearing walls shall not exceed the spans set forth in Table 602.6. The table is not to be used where concentrated loads are supported by the headers.[19]

The header requirement is modified for non-load-bearing walls, as follows:

Load-bearing headers are not required in interior or exterior nonbearing walls. A single flat 2 in. × 4 in. member may be used as a header in interior or exterior nonbearing walls for openings up to 8 ft in width if the vertical distance to the parallel nailing surface above is not more than 24 in. For such nonbearing headers, no cripples or blocking are required above the header.[20]

TABLE 10.5   Maximum Spans for Headers Located over Openings in Walls (ft)

Size of Header[a,b]	Headers in Bearing Walls[b]			Headers in Walls Not Supporting Floors or Roofs
	Supporting Roof Only	One Story Above	Two Stories Above	
2×4	4	—	—	—
2×6	6	4	—	—
2×8	8	6	—	10
2×10	10	8	6	12
2×12	12	10	8	16

For SI: 1 in. = 25.4 mm, 1 in. = 304.8 mm.
[a]Nominal 4-in.-thick single headers may be substituted for double members.
[b]Spans are based on No. 2 Grade Lumber with 10-ft. tributary floor and roof loads.

Source: Reproduced from the 1995 edition of the *CABO One and Two Family Dwelling Code*™, copyright © 1995, with the permission of the publisher, The International Codes Council, Whittier, CA, Table 602.6, p. 68.

Case Study House Example

The set of working drawings for the Case Study House (Appendix A) can be examined to determine if the exterior and interior wall systems that are specified comply with the requirements of the 1995 CABO Code.

STUD SIZE/SPACING: EXTERIOR WALL

The material specification information in Appendix C in "Item 5: Exterior Walls" indicates that No. 2 Hem–Fir lumber is used. In addition, the following exterior wall information is provided on Sheet A-5:

- Exterior wall between the second floor and the roof:
    2×4 in. studs @ 16 in. o.c.
    1/2 in. insulated sheathing

Table 10.1 indicates that for this *supporting roof and ceiling only situation*, 2×4 in. studs @ 24 in. o.c. would have been acceptable.

- Exterior wall between the first floor and the second floor
    2×4 in. studs @ 16 in. o.c.
    1/2 in. insulated sheathing

Table 10.1 indicates that for this *supporting one floor, roof and ceiling situation*, 2×4 in. studs @ 16 in. o.c. are acceptable.

It should be noted that Sheet A-1 indicates that the rear exterior wall of the family room, between the first floor and the second floor, is a 6-in. wall. No information about the size and spacing of the studs in that wall is specified, but it is reasonable to assume that 2×6 in. studs at 16 in. o.c. were used. Table 10.1 indicates that for this *supporting one floor, roof and ceiling situation*, 2×6 in. studs at 24 in. o.c. would have been acceptable.

STUD SIZE/SPACING: INTERIOR WALLS

Footnote 1 on Sheet A-1 indicates that all interior partitions are 2×4 in. studs unless otherwise noted. An examination of Sheet A-1 indicates that the only interior partition that is load bearing is the 2×6 in. wall that is the interior wall of the dining room and living room. The material specification information in Appendix C, in "Item 9: Partition Framing", indicates that 2×6 in. No. 2 Hem–Fir studs spaced at 16 in. o.c. are used in this wall.

Table 10.1 indicates that for this *supporting one floor situation*, 2×6 in. studs at 24 in. o.c. would have been acceptable. It should be noted that because pre-engineered roof trusses are used, none of the interior partition walls on the second floor (Sheet A-2) are load bearing.

EXTERIOR WALL BRACING

A review of the working drawings for the Case Study House indicates that the type of exterior wall bracing that is used is not mentioned. A similar review of the material specification information in Appendix C, in "Item 5: Exterior Walls", indicates that corner bracing will be provided, but that the type to be used is not specified.

It would appear, therefore, that additional information about the exterior wall bracing must be added to the working drawings. As shown in Table 10.3, since the Case Study House is assumed to be located in central Pennsylvania (seismic zone = 0), 1×4 in. let-in bracing is required at each end of the wall and at least every 25 ft of wall length. Alternatively, structural sheathing panels, each 48 in. in width, can be installed at each end of the wall and at least every 25 ft of wall length. As noted in Table 10.2, 5/16-in. or 3/8-in.-thick structural panels with a 12/0, 16/0, 20/0, or Wall—16 o.c. rating can be used to provide the needed exterior wall bracing.

HEADER SELECTION

For the Case Study House, Sheet A-4 indicates the location of all of the openings on the four elevation views of the house. Sheet A-6 presents a generalized Window and Door Schedule for all of the houses which are built by S&A Custom Built Homes. The specific information about the type of window or door used for each opening, as well as the size of other openings, is provided on Sheets A-1, A-2, and A-3.

The actual size of the headers used for all of the wall openings is indicated in Note #7 on Sheet A-1 as two 2×10 in. members unless otherwise noted. According to Table 10.5, the allowable spans are:

Two 2×10 in. headers

Supporting roof only	10 ft
One story above	8 ft
Two stories above	6 ft

The locations of headers corresponding to these loading situations are shown in Fig. 10.2:

- Exterior wall between the second floor and the roof
  For this *supporting roof only situation*, the largest wall opening size for the two 2852 windows (Sheet A-2) is 67-13/16 in. = 5 ft 7-13/16 in. (Sheet A-6). Thus, two 2×10 in. headers are adequate.

- Exterior wall between the first floor and the second floor
  For this *one story above situation*, the largest wall opening size for the 6/0 × 6/8 patio door (Sheet A-1) is 72 in. = 6 ft (Sheet A-6). Thus, two 2×10 in. headers are adequate.

- Exterior wall in breakfast nook extension
  For this *supporting roof only situation*, the opening size for three 2856 windows (Sheet A-1) is 101-1/2 in. = 8 ft 5-1/2 in. (Sheet A-6). Thus, two 2×10 in. headers are adequate.

With regard to the breakfast nook extension, Sheet A-1 indicates that two 1-3/4 in. × 11-7/8 in. × 10 ft, 11 in. LVL headers are used to carry the load from above, around the required opening in the exterior bearing wall.

- Interior wall on the first floor
  The largest opening in the 2×6 in. interior bearing wall is 3 ft, 4-1/2 in. (Sheet A-1). For this *one story above situation*, two 2×10 in. headers are adequate.

Additional Wood Wall Framing Details

In addition to establishing the requirements for: (1) the size and spacing of studs in both load-bearing and non-load-bearing walls, (2) exterior wall bracing, (3) fastening, and (4) headers for wall openings, Chapter 6 of the 1995 CABO Code also provides detailed information about the following specific aspects of framing:

- Top plates—the use of single top plates
- Bearing studs—spacing requirements when the spacing of floor and roof members is not the same as the stud spacing
- Drilling and notching—studs
- Drilling and notching—top plate
- Firestopping requirements
- Cripple walls[21]

Requirements such as these, although they seldom appear directly on a set of working drawings, should become part of a homebuilder quality management checklist for the wall framing phase of the construction process.

Masonry Wall Systems

The load-bearing and non-load-bearing walls of a house also can be constructed using either **clay masonry units** or **concrete masonry units**. A masonry wall assembly typically consists of the following interdependent components:

- Masonry units
    Clay: solid or hollow
    Concrete: solid or hollow
- Masonry mortar
- Flashing materials
- Masonry anchors, ties, and joint reinforcement
- Quality workmanship

When all of these components are properly integrated, the desired properties of the wall, namely **strength** (to carry the applied vertical and lateral loads), **water tightness** (to prevent rain penetration), and **durability** (wear and weather resistance), will be met. Proper wall assembly design can provide resistance to heat transfer, fire resistance, sound resistance, and radiation protection.[22]

The failure of a masonry wall system does not necessarily imply total structural collapse. Failure can take many forms, depending upon which of the previous components is at fault.

Clay Masonry Units

Structural clay products consist of a mixture of surface clays, shales, fire clays, and burned clays which is burned in a kiln for two to five days to produce bricks which have the desired properties. These products can be classified into the following categories: (1) solid masonry units (brick), (2) hollow masonry units (hollow brick, structural clay tile, and structural facing tile), and (3) architectural terra cotta (anchored ceramic veneer, adhesion ceramic veneer, and ornamental).[23]

Brick, which is the common usage term for a **clay solid masonry unit**, is the structural clay product which is most often used in housing construction. It is considered to be a solid masonry unit if the cored area does not exceed 25% of the total cross-sectional area of the unit. Brick can be classified further into: (1) **building or common brick**, which is used chiefly as a structural material where appearance is not as important as strength and durability, and (2) **facing brick**, which is used where appearance is the most important function.

GRADING SYSTEM

ASTM C62[24] defines the following three grades for building brick based upon resistance to weathering: **SW** (severe weathering), which should be used where high and uniform resistance to damage caused by cyclic freezing is desired and where brick may be frozen when saturated with water; **MW** (moderate weathering), which should be used where moderate resistance to cyclic freezing damage is permissible or where brick may be damp but not saturated with water when freezing occurs; and **NW** (negligible weathering), which is used where the brick needs little resistance to cyclic freezing damage and is protected from water absorption and freezing, such as for interior partitions. Different physical properties (i.e., compressive strength, absorption characteristics, and saturation coefficient) are specified for each grade.

SIZE CONVENTION

Many different sizes and shapes of brick are produced in the United States. Figure 10.5 indicates some of the commonly used **non-modular** and **modular** bricks. The **actual dimensions** of **non-modular brick** are indicated, with the standard brick size being 3-3/4 in. $\times$ 8 in. $\times$

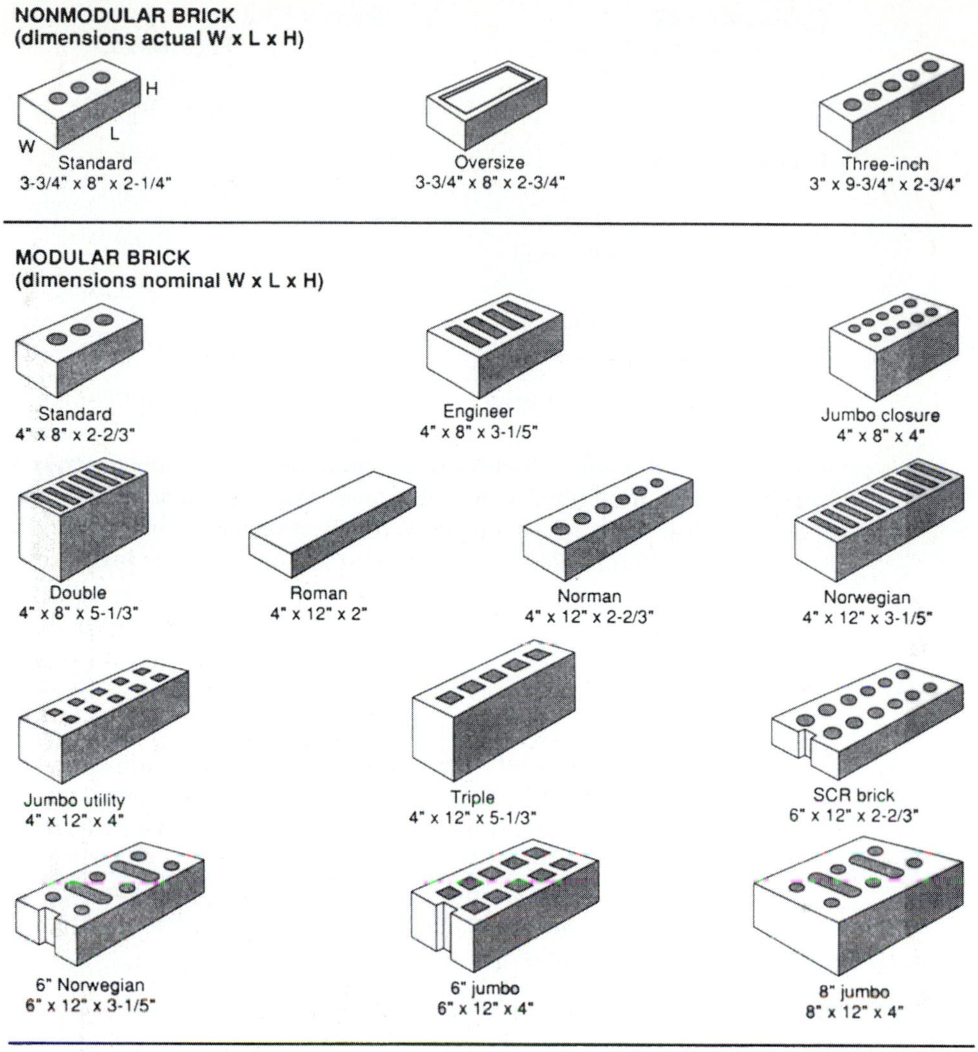

NONMODULAR BRICK
(dimensions actual W x L x H)

Standard
3-3/4" x 8" x 2-1/4"

Oversize
3-3/4" x 8" x 2-3/4"

Three-inch
3" x 9-3/4" x 2-3/4"

MODULAR BRICK
(dimensions nominal W x L x H)

Standard
4" x 8" x 2-2/3"

Engineer
4" x 8" x 3-1/5"

Jumbo closure
4" x 8" x 4"

Double
4" x 8" x 5-1/3"

Roman
4" x 12" x 2"

Norman
4" x 12" x 2-2/3"

Norwegian
4" x 12" x 3-1/5"

Jumbo utility
4" x 12" x 4"

Triple
4" x 12" x 5-1/3"

SCR brick
6" x 12" x 2-2/3"

6" Norwegian
6" x 12" x 3-1/5"

6" jumbo
6" x 12" x 4"

8" jumbo
8" x 12" x 4"

Figure 10.5
Sizes of Non-Modular and Modular Brick
Source: Charlie Wing, *The Visual Handbook of Building and Remodeling* (Emmaus, PA: Rodale Press, 1990), p. 54.

2-1/4 in. **Modular** bricks are sized according to their nominal dimensions, with the **standard brick size** being 4 in. × 8 in. × 2-2/3 in. The actual dimensions of that type of brick are 3-1/2 in. × 7-1/2 in. × 2-1/6 in. Thus, when a 1/2 in.-thick-mortar joint is used, three courses of modular standard brick result in 8 in. (3 × 2-2/3 in.) of wall.

The modular **standard** brick satisfies the need for a basic 4 in. × 8 in. brick, while the **Roman** and **Norman** brick satisfy the need for a 4 in. × 12 in. brick. The SCR brick is formed with a 3/4 in. × 3/4 in. jamb slot at one end and is intended for use in 6-in.-wide solid, load-bearing masonry walls for all types of single-story situations. The walls of such structures should not exceed a height of 9 ft to the eaves or 15 ft to the gable peak.[25]

Concrete Masonry Units

Concrete masonry units are manufactured using a low-slump mixture of Portland cement, suitable aggregates, water, and often admixtures. These materials are machine molded into the desired shape using compaction and vibration and steam cured in a drying kiln. An aging period is required for the units to achieve the required strength, moisture content, and other properties.

Concrete masonry units can be classified into the following categories:[26]

Concrete Brick
  - Building Brick
  - Slump Brick

Special Units
  - Split Face Block
  - Faced Block
  - Decorative Block

Concrete Block
  - Solid
    —Load bearing
    —Non-load bearing
  - Hollow
    —Load bearing
    —Non-load bearing

Although **concrete brick** is completely solid (it does not contain any cores), **solid concrete block** can have a cored area which does not exceed 25% of the total cross-sectional area of the unit. **Hollow load-bearing and non-load-bearing concrete blocks**, which are the products most often used in housing construction, have an empty core area which typically ranges from 40% to 50% of the total cross-sectional area of the unit. Hollow non-load-bearing blocks are thin shelled and lightweight and are typically used as interior non-load-bearing partitions.

Special concrete masonry units are made to create particular architectural effects.

GRADING SYSTEM

The two ASTM standards which most directly relate to the uses of concrete block in house construction are ASTM C90,[27] which addresses both solid and hollow load-bearing concrete block, and ASTM C129,[28] which addresses both solid and hollow non-load-bearing concrete block. Three weight classes are specified: (1) normal weight, (2) medium weight, and (3) light weight. In addition, two types are specified: (1) Type I, moisture controlled and (2) Type II, non-moisture controlled. Moisture limits are established in order to minimize in-place shrinkage of units and a later tendency to wall cracking that can occur in the dryer areas of the United States.

SIZE CONVENTION

Concrete block is manufactured in modular sizes which conform to the basic 4-in. architectural module. Typical wall thicknesses of 4, 6, and 8 in. are used for interior walls, and 8, 10, and 12 in. are used for exterior walls. The 8 in. × 16 in. × 8 in. **stretcher block** shown in Fig. 10.6, which is the basic unit in a wall, is probably the most commonly used in house construction. The actual dimensions of this block are 7-5/8 in. × 15-5/8 in. × 7-5/8 in. to allow for 3/8-in. mortar joints. The equivalent dimensions for a stretcher block in a 12-in.-wide concrete block wall are shown.

The **regular block** (also called a corner block) is used for squared corners, and the **bullnose block** is used for rounded corners. **Jamb blocks** are installed around window and door openings. Although concrete block is typically 8 in. high, it is possible to use 4-in.-high block through special order or by cutting 8-in. block. Three-core concrete blocks, instead of the two-core variety shown in Fig. 10.6, are also readily available in most areas.

Masonry Mortar

Masonry mortar serves a number of purposes in masonry wall systems, including: (1) bonding the masonry units together and sealing the spaces between them, (2) compensating for the size variations in masonry units, (3) facilitating the integral action of the metal ties and reinforcement with the wall, and (4) providing aesthetic quality to the wall by creating shadow lines and/or color effects.[29] Masonry mortar combines one or more cementitious materials (Portland cement, hydrated lime, masonry cement), clean well-graded sand, and enough clean water to create a plastic, workable mixture.

(dimensions actual W x L x H)

Partition
3-5/8"x7-5/8"x3-5/8"

1/2 partition
3-5/8"x15-5/8"x3-5/8"

Partition
3-5/8"x15-5/8"x7-5/8"

Double ends
3-5/8"x7-5/8"x7-5/8"

Double end
7-5/8"x15-5/8"x7-5/8"

Regular
7-5/8"x15-5/8"x7-5/8"

Stretcher
7-5/8"x15-5/8"x7-5/8"

Steel sash jamb
7-5/8"x15-5/8"x7-5/8"

Figure 10.6
Actual
Dimensions of
Concrete
Masonry Units
Source: Charlie
Wing, *The Visual*
Handbook of
Building and
Remodeling
(Emmaus, PA:
Rodale Press,
1990), pp. 60–61
(modified).

Double ends
11-5/8"x15-5/8"x7-5/8"

Regular
11-5/8"x15-5/8"x7-5/8"

Stretcher
11-5/8"x15-5/8"x7-5/8"

Single bull nose
11-5/8"x15-5/8"x7-5/8"

Steel sash jamb
11-5/8"x15-5/8"x7-5/8"

Jamb
11-5/8"x15-5/8"x7-5/8"

Twin
1-5/8"x15-5/8"x7-5/8"

Solid-bottom U-block
11-5/8"x15-5/8"x7-5/8"

Mortars have two distinct sets of properties:

1. Plastic (wet) mortar properties
   a. Workability
   b. Water reventivity
   c. Initial flow
   d. Flow—after suction
2. Hardened mortar properties
   a. Bond strength
   b. Durability
   c. Compressive strength
   d. Volume change
   e. Watertightness
   f. Rate of hardening
   g. Appearance[30]

The proportions of each of the materials used determines the predominant properties of the mortar and hence its use on a project. These predominant properties must be matched to the predominant forces acting on the masonry structure. Walls that are predominantly load bearing, for instance, require a mortar with a high compressive strength, while a mortar possessing a high tensile bond strength should be used on walls that must resist strong winds.

Four different types of mortars (M, S, N, and 0) that are suitable for both clay masonry units and concrete masonry units are specified in ASTM C270.[31] The recommended uses for these types of mortars are:

- Type M Mortar. A high-strength mortar suitable for general use and recommended specifically for masonry below grade or in contact with earth such as foundations, retaining walls, or paving.
- Type S Mortar. A high-strength mortar suitable for general use and specifically where high transverse strength of masonry is desired, for reinforced masonry, where mortar bonds the facing and backing, and areas subject to winds greater than 80 mph (129 kmph).
- Type N Mortar. A medium-strength mortar suitable for general use in exposed masonry above grade and recommended specifically where high compressive or transverse masonry strengths are not required.
- Type O Mortar. A low-strength mortar suitable for use in non-bearing applications in walls of low axial compressive strength and where masonry is not subject to severe weathering.[32]

Types of Masonry Walls

In many parts of the United States, it is standard practice to build the foundation wall system of a house using concrete masonry units such as shown in Fig. 7.5. Such a wall system is **pre-approved** in "Chapter 4: Foundations" of the 1995 CABO Code. Table 9.12 indicates how the various unreinforced masonry foundation wall alternatives (ungrouted, fully grouted, and composed of solid masonry units) can be used in Seismic Zones 0, 1, and 2 when favorable soil and ground water conditions exist. Table 9.13 indicates the reinforcement and structural design requirements for masonry walls when less favorable soil conditions exist or the house is located in Seismic Zone 3 or 4.

Clay masonry units, as well as concrete masonry units, can be used for both exterior and interior walls in houses that have been designed to take advantage of these products in both the structural systems that support the house and the architectural systems that enhance the appearance of the house. Masonry walls commonly used in both below-grade and above-grade applications include: (1) plain concrete masonry walls, (2) solid concrete masonry walls, (3) cavity walls, (4) veneered walls, and (5) reinforced concrete masonry walls (Fig. 10.7).

BASIC TERMINOLOGY

The ability of a masonry wall to resist the vertical and horizontal loads imposed upon it is a function of both the type of masonry units and masonry mortar that are used and the interaction between individual masonry units that is created when these units are joined together to form a wall. Masonry walls can be designed as single-wythe walls, which have a thickness equal to the width of the masonry unit that is used, or as multiple (usually two) wythe walls that are closely spaced and structurally joined together in one of a number of commonly accepted ways.

Figure 10.8 illustrates some of the terms that are used in masonry wall construction and some of the brick positions found in single- and double-wythe brick walls. The ultimate performance of a brick wall depends upon whether or not the **bond** that is used is appropriate for the loads that the wall must carry.

Three different types of bond are commonly recognized. The first, **structural bond**, refers to the method that is used to interlock or tie together the individual masonry units. This type of bond is typically achieved by: (1) the overlapping of the masonry units between courses and the interlocking of masonry units between wythes, (2) the use of ties that are embedded in the mortar joints, and (3) the use of masonry grout between wythes. **Pattern bond** refers to the pattern that results on the face of the wall due to the arrangement of the

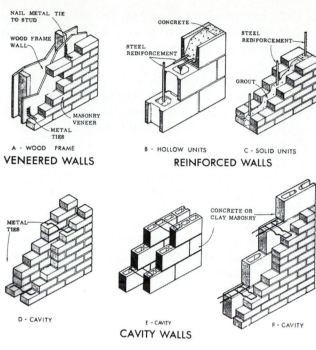

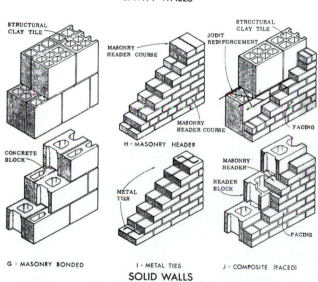

Figure 10.7
Typical Masonry
Wall Assemblies
Source: Harold B.
Olin, John L.
Schmidt, and Walter
H. Lewis.
*Construction
Principles,
Materials, and
Methods,* 4th ed.
(Chicago, IL: The
Institute of
Financial
Education, 1980),
p. 343.5.

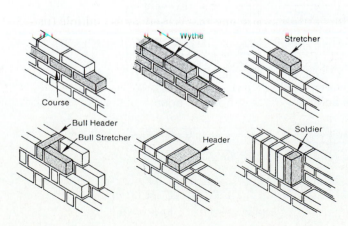

Figure 10.8
Basic
Terminology of
Masonry Units in
a Wall
Source: Gordon A.
Sanders, *Light
Building
Construction*
(Reston, VA:
Reston Publishing
Company, Inc.,
1985), p. 233.

individual masonry units and the masonry mortar. The pattern bond can serve either (or both) a structural or decorative function. The third type of bond, called **mortar bond**, results from the adhesion of the mortar to the masonry units and the reinforcing steel if it is used in the wall.

The basic types of **pattern bond** for brick walls are shown in Fig. 10.9. Both brick and concrete masonry walls are often constructed in **running bond**, which is the simplest of the basic pattern bonds because it consists of only stretcher units. Used primarily in cavity wall and veneer wall construction, it is built with each head joint positioned with either a one-half or one-third overlap over the unit in the next course. **Common (or American) bond** uses full-length headers every fifth to seventh course to provide an interlocking between double wythe walls (see Fig. 10.8), with the remaining courses being in a running bond pattern. Figure 10.9 also indicates a variation of the common bond pattern which uses Flemish headers positioned between stretcher units. Both the **Flemish bond** and the **English bond** patterns also rely on header units to provide an interlocking between double wythe walls. **Stack (or block) bond** is an essentially non-structural pattern bond.

Concrete masonry unit walls also are built using some commonly accepted bond patterns. Several of these are shown in Fig. 10.10. Decorative pattern bonds include: (1) coursed ashlar, (2) diagonal bond, (3) basket weave, and (4) diagonal basket weave.

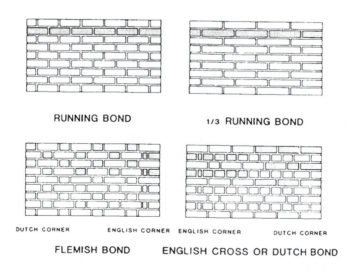

Figure 10.9
Traditional
Pattern Bonds for
Brick Walls

*Source: Technical
Notes on Brick
Construction, No. 30
(Reston, VA: Brick
Institute of
America, 1988),
p. 2.*

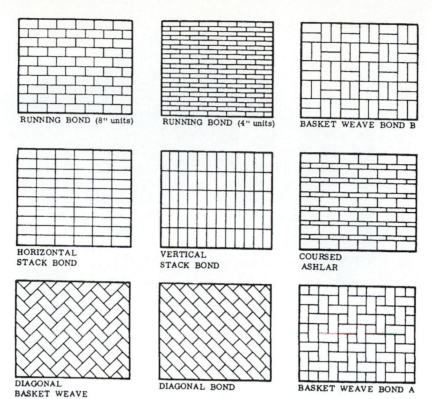

RUNNING BOND (8" units)   RUNNING BOND (4" units)   BASKET WEAVE BOND B

HORIZONTAL STACK BOND   VERTICAL STACK BOND   COURSED ASHLAR

DIAGONAL BASKET WEAVE   DIAGONAL BOND   BASKET WEAVE BOND A

Figure 10.10 Traditional Pattern Bonds for Concrete Masonry Walls

Source: Harold B. Olin, John L. Schmidt, and Walter H. Lewis. *Construction Principles, Materials and Methods*, 4th ed. (Chicago, IL: The Institute of Financial Education, 1980), pp. 343–10.

PLAIN CONCRETE MASONRY SINGLE WYTHE WALL

A plain concrete masonry, single wythe wall consists of a single vertical tier or stack of masonry units which does not contain steel reinforcement. The cores of the unit may or may not be filled with grout. This basic type of wall system has been traditionally used to construct foundation walls for houses. The basic components for such an application are shown in Fig. 10.11.

As indicated in Table 9.12, when ungrouted, a 12-in.-thick hollow masonry unit wall can be used to resist a depth of unbalanced fill of only 6 ft. in Seismic Zone 0, 1, or 2 where unstable soil or groundwater conditions do not exist. This limits the usage to the crawl space situation in Fig. 9.11(b), since most basements require at least 8-ft-high walls. In order to resist a depth of unbalanced fill of 8 ft, both a solid or hollow unit plain concrete masonry wall must be fully grouted and either 10 in. or 12 in. thick.

The National Concrete Masonry Association (NCMA) also requires that pilasters, crosswalls, or supporting stoops[33] (see Fig. 10.12) be installed wherever a horizontal section of a plain concrete masonry foundation wall exceeds three times its wall height. The 1995 CABO Code does not put a similar restriction on Table 9.12.

SOLID MASONRY WALLS

Solid masonry walls, as shown in Fig. 10.7, typically consist of two wythes that act monolithically because they are spaced close together; structurally connected by header units, metal ties, or joint reinforcement; and have the collar joint between them solidly filled with mortar or grout. They possess greater strength and moisture penetration resistance than single-wythe walls because of their greater thickness.

CAVITY WALLS

A **cavity wall** (Fig. 10.7) consists of two wythes of solid or hollow masonry units that are deliberately separated into an inner and outer wall by a continuous air space not less than 2 in. or more than 4-1/2 in. wide. The two wythes are bonded together with metal ties or joint

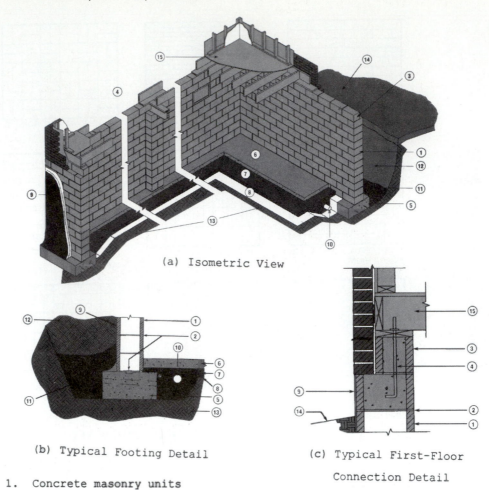

(a) Isometric View

(b) Typical Footing Detail

(c) Typical First-Floor Connection Detail

Figure 10.11
A Typical Plain Concrete Masonry Foundation Wall
Source: NCMA Guide for Home Owners and Home Builders on Residential Concrete Masonry Basement Walls (Herdon, VA: National Concrete Masonry Association, 1994), p. 18.

1. Concrete masonry units
2. Type M or S mortar
3. Solid top course
4. Anchor bolts
5. Concrete or solid concrete masonry footing
6. Concrete slab
7. Aggregate base
8. Vapor barrier
9. Water proof or damproof membrane
10. Foundation drain
11. Free draining backfill
12. Backfill
13. Undisturbed soil
14. Top of grade
15. Floor diaphragm

reinforcement. Often, the 4-in. outer wall is composed of non-load-bearing clay brick and the inner wall is composed of 4-, 6-, or 8-in. solid or hollow concrete masonry units that carry the weight of the floors and the roof.

VENEERED WALLS

Clay brick, concrete brick, and other architectural facing units are often used as a non-structural cladding which is attached to backing material that could be a masonry, concrete, wood stud, or metal stud load-bearing wall. Figures 7.42 and 10.7 indicate such an application on a wood stud wall system.

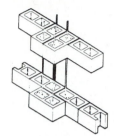

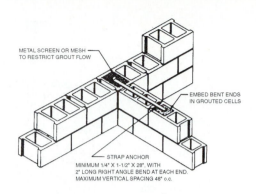

METAL SCREEN OR MESH
TO RESTRICT GROUT FLOW

EMBED BENT ENDS
IN GROUTED CELLS

STRAP ANCHOR
MINIMUM 1/4" X 1-1/2" X 28", WITH
2" LONG RIGHT ANGLE BEND AT EACH END.
MAXIMUM VERTICAL SPACING 48" o.c.

Figure 10.12
Acceptable
Lateral Support
Options for Plain
Concrete
Masonry
Foundation Walls
*Source: NCMA
TEK 14-10A-Lateral
Support of Concrete
Masonry Walls*
(Herdon, VA:
National Concrete
Masonry
Association, 1994),
p. 2.

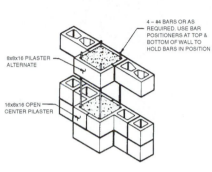

4 – #4 BARS OR AS
REQUIRED. USE BAR
POSITIONERS AT TOP &
BOTTOM OF WALL TO
HOLD BARS IN POSITION

8x8x16 PILASTER
ALTERNATE

16x8x16 OPEN
CENTER PILASTER

(a) Pilaster Connection

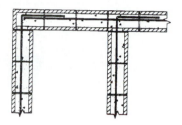

(b) Intersecting Wall Connection

REINFORCED CONCRETE MASONRY WALLS

A concrete masonry wall (Fig. 10.7) must often resist a combination of loads which are imposed in both the vertical direction (i.e., dead load, live load, etc.) as well as the lateral direction (i.e., wind load, seismic load, lateral earth pressure, etc.). At a certain magnitude of loading, **steel reinforcement** must be installed in the wall because it becomes structurally inefficient for the masonry units to act by themselves. A reinforced concrete masonry wall can more efficiently resist the loads when it relies on both the tensile strength of the reinforcement and the compressive strength of the masonry units.

Figure 10.13 indicates both the lateral loads that act on a concrete masonry foundation and a simplified representation of the deflected shape of the wall due to the imposed lateral earth pressure. The vertical reinforcement is placed toward the inside of the wall because that side experiences tensile stresses due to the deflected shape.

The triangular-shaped lateral pressure diagram represents an equivalent fluid pressure which, for free draining soil, is usually assumed to be 30 pcf. The placement of the vertical and horizontal reinforcement to counteract this loading is shown in Fig. 10.14.

The reinforced concrete masonry foundation wall requirements of the 1995 CABO Code are provided in both Tables 9.12 and 9.13. For the more favorable soil and seismic options of Table 9.12, a maximum depth of unbalanced fill of 7 ft can be resisted by an 8-in.-thick hollow concrete block wall if vertical No. 4 reinforcing bars are spaced at 24 in. o.c. and the cells containing these bars are fully grouted. There is also a location restraint placed on the reinforcement.

Table 9.13 addresses the more unfavorable condition of either Seismic Zone 3 or 4 or unstable soil conditions when the equivalent fluid weight of the soil does not exceed 30 lbs/ft^3. Steel reinforcement, and a spacing requirement for lateral wall support, must be designed for hollow masonry walls if the height of unbalanced fill exceeds 4 ft. For solid masonry walls, the horizontal reinforcement requirement in the upper 12 in. of the wall and the vertical reinforcement requirement are specified for various depths of unbalanced fill/lateral wall support combinations. If the height of unbalanced fill exceeds 8 ft, a structural design of the wall system is required.

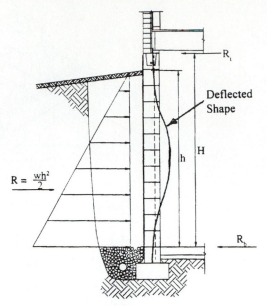

Figure 10.13 Lateral Forces Acting on a Reinforced Concrete Masonry Foundation Wall

Source: NCMA TEK 15-3-Reinforced Eight-Inch Basement Walls (Herdon, VA: National Concrete Masonry Association, 1991) (modified), p. 4.

$$M_{max} = \frac{wh^3}{6H} \left(H + \frac{2h}{3} \sqrt{\frac{h}{3H}} - h \right) \text{ ft. lbs./ft.}$$

$$R_b = \frac{wh^2}{2} - \frac{wh^3}{6H} \text{ lbs./ft.}$$

$$R_t = \frac{wh^3}{6H} \text{ lbs./ft.}$$

Where:

H = Wall Height, ft.
h = Backfill Height, ft.
R = Resultant Lateral Force, lb./ft.
w = Equivalent Fluid Weight of Soil, pcf
M_{max} = Maximum Moment Due to Lateral Soil Pressure
R_t = Reaction at top of wall, lbs./ft.
R_B = Reaction at bottom of wall, lbs./ft.

ADDITIONAL ASPECTS

Several additional aspects of masonry wall construction should be mentioned because they are important factors in the overall performance of the wall system and the house.

Moisture Control—The presence and movement of water in the liquid, solid, and vapor states must be recognized and considered in the design of masonry walls. The requirements of the 1995 CABO Code,[34] if followed during the construction stage, should reduce the potential moisture damage below grade.

A combination of flashing and weepholes is typically used to control moisture in masonry walls above grade. A detailed discussion of flashing is found in NCMA TEK 19-5.[35] As noted there, flashing consists of a sheet of impervious material which is built into the wall system to intercept the flow of water through the masonry units and direct it to the exterior of the house. Weepholes are typically placed in the mortar joints immediately above the flashing level so that moisture can escape (Fig. 10.15).

The flow of moist air through a wall system is typically controlled by using air barriers and vapor barriers.

Wall Movement Control—Temperature changes, moisture changes, and differential settlement can create movements in the wall which can cause detrimental cracking if the concrete masonry places a restraint against that movement. As noted in NCMA TEK 10-2,[36]

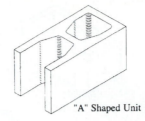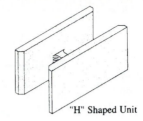

Bond Beam Unit   "A" Shaped Unit   "H" Shaped Unit

(a) Special Block Shapes for Reinforced Construction

Figure 10.14
Placement of
Reinforcement in
Reinforced
Concrete
Masonry
Construction
*Source: NCMA
TEK 14-2-
Reinforced Concrete
Masonry* (Herdon,
VA: National
Concrete Masonry
Association, 1995),
p. 5.

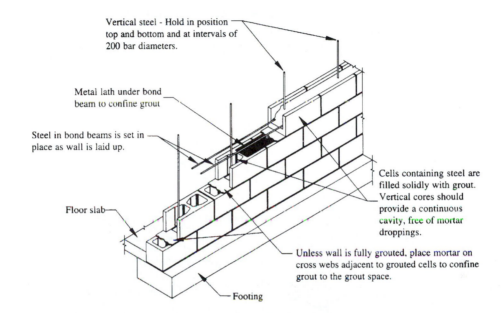

Vertical steel - Hold in position
top and bottom and at intervals of
200 bar diameters.

Metal lath under bond
beam to confine grout

Steel in bond beams is set in
place as wall is laid up.

Floor slab

Cells containing steel are
filled solidly with grout.
Vertical cores should
provide a continuous
cavity, free of mortar
droppings.

Unless wall is fully grouted, place mortar on
cross webs adjacent to grouted cells to confine
grout to the grout space.

Footing

(b) Location of Steel Reinforcement

such movements are mitigated by using a combination of **control joints**, **horizontal joint reinforcement**, and **bond beams**.

Connectors and Joint Reinforcement—The three types of connectors that are commonly used in masonry construction are **wall ties**, **anchors**, and **fasteners**. Wall ties are used to connect one masonry wythe to an adjacent one, and anchors are used to connect masonry to a structure or frame (Fig. 10.16). Fasteners are used to connect a particular appliance to masonry. NCMA TEK 12-1[37] provides some design guidelines and additional descriptive information for anchors and ties.

In addition to serving as a wall tie, the truss-type and ladder-type prefabricated wire ties shown in Fig. 10.16 can also be used as joint reinforcement to provide crack control in unreinforced concrete masonry walls and horizontal reinforcement in single-wythe reinforced concrete masonry walls. This type of joint reinforcement also can be used to reinforce stack bond masonry and improve the structural bonding of intersecting concrete masonry walls.

Energy Considerations—A number of insulation strategies are available to the homebuilder who wants to improve the thermal performance of a house by taking advantage of the inherent energy advantages of masonry construction.

Figure 10.15
The Use of
Flashing and
Weepholes in
Cavity Wall
Construction
*Source: NCMA
TEK 19-5-Use of
Flashing in Concrete
Masonry* (Herdon,
VA: National
Concrete Masonry
Association, 1993),
p. 3.

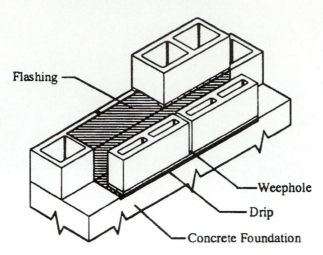

(a) Typical Wall Ties

Figure 10.16
Typical Concrete
Masonry
Connector Details
*Source: NCMA
TEK 12-1-Anchors
and Ties for
Masonry* (Herdon,
VA: National
Concrete Masonry
Association, 1995),
p. 3.

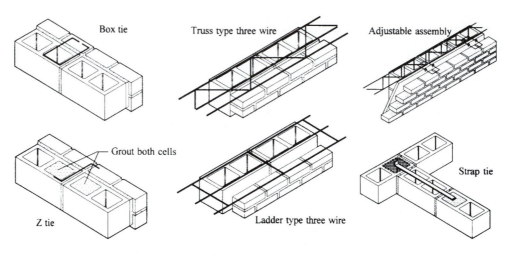

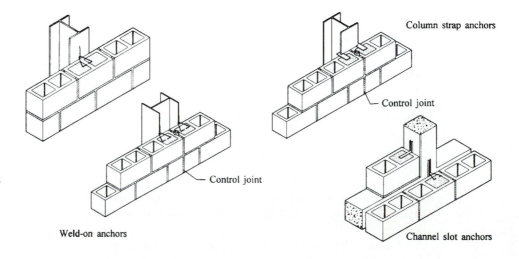

(b) Typical Anchors

Three different locations are used to insulate above-grade concrete masonry walls. **Exterior insulated masonry walls** are walls that have insulation placed outside the thermal mass. Such an insulation strategy keeps the masonry wall directly in contact with interior conditioned air. Both single-wythe and multiple-wythe walls can be insulated this way [Fig. 10.17(a)]. A single-wythe wall, which has a layer of rigid board insulation fastened to the masonry wall with mechanical non-conducting fasteners, requires a protective finish (i.e., stucco or plaster applied to a reinforcing mesh) to maintain the integrity of the insulation. A multiple-wythe wall can have the cavity filled with a rigid insulation board, granular insulation fill, or foam insulation.

Integral insulation is placed in the ungrouted core spaces of the concrete block units [Fig. 10.17(b)]. A number of manufacturers currently produce molded polystyrene inserts, expanded perlite, or vermiculite granular fills or insulation foams that can be used. **Interior insulation** options include fibrous batts, rigid polystyrene, or polyisocyanurate boards or fibrous blown in insulation which is installed between wood or metal studs that are installed directly on or adjacent to the concrete masonry walls [Fig. 10.17(c)]. Interior insulation can also be attached directly to the wall using specially designed clips and channels. The interior wall of the house is then finished with conventional gypsum wallboard, paneling, or lath and plaster.

Interior Wall Finish

The design and construction requirements which control the type of interior wall covering materials which can be used on the interior walls of a house are presented in Chapter 7 of the 1995 CABO Code.[38] Information is provided about: (1) gypsum plaster and Portland cement plaster, (2) gypsum wallboard, (3) ceramic tile, (4) wood veneer and hardboard paneling, and (5) wood shakes and shingles. Although plaster walls were once the wall covering

Figure 10.17
Insulation
Alternatives for
Concrete
Masonry Walls
*Source: NCMA
TEK 6-11-Insulating
Concrete Masonry
Walls* (Herdon, VA:
National Concrete
Masonry
Association, 1995),
pp. 1–3.

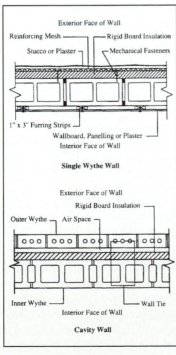

(a) Typical exterior insulation

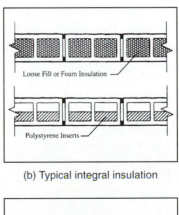

(b) Typical integral insulation

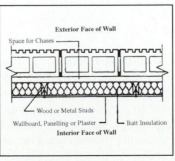

(c) Typical interior insulation

of choice for new house construction, the comparative ease of installation and resultant lower cost of gypsum board have made it the most commonly used interior wall covering material used in houses.

Overview of Gypsum Board

Gypsum board, also known as drywall, plaster board or wallboard, consists of a noncombustible core of gypsum with a paper surface on both faces and on all edges. It is an excellent fire-resistive material, provides good sound isolation properties, is dimensionally stable and durable, and is easy to install and repair.

Gypsum board has become an economical commodity product which is available in 4-ft.-wide and 8-, 10-, and 12-ft long sizes which are compatible with the standard 16-in. and 24-in. o.c. spacing of studs and joists. It is typically available with edges that are tapered, square edge, beveled, rounded, or tongue-and-grooved.

Regular gypsum board is the most widely used, although **Type X gypsum board** is used when improved fire resistance is required. Typical applications of Type X gypsum board include the finish on the garage side of the wall which separates the living space from the garage in single-family detached housing and on the separation wall between individual housing units in townhouse construction. Water-resistant gypsum board, which has a water-resistant core and is covered with water-repellant paper, is used as a backer board for the application of ceramic or plastic wall tile or plastic finish panels in bath, shower, kitchen, and laundry areas.[39]

Figures 7.45 and 10.18 indicate how gypsum wallboard is typically installed over a wood stud wall using nails which are specifically designed for such use. Wallboard may also be screwed to either a wood or steel stud wall system, or adhesives may be used to bond single layers of gypsum board directly to framing, furring, masonry, or concrete.

The quality of a gypsum wallboard installation is often determined by the care that is taken to cement and tape the joints, cover up the nail or screw locations, and in the treatment of interior and exterior corners and intersections of walls and ceilings. A number of excellent references provided by the Gypsum Association present the design and construction details which result in a quality gypsum board wall installation.[40] Homebuilders are encouraged to use these references in their total quality management programs for the specialty contractors that are used for this phase of the construction process.

Other Wall Covering Materials

Basements, family rooms, and other locations in a house are often finished with plywood, hardboard, or particleboard panels. As noted by Sherwood and Stroh, "hardboard and particleboard imprinted with a wood grain pattern are generally less expensive than plywood. A photograph of wood is used to imprint a facing material, which produces a very realistic pattern. Both smooth and textured facings also are available in solid colors as well as designs. Facing materials usually are easy to clean."[41]

Exterior Wall Finish

The design and construction requirements which control the type of exterior wall finishes on houses are presented in the 1995 CABO Code.[42] As noted, "all exterior walls shall be covered with approved materials designed and installed to provide a barrier against the weather and insects to enable environmental control of the interior spaces."[43] A wide variety of wood, masonry veneer, vinyl, and metal siding products can be used to meet these objectives while at the same time meeting the architectural design objectives noted in Chapter 5 of this book. Some of these will be discussed briefly. Manufacturers' literature should be consulted in order to obtain additional information about a specific product.

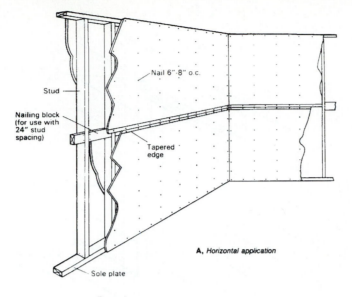

A, *Horizontal application*

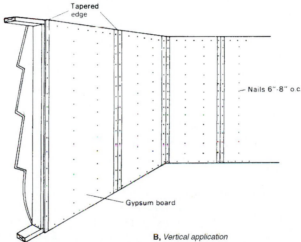

B, *Vertical application*

Figure 10.18
Installation Detail
of Gypsum Board
on Walls
Source: Gerald E.
Sherwood and
Robert C. Stroh,
*Wood Frame House
Construction*
(Mineola, NY:
Dover Publications,
Inc., 1989), p. 177.

Overview of Exterior Wall Finishes

SOLID WOOD SIDING

Solid wood siding is available in many styles for both horizontal and vertical applications, including beveled (see Fig. 10.19), dropped, and beaded planks for horizontal use and V-groove, tongue-and-groove, board-and-batten, and channel configurations for vertical siding usage. Almost any type of wood can be used for solid wood siding, including such species as cedar, redwood, fir, cypress, pine, spruce, and hemlock. All of these sidings should be weatherproofed with water-repellent treatments that will have to be periodically renewed as needed.

PLYWOOD SIDING

Plywood siding also is available in many varieties of wood finish and texture at varying costs. Plywood panels are 4 ft wide and range from 8 to 12 ft in length. When using plywood siding, the correct length of panel should be matched to the house dimensions in order to min-

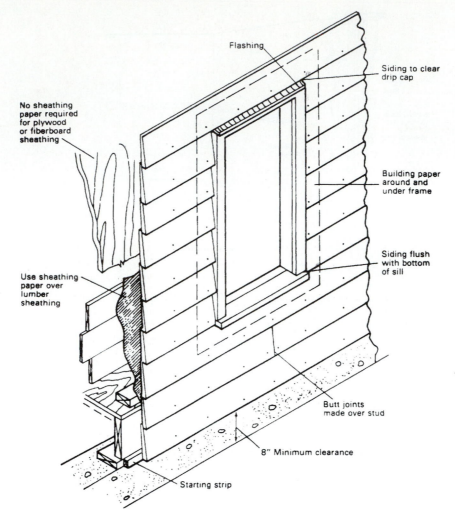

Figure 10.19

Installation of Bevel Pattern Wood Siding

Source: Gerald E. Sherwood and Robert C. Stroh, *Wood Frame House Construction* (Mineola, NY: Dover Publications, Inc., 1989), p. 121.

imize the number of horizontal joints. All plywood siding must be of the exterior type which is manufactured with waterproof glue.

WOOD SHINGLES AND SHAKES

Wood shingles and shakes typically are manufactured from red or white cedar. The difference between the two is that both sides of shingles are sawn smooth, while shakes have at least one rough-textured side. This siding is naturally resistant to decay and is an excellent insulator. The cedar is originally a golden brown color and will eventually weather to silver or dark gray. The speed of change and final shade depend primarily on local atmospheric and climatic conditions. Wood shingles and shakes are expensive and time consuming to install.

BRICK OR MASONRY

Solid brick or masonry exterior walls, as well as brick or masonry veneer walls, provide a practically maintenance-free exterior siding. A brick or masonry veneer wall, as noted earlier, is not a structural part of the house; it is merely a facing supported by the foundation and tied to the wood frame walls. Brick and masonry units are available in a variety of colors and textures. The wall pattern and type of mortar joint also can be modified (see Figs. 10.9 and 10.10). A more attractive exterior appearance can be obtained than is available with other siding materials, although brick and masonry are more expensive to purchase and install.

STONE

Stone creates an attractive exterior appearance that is durable and practically maintenance free. The disadvantage is that stone generally is more expensive than a brick or masonry veneer. All three of these options offer very low insulating values; thus, houses with these sidings tend to be colder in the winter.

ALUMINUM SIDING

Aluminum siding is a low-maintenance exterior wall covering that will not rot, split, warp, or crack. Its baked-on finish can last between 20 and 40 years, and it can be obtained in a variety of colors, both light and dark. Panels with 4- and 8-in. wide exposures are available. One of the disadvantages of aluminum siding is that it dents and scratches very easily. A damaged panel would have to be replaced.

VINYL SIDING

Vinyl siding is another popular low-maintenance wall covering that essentially has replaced aluminum siding in many areas of the United States because of its cost advantage. The color is molded throughout the entire thickness of the material; therefore, a scratch will do little damage. Vinyl siding, due to its resilient nature, does not dent. A disadvantage of vinyl siding is that it can buckle or ripple if it is not installed correctly. It also is not available in as many colors as aluminum siding. Figure 7.41 provides an example of how the exterior appearance of the Case Study House was enhanced by using vinyl siding.

STUCCO

Stucco is an exterior wall covering that requires little or no maintenance. It is a plaster-like material made of a mixture of cement, lime, sand, and water. Stucco is water and fire resistant. It is troweled onto either masonry or frame walls, with no seams or joints. The natural color of stucco is white, but different colors can be added to the mixture. Different finish textures also are available, from smooth to rough to a sandy or pebbled surface.

EIFS EXTERIOR WALLS

A stucco appearance as well as superior insulating advantages can be achieved by using an Exterior Insulation and Finish System (EIFS), which uses a combination of synthetic materials in the wall system. A number of product manufacturers now produce such systems, and they are becoming increasingly popular with homebuilders because of the diversity of architectural features that can be obtained using this innovative system.

1995 CABO Code Requirements

Specific design and construction requirements for a number of the exterior wall finish materials noted above are provided in the 1995 CABO Code.[44] The major guidelines relate to the thickness of siding, joint treatment, the requirement for an underlayment of sheathing paper, and the types of supports and fasteners that are required. An example of the type of requirements for vinyl siding and wood siding is shown in Table 10.6.

TABLE 10.6 Vinyl and Wood Siding Attachment and Minimum Thickness Requirements

Siding Material	Nominal Thickness[a] (in.)	Joint Treatment	Sheathing Paper Required	Type of Supports for the Siding Material and Fasteners[b,c,d]				
				Wood or Wood Structural Panel Sheathing	Fiberboard Sheathing Into Stud	Gypsum Sheathing Into Stud	Direct to Studs	Number or Spacing of Fasteners
Vinyl Siding[e]	0.035	Lap	No	0.120 nail 1-1/2" Staple 1-3/4"	0.120 nail 2" Staple 2-1/2"	0.120 nail 2" Staple 2-1/2"	Not Allowed	Same as stud spacing
Wood[f] Rustic, drop Shiplap	3/8 Min. 19/32 Avg.	Lap	No	Fastener penetration into stud—1"			0.113 nail— 2-1/2", Staple 2"	Face nailing up to 6" widths, 1 nail per bearing; 8" widths and over, 2 nails per bearing
Bevel	7/16	Lap	No					
Butt tip	3/16	Lap	No					

[a]Based on stud spacing of 16 in. o.c. Where studs are spaced 24 in., siding may be applied to sheathing approved for that spacing.
[b]Nail is a general description and may be T-head, modified round head, or round head with smooth or deformed shanks.
[c]Staples shall have a minimum crown width of 7/16-in. O.D. and be manufactured of minimum No. 16 gage wire.
[d]Nails or staples must be aluminum, galvanized, or rust-penetrative coated and shall be driven into the studs for fiberboard or gypsum backing.
[e]Vinyl siding shall comply with ASTM D 3679.
[f]Woodboard sidings applied vertically shall be nailed to horizontal nailing strips or blocking set 24 in. o.c. Nails shall penetrate 1-1/2 in. into studs, studs and wood sheathing combined, or blocking. A weather-resistant membrane shall be installed weatherboard fashion under the vertical siding unless the siding boards are lapped or battens are used.

Source: Reproduced from the 1995 edition of the CABO One and Two Family Dwelling Code,™ copyright © 1995, with the permission of the publisher, The International Codes Council, Whittier, CA, excerpt from Table 703.4, p. 85.

ROOF–CEILING SYSTEMS

Both wood wall and masonry wall systems, as well as interior and exterior wall finish systems, are discussed in the beginning of this chapter. The remaining primary structural element of a platform frame house that must be examined is the **roof–ceiling system**. Closely associated with that system are the roof covering systems that can be used.

Figure 9.4 indicates that one of the commonly used roof–ceiling systems is the **roof rafter/ceiling joist** option. Another is the **pre-engineered roof truss** option (Fig. 9.8). The loads that must be carried by the roof–ceiling system consist of a combination of dead load, live load, snowload, and wind load (Fig. 9.6).

Chapter 8 of the 1995 CABO Code,[45] presents the requirements which ensure that the loads imposed on the roof–ceiling system are supported and also properly transmitted to the rest of the structural frame. Figure 10.20 provides additional details about the roof–ceiling system. The primary emphasis in Chapter 8 is placed on the roof rafter/ceiling joist option. Only the following brief reference is made to the pre-engineered roof truss option:

> Wood trusses shall be designed in accordance with approved engineering practice. Truss components may be joined by nails, glue, timber connectors, or other approved fastening devices.[46]

Reference to the appropriate industry specifications for metal-plate-connected wood trusses also is provided, and it is noted that truss members cannot be cut or altered unless so designed.

Homebuilders desiring to use alternative roof–ceiling systems must follow the procedures provided in the 1995 CABO Code related to the approval of alternative materials and systems.[47]

The roof–ceiling system must be able to accommodate the climatic, dead, and live loads imposed on it. In addition, water draining from the roof must be collected and discharged to the ground surface at least 5 ft from foundation walls or to an approved drainage system in areas where expansive or collapsible soils are known to exist.[48]

Figure 10.20
Roof-Ceiling
System with the
Braced Rafter
Option Indicated
Source: Reproduced
from the 1995 edition of the *CABO
One and Two
Family Dwelling
Code,*™ copyright ©
1995, with the permission of the publisher, The
International Codes
Council, Whittier,
CA, Fig. 802.4.1,
p. 90.

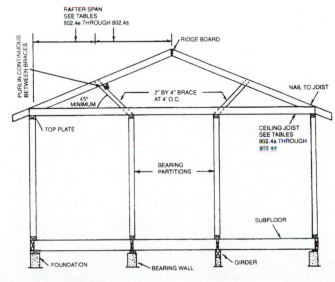

For SI: 1 inch = 25.4 mm, 1 foot = 304.8 mm.

Note:
Where ceiling joists run perpendicular to the rafters, rafter ties shall be nailed to the rafters near the plate line and spaced not more than 4 feet on center.

Introduction to Roof Framing

Essential information about the design of the roof rafter/ceiling joist option is provided in the 1995 CABO Code.[49] One of the requirements is that a grade mark or a certificate of inspection by an approved agency must be provided which includes adequate information to determine values for F_b, the allowable stress in bending, and E, the modulus of elasticity of the lumber being used. Acceptable design approaches include the requirements in Chapter 8 of the 1995 CABO Code or a series of listed industry standards.

These design approaches typically select the size and spacing of: (1) the roof rafters so that the vertical dead, live, and snow loads can be supported safely, and (2) the ceiling joists so that attic loads can be supported safely. The design procedure is similar to the one that is used for floor joists. Although these vertical loads must certainly be supported, there are also major problems created for the roof–ceiling framing system in certain locations in the United States (see Fig. 9.9) by the **negative pressure** or **uplift** caused by **high winds**.

For wood-framed houses, the design of the roof–ceiling system must conform to the framing information provided in Fig. 10.20 and the **fastening requirements** indicated in the 1995 CABO Code in the expanded version of Table 10.4. For masonry houses, the design of the roof–ceiling system must conform to the **anchorage requirements** that are provided for various wind load/seismic zone conditions.[50]

Additional attention is given to the roof tie-down requirements related to the effects of wind by indicating that roof assemblies subject to wind uplift pressures of 20 psf or greater must have tie-down connectors that can resist the forces indicated in a "wind uplift force" table that is provided. The uplift forces must be transmitted from the rafter or truss ties to the foundation.[51]

Allowable Spans for Ceiling Joists

Figure 9.4 indicates that the ceiling joists supporting the attic space and the ceiling on the top level of the house typically span from the exterior bearing walls to the interior bearing wall. Table 9.7 specifies two possible attic live loads: 10 psf for houses with roof slopes not steeper than 3 in 12 that will not have storage capability, and 20 psf for houses which have limited storage in the attic. In addition, a uniform dead load of 10 psf commonly is used to represent the weight of the ceiling framing and the attic and ceiling covering components.

Description of Span Tables

Chapter 8 of the 1995 CABO Code provides four allowable span tables for ceiling joists which take into account the loading conditions most commonly encountered in houses. These conditions, as shown in Table 10.7, include both plaster and gypsum ceilings on the upper level of the house. The deflection criteria for a plaster ceiling are more rigorous than for a gypsum ceiling because of the greater tendency of plaster ceilings to experience cracking under live load deflection.

TABLE 10.7   Types of Allowable Span Tables for Ceiling Joists

Table No.	Attic Storage	Live Load (psf)	Dead Load (psf)	Type of Ceiling	Deflection Limitation
802.4a	Limited	20	10	Plaster	$\ell/360$
802.4b	Limited	20	10	Gypsum	$\ell/240$
802.4c	None	10	5	Plaster	$\ell/360$
802.4d	None	10	5	Gypsum	$\ell/240$

Source: Reproduced from the 1995 edition of the *CABO One and Two Family Dwelling Code*™, copyright© 1995, with the permission of the publisher, the International Codes Council, pp. 93–100.

An excerpt from the indicated Table 802.4b for the specific ceiling joist size of 2 in. × 8 in. is provided in Table 10.8 to indicate how the span tables for ceiling joists are used. It is important to note that *Table 10.8 can be used only for the case of simply supported members which carry a uniformly distributed load*. In addition, the "repetitive member use" F_b value can be used only if the ceiling joists are spaced no more than 24 in. o.c.

A comparison of Table 10.8 for ceiling joists with Tables 9.15 and 9.16 for floor joists indicates that its format is similar and that it can therefore be used in a similar fashion.

Case Study House Example

A review of Sheet A-5 of the working drawings for the Case Study House (Appendix A) indicates that pre-engineered roof trusses, spaced at 24 in. o.c., are used as the roof–ceiling system over the main part of the house. This is confirmed on Sheet A-2 (i.e., the second-floor plan). Sheet A-1 also indicates that engineered wood roof trusses, spaced at 24 in. o.c., are used as the roof–ceiling system over the garage.

In order to illustrate how a roof rafter/ceiling joist system is selected, the Case Study House must therefore be redesigned with a load-bearing wall on the second level, as shown in Fig. 9.4. The following redesign assumptions must be made:

1. The existing 2 in. × 4 in. non-load-bearing wall on the second level (see Sheet A-2) that serves as the interior wall for Bedrooms #2 and #3 will be changed to a 2 in. × 6 in. load-bearing wall.

2. Since the new load-bearing wall is directly above the existing 2 in. × 6 in. bearing wall on the first level, the spans of the ceiling joists above the second level are 12 ft and 16 ft.

3. A condition of limited attic storage will be assumed, and a gypsum ceiling will be used on the second level.

TABLE 10.8 Allowable Spans for Ceiling Joists with 20 lbs per Square Foot Live Load (Limited Attic Storage where Development of Future Rooms is not Possible) (Gypsum Ceiling)

DESIGN CRITERIA:
Deflection—For 20 lbs per sq ft live load. Limited to span in inches divided by 240.
Strength—Live load of 20 lbs per sq ft plus dead load of 10 lbs per sq ft determines fiber stress value.

Joist Size and Spacing		Modulus of Elasticity, "E," in 1,000,000 psi						
(in.)	(in.)	0.4	0.6	0.8	1.0	1.1	1.2	1.4
	12.0	12-10	14-8	16-2	17-5	18-0	18-6	19-6
		560	740	900	1,040	1,110	1,170	1,300
	13.7	12-3	14-1	15-6	16-8	17-2	17-9	18-8
		590	770	940	1,090	1,160	1,230	1,360
2×8	16.0	11-8	13-4	14-8	15-10	16-4	16-10	17-9
		620	810	990	1,140	1,220	1,290	1,430
	19.2	11-0	12-7	13-10	14-11	15-5	15-10	16-8
		660	870	1,050	1,220	1,300	1,370	1,520
	24.0	10-2	11-8	12-10	13-10	14-3	14-8	15-6
		710	930	1,130	1,310	1,400	1,480	1,640

For SI: 1 in. = 25.4 mm, 1 pound per sq in. = 6.895 kPa, 1 pound per sq ft = 0.0479 kN/m².

NOTE: The extreme fiber stress in bending, "F_b," in pounds per sq in. is shown below each span.

Source: Reproduced from the 1995 edition of the *CABO One and Two Family Dwelling Code*™, copyright © 1995, with the permission of the publisher, the International Codes Council, excerpt from Table 802.4b, pp. 95–96.

CEILING JOISTS

The selection of the ceiling joist size and spacing follows the "Checklist for Design" steps presented when floor joist design was discussed.

    I. Checklist for Design

        A. Determine

            1. Span

                Actual spans: 12 ft and 16 ft

                Clear spans:

                      12—(3.5 in. + 5.5 in./2) = 11 ft, 5-3/4 in.

                      16—(3.4 in. + 5.5 in./2) = 15 ft, 5-3/4 in.

                Since it is more efficient, from a framing standpoint, to have the ceiling joists all one size, only the 16 ft span situation will be considered in this illustration. Thus, Span = 15 ft, 5-3/4 in.

            2. Joist spacing desired

                The two trial spacing distances will be 16 in. o.c. (because it matches the spacing of the wood studs on the exterior load-bearing walls) and 24 in. o.c. (because it matches the spacing of the pre-engineered roof trusses that were actually used). Thus, Spacing = 16 in. or 24 in.

        B. Enter standard span tables and determine:

            1. Size of joist

            2. Required F_b (i.e., the actual F_b incurred)

            3. E

               The allowable span tables in the 1995 CABO Code provide information about 2 in. × 4 in., 2 in. × 6 in., 2 in. × 8 in., and 2 in. × 10 in. ceiling joists. Table 10.8, which presents only 2 in. × 8 in. ceiling joist information, is used to illustrate the selection process.

                      Table 10.8 indicates that for 2 in. × 8 in. ceiling joists:

               Spacing = 16 in. o.c.; for a 15-ft, 10-in. span, the required $E = 1.0 \times 10^6$ psi and the required $F_b = 1140$ psi

               Spacing = 24 in. o.c.; for a 15-ft, 6-in. span, the required $E = 1.4 \times 10^6$ psi and $F_b = 1640$ psi

        C. Enter Table 9.14 with

            1. Required F_b

            2. Required E

        D. Read grade and species of various sizes of lumber permitted

               There are a number of species/grade combinations of lumber that could meet the above requirements. For illustrative purposes, it will be assumed that No. 2 Hem-Fir, Surface Dry, will be used for the ceiling joists to match what is used in the remainder of the house. Table 9.14 indicates:

            2 in. × 8 in., No. 2 Hem-Fir:

            $F_b = 1175$ psi (normal duration load)

            $E = 1.3 \times 10^6$ psi

               As a result, the 24-in. o.c. selection cannot be used because the required E and F_b values exceed the allowable E and F_b values for No. 2 Hem-Fir. The final selection, therefore, is:

            Ceiling Joists:

            2 in. × 8 in. No. 2 Hem-Fir at 16 in. o.c.

Allowable Spans for Roof Rafters

Figure 9.4 indicates that the roof rafters span from the exterior bearing wall to the ridge board at the top of the roof. The **span of a roof rafter** is considered to be the **horizontal projection** of that included distance. The vertical loads are considered to be applied on that horizontal projection. Purlins, as shown in Fig. 10.20, can be installed to reduce the span that is used to design the roof rafters. As noted in the 1995 CABO Code:

> Purlins may be installed to reduce the span of rafters . . . Purlins shall be sized no less than the required size of the rafters that they support. Purlins shall be continuous and shall be supported by 2 × 4 (51 mm by 102 mm) struts installed to bearing walls at a slope not less than 45° from the horizontal. The struts shall be spaced not more than 4 ft (1219 mm) o.c., and the unbraced length of struts shall not exceed 8 ft (2438 mm).[52]

The appropriate dead, live, snow, and wind loads that must be supported by the roof rafter system and transmitted to the exterior bearing walls are determined from Table 9.1 for the geographical area and building code jurisdiction in which the house is built.

Description of Span Tables

Chapter 8 of the 1995 CABO Code provides 15 allowable span tables for roof rafters which take into account the most commonly encountered roof loading conditions in houses. These conditions, as shown in Table 10.9, are divided into three categories. The first: low or high slope rafters, has the two subcategories of gypsum and plaster ceilings with their required deflection limitations. Separate tables are provided for each subcategory based upon 20 psf, 30 psf, and 40 psf live-load values. The second category, low slope rafters, is for houses with

TABLE 10.9   Types of Allowable Span Tables for Roof Rafters

I. Low or High Slope Rafters

Table No.	Live Load (psf)	Dead Load (psf)	Type of Ceiling	Deflection Limitation
802.4e	20	15	Gypsum	$\ell/240$
802.4f	30	15	Gypsum	$\ell/240$
802.4g	40	15	Gypsum	$\ell/240$
802.4h	20	15	Plaster	$\ell/360$
802.4i	30	15	Plaster	$\ell/360$
802.4j	40	15	Plaster	$\ell/360$

II. Low Slope Rafters (slope of 3 in 12 or less)

Table No.	Live Load (psf)	Dead Load (psf)	Type of Ceiling	Deflection Limitation
802.4k	20	10	No Finished Ceiling	$\ell/240$
802.4l	30	10	No Finished Ceiling	$\ell/240$
802.4m	40	10	No Finished Ceiling	$\ell/240$

III. High Slope Rafters (slope over 3 in 12)

Table No.	Live Load (psf)	Dead Load (psf)	Type of Roof Covering	Deflection Limitation
802.4n	20	15	Heavy Roof Covering	$\ell/180$
802.4o	30	15	Heavy Roof Covering	$\ell/180$
802.4p	40	15	Heavy Roof Covering	$\ell/180$
802.4q	20	7	Light Roof Covering	$\ell/180$
802.4r	30	7	Light Roof Covering	$\ell/180$
802.4s	40	7	Light Roof Covering	$\ell/180$

no finished ceilings. Again, separate tables are provided for three levels of live load. The third category, high slope rafters, has the two subcategories of heavy roof covering and light roof covering with their related dead load values. Separate tables are provided for three levels of live load.

An excerpt from the indicated Table 802.4q for the specific roof rafter size of 2 in. × 6 in. is provided in Table 10.10 to indicate how the span tables for roof rafters are used. Table 802.4q was selected because earlier in this chapter a design roof load of 21 psf was determined for the Case Study House (Appendix A), and because a light roof covering (fiberglass shingles) actually is used on the Case Study House.

Case Study House Example

The selection of the roof rafter size and spacing also follows the "Checklist Design" steps because the actual roof–ceiling system over the main part of the Case Study House consisted of pre-engineered roof trusses.

ROOF RAFTERS

I. Checklist for Design
    A. Determine
        1. Span
            Sheet A-5 indicates that the horizontal projection of the roof rafter is 14 ft. Thus, Span = 14 ft.
        2. Rafter spacing desired
            Since the assumed roof rafters must be nailed directly to the ceiling joists, a spacing of 16 in. o.c. will be used to match the selected spacing of the ceiling joists. Thus, Spacing = 16 in. o.c.

TABLE 10.10   Allowable Spans for High Slope Roof Rafters
Slope over 3 in 12 - 20 lbs per sq ft Live Load (Light Roof Covering)

DESIGN CRITERIA:
Strength—7 lbs per sq ft dead load plus 20 lbs per sq ft live load determines fiber stress.
Deflection—For 20 lbs per sq ft live load. Limited to span in inches divided by 180.

Rafter Size and Spacing		Allowable Fiber Stress in Bending, "F_b," (psi)						
(in.)	(in.)	600	800	1000	1200	1300	1400	1500
	12.0	10-7 0.38	12-3 0.59	13-8 0.83	15-0 1.09	15-7 1.23	16-2 1.37	16-9 1.52
	13.7	9-11 0.36	11-5 0.55	12-9 0.77	14-0 1.02	14-7 1.15	15-1 1.28	15-8 1.42
2 × 6	16.0	9-2 0.33	10-7 0.51	11-10 0.72	13-0 0.94	13-6 1.06	14-0 1.19	14-6 1.32
	19.2	8-4 0.30	9-8 0.47	10-10 0.65	11-10 0.86	12.4 0.97	12-9 1.08	13-3 1.20
	24.0	7-6 0.27	8-8 0.42	9-8 0.59	10-7 0.77	11-0 0.87	11-5 0.97	11-10 1.08

For SI: 1 in. = 25.4 mm, 1 pound per sq in. = 6.895 kPa, 1 pound per sq ft = 0.0479 kN/m².

NOTE: The modulus of elasticity, "E," in 1,000,000 pounds per sq in. is shown below each span.

RAFTERS: Spans are measured along the horizontal projection, and loads are considered as applied on the horizontal projection.

Source: Reproduced from the 1995 edition of the *CABO One and Two Family Dwelling Code™*, copyright © 1995, with the permission of the publisher, the International Codes Council, excerpt from Table 802.4q, pp. 125–126.

B. Enter standard span tables and determine:

  1. Size of joist

  2. Required F_b (i.e., the actual f_b incurred)

  3. E

  The roof rafter tables described in Table 10.9 provide information about roof rafters which range in size from 2 in. $\times$ 4 in. to 2 in. $\times$ 12 in. Table 10.10 only presents information about 2 in. $\times$ 6 in. roof rafters.

  An examination of Table 10.10 will indicate that it is constructed a little differently than Table 10.8. In Table 10.10, the columns refer to different values of F_b while the bottom entry in each row refers to the required values of E. The upper entry in each row indicates the span length which results in these values of F_b and E.

  Table 10.10 indicates that for 2 in. $\times$ 6 in. roof rafters spaced at 16 in. o.c., a span of 14 ft requires a structural member that has at least F_b = 1400 psi and E = 1.19 $\times$ 10^6 psi.

C. Enter Table 9.14 with

  1. Required F_b

  2. Required E

D. Read grade and species of various sizes of lumber permitted

  For illustrative purposes, No. 2 Hem-Fir Surface Dry will be used for the roof rafters. It is important to note that because the Case Study House is located in central Pennsylvania, the F_b value corresponding to the snow loading condition is used.

  Table 9.14 indicates

  2 in. $\times$ 6 in. No. 2 Hem-Fir:

  F_b = 1460 psi (snow loading)

  E = 1.3 $\times$ 10^6 psi

  Since the actual values of F_b and E for No. 2 Hem-Fir exceed the required values of F_b = 1400 psi and $E \times$ 1.19 $\times$ 10^6 psi, the final selection is

  Roof Rafters:

  2 in. $\times$ 6 in. No. 2 Hem-Fir at 16 in. o.c.

Additional Roof Framing Details

In addition to establishing the requirements related to the allowable spans for ceiling joists and roof rafters, the 1995 CABO Code provides detailed information about specific aspects of roof framing. Included are requirements related to:

- Fire-retardant-treated lumber
- Cathedral ceilings
- Framing details
- Ceiling joists lapped
- Bearing
- Finished ceiling material
- Cutting and notching
- Bored holes
- Lateral support
- Bridging
- Framing of openings
- Headers[53]

These detailed requirements, although they seldom appear directly on a set of working drawings, should become part of a homebuilder's quality management checklist for the roof–ceiling system phase of the construction process.

Roof Sheathing

The 1995 CABO Code provides basic requirements for: (1) lumber sheathing, (2) plywood sheathing, and (3) particleboard sheathing.[54] Allowable span information for lumber sheathing is provided. It is noted that the grade and allowable span information for plywood and wood structural panels which appears in Table 9.21 is applicable to roof sheathing. Appropriate industry standards which govern the selection and installation of plywood and wood structural patterns also are cited. Similar information is provided in the 1995 CABO Code for particleboard sheathing.[55]

Other Considerations

The 1995 CABO Code also provides brief information about Metal Framing, Ceiling Finishes, and Attic Access.[56] In addition, the important issue of providing proper ventilation through the enclosed attic and rafter spaces is addressed. A total net free ventilating area not less than 1 to 150 of the area of the space ventilated is required unless certain specified conditions apply.[57]

Roof Coverings

The design and construction requirements which control the type of roof coverings that can be used on houses are presented in Chapter 9 of the 1995 CABO Code.[58] Information is provided about: (1) Deck Preparation, (2) Asphalt Shingles, (3) Slate Shingles, (4) Metal Roofs, (5) Tile, Clay or Concrete Shingles, (5) Built-Up Roofing, (6) Wood Shingles, (7) Wood Shakes, and (8) Re-roofing. Homebuilders desiring to use alternative roof coverings must follow the provided procedures related to the approval of alternate materials and systems.[59]

Overview of Roof Coverings

ASPHALT SHINGLES

Asphalt shingles are the most common roofing material for houses that is used today. These durable shingles generally have a life expectancy of 15 to 30 years. They are constructed of a heavy paper known as felt, coated with hot liquid asphalt, and then covered with fine rock granules. Asphalt shingles vary in weight, ranging from approximately 160 to 350 pounds per square (100 sq ft). Generally speaking, the heavier shingles are more expensive and have a longer life. Asphalt shingles should not be installed on a roof with a slope of less than 2 in 12 due to the potential for water seepage under the shingles.

FIBERGLASS SHINGLES

These shingles are similar in appearance to the asphalt type, but are more resistant to fire and lighter in weight. Fiberglass shingles resist warping and curling and will not absorb moisture, which causes rotting. As with the asphalt shingle, heavier fiberglass shingles are more expensive and durable and the same slope restrictions apply.

Figure 10.21 provides an example of the components of an asphalt or fiberglass shingle roof.

SLATE SHINGLES

Slate is one of the best, and one of the most expensive, roofing materials available. It can be purchased with a rough or smooth texture and in a variety of colors depending on the source of the slate. These shingles are so durable that they usually will outlast the life of the build-

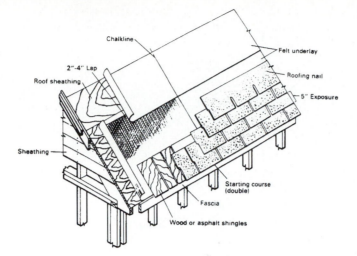

Figure 10.21
Installation of Fiberglass or Asphalt Shingle Roof Covering
Source: Gerald E. Sherwood and Robert C. Stroh, *Wood Frame House Construction* (Mineola, NY: Dover Publications, Inc., 1989), p. 93.

ing. Due to their extreme weight (approximately 3,000 pounds per square), a stronger roof framing system is required. Generally, these shingles should not be used on a roof with a slope of less than 6 in 12.

METAL ROOF

Metal roofs can be built of any non-rusting metal, such as aluminum, copper, galvanized metal, or terne (tin-coated) and are very durable. However, they are relatively expensive and can be extremely noisy. They are generally lightweight (40–60 pounds per square) and come in a variety of colors and styles. This form of roofing is becoming more popular in the residential construction industry.

TILE SHINGLES

Clay or cement tile shingles come in a variety of shapes, colors, and textures. They are easy to install, but as with the slate shingles, can be quite heavy (800–2600 pounds per square). Due to their weight, a stronger roof framing system also is required. These shingles are durable and have long life expectancies; however, they can be expensive.

BUILT-UP ROOFS

Built-up roofs require a solid deck of sheathing which may or may not be insulated. They typically are composed of a base ply of roofing material as well as three, four, or five layers of roofing felt, each mopped down with tar or asphalt. The final surface usually is coated with asphalt and covered with gravel. Such roofs typically are referred to as 10-year, 15-year, or 20-year roofs, according to the manufacturer's specifications.

WOOD SHAKES AND SHINGLES

The construction of this type of covering is identical to that described for wood shakes and shingles in the exterior wall finish section. These coverings tend to be much more expensive than standard asphalt or fiberglass shingles. Some homeowners, however, consider the resulting roof quality and appearance to be much higher. The minimum recommended roof slope for wood shakes is 4 in 12, and for wood shingles it is 3 in 12. The main drawback to wood shakes and shingles is flammability.

PROBLEMS WITH ROOF COVERINGS

A major problem related to roof coverings is the damage that can be done by either **water** or **ice** if the roof is not properly installed. Figure 10.22 indicates the type of flashing that is required to ensure that water is properly channeled whenever there are intersecting surfaces.

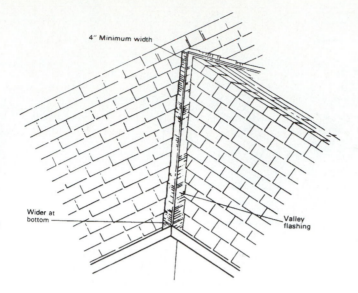

Figure 10.22
Roof Valley
Flashing Detail
Source: Gerald E.
Sherwood and
Robert C. Stroh,
*Wood Frame House
Construction*
(Mineola, NY:
Dover Publications,
Inc., 1989), p. 104.

Figure 10.23 indicates the damage that is caused in colder climates due to ice dams that can form along roof edges at the cornice overhang. As noted by Sherwood:

> Ice dams form as a result of the melting snow that has fallen on the warmer attic areas of the roof. The water from the melted snow runs down the roof to the colder cornice area where it freezes again, forming the ice dam. As more water runs down the roof, ice gradually becomes deeper, forming a trough that catches water, causing water to back up under the roof covering material and leak through the ceiling and walls.[60]

The damage caused by ice dams can be reduced by increasing the distance the backed-up water must travel to reach the inside of the house through the use of flashing as well as improving the soffit ventilation and insulation details.[61]

Roof Covering Requirements

The basic technical requirements for each of the types of roof covering systems described above are provided in the 1995 CABO Code.[62] As an example, Table 10.11 provides the pertinent information related to asphalt shingle roofs. Similar tables for wood shingle and shake roofs also are provided.

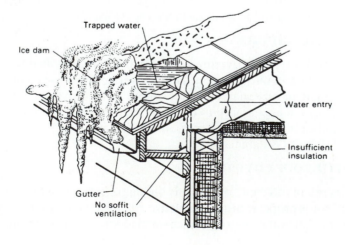

Figure 10.23
Snow and Ice
Dam Formation
Source: Gerald E.
Sherwood and
Robert C. Stroh,
*Wood Frame House
Construction*
(Mineola, NY:
Dover Publications,
Inc., 1989), p. 91.

TABLE 10.11   Technical Requirements for Asphalt Shingle Roofs

Roof Slope	Not Permitted Below 2:12	
	2:12 to less than 4:12	4:12 and over
DECK REQUIREMENT	Asphalt shingles shall be fastened to solidly sheathed roofs. Sheathing shall conform to Tables 503.2.1.1a and 803.3.2.	
UNDERLAYMENT Temperate climate	Asphalt strip shingles may be installed on slopers as low as 2″ in 12″, provided the shingles are approved self-sealing or are hand-sealed and are installed with an underlayment consisting of two layers of nonperforated Type 15 felt applied shingle fashion. Starting with an 18″-wide sheet and a 36″-wide sheet over it at the eaves, each subsequent sheet shall be lapped 19″ horizontally.	One layer nonperforated Type 15 felt lapped 2″ horizontally and 4″ vertically to shed water.
Severe climate: In areas subject to wind-driven snow or roof ice buildup.	Same as for temperate climate, and additionally the two layers shall be solid cemented together with approved cementing material between the plies extending from the eave up the roof to a line 24″ inside the exterior wall line of the building.	Same as for temperate climate, except that one layer No. 40 coated roofing or coated glass base sheet shall be applied from the eaves to a line 12″ inside the exterior wall line with all laps cemented together.
ATTACHMENT Type of fasteners	Corrosion-resistant nails, minimum 12-gage 3/8″ head, or approved corrosion-resistant staples, minimum 16-gage 15/16″ crown width.	
	Fasteners shall be long enough to penetrate into the sheathing 3/4″ or through the thickness of the sheathing, whichever is less.	
No. of fasteners[a]	4 per 36-40″ strip 2 per 9-18″ shingle	
Exposure Field of roof	Per manufacturer's instructions included with packages of shingles	
Hips and ridges	Hip and ridge weather exposures shall not exceed those permitted for the field of the roof	
Method	Per manufacturer's instructions included with packages of shingles.	
FLASHINGS Valleys Other flashings	Per Section 903.5 Per Sections 903.6 and 903.7	

For SI: 1″ = 25.44 mm.
[a]Figures shown are for normal application. For special conditions such as mansard application and where roofs are in special wind regions, shingles shall be attached per manufacturer's instructions.

Source: Reproduced from the 1995 edition of the *CABO One and Two Family Dwelling Code*™, copyright © 1995, with the permission of the publisher, the International Codes Council, Table 903.4, p. 132.

The 1995 CABO Code does not, however, provide all of the information that is required to ensure that a satisfactory roof installation is obtained. The building code official or the homebuilder must refer to the technical literature provided by each of the product manufacturers in order to ensure that the trade contractor involved is installing the roof properly.

CHIMNEYS AND FIREPLACES

As noted in Chapter 1 of this book, one of the features most desired by homebuyers is one or more fireplaces to provide warmth as well as to enhance the decorative features of the interior living space of the house. Some homebuyers also want a wood stove installed to provide an additional source of heat for the house. Both of these devices require chimneys to remove the smoke from the house, and both also involve significant structural considerations if they are to operate efficiently and, above all, safely.

The 1995 CABO Code provides the requirements which ensure that these considerations are addressed.[63] The topics covered are: (1) Masonry Chimneys, (2) Factory-Built Chimneys, (3) Masonry Fireplaces, and (4) Factory-Built Fireplaces.

The essential features of a typical fireplace are shown in Fig. 10.24. Some of the critical features that must be considered are: (1) the relationship between the depth of the fireplace to the height of the opening; (2) the relationship between the flue area and the area of the fireplace opening; (3) separation and clearance between the fireplace and the wood stud wall; (4) structural support of the fireplace; and (5) design of the damper, cleanout opening, and ash dump.

The heat production efficiency of fireplaces is very low. Wood stoves with controlled air intakes create an efficient, slow, and more effective combustion environment. Both freestanding stoves, as well as ones which are designed to be inserted directly into fireplace openings, are widely available in many designs, some of which include fans and water circulating coils. As the popularity of wood stoves has increased, so has the frequency of fires and accidents. This often results because the manufacturer's fire safety precautions have not been followed during construction and installation.

Homebuilders must ensure that the manufacturer's instructions as well as the code requirements are followed properly during installation to ensure that these devices truly enhance the lifestyle of the homebuyer.

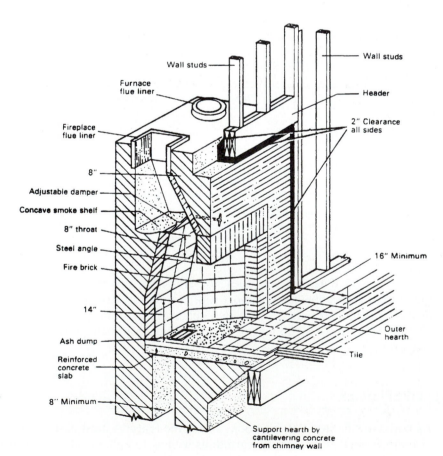

Figure 10.24

Components of a Masonry Fireplace

Source: Gerald E. Sherwood and Robert C. Stroh, *Wood Frame House Construction* (Mineola, NY: Dover Publications, Inc., 1989), p. 145.

CHAPTER SUMMARY

Wall systems can be divided into two major parts: **wood wall systems** and **masonry wall systems**. The section of this chapter on wood wall systems indicated that the exterior and interior bearing and non-bearing walls actually are not "designed" in a conventionally framed house because the basic 2×4 or 2×6 stud wall, within the limits established by the 1995 CABO Code, will satisfactorily carry the required loads in a typical house. The design of the headers that must be installed over all openings in load-bearing walls was explained, and examples of wood wall design related to the Case Study House were provided to indicate how the 1995 CABO Code requirements are implemented.

The part of the chapter dealing with masonry wall systems began with a fairly extensive introduction to masonry construction. Without such an introduction, it is difficult to understand the 1995 CABO Code requirements for this alternative structural framing approach. A discussion of those requirements immediately followed the introductory material.

A discussion of both the interior as well as the exterior wall finish systems that typically are used by homebuilders concluded the section on wall systems. In each case a brief introduction to the 1995 CABO Code requirements related to these wall finish systems was provided.

The final structural system that was addressed was the **roof–ceiling system**. Although quite a few homebuilders have adopted the pre-engineered roof truss option, very little guidance is provided directly in the 1995 CABO Code about this approach. The primary emphasis is on the conventional rafter/ceiling joist system. A major focus of this part of the chapter was an explanation of how the allowable span tables for ceiling joists and roof rafters are used. The Case Study House context again was utilized to explain how these structural elements are designed.

The section on the roof–ceiling system also included a discussion of the **roof covering systems** that typically are used by homebuilders. After an introductory presentation of these systems, the 1995 CABO Code requirements for roof covering systems was described briefly.

The final section of this chapter provided a brief introduction to **chimneys** and **fireplaces**. Some of the critical features which affect the design of these heat-producing devices were presented.

The fairly comprehensive treatment of the previously noted structural systems should provide homebuilders and building code officials with a more organized understanding of this important phase of house construction.

NOTES

[1] *CABO One and Two Family Dwelling Code, 1995 Edition*, pp. 63–142. (Falls Church, VA: The Council of American Building Officials, 1995).

[2] Ibid., Chapter 6, pp. 63–80.

[3] Ibid., Section 108, p. 2.

[4] Ibid., Chapter 7, pp. 81–88.

[5] Ibid., Chapter 6, pp. 63–80.

[6] Ibid., Section 602.2, p. 63.

[7] Ibid., Section 602.3.1, p. 63.

[8] Ibid., Section 602.3.2, p. 63.

[9] Ibid., Sections 602.3.3 and 602.3.4, p. 63.

[10] Ibid., Sections 602.4 and 602.4.1, p. 63.

[11] Ibid., Section 602.9, p. 70.

[12] Ibid., Table 602.3c, p. 66.

[13] Ibid., Sect. 602.9, p. 70.

[14]Ibid., Tables 602.3a and 602.3a(1), pp. 64–65.

[15]Ibid., Section 602.6, p. 66.

[16]Ibid., Section 602.6.1, p. 66.

[17]Ibid., Section 602.6.2, p. 66.

[18]*CABO One and Two Family Dwelling Code, 1992 Edition*, pp. 54–58.1.   (Falls Church, VA: The Council of American Building Officials, 1995).

[19]*1995 CABO Code*, Section 602.6, p. 66.

[20]Ibid., Section 602.6.3, p. 66.

[21]Ibid., Section 602, pp. 63–70.

[22]Harold B. Olin, John L. Schmidt and Walter H. Lewis, *Construction Principles, Materials and Methods*, 4th Ed., pp. 205.3–205.5.   (Chicago, IL: The Institute of Financial Education, 1980).

[23]Ibid., pp. 205.7.

[24]*ASTM C62-92c—Standard Specification for Building Brick (Solid Masonry Units Made from Clay or Shale)* (Philadelphia, PA: American Society for Testing and Materials, 1992).

[25]Ernest R. Weidhaas, *Architectural Drafting and Construction*, 4th ed., p. 481.   (Boston, MA: Allyn and Bacon, 1989).

[26]Olin et al., *Construction Principles*, pp. 205.18.

[27]*ASTM C90-93, Standard Specification for Load Bearing Concrete Masonry Units*   (Philadelphia, PA: American Society for Testing and Materials, 1993).

[28]*ASTM C129-85, Non-Load Bearing Concrete Masonry Units*   (Philadelphia, PA: American Society for Testing and Materials, 1985).

[29]Olin, et al., *Construction Principles*, p. 205.23 (modified).

[30]Ibid., pp. 205.23–205.25.

[31]*ASTM C270-92a—Standard Specifications for Mortar for Unit Masonry*   (Philadelphia, PA: American Society for Testing and Materials, 1992).

[32]*Pocket Guide to Brick Construction*   (Reston, VA: Brick Institute of America, 1990), p. 36.

[33]*NCMA Guide for Home Owners and Home Builders on Residential Concrete Masonry Basement Walls*, p. 19. (Herdon, VA: National Concrete Masonry Association, 1994).

[34]*1995 CABO Code*, Sections 405 and 406, pp. 32–34.

[35]*NCMA TEK 19-5—Use of Flashing in Concrete Masonry Walls*   (Herdon, VA: National Concrete Masonry Association, 1993).

[36]*NCMA TEK 10.2—Control of Wall Movement with Concrete Masonry*   (Herdon, VA: National Concrete Masonry Association, 1972).

[37]*NCMA TEK 12-1—Anchors and Ties for Masonry*   (Herdon, VA: National Concrete Masonry Association, 1995).

[38]*1995 CABO Code*, Section 702, pp. 81–83.

[39]*GA-201-90, Using Gypsum Board for Walls and Ceilings*, p. 3.   (Washington, D.C.: The Gypsum Association, 1990).

[40]*GA 505-91, Gypsum Board Terminology*   (Washington, D.C.: The Gypsum Association, 1991); *GA 201-90, Using Gypsum Board for Walls and Ceilings*; *GA 216-93, Application and Finishing of Gypsum Board* (Washington, D.C.: The Gypsum Association, 1993); *GA 600-94, Fire Resistance Design Manual*   (Washington, D.C.: The Gypsum Association, 1994).

[41]Gerald E. Sherwood and Robert C. Stroh, *Wood Frame House Construction*, p. 176.   (Mineola, NY: Dover Publications, Inc., 1989).

[42]1995 CABO Code, Section 703, pp. 83–88.

[43]Ibid., Section 703.1, p. 83.

[44]Ibid., Section 703, pp. 83–88.

[45]Ibid., Chapter 8, pp. 89–130.

[46]Ibid., Section 802.11, p. 91.

[47]Ibid., Section 108, p. 2.

[48]Ibid., Sections 801.2 and 801.3, p. 89.

[49]Ibid., Section 802, pp. 89–91.

[50]Ibid., Section 604.10 and Figures 604.10a, 604.10b, and 604.10c, pp. 73–76.

[51]Ibid., Section 802.12 and Table 802.12, p. 91.

[52]Ibid., Section 802.4.1, p. 89.

[53]Ibid., Section 802, pp. 89–91.

[54]Ibid., Section 803, pp. 91–92.

[55]Ibid., Table 803.3.2, p. 92.

[56]Ibid., Sections 804, 805, and 807, p. 92.

[57]Ibid., Section 806, p. 92.

[58]Ibid., Chapter 9, pp. 131–136.

[59]Ibid., Section 108, p. 2.
[60]Sherwood and Stroh, *Woodframe House Construction*, p. 88.
[61]Ibid., p. 88 and Fig. 80b.
[62]Ibid., Sections 903 to 909, pp. 131–135.
[63]Ibid., Chapter 10, pp. 137–142.

BIBLIOGRAPHY

ASTM C62-92c—Standard Specification for Building Brick (Solid Masonry Units Made from Clay or Shale). Philadelphia, PA: American Society for Testing and Materials, 1992.

ASTM C90-93, Standard Specification for Load Bearing Concrete Masonry Units. Philadelphia, PA: American Society for Testing and Materials, 1993.

ASTM C129-85, Non-Load Bearing Concrete Masonry Units. Philadelphia, PA: American Society for Testing and Materials, 1985.

ASTM C270-92a, Standard Specifications for Mortar for Unit Masonry. Philadelphia, PA: American Society for Testing and Materials, 1992.

Building Code Requirements for Masonry Structures (ACI 530-92/ASCE 5-92/TMS 402-92). New York: American Concrete Institute/American Society of Civil Engineers/The Masonry Society, 1992.

CABO One and Two Family Dwelling Code, 1992 Edition. Falls Church, VA: The Council of American Building Officials, 1992.

CABO One and Two Family Dwelling Code, 1995 Edition. Falls Church, VA: The Council of American Building Officials, 1995.

Cost-Effective Home Building: A Design and Construction Handbook. Washington, D.C.: The Home Builder Press, The National Association of Home Builders, 1994.

Drysdale, Robert G., Ahmad A. Hamind, and Lawrie R. Baker. *Masonry Structures: Behavior and Design*. Englewood Cliffs, NJ: Prentice Hall, 1994.

GA-201-90, Using Gypsum Board for Walls and Ceilings. Washington, D.C.: The Gypsum Association, 1990.

GA 216-93, Application and Finishing of Gypsum Board. Washington, D.C.: The Gypsum Association, 1993.

GA 505-91, Gypsum Board Terminology. Washington, D.C.: The Gypsum Association, 1991.

GA 600-94, Fire Resistance Design Manual. Washington, D.C.: The Gypsum Association, 1994.

"Houston Builder Gets Back to Basics with Concrete," *Professional Builder*, vol. 60, no. 4, February 1995.

NCMA Guide for Home Owners and Home Builders on Residential Concrete Masonry Basement Walls. Herdon, VA: National Concrete Masonry Association, 1994.

NCMA TEK 3-6A-Concrete Masonry Veneers. Herdon, VA: National Concrete Masonry Association, 1995.

NCMA TEK 6-11-Insulating Concrete Masonry Walls. Herdon, VA: National Concrete Masonry Association, 1995.

NCMA TEK 10.2-Control of Wall Movement with Concrete Masonry. Herdon, VA: National Concrete Masonry Association, 1972.

NCMA TEK 12-1-Anchors and Ties for Masonry. Herdon, VA: National Concrete Masonry Association, 1995.

NCMA TEK 14-2-Reinforced Concrete Masonry. Herdon, VA: National Concrete Masonry Association, 1995.

NCMA TEK 14-10A-Lateral Support of Concrete Masonry Walls. Herdon, VA: National Concrete Masonry Association, 1994.

NCMA TEK 15-3-Reinforced Eight-Inch Basement Walls. Herdon, VA: National Concrete Masonry Association, 1991.

Olin, Harold B., J. L. Schmidt and Walter H. Lewis. *Construction Principles, Materials and Methods*, 4th Ed. Chicago, IL: The Institute of Financial Education, 1980.

Research Means Better, More Affordable Houses. Upper Marlboro, MD: NAHB Research Center, 1993.

Sanders, Gordon A. *Light Building Construction*. Reston, VA: Reston Publishing Company, Inc., 1985.

Sherwood, Gerald E. and Robert C. Stroh. *Wood Frame House Construction*. Mineola, NY: Dover Publications Inc., 1989.

Technical Notes on Brick Construction, No. 30. Reston, VA: Brick Institute of America, 1988.

Weidhaas, E. R. *Architectural Drafting and Construction*, 4th Ed. Boston, MA: Allyn and Bacon, 1989.

Wing, Charlie. *The Visual Handbook of Building and Remodeling*. Emmaus, PA: Rodale Press, 1990.

11

Structural Analysis

INTRODUCTION

This chapter concludes the four-chapter sequence which addresses the design and construction of the structural systems in a house. Chapter 8 introduces the basic structural characteristics of wood, the material which is still the most widely used by builders to frame a house. The fundamental structural loading conditions of compression, tension, bending, shear, and deflection are presented, and the grading rules related to lumber are introduced. Selected design tables appearing in the National Design Specification (NDS®) for Wood Construction,[1] which presents the structural design procedures and allowable values for the various species and grades of lumber which a builder typically uses, also are discussed.

Chapters 9 and 10 are devoted to a presentation of the design approach which may be used by the builder if the house being built conforms to one of the two "conventional systems" used in the United States (i.e., platform framing and balloon framing). As noted, there are a number of chapters in the 1995 CABO Code which provide guidance to the builder regarding the selection of the sizes of the various wood structural members that are used to frame a house. In addition, the 1995 CABO Code coverage of foundation design and construction, as well as masonry wall construction, is presented.[2] The various design procedures are illustrated in relation to the "Case Study House Example" plans and specifications which appear in Appendix A.

Builders who do not have either an engineering or an architectural background are limited to the "structural design by table" approach which is presented in the 1995 CABO Code. This approach, while often conservative, is perfectly satisfactory as long as the underlying assumptions upon which the structural design tables are developed (i.e., uniform loading on simply supported rafters and joists, etc.) are not violated in the house being analyzed.

Chapter 11, "Structural Analysis", has been included in this text to serve several purposes. The first is to illustrate the types of structural engineering calculations which a licensed engineer or architect should perform when a builder is asked to provide structural framing systems which no longer satisfy the simplifying assumptions. This is particularly important in custom and luxury houses where the division of the house into the desired living spaces often results in the use of cantilevered beams, concentrated loads imposed on other structural members, etc. In such situations, the prudent homebuilder must turn to a licensed engineer or architect for professional advice.

The second reason for including this chapter is to introduce the homebuilder to the type of structural calculations which can actually save money as the structural framing system for a house is being developed. It is not possible to optimize the solutions to the structural framing of a house without performing the types of structural calculations illustrated in this chapter. The solutions provided by the 1995 CABO Code, which are based on conservative assumptions, often can be improved if a licensed engineer or architect performs the appropriate structural analysis.

This chapter begins with a presentation of the general procedures for analyzing or designing timber tension, compression, and flexural members as well as members which are subjected to combined loading conditions. The design procedures are based upon the provisions of the National Design Specification (NDS®) for Wood Construction.[3] This chapter also includes an analysis of the resolution of horizontal and vertical loads on roof rafter members. The structural theory presented in the beginning of the chapter is then applied to the various structural members of the Case Study House Example (Appendix A).

DESIGN OF TIMBER LOAD-CARRYING MEMBERS*

General

The design values listed in Tables 8.4, 8.5, and 8.7 for solid sawn visually stress rated (VSR) Spruce-Pine-Fir, Hem-Fir, and Southern Pine, and for machine stress rated (MSR) lumber are valid only for a specific set of conditions, e.g., normal load duration (10 years), moisture content less than 19%, specific member size, normal temperatures (<100°F), single-member use, and members which are sufficiently braced such that they do not buckle. Table 11.1 summarizes the adjustment factors important to each of the base design values, e.g., F_b, F_t, F_v, E.

The size factor (C_F), flat use factor (C_{fu}), and wet service factor (C_M) are summarized in Tables 8.6a, 8.6b, and 8.6c, respectively. The repetitive member factor (C_r) is defined in footnote d of Table 8.4. The load duration factor (C_D) is summarized in Fig. 8.18. The temperature factor (C_t) is 1.0, unless temperatures are sustained above 100°F. (If this is a concern, refer to Table 2.3.4 of the NDS®.[4]) The beam stability (C_L) and buckling stability factor (C_p) will be discussed in detail in later sections of this chapter. The volume factor (C_V) is used only for glued-laminated beams larger than a standard size (5.125 in. × 12 in. × 21 ft) beam and will be defined as the need arises in examples. The form factor (C_f), curvature factor (C_c), shear stress factor (C_H), and buckling stiffness factor (C_T) are for special cases not of concern to the scope of this text. The bearing area factor (C_b) equals $(\ell_b + 0.375)/\ell_b$, where ℓ_b is the length of bearing and is applied only to bearing areas less than 6 in. in length and not nearer than 3 in. from the end of the member. In this chapter, basic design values will be denoted by an unprimed capital F (e.g., F_b) and adjusted design values by a primed capital F (e.g., F_b').

Tension Members

Tension members in the structural system in houses are commonly encountered in the lower chords and some webs of trusses and in bracing members. The governing design equation is:

---

*Appropriate references should be made to Chapters 8–10 of this text for additional information related to the topics which are covered in Chapter 11.

TABLE 11.1 Applicability of Adjustment Factors

	Load Duration Factor	Wet Service Factor	Temperature Factor	Beam Stability Factor[a]	Size Factor[b]	Volume Factor[a,c]	Flat Use Factor[d]	Repetitive Member Factor[e]	Curvature Factor[f]	Form Factor	Column Stability Factor	Shear Stress Factor[g]	Buckling Stiffness Factor[h]	Bearing Area Factor
$F'_b = (F_b)$	(C_D)	(C_M)	(C_t)	(C_L)	(C_F)	(C_V)	(C_{fu})	(C_r)	(C_c)	(C_f)	•	•	•	•
$F'_t = (F_t)$	(C_D)	(C_M)	(C_t)	•	(C_F)	•	•	•	•	•	•	•	•	•
$F'_v = (F_v)$	(C_D)	(C_M)	(C_t)	•	•	•	•	•	•	•	•	(C_H)	•	•
$F'_{c\perp} = (F_{c\perp})$	•	(C_M)	(C_t)	•	•	•	•	•	•	•	•	•	•	(C_b)
$F'_c = (F_c)$	(C_D)	(C_M)	(C_t)	•	(C_F)	•	•	•	•	•	(C_P)	•	•	•
$E' = (E)$	•	(C_M)	(C_t)	•	•	•	•	•	•	•	•	•	(C_T)	•
$F'_g = (F_g)$	(C_D)	•	(C_t)	•	•	•	•	•	•	•	•	•	•	•

[a]The beam stability factor, C_L, shall not apply simultaneously with the volume factor, C_v, for glued laminated timber bending members. Therefore, the lesser of these adjustment factors shall apply.

[b]The size factor, C_F, shall apply only to visually graded sawn lumber members and to round timber bending members.

[c]The volume factor, C_v, shall apply only to glued laminated timber bending members.

[d]The flat use factor, C_{fu}, shall apply only to dimension lumber bending members 2 in. to 4 in. thick and to glued laminated timber bending members.

[e]The repetitive member factor, C_r, shall apply only to dimension lumber bending members 2 in. to 4 in. thick.

[f]The curvature factor, C_c, shall apply only to curved portions of glued laminated timber bending members.

[g]Shear design values parallel to grain, F_v, for sawn lumber members shall be permitted to be multiplied by the shear stress factors, C_H, specified in Tables 4A, 4B, 4C, and 4D of the NDS® Supplement.

[h]The buckling stiffness factor, C_T, shall apply only to 2 in. × 4 in. or smaller sawn lumber truss compression chords subjected to combined flexure and axial compression when 3/8-in. or thicker plywood sheathing is nailed to the narrow face.

Source: *National Design Specification® for Wood Construction and Supplement*, p. 5 (American Forest & Paper Association, 1991).

$$F_t' > f_t = \frac{P}{A}$$ (Eq. 11.1)

where: F_t' = adjusted design tensile stress
$= F_t(C_D)(C_M)(C_t)(C_F)$
C_D = load duration factor
C_M = moisture factor
C_t = temperature factor
C_F = size (depth) factor
f_t = actual tensile stress
P = tensile load
A = cross-sectional area

EXAMPLE NO. 1: Tension Members

Evaluate the allowable tensile load which an 8-ft-long No. 2 Southern Pine (SP) dressed 2×8 can carry if the load is induced by a wind load. The lumber is dressed and surfaced at 15% moisture content.

SOLUTION

a. The adjusted design stress for the member is the product of F_t (Table 8.5), C_D (Fig. 8.18), C_M (Table 8.6c) and C_F (for Southern Pine); C_F is included in tabulated base design values).

$F_t' = 650$ psi (1.60) (1.0) (1.0) $(1.0) = 1,040$ psi

b. From Eq. 11.1 and the section properties in Table 8.8, the allowable load is simply

$P = F_t'(A) = 1,040 \ (10.88) = 11,310$ lb.

If the member is connected by bolts, the load capacity of the member is: $P = F_t'$ (net area), where net area equals the area of the 2×8 minus the projected area of the bolt hole.

Compression Members—Columns

The structural member which is in a state of pure compression is a rarity. That is, axial compressive stresses are usually accompanied by flexural stresses due to eccentric loading, lateral loading, and end moments. Nonetheless, the primary design method for compression webs in trusses, bracing members, and some vertical framing members assumes that the column is loaded in pure compression. Furthermore, the response of a timber column in pure compression is an integral component of the design procedure for beam columns (members subjected to combined axial compressive and flexural stresses).

A timber column is a straight member loaded by a centrically applied compressive force (Fig. 11.1). The ends of the column are assumed to be pinned (i.e., rotation is not constrained). The x and y axes of the rectangular cross section are the strong and weak axes, respectively. The strong axis has the larger moment of inertia. The slenderness ratio of a column is defined as L/r, where L equals the unsupported length of the column and r is the radius of gyration $(I/A)^{1/2}$ of the section. Two slenderness ratios are defined for the column in Fig. 11.1, namely L/r_x and L/r_y, where $r_x = (I_x/A)^{1/2} = h/(12)^{1/2}$ and $r_y = b/(12)^{1/2}$. In solid rectangular columns, the slenderness factors for the section in Fig. 11.1 are taken as L/h about the x axis and L/b about the y axis. In general discussions the slenderness factor will be designated as L/d, and it is assumed that it refers to the largest slenderness factor for the column.

The allowable stresses for columns with intermediate bracing or nonpinned end restraints may also be calculated by using the equations for pinned columns (i.e., Eqs. 11.2 and 11.3), provided that the column length is replaced by an appropriate theoretical effective length, L_e. The effective column length of a compression member is the distance

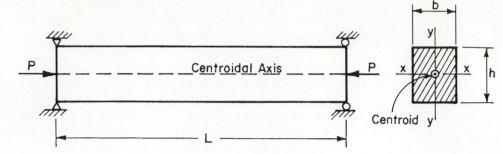

Figure 11.1
Compression
Member
Definition Sketch

between the two points along its length at which the member is assumed to buckle in the shape of a sine wave, i.e., where there is a reversal of curvature and zero bending moment. The effective column length is dependent on the values of end fixity and lateral translation (deflection) associated with the ends of columns and the points of lateral support between the ends of columns. Lower values of effective length are associated with more end fixity and less lateral translation, while higher values are associated with less end fixity and more lateral translation.

The effective lengths for columns with various end constraints are summarized in Table 11.2. The influence of lateral braces upon L/d is demonstrated in Example 2.

TABLE 11.2   Effective Column Length

Buckling Modes						
Theoretical K_e value	0.5	0.7	1.0	1.0	2.0	2.0
Recommended design K_e when ideal conditions are approximated	0.65	0.80	1.2	1.0	2.10	2.4
End condition code	Rotation fixed, translation fixed					
	Rotation free, translation fixed					
	Rotation fixed, translation free					
	Rotation free, translation free					

Source: *National Design Specification® for Wood Construction and Supplement,* p. 116 (American Forest & Paper Association, 1991.

Timber columns will buckle about the cross-sectional axis with the largest slenderness ratio, L_e/d. Buckling does not always occur about the weak axis of the section. The column's unsupported length may be less than the total column length. Furthermore, the unsupported length may be different with respect to buckling about the x or y axis.

Prior to the 1991 edition of the NDS®,[5] timber columns were classified as either short, intermediate, or long, depending upon their slenderness factor, L_e/d. A column with $L_e/d \leq 11$ was defined as short; a column with $11 < L_e/d < K$ (where $K = 0.671 [E/(F_c \times C_D)]^{1/2}$) was defined as intermediate; and a column with $L_e/d > K$ was defined as long. A short column fails by crushing of the wood fibers. An intermediate column fails by buckling inelastically; that is, buckling commences before yielding, but yielding occurs during buckling. Long columns buckle elastically; that is, they do not yield before or during buckling.

In the 1991 Specification, the equation format for predicting wood column behavior was significantly changed. Columns are no longer classified as short, intermediate, or long. Instead, the behavior is defined by a single equation for L_e/d ratios ranging from 0 to 50. Columns with L_e/d ratios greater than 50 are not permitted and are assigned an adjusted compressive stress of zero. (One exception is that L_e/d ratios up to 75 are permitted during construction, but not in service.)

The governing condition for the analysis and design of timber columns is:

$$F'_c > f_c = \frac{P}{A} \qquad \text{(Eq. 11.2)}$$

where: F'_c = adjusted design compressive stress parallel to the grain
   $= F_c(C_D)(C_M)(C_t)(C_F)(C_P)$
   f_c = calculated uniform compressive stress
   P = compressive load
   A = cross-sectional area

The column stability factor (C_P) is defined by Eq. 11.3[6]

$$C_p = \frac{(1 + F_{cE}/F^*_c)}{2c} - \sqrt{\left[\frac{1 + F_{cE}/F^*_c}{2c}\right]^2 - \frac{F_{cE}/F^*_c}{c}} \qquad \text{(Eq. 11.3)}$$

where: $F^*_c = F_c(C_D)(C_M)(C_T)(C_F)$

   $F_{cE} = \dfrac{K_{cE} E'}{(L_e/d)^2}$

   K_{cE} = 0.3 for VSR lumber
      = 0.418 for MSR lumber, glulam, etc., with coefficient of variation (COV) of $E \leq 0.11$.
   c = 0.8 for sawn lumber
      = 0.85 for round timber piles
      = 0.90 for glued laminated timber
   $E' = E(C_M)(C_t)(C_T)$

The higher values for K_{cE} and c for manufactured lumber products result in larger C_p values. This reflects the reduced variability of the structural properties (e.g., F_b, F_c, E) of such products. The generalized variation of F'_c with L_e/d is shown graphically in Fig. 11.2.

Centrically loaded compression members are seldom found in practice. Although many columns are analyzed or designed assuming centric loading, it is better practice to assume an accidental eccentricity of the larger of 1 in. or 0.1h about the strong axis (see Fig. 11.1) and 1 in. or 0.1b about the weak axis, and to analyze or design the member as a beam column. Analysis of beam columns is discussed later in this chapter.

The NDS® prohibits the use of compression members with $L_e/d > 50$.[7] In cases where $L_e/d > 50$, the slenderness ratio can be reduced by installing intermediate bracing or by increasing the minimum thickness with properly fastened scabbing in the middle 1/2 to 1/3

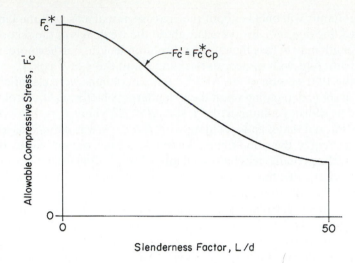

Figure 11.2
Variation of
Allowable
Compressive
Stress with
Slenderness
Factor for Solid
Rectangular
Timber Columns

of the critical unsupported length. Another alternative is to replace the solid timber member with a spaced column. The spaced column consists of two compression members spaced approximately 1.5 in. apart and fastened together with special timber connectors. Slenderness factors up to 80 are then permissible, and design compressive stresses are 2.5 to 3 times as great as for single compression members. The reader is referred to the NDS® for additional details.[8] For timber columns with non-rectangular cross sections, the design equations presented above are easily modified. For such situations, d is replaced with $r(12)^{1/2}$ in all of the design equations for F'_c (r = slenderness ratio). The limiting slenderness ratio becomes 173 (corresponding to $L_e/d = 50$).

EXAMPLE NO. 2:   Effective Length of Compression Members

      a. Evaluate the slenderness ratio of the timber column in Fig. 11.3(a) using the theoretical K_e values shown in Table 11.2. The brace prevents displacement in the plane of the page, but not

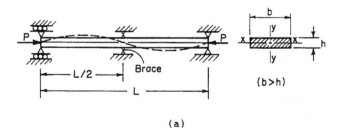

(a)

Figure 11.3
Example No. 2:
Effective Length
of Compression
Members

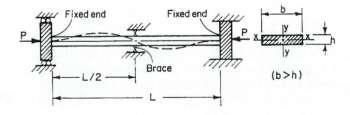

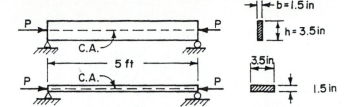

Figure 11.4
Example No. 3:
Compression
Member Analysis

in any other plane. The slenderness ratio about the strong axis (the axis with greatest moment of inertia) is $(L_e/d)_y = L/b$. The slenderness ratio about the weak axis (the axis with least moment of inertia) is $(L_e/d)_x = L/(2h)$.

b. Evaluate the slenderness ratio of the fixed-fixed column in Fig. 11.3(b) using the theoretical K_e values shown in Table 11.2. The bracing prevents buckling only in the plane of the page. The buckling mode is shown by the dashed line. The brace is essentially a pin for effective length considerations. About the strong axis, $(L_e/d)_y = 0.5L/b$. About the weak axis, $(L_e/d)_x = 0.7(L_e/2)/h = 0.35L/h$.

EXAMPLE NO. 3:   Compression Member Analysis

Evaluate the allowable load for the 5 ft long, dressed No. 2 Southern Pine 2×4 (surfaced dry and used at 19% moisture content) column shown in Fig. 11.4 if the ends are pinned with respect to buckling about both the strong and weak axes. Assume that the load is caused by snow.

SOLUTION

Step 1: Governing Equation and Parameters

$$f_c = P/A \leq F'_c = F_c(C_D)(C_M)(C_t)(C_F)(C_P)$$

$C_D = 1.15$ (snow load —Fig. 8.18)
$C_M = C_t = C_F = 1.0$
$L_e = 1.0\ L$ (Table 11.2)
$F_c = 1650$ psi (Table 8.5)
$E = 1.6 \times 10^6$ psi (Table 8.5)
$A = 5.25$ in^2 (Table 8.8)

Step 2: Evaluate C_p (Eq. 11.3)

$(L_e/d)_{max} = 1.0\ L/d = (1.0)(5\ \text{ft})(12\ \text{in./ft})/1.5\ \text{in.}$
$= 40 < 50$
$c = 0.8$

$F_c^* = 1650\ (1.15) = 1,897.5$ psi

$$F_{cE} = \frac{K_{cE}\ E'}{(L_e/d)^2}, \quad E' = 1.6 \times 10^6\ \text{psi},\ K_{cE} = 0.3$$

$$= \frac{.3(1.6 \times 10^6)}{(40)^2} = 300\ \text{psi}$$

$$C_p = \frac{1 + \left(\dfrac{300}{1,897.5}\right)}{1.6} - \sqrt{\left[\frac{1 + \left(\dfrac{300}{1,897.5}\right)}{1.6}\right]^2 - \frac{\dfrac{300}{1,897.5}}{.8}} = 0.153$$

$$F'_c = 1,897.5\ (0.153) = 290.3\ \text{psi}$$

Step 3: Allowable Load

$$P = 290.3\ (1.5)\ (3.5) = 1,524\ \text{lbs}$$

EXAMPLE NO. 4:   Compression Member Design

Design the smallest square column to support a concentric compressive load of 16,000 lb. Use No. 1 Spruce-Pine-Fir and assume that the load is due to wind. The unsupported length of the column is 8 ft.

SOLUTION

Step 1: Governing Equation and Parameter Definition

$$P = F'_c A$$
$$F'_c = (F_c)(C_D)(C_M)(C_t)(C_F)(C_p)$$

Step 2: For first estimate, assume $C_p = 1.0$ and use properties for timbers

$$F_c = 700 \text{ psi (Table 8.4)}$$
$$C_D = 1.6 \text{ (Fig. 8.18)}$$
$$C_t = 1.0$$
$$C_m = 1.0$$
$$C_F = 1.0 \text{ (assumed)}$$
$$F'_c = 700 (1.6) = 1,120 \text{ psi}$$
$$\therefore A = 16,000/1,120 = 14.28 \text{ in}^2$$
$$b = \sqrt{14.28} = 3.78 \text{ in.}$$

Thus try a 6×6 and check stability effects.

Step 3: Try a 6×6

$$F_c = 700 \text{ psi}$$
$$E = 1.3 \times 10^6 \text{ psi}$$
$$C_F = 1.0 \text{ (Table 8.4)}$$
$$L_e/d = 96/5.5 = 17.45$$
$$F_{cE} = \frac{.3 (1.3 \times 10^6)}{(17.45)^2} = 1,280 \text{ psi}$$
$$F^*_c = 700 (1.6) = 1,120 \text{ psi}, F_{cE}/F^*_c = 1.143$$
$$c = 0.8$$
$$C_p = \frac{1 + 1.143}{1.6} - \sqrt{\left[\frac{1 + 1.143}{1.6}\right]^2 - \frac{1.143}{.8}} = 0.735$$
$$F'_c = 1,120 (0.735) = 823 \text{ psi}$$
$$P_{all} = (823) (5.5)^2 = 24,895 \text{ lbs} > 16,000 \text{ lbs}$$

A quick check clearly shows that a 4×4, the next smaller standard size, would be inadequate (i.e., 823 (3.5)² = 10,080 lbs. and F'_c would be lower than 823 because L_e/d for a 4×4 is greater than L_e/d for the 6×6). Thus, use a 6×6.

Flexural Members—Beams

Figure 11.5 is a sketch of a typical timber flexural member. Such members appear in houses as roof rafters, ceiling joists, and floor joists. The length of the member is much greater than either of its cross-sectional dimensions, and the applied load is either a force system which has a component perpendicular to the longitudinal axis or a moment couple which is applied at some location along the member. The resulting shear and moment diagrams for a number of beam loading conditions are illustrated in Appendix E. There are two basic types of timber beams: laterally braced and laterally unbraced beams. Laterally braced beams have sufficient lateral supports, such as attached decking or bridging, to permit the beam to reach its allowable flexural stress before the beam buckles laterally. For such conditions, $C_L = 1.0$ and the full design value for extreme fiber stress in bending (F'_b) for the species and grade of lumber may be used. An unbraced beam may buckle laterally at a stress level below the allowable flexural stress, so the design flexural stress must be reduced by the stability factor C_L. The criteria for braced and unbraced beams are discussed in this section.

The design or analysis of a timber beam may be controlled by a number of factors. Chief among these are the external load-induced flexural (i.e., bending) and transverse shear stresses, the bearing stresses at end supports and at concentrated loads, and the deflec-

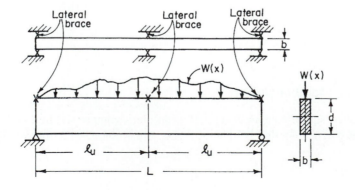

Figure 11.5
Definition Sketch for a Laterally Braced Timber Beam

tion of the beam. For many timber beams, flexural stresses will control the design. Thus, the first step in nearly all beam analysis procedures focuses on an evaluation of flexural stresses. Some exceptions to this sequence of analysis occur in the case of very short or heavily loaded beams where shear is likely to control the design or in the case of beams with severe deflection limitations.

In all of the ensuing discussions, it is assumed that the beams are of solid sawn cross section and that flexural loads are applied through the shear center of the section so that no torsion (twisting) of the beam occurs. The scope of the presentation does not include laminated beams or torsional stresses.

Laterally Braced Beams ($C_L = 1.0$)

In order to qualify as a laterally braced beam, ℓ_u (Fig. 11.5) must be equal to zero. This condition requires continuous lateral bracing of the compression flange. For all other cases, $\ell_u > 0$ and the beam stability factor (C_L) is less than 1.0.

1. Flexural Stresses

The governing equation for flexural stresses in beams with adequate lateral bracing is:

$$f_b = M/S < F_b' = F_b(C_D)(C_M)(C_t)(C_F)(C_{fu})(C_r)(C_f)(C_L) \qquad \text{(Eq. 11.4)}$$

where:
f_b = calculated flexure stress
F_b' = adjusted design flexure stress
F_b = base flexure stress (Table 8.4 or 8.5)
M = the maximum bending moment along the beam
S = section modulus of the beam
$C_D \dots C_f$ = adjustment factors (Table 11.1)
C_L = beam stability factor = 1.0 for braced beams

2. Shear Stresses

The basic governing equation for transverse or horizontal shear (Fig. 11.6) is:

$$f_v = VQ/Ib \leq F_v' = F_v(C_D)(C_M)(C_t)(C_H) \qquad \text{(Eq. 11.5)}$$

where:
F_v' = adjusted design horizontal shear stress
F_v = base design horizontal shear stress
f_v = calculated transverse shear stress
V = the maximum transverse shear force along the beam
I = 2nd moment of area about the neutral axis
b = width of the beam at the point where shear stress is evaluated
Q = 1st moment of the cross-sectional area between the outer fiber and the point at which the shear stress is evaluated, taken about the neutral axis
$C_D \dots C_H$ = adjustment factors (Table 11.1)

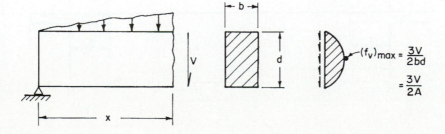

Figure 11.6
Shear Stresses in
a Rectangular
Section

For a rectangular section, it is easily shown, when I and Q are calculated, that $f_v = 3V/2A$. The assumed distribution of the shear stress across a beam section is shown in Fig. 11.6. For most solid sawn beams the shear requirements will be satisfied using the basic criterion of Eq. 11.5. For cases where Eq. 11.5 is not satisfied, consult the NDS® for allowable reductions in the computation of f_v.[9]

3. Bearing Stress

High concentrated loads at end supports or at point loads may cause compressive (i.e., wood fiber crushing) failures normal to the grain. To prevent this occurrence, adequate bearing area under each load is required (Fig. 11.7). The governing design criteria for end supports and point loads are:

$$f_{c\perp} = R/A < F'_{c\perp} = F_{c\perp}(C_M)(C_t)(C_b) \text{ at supports} \qquad \text{(Eq. 11.6)}$$

and

$$f_{c\perp} = P/A \le F'_{c\perp} = F_{c\perp}(C_M)(C_t)(C_b) \text{ at other concentrated loads}$$

where: $F_{c\perp}$ = base design compressive stress normal to the grain
$F'_{c\perp}$ = adjusted design compressive stress normal to the grain
R = reactive force
P = concentrated load
A = bearing area at the load or support
($b \times z$ = width × bearing length)
$f_{c\perp}$ = calculated bearing stress
$C_M \ldots C_b$ = adjustment factors (Table 11.1)

4. Deflection

Beam deflections are evaluated according to standard methods of elastic analysis. For short duration loads, creep is not a factor. However, for long-term loads, creep may increase deflections considerably. Thus, total deflection, Δ, may be taken as the live load deflection, Δ_L, plus twice the dead load deflection, Δ_D (Eq. 11.7). (Note that if the structural component has a long-term live load (Δ_{LL}), then Eq. 11.7 becomes $\Delta = \Delta_L + 2(\Delta_D + \Delta_{LL})$.)

$$\Delta_{all} = \Delta_L + 2\Delta_D \qquad \text{(Eq. 11.7)}$$

On the other hand, the initial deflection due to the dead load occurs before the plaster or gypsum board ceiling underneath the beam is applied. The dead load, therefore, is

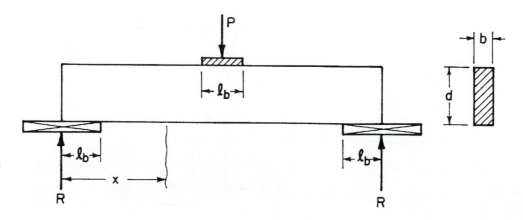

Figure 11.7
Definition Sketch
for Allowable
Bearing Stresses
in Timber Beams

sometimes omitted when computing deflection. The 1995 CABO Code has adopted this assumption for analyzing beams.[10] In order for the assumptions for structural analysis in this chapter to be the same as those used in the 1995 CABO Code analysis chapter, dead load also is omitted from all of the deflection calculations.

The design criteria for deflection are:

$$\Delta_{all} > \Delta = kwL^4/EI \text{ for distributed loads} \qquad \text{(Eq. 11.8a)}$$

$$\Delta_{all} > \Delta = kPL^3/EI \text{ for concentrated loads} \qquad \text{(Eq. 11.8b)}$$

where: k = coefficient dependent upon the nature and distribution of loads. It may be obtained by elastic beam deflection analysis or by reference to standard engineering handbooks.

The allowable deflections of beams are specified in design and building codes. Typical deflection limits include $\Delta_{all} = L/360$ for beams supporting plastered ceilings and $\Delta_{all} = L/240$ for beams supporting unplastered ceilings, where L is the span of the beam.

EXAMPLE NO. 5: Laterally Braced Beam

Select the smallest standard size No. 2 Southern Pine dressed member surfaced and used at 15% moisture content to carry a uniformly distributed snow load of 100 lb/ft over a 10–ft simply supported span. Assume that the compression flange is adequately braced ($\ell_u = 0$) and that the allowable deflection has been established as L/240.

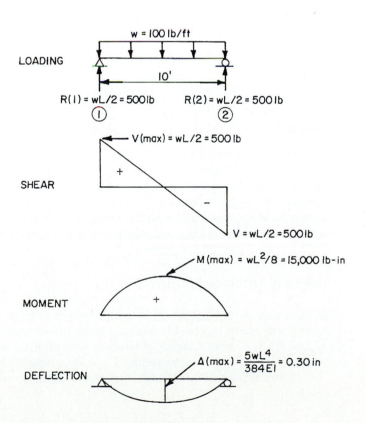

SOLUTION

Step 1: Define the parameters (assume a size, say a 2×8).

$$F_b = 1{,}200 \text{ psi (Table 8.5)}$$
$$E = 1.6{\times}10^6 \text{ psi (Table 8.5)}$$
$$F_v = 90 \text{ psi (Table 8.5)}$$
$$F_{c\perp} = 565 \text{ psi (Table 8.5)}$$

Step 2: Determine the maximum shear and moment.

$$V_{max} = wL/2 = 100(10)/2 = 500 \text{ lb}$$
$$M_{max} = wL^2/8 = 100(10)^2(12)/8 = 15{,}000 \text{ lb-in.}$$

Step 3: Select a section based upon the flexural stress criteria.

$$F_b' = F_b(C_D)(C_M)(C_t)(C_F)(C_{fu})(C_r)(C_f)(C_L)$$
$$= 1{,}200(1.15)(1)(1)(1)(1)(1)(1) = 1{,}380 \text{ psi}$$
$$f_b = M/S < F_b'$$

$S_{req'd} > M/F_b' = 15{,}000 \text{ lb-in.}/1{,}380 \text{ psi} = 10.87 \text{ in}^3$. Entering Table 8.8, it appears that a 2×8 is indeed the smallest adequate member with $S > 10.87 \text{ in}^3$. The section properties of the 2×8 are:

$$S = 13.14 \text{ in}^3$$
$$I = 47.64 \text{ in}^4$$
$$\text{Dead load} = 2.6 \text{ lb/ft}$$

Since the beam dead load is only 2.6% of the live load, it can be neglected in the design. If the dead load exceeds 5–10% of the live load, the adequacy of the selection should be rechecked.

Step 4: Evaluate the resulting horizontal shear stress.

$$F_v' = 90 \times 1.15 = 103.5 \text{ psi}$$
$$f_v = 3V/2A = 3(500 \text{ lb})/(2(10.88 \text{ in}^2)) = 69 \text{ psi}$$

Since $f_v < F_v'$, the shear strength of the 2×8 beam is adequate.

Step 5: Evaluate the resulting bearing requirements.

$$F_{c\perp}' = 565 \text{ psi}$$
$$f_{c\perp} = R/bz = 500 \text{ lb}(1.5 \text{ in.})(z)$$

Since $F_{c\perp}' \geq f_{c\perp}$, it follows that $z \geq 500/(565 \times 1.5)$ = 0.59 in. A minimum bearing length z of 0.59 in. is therefore required in order to provide enough bearing area. For the 1995 CABO Code (Sections 502.4[11] and 802.5[12]), the bearing distance at the end of a member must be at least 1½ in. of bearing on wood or metal and at least 3 in. on masonry.

Step 6: Check the deflection requirements of a uniformly loaded simply supported beam (live load analysis only).

$$\Delta = \Delta_L = (5w_L L^4)/(384EI)$$
$$\Delta = [5(100)(10)^4/(384(1.6{\times}10^6)(47.64))] \times 1{,}728$$
$$\Delta = 0.30 \text{ in.}$$

Since $\Delta_{all} = L/240 = 10 \text{ ft} \times (12 \text{ in./ft})/240 = 0.5 \text{ in.}$, the deflection requirement is met and a 2×8 is adequate to carry the specified load.

Summary

The design of most of the roof rafters, ceiling joists, and floor joists in a house (as well as individual beams which carry uniformly distributed loads) whose loads are transmitted to the members through a sheathing system which is firmly attached to the member will be designed using the procedure presented. This procedure uses the full value of the adjusted design bending stress (F_b') with $C_L = 1.0$ in the calculations. The discussion in the sections which follow presents the procedure which is required both for more general loading and for laterally unbraced conditions.

Laterally Unbraced Beams

If the unsupported length of the compression edge of a timber beam is large such that $\ell_u > 0$, the beam will buckle laterally before reaching the full allowable bending moment, $M_{all} = F_b^* \times S$. In the design of laterally unsupported beams, the adjusted design bending stress, F_b^*, must therefore be reduced to a level below which lateral buckling will not occur.

The adjusted design flexure stress for a laterally unbraced beam is defined by Eq. 11.9 or 11.10.

$$F_b' = F_b(C_D)(C_M)(C_t)(C_F)(C_v)(C_{fu})(C_r)(C_c)(C_f)(C_L) \qquad \text{(Eq. 11.9)}$$

$$F_b' = F_b^*(C_L) \qquad \text{(Eq. 11.10)}$$

The beam stability factor is defined by Eq. 11.11.[13]

$$C_L = \frac{(1 + F_{bE}/F_b^*)}{1.9} - \sqrt{\left[\frac{1 + F_{bE}/F_b^*}{1.9}\right]^2 - \frac{F_{bE}/F_b^*}{0.95}} \qquad \text{(Eq. 11.11)}$$

where: $F_{bE} = K_{bE}E'/R_B^2$

$K_{bE} = 0.438$ for VSR lumber

$\quad\; = 0.609$ for MSR lumber and engineered wood products with low coefficient of variability (COV < 0.11) of structural properties

$E' = E(C_M)(C_t)(C_T)$

$R_B = $ slenderness factor $= \sqrt{\ell_e d/b^2}$

$\ell_e = $ effective unbraced length for bending members (Table 11.3)

$d = $ beam dimension in direction of applied flexural load

$b = $ beam dimension perpendicular to direction of applied flexural load

TABLE 11.3 Effective Length, ℓ_e, for Bending Members

Cantilever[a]	when $\ell_u/d < 7$	when $\ell_u/d \geq 7$
Uniformly distributed load	$\ell_e = 1.33\,\ell_u$	$\ell_e = 0.90\,\ell_u + 3d$
Concentrated load at unsupported end	$\ell_e = 1.87\,\ell_u$	$\ell_e = 1.44\,\ell_u + 3d$
Single Span Beam[a]	**when $\ell_u/d < 7$**	**when $\ell_u/d \geq 7$**
Uniformly distributed load	$\ell_e = 2.06\,\ell_u$	$\ell_e = 1.63\,\ell_u + 3d$
Concentrated load at center with no intermediate lateral support	$\ell_e = 1.80\,\ell_u$	$\ell_e = 1.37\,\ell_u + 3d$
Concentrated load at center with lateral support at center	$\ell_e = 1.11\,\ell_u$	
Two equal concentrated loads at 1/3 points with lateral support at 1/3 points	$\ell_e = 1.68\,\ell_u$	
Three equal concentrated loads at 1/4 points with lateral support at 1/4 points	$\ell_e = 1.54\,\ell_u$	
Four equal concentrated loads at 1/5 points with lateral support at 1/5 points	$\ell_e = 1.68\,\ell_u$	
Five equal concentrated loads at 1/6 points with lateral support at 1/6 points	$\ell_e = 1.73\,\ell_u$	
Six equal concentrated loads at 1/7 points with lateral support at 1/7 points	$\ell_e = 1.78\,\ell_u$	
Seven or more equal concentrated loads, evenly spaced, with lateral support at points of load application	$\ell_e = 1.84\,\ell_u$	
Equal end moments	$\ell_e = 1.84\,\ell_u$	

[a]For single span or cantilever bending members with loading conditions not specified in Table 11.3:

$\qquad \ell_e = 2.06\,\ell_u \qquad$ when $\ell_u/d < 7$

$\qquad \ell_e = 1.63\,\ell_u + 3d$ when $7 \leq \ell_u/d \leq 14.3$

$\qquad \ell_e = 1.84\,\ell_u \qquad$ when $\ell_u/d > 14.3$

Source: *National Design Specification® for Wood Construction and Supplement*, p.11 (American Forest & Paper Association 1991).

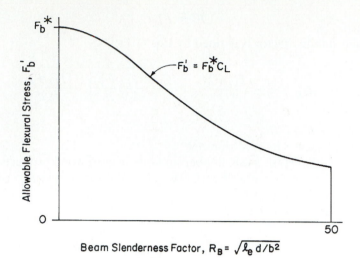

Figure 11.8
Variation of
Allowable
Flexure Stress
with Slenderness
Factor

Figure 11.8 graphically illustrates the variation of F_b' with R_B. A laterally unsupported beam must satisfy the same requirements for shear, bearing, and deflection as a braced beam. The procedures for evaluating these stress levels and the allowable stresses in unbraced beams are identical to those for braced beams. Finally, the maximum allowable value for R_B is 50. If $R_B > 50$, $F_b' = 0$ psi.

EXAMPLE NO. 6:   Laterally Unbraced Beam

Evaluate the allowable flexural stress for a dressed 2×10 No. 2 Southern Pine (surfaced and used at 15% moisture content) beam which carries a uniformly distributed load over an unbraced simply supported span of 10 ft. Assume VSR lumber and a snow load duration.

SOLUTION

Step 1: Define the governing equations and key parameters

$$F_b' = F_b(C_D)(C_M)(C_t)(C_F)(C_r)(C_f)(C_L) = F_b^*(C_L)$$
$C_D = 1.15$ (Fig. 8.18); $C_M = C_t = C_F = C_r = C_f = 1.0$
$E = 1.6 \times 10^6$ psi (Table 8.5)
$F_b = 1,050$ psi (Table 8.5)
$F_b^* = 1,050(1.15) = 1207.5$ psi

Step 2: Evaluate the slenderness factor, R_B

$\ell_e = 1.63\ell_u + 3d$
$\quad = 1.63(10)(12 \text{ in./ft}) + 3(9.25 \text{ in.}) = 223.35$ in.
$R_B = (\ell_e d/b^2)^{1/2} = [(223.35)(9.25)/(1.5)^2]^{1/2} = 30.3$

Step 3: Evaluate the beam stability factor (C_L)

$$F_{bE} = 0.438(1.6 \times 10^6)/(30.3)^2 = 763.3 \text{ psi}$$

$$C_L = \frac{1 + 763.3/1207.5}{1.9} + \sqrt{\left[\frac{1 + 763.3/1207.5}{1.9}\right]^2 - \frac{763.3/1207.5}{0.95}}$$
$$= 0.589$$

Step 4: Evaluate the allowable flexural stress.

$$F_b' = 1,207.5(0.589) = 711 \text{ psi}$$

Notice that F_b' is approximately 60% of the allowable flexure stress without lateral buckling ($F_b^* = 1,207$ psi). If f_b exceeds F_b', the 2×10 will buckle laterally. The objective of the designer, therefore, is either to brace the 2×10 to increase F_b' or to assure that the flexural stress (f_b) never exceeds $F_b' = 711$ psi in this beam.

Combined Loading Conditions

Many structural members are subject to a combination of flexural, tensile, and compressive loads. The more common situations encountered include axial tension and flexure, centric

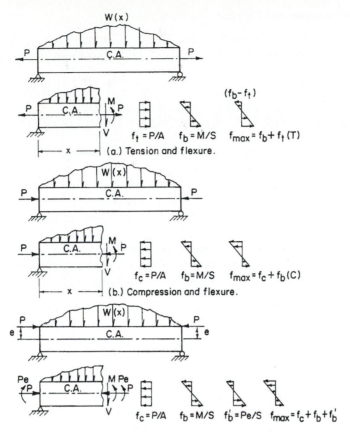

Figure 11.9
Stress
Distributions for
Combined
Loadings

(a.) Tension and flexure.

(b.) Compression and flexure.

(c.) Eccentric compression and flexure.

axial compression and flexure, and eccentric axial compression and flexure. The combined states of stress for each are shown in Fig. 11.9.

The fundamental design criterion for timber members under combined states of loading is represented by the "Interaction Equation." Equation 11.12 is the generalized interaction equation for all combined loadings.

$$\frac{\text{Calculated Axial Stress}}{\text{Allowable Axial Stress}} + \frac{\text{Calculated Flexural Stress}}{\text{Allowable Flexural Stress}} \leq 1.0 \qquad \text{(Eq. 11.12)}$$

By referring to the stress distributions in Fig. 11.9, it can be noted that the generalized interaction equation is quite rational if the allowable axial and flexural stresses are equal. Reference to design value tables in Chapter 8, however, shows that equality generally does not exist between the two allowable stresses. Thus, the interaction equation is an empirical approach. Nonetheless, it is relatively simple to use and it yields reasonable results in practice.

1. Axial Tension plus Flexure

Eqs. 11.13a and 11.13b are the NDS® design criteria for combined axial tension plus flexure.[14]

$$\frac{f_t}{F'_t} + \frac{f_b}{F^*_b} \leq 1.0 \qquad \text{(Eq. 11.13a)}$$

$$\frac{f_b - f_t}{F'_b} \leq 1.0 \qquad \text{(Eq. 11.13b)}$$

2. Axial Compression and Flexure

Members which carry both flexural and axial compressive loads simultaneously are called beam columns. Eq. 11.14 is the NDS® governing design criterion for beam columns.[15]

$$\left(\frac{f_c}{F_c'}\right)^2 + \frac{f_{b1}}{F_{b1}'[1 - (f_c/F_{cE1})]} + \frac{f_{b2}}{F_{b2}'[1 - f_c/F_{cE2} - (f_{b2}/F_{bE})^2]} \leq 1.0 \quad \text{(Eq. 11.14)}$$

where: F_c' = adjusted design compressive stress (adjusted for slenderness effects using maximum L_e/d)

F_{b1}' = adjusted design flexural stress (adjusted for slenderness effects with respect to the strong axis of bending [$C_L \leq 1.0$])

F_{b2}' = adjusted design flexural stress with respect to the weak axis of bending ($C_L = 1$)

f_c = calculated axial compressive stress

f_{b1} = calculated bending stress for load applied parallel to wide face

f_{b2} = calculated bending stress for load applied perpendicular to the wide face

$F_{cE1} = K_{cE}E'/(L_{e1}/d_1)^2$

$F_{cE2} = K_{cE}E'/(L_{e2}/d_2)^2$

$F_{bE} = K_{bE}E'/(R_B)^2$

d_1 = wide face dimension

d_2 = narrow face dimension

Eq. 11.14 is for the general case of biaxial bending coupled with axial compression. In the common case of axial compression plus bending about the strong axis of the cross section, Eq. 11.14 reduces to the first two terms.

3. Eccentric Axial Compression and Flexure

Eq. 11.15 is the governing interaction equation for the general case of biaxial bending and eccentricity of the axial compressive load in both directions.[16]

$$\left(\frac{f_c}{F_c'}\right)^2 + \frac{f_{b1} + f_c(6e_1/d_1)[1 + 0.234(f_c/F_{cE1})]}{F_{b1}'[1 - (f_c/F_{cE1})]}$$

$$+ \frac{f_{b2} - f_c(6e_2/d_2)[1 + 0.234(f_c/F_{cE2})]}{F_{b2}'[1 - f_c/F_{cE2} - (f_b/F_{bE})^2]} \leq 1.0 \quad \text{(Eq. 11.15)}$$

where e_1 = eccentricity of axial load measured parallel to the wide face of cross section

e_2 = eccentricity of axial load measured parallel to the narrow face of cross section

Eq. 11.15 reduces to a simpler form for eccentricity about only one axis or flexure about only one axis. Indeed, Eq. 11.14 is a special case of Eq. 11.15 for $e_1 = e_2 = 0$.

The terms $[1 + 0.234(f_c/F_{cE1})]$, $[1 - f_c/F_{cE1}]$, and $[1 - f_c/F_{cE2} - (f_b/F_{cE})^2]$ in Eqs. 11.14 and 11.15 are factors to account for secondary stresses in beam columns. They are sometimes called **moment magnification factors**. Figure 11.10(a) will help explain this phenomenon for the combined flexure and centric compression case. The total moment in the beam column is the moment due to loads, M_w, plus the secondary moment, $P\Delta$. In short members, $P\Delta$ is much smaller than M_w; as the member lengthens, Δ is larger for the same loads and $P\Delta$ becomes more significant. In Eq. 11.14, this effect is incorporated by reducing the adjusted design flexural strength. The reduction increases as the axial compression stress increases and as the beam length increases (i.e., F_{cEi} decreases).

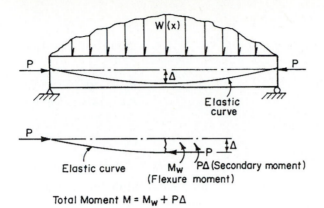

Total Moment $M = M_w + P\Delta$

(a.) Centric loads.

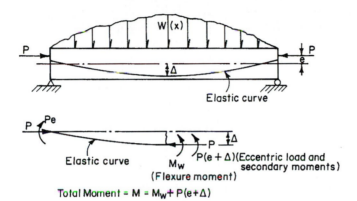

Figure 11.10
Secondary
Moments in
Beam Columns

Total Moment = $M = M_w + P(e+\Delta)$

(b.) Eccentric loads.

For the case of a beam column with eccentric loading about one axis, the maximum moment $[M_w + P(e + \Delta)]$ is illustrated in Fig. 11.10(b). For shorter members, Δ is generally much smaller than e and $M_{\max} \approx M_w + Pe$. For longer members, Δ can approach the order of magnitude of e and $M_{\max} = M_w + P(e + \Delta)$. The design specification (Eq. 11.15) incorporates the secondary moment effect by increasing the bending stress due to eccentricity by $0.234 \, f_c/F_{cEi}$ and reducing the adjusted design bending stress.

EXAMPLE NO. 7:   Beam-Column Analysis

Evaluate the magnitude of the maximum axial compressive load which may be superimposed on the No. 2 Southern Pine dressed 2×4 (surfaced and used at 15% moisture content) shown in Fig. 11.11. Assume normal duration loading and pinned ends.

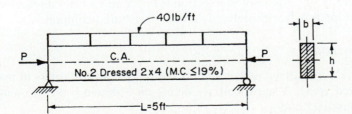

Figure 11.11
Illustration for
Example No. 7

SOLUTION

Step 1: Obtain the section and structural properties from Tables 8.5 and 8.8.

$$A = 5.25 \text{ in}^2 \text{ (Table 8.8)}$$
$$b = 1.5 \text{ in. (Table 8.8)}$$
$$d = 3.5 \text{ in. (Table 8.8)}$$
$$S = 3.06 \text{ in}^3 \text{ (Table 8.8)}$$
$$F_b = 1,500 \text{ psi (Table 8.5)}$$
$$F_c = 1,650 \text{ psi (Table 8.5)}$$
$$E = 1.6 \times 10^6 \text{ psi (Table 8.5)}$$
$$C_D = 1.15$$

All other C_i's = 1.0
$$F_b^* = 1,500(1.15) = 1,725 \text{ psi}$$
$$F_c^* = 1,650(1.15) = 1,898 \text{ psi}$$

Step 2: Governing Interaction Equation

The beam is subject to axial compression and bending about the strong axis; thus, the governing equation is

$$\left(\frac{f_c}{F_c'}\right)^2 + \frac{f_b}{F_b'[1 - (f_c/F_{cE})]} \leq 1.0$$

Step 3: Evaluate terms in the interaction equation.

$$f_c = P/A = P/5.25$$
$$f_b = M/S = wL^2/8S = 3.33(60)^2/[(8)(3.06)]$$
$$= 490 \text{ psi}$$
$$F_c' = F_c^* C_p$$
$$L_e = K_e L = (1.0)(60) = 60 \text{ in.}$$
$$L_e/d = 60/1.5 = 40$$
$$K_{cE} = 0.30$$
$$F_{cE} = 0.3(1.6 \times 10^6)/(40)^2 = 300 \text{ psi}$$
$$F_{cE}/F_c^* = 0.158$$

$$C_p = \frac{1.158}{1.6} - \sqrt{\left(\frac{1.158}{1.6}\right)^2 - \frac{0.158}{0.8}} = 0.153$$

$$F_c' = 1,898(0.153) = 290.3 \text{ psi}$$
$$F_b' = F_b^* C_L$$
$$\ell_e = 1.63(60) + 3(3.5) = 108.3 \text{ in.}$$
$$R_B = \sqrt{108.3(3.5)/(1.5)^2} = 12.98$$
$$F_{bE} = 0.438(1.6 \times 10^6)/(12.98)^2 = 4,160 \text{ psi}$$
$$F_{bE}/F_b^* = 2.41$$

$$C_L = \frac{3.41}{1.9} - \sqrt{\left(\frac{3.41}{1.9}\right)^2 - \frac{2.41}{0.95}} = 0.968$$

$$F_b' = 1,725(0.968) = 1,670 \text{ psi}$$

Step 4: Substitute terms into the interaction equation and solve for P.

$$\left(\frac{P/5.25}{290.3}\right)^2 + \frac{490}{1,670\left[1 - \dfrac{P/5.25}{300}\right]} \leq 1.0$$

$$\frac{P^2}{2.323 \times 10^6} + \frac{0.293}{\left[1 - \dfrac{P}{1,575}\right]} \leq 1.0$$

Using a trial-and-error solution approach with a first estimate of 1,000 lb, the following solution sequence is obtained:

Load Estimate (lbs)	LHS of Eq.
1,000	1.234
800	0.871
900	1.033
895	1.024
880	0.997 ≈ 1.0

Thus, $P = 880$ lbs.

ANALYSIS OF EXTERNAL LOADS ON ROOF RAFTER MEMBERS

Resolution of the Roof Loads into Vertical and Horizontal Components

Roof loads (snow, wind, and dead) are not all prescribed in the same terms. For example, snow loads are defined as lbs per square foot of horizontal projected area, whereas wind loads are usually defined as lbs per square foot of surface area. These loads are not always directly additive. The purpose of this section is to present a brief discussion on the transformation of roof loads to horizontal and vertical distributed components.

Vertical Snow Load

Snow load (w_s) is always defined as a vertical gravity load with units of psf_h (lbs/ft^2 of horizontal projected area). A snow load, w_s, acting on a sloped roof with length L produces a resultant load of $w_s(L\cos\theta)$.

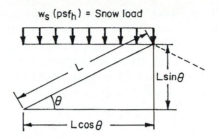

Vertical Dead Load

The weight of the roof, which is a dead load, includes the weight of the shingles, sheathing, and roof rafters. It is usually expressed as a unit weight (w_d in lb/ft^2) on the **sloping roof area**. It must, therefore, be converted to an equivalent unit weight on the horizontal projection of the roof. The appropriate procedure for making this conversion is shown below.

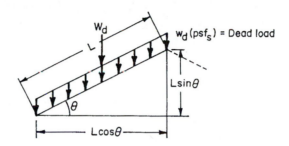

The resultant load W_d (i.e., the total weight in pounds) on a one-foot-wide strip (normal to plane of the page) of roof is:

$$W_d = w_d(L)(1) = w_d L$$

The resultant load can now be expressed as a unit weight (w_d' in psf$_h$) on a one-foot-wide strip of horizontal projection of the roof, as follows:

$$w_d' = W_d/L\cos\theta = w_d L/L\cos\theta = w_d/\cos\theta$$

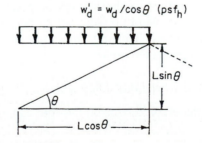

Wind Load

Wind loads act normal to the surface and are normally expressed in units of lbs/sq foot of surface area (psf$_s$).

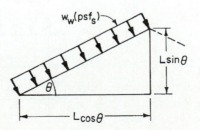

Before wind loads can be added to snow loads, it is necessary to convert the wind pressure shown above into its equivalent vertical component (acting on the horizontal projection of the roof) and its equivalent horizontal component (acting on the vertical projection of the roof). This can be accomplished by first determining the value of the resultant wind load W_w (i.e., the total load in pounds) on a one-foot-wide strip (normal to the plane of the page) of roof and then expressing that resultant as equivalent vertical and horizontal loads.

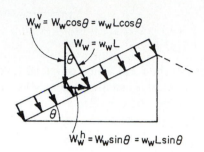

The vertical (W_w^v) and horizontal (W_w^h) components of the resultant force W_w are now expressed as the wind pressures (w_w^v and w_w^h in psf) on a one-foot strip of the roof, as follows:

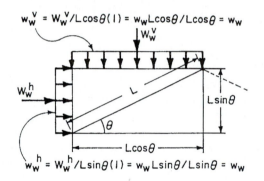

The vertical and horizontal components of the wind pressure have the same magnitude as the original wind pressure acting perpendicular to the roof surface. However, the area base has changed. The units on w_w^v are in psf_h and on w_w^h are psf_v ($\text{psf}_v \equiv$ lbs per square foot of vertical projected area).

Combined Unit Load Diagram

The combined unit loads on the roof can, therefore, be represented as follows:

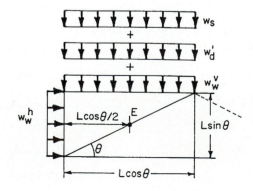

Experience indicates that a full snow load and a full wind load rarely occur at the same time. When a strong wind is blowing, one which results in the full wind load being active, the snow load is reduced by 50% because snow is being blown off the roof. Also, if the full snow load is acting, then only one-half the design wind load is used. Many building codes use an "Equivalent Vertical Load" to represent the combined effect of snow load and wind load on a roof. This equivalent load is often assumed to act on the horizontal projection of the roof area.

Calculation of the Maximum Moment on a Roof Rafter

The combined unit loads on the roof which have been determined above can be used to calculate the maximum bending moment, and the maximum shear, on the roof rafter of the house so that the proper sized member can be selected.

EXAMPLE NO. 8:   Calculate the Maximum Bending Moment in a Roof Rafter Using the Vertical and Horizontal Representation of Loads

Determine the maximum bending moment in the roof rafter (i.e., at Point E) specified below:
Given:

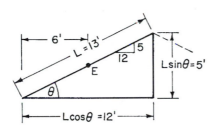

1. $L = 13$ ft, $\theta = 22.6°$
2. $w_s = 20$ psf$_h$
3. $w_d = 11$ psf
   (shingles = 3 psf$_s$, 5/8 in. sheathing = 2 psf$_s$, roof rafters = 6 psf$_s$)
4. $w_w = 10$ psf$_s$
5. Rafter Spacing = 2 ft o.c.

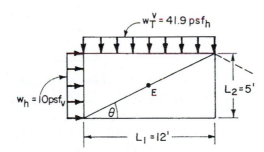

- $w_T^v = w_s + w_d' + w_w^v$
  $= w_s + w_d/\cos\theta + w_w$
  $= 20 + 11(12/13) + 10 = 20 + 11.9 + 10$
  $= 41.9$ lb/ft^2
- $w^h = w_w = 10$ lb/ft^2

The total bending moment at $E = (w_T^v)(2)L_1^2/8 + w^h(2)L_2^2/8$

$$\text{Total } M_E = (41.9)(2)(12)^2/8 + (10)(2)(5)^2/8$$
$$= 1{,}508.4 + 62.5$$
$$= 1{,}570 \text{ ft-lbs}$$

Roof Load Components Perpendicular and Parallel to the Roof Line

The bending moment result at the centerline of the roof rafter above is correct. The only problem is that the roof rafter is not simply a beam; rather, it is a beam column since there are components of the roof loads which are both perpendicular and parallel to the roof line.

Resolution of wind, snow, and dead loads into components parallel and perpendicular to the rafter is presented next.

Vertical Snow Load and Dead Load

The combination of the snow load and the dead load can be represented as follows:

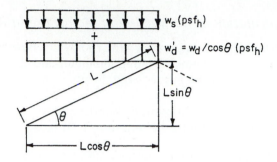

The resultant vertical load on a one-foot strip of roof due to these two unit loads is:

$$W^v = (w_s + w_d')(L)(1)(\cos\theta) = (w_s + w_d')L\cos\theta$$

W^v can then be resolved into components parallel and perpendicular to the roof line, as follows:

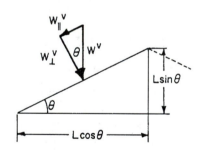

$$W_{\parallel}^v = W^v\sin\theta = (w_s + w_d')L\cos\theta\sin\theta$$
$$W_{\perp}^v = W^v\cos\theta = (w_s + w_d')L\cos^2\theta$$

The parallel ($W_{\parallel}^v$) and perpendicular ($W_{\perp}^v$) components of the resultant force W^v can be expressed as the unit loads ($w_{\parallel}^v$ and $w_{\perp}^v$ in psf_s) on a one-foot strip of roof as follows:

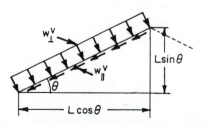

$$w_{\parallel}^v = (w_s + w_d')L\cos\theta\sin\theta/[L(1)] = (w_s + w_d')\cos\theta\sin\theta$$
$$w_{\perp}^v = (w_s + w_d')L\cos^2\theta/[L(1)] = (w_s + w_d')\cos^2\theta$$

Wind Load

The wind load acts in a direction perpendicular to the roof line and has units of psf$_s$. Thus, it can be added directly to $w_\perp^y$.

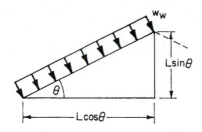

Combined Unit Load Diagram

The combined unit loads on the roof can, therefore, be represented as follows:

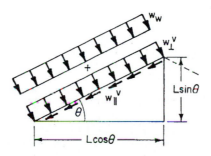

Calculation of Loads and Maximum Moments on a Roof Rafter

The parallel and perpendicular components of roof unit loads can also be used to calculate the maximum bending moment and the maximum shear on the roof rafter.

EXAMPLE NO. 9: Calculate the Maximum Bending Moment and Axial Load in a Roof Rafter Using the Perpendicular and Parallel Representation of Loads

For the roof rafter analyzed in Example No. 8, determine the maximum bending moment in the roof rafter (i.e., at Point E).

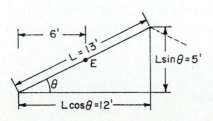

Given:

1. $L = 13$ ft, $\theta = 22.6°$
2. $w_s = 20$ psf$_h$
3. $w_d = 11$ psf$_s$
4. $w_w = 10$ psf$_s$
5. Rafter Spacing = 2 ft o.c.
6. $w_d' = w_d/\cos\theta = 11.9$ psf$_h$

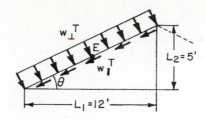

- $w_\perp^T = w_w + w_\perp^y = 10 + (w_s + w_d')\cos^2\theta$
  $= 10 + (20 + 11.9)(12/13)^2 = 10 + (31.9)(0.85)$
  $= 10 + 27.1$
  $= 37.1 \text{ lb/ft}^2$
- $w_\parallel^T = (w_s + w_d')\cos\theta\sin\theta = (20 + 11.9)(12/13)(5/13)$
  $= (31.9)(0.36)$
  $= 11.5 \text{ lb/ft}^2$

The total bending moment at $E = (w_\perp^T)(2)L^2/8$.

$$\text{Total } M_E = (37.1)(2)(13)^2/8 = 1{,}568 \text{ ft-lbs}$$

As expected, the bending moment at E in Example No. 9 is the same (except for round-off error) as the value obtained in Example No. 8.

The resolution of the forces on the roof rafter into perpendicular and parallel components does, however, indicate that the roof rafter is also subjected to an axial load which has a maximum value of:

$$\text{Maximum axial load} = w_\parallel^T(L)(2)$$
$$= (11.5)(13)(2) = 299 \text{ lb}$$

This axially directed load, which is not a constant value along the roof rafter (it accumulates to a maximum value of 299 lbs at the outside edge of the roof rafter), results in the roof rafter being a beam-column type of member.

MODIFIED CASE STUDY HOUSE EXAMPLE

The first step in the structural analysis of the Case Study House requires that a "Load Distribution" be developed. This distribution can then be used to determine the required size of the roof rafters, ceiling joists, floor joists, wall studs, girders, columns, and footers. It will again be assumed that No. 2 Hem-Fir, surfaced dry and used at less than 19% moisture content, is being used for all lumber and that the live loads are as determined in Chapter 9. It should be noted that the real Case Study House is constructed with roof trusses, thereby not requiring a bearing wall between the second floor and the roof trusses. This modified Case Study House example assumes a roof rafter and ceiling joist system with a bearing wall located as shown in Fig. 11.12 in order to illustrate the necessary calculations.

Load Distribution

Step 1: Determine the live loads (Chapter 9)

1. Roof snow load = 20 lb/ft² (rounded off from the value used in Chapter 9 since the 1995 CABO roof rafter tables are based on 20 lb/ft², 30 lb/ft², or 40 lb/ft²)

2. Roof wind load = 0 lb/ft² (as determined in Chapter 9 when the 1995 CABO Code guidelines were used)

3. Attic (limited storage) = 20 lb/ft²

4. Second floor (sleeping) = 30 lb/ft²

5. First floor (dwelling) = 40 lb/ft²

Step 2: Determine the total loads (representative dead load values are obtained from either Table 8.8 or Table 9.6)

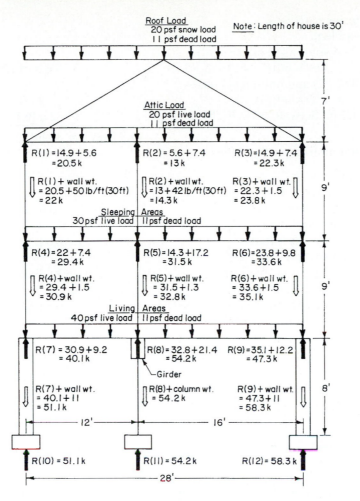

Figure 11.12
Load Distribution
on Modified Case
Study House

1. Roof loads (on horizontal projection)

Snow load	20 lb/ft^2
Dead load	
Fiberglass shingles	2.0 lb/ft^2
1/2 in. plywood = [0.4 lb/ft^2/1/8 in.](4)	1.6 lb/ft^2
Framing—2×6 rafters at 12 in. o.c.	6.0 lb/ft^2
	9.6 lb/ft^2
	≈ 10 lb/ft^2 (on sloped surface)
Dead load on horizontal projection	= 10/cos26.6°
	= 11 lb/ft^2
Total roof load	= 31 lb/ft^2

2. Attic floor

Limited attic storage	20 lb/ft^2
Dead load	
1/2 in. plywood = [0.4 lb/ft^2/1/8 in.](4)	1.6 lb/ft^2
Framing—2×8 joists at 12 in. o.c.	6.0 lb/ft^2 (estimated framing)
Insulation—12 in. (blown-in)	1.1 lb/ft^2
1/2 in. drywall = [0.55 lb/ft^2/1/8 in.](4)	2.2 lb/ft^2
10.9 lb/ft^2 ≈	11 lb/ft^2
Total attic floor load	= 31 lb/ft^2

3. Interior partitions
   Dead load
      2×4 studs at 16 in. o.c.
        = 1.094 lb/ft(12 in./16 in.)(8 ft)   6.6 lb/ft (estimated framing)
      Top (2) and bottom (1) plates
        = 1.094 lb/ft(3)   3.3 lb/ft
      1/2 in. drywall = 2(2 lb/ft^2)(8 ft) $\underline{32.0\ \text{lb/ft}}$

                              41.9 lb/ft ≈ 42 lb/ft wall

      Total interior partition load   42 lb/ft wall

4. Exterior walls
   Dead load
      Vinyl siding = 2 lb/ft^2(8 ft)   16.0 lb/ft
      1/2 in. insulating sheathing
        = 0.75 lb/ft^2(8 ft)   6.0 lb/ft
      3 1/2 in. batt insulation
        = 1.1(3.5 in./12 in.)(8 ft)   2.6 lb/ft
      2×4 studs at 16 in. o.c.
        = 1.094 lb/ft(12 in./16 in.)(8 ft)   6.6 lb/ft (estimated framing)
      Top (2) and bottom (1) plates
        = 1.094 lb/ft(3)   3.3 lb/ft
      1/2 in. drywall = (2 lb/ft^2)(8 ft) $\underline{16.0\ \text{lb/ft}}$

                              50.5 lb/ft ≈ 50 lb/ft wall

      Total exterior wall load   50 lb/ft wall

5. Second floor (sleeping area)
   Live load   30.0 lb/ft^2
   Dead load
      Carpet   0.5 lb/ft^2
      3/4 in. plywood
        = 0.4 lb/ft^2/1/8 in. (6)   2.4 lb/ft^2
      2×10 joists at 16 in. o.c.   6.0 lb/ft^2 (estimated framing)
      1/2 in. drywall = 0.55 lb/ft^2/1/8 in.(4) $\underline{2.2\ \text{lb/ft}^2}$

                            11.1 lb/ft^2 ≈ 11 lb/ft^2

      Total second floor load   41 lb/ft^2

6. First floor (living area)
   Live load   40 lb/ft^2
   Dead load (same as above) $\underline{11\ \text{lb/ft}^2}$
      Total first-floor load   51 lb/ft^2

7. Basement wall
   Dead load
      10 in. concrete block (light aggregate)   46 lb/ft^2

8. Basement
   Dead load
      Girder—(2) 1-3/4 in. × 9-1/4 in. LVLs (neglect)
      Columns (neglect)
      Concrete floor (neglect)

The corresponding load distribution that results is illustrated in Fig. 11.12.

The calculations necessary to determine the floor reactions in Fig. 11.12 are summarized:

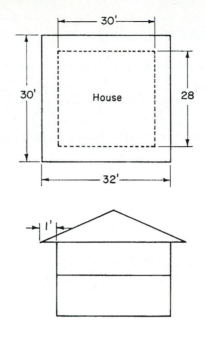

$R(1) = 31 \text{ lb/ft}^2(30 \text{ ft})(32 \text{ ft})/2 + 31 \text{ lb/ft}^2(30 \text{ ft})(12 \text{ ft}/2)$
$\quad = 14.9 \text{ k} + 5.6 \text{ k} = 20.5 \text{ k}$

$R(2) = 31 \text{ lb/ft}^2(12 \text{ ft}/2)(30 \text{ ft}) + 31 \text{ lb/ft}^2(16 \text{ ft}/2)(30 \text{ ft})$
$\quad = 5.6 \text{ k} + 7.4 \text{ k} = 13 \text{ k}$

$R(3) = 31 \text{ lb/ft}^2(30 \text{ ft})(32 \text{ ft})/2 + 31 \text{ lb/ft}^2(30 \text{ ft})(16 \text{ ft}/2)$
$\quad = 14.9 \text{ k} + 7.4 \text{ k} = 22.3 \text{ k}$

$R(4) = [R(1) + \text{wall}] + (\text{floor})$
$\quad = 22 \text{ k} + 41 \text{ lb/ft}^2(30 \text{ ft})(12 \text{ ft}/2)$
$\quad = 22 \text{ k} + 7.4 \text{ k} = 29.4 \text{ k}$

$R(5) = [R(2) + \text{wall}] + (\text{floor})$
$\quad = 14.3 \text{ k} + 41 \text{ lb/ft}^2(12 \text{ ft}/2)(30 \text{ ft}) + 41 \text{ lb/ft}^2 (16 \text{ ft}/2)(30 \text{ ft})$
$\quad = 14.3 \text{ k} + 7.4 \text{ k} + 9.8 \text{ k} = 31.5 \text{ k}$

$R(6) = 23.8 \text{ k} + 41 \text{ lb/ft}^2(30 \text{ ft})(16 \text{ ft}/2)$
$\quad = 23.8 \text{ k} + 9.8 \text{ k} = 33.6 \text{ k}$

$R(7) = 30.9 \text{ k} + 51 \text{ lb/ft}^2(30 \text{ ft})(12 \text{ ft}/2)$
$\quad = 30.9 \text{ k} + 9.2 \text{ k} = 40.1 \text{ k}$

$R(8) = 32.8 \text{ k} + 51 \text{ lb/ft}^2(12 \text{ ft}/2)(30 \text{ ft}) + 51 \text{ lb/ft}^2 (16 \text{ ft}/2)(30 \text{ ft})$
$\quad = 32.8 \text{ k} + 9.2 \text{ k} + 12.2 \text{ k} = 54.2 \text{ k}$

$R(9) = 35.1 \text{ k} + 51 \text{ lb/ft}^2(30 \text{ ft})(16 \text{ ft}/2)$
$\quad = 35.1 \text{ k} + 12.2 \text{ k} = 47.3 \text{ k}$

Masonry wall weight $= 46 \text{ lb/ft}^2(30 \text{ ft})(8 \text{ ft}) = 11 \text{ k}$

Figure 11.13 summarizes the load placed on the wall footings and the individual column footings. The reactions at the individual footings that support the main girder are determined as follows:

Load per foot of girder $= 54.2 \text{ k}/30 \text{ ft} = 1.81 \text{ k/ft}$
Weight on each interior column $= 1.81 \text{ k/ft}(7.5 \text{ ft}) = 13.5 \text{ k}$
Weight on the end foundation wall due to the girder load $=$
$1.81 \text{ k/ft}(7.5 \text{ ft})/2 = 6.8 \text{ k}$
Weight on the end foundation wall due to the weight of the end of the house $=$
$50 \text{ lb/ft}(28 \text{ ft})(2) = 2.8 \text{ k}$
Total load on end wall $= 6.8 \text{ k} + 2.8 \text{ k} = 9.6 \text{ k}$

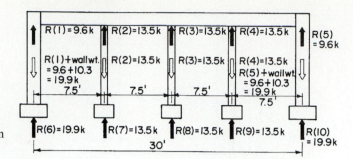

Figure 11.13
Load Distribution
on Footings

Roof Rafter Design: Beam-Column Analysis

Design a typical No. 2 Hem-Fir roof rafter, surfaced dry and used at less than 19% moisture content, as a beam column for the conditions as listed.

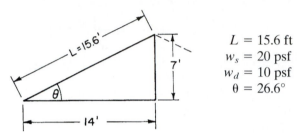

$L = 15.6$ ft
$w_s = 20$ psf
$w_d = 10$ psf
$\theta = 26.6°$

Assume rafter spacing = 1 ft o.c.

Solution

Step 1: Convert the dead load to a horizontal projected area.

$$w_d' = w_d/\cos\theta = 10 \text{ psf}/\cos 26.6° = 11.2 \text{ psf}$$

Step 2: Determine the total vertical load

$$w_t = w_d' + w_s = 11.2 \text{ psf} + 20 \text{ psf} = 31.2 \text{ psf}$$

Step 3: Resolve into $\perp$ and $\parallel$ loads on the rafter.

$$\begin{aligned}
w_\perp^v &= w_t\cos^2\theta \\
&= 31.2 \text{ psf } (1 \text{ ft})(\cos^2 26.6°) \\
&= 24.9 \text{ lb/ft} \\
w_\parallel^v &= w_t\sin\theta\cos\theta \\
&= 31.2 \text{ psf } (1 \text{ ft})(\sin 26.6°)(\cos 26.6°) \\
&= 12.5 \text{ lb/ft}
\end{aligned}$$

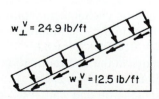

Step 4: Assume a 2×10 roof rafter and check the interaction equation. Obtain the section and structural properties from Tables 8.4 and 8.8.

$A = 13.875$ in^2
$b = 1.5$ in.
$d = 9.25$ in.
$S = 21.391$ in^3
$F_b = 850$ psi
$F_c = 1,250$ psi
$E = 1.3 \times 10^6$ psi
$C_D = 1.15$
$C_F = 1.10$ (bending)
$\quad\;\; = 1.00$ (compression)

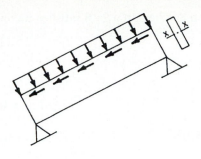

Step 5: Identify the governing interaction equation.

$$\left(\frac{f_c}{F'_c}\right)^2 + \frac{f_b}{F'_{b1}[1 - f_c/F_{cE1}]} \leq 1.0$$

Step 6: Check if a 2×10 satisfies the interaction equation.

$$f_c = P/A = (12.5 \text{ lb/ft})(15.6 \text{ ft})/13.875 = 14.1 \text{ psi}$$
$$f_b = M_{max}/S_x = wL^2/(8S_x)$$
$$\quad = 24.9(15.6)^2(12)/(8 \times 21.391) = 425 \text{ psi}$$

$F'_c = F^*_c C_p$
$\quad F^*_c = Fc(C_D)(C_F) = 1,250(1.15)(1.0) = 1,438$ psi
$\quad K_{cE} = 0.30$
$\quad (L_e/d)_x = 15.6 \times 12/9.25 = 20.2;\ (L/d)y \approx 0$
$\quad F_{cE} = 0.3(1.3\times10^6)/(20.2)^2 = 956$ psi
$\quad c = 0.8$
$\quad F_{cE}/F^*_c = 956/1,438 = 0.665$

$$C_p = \frac{1.665}{1.6} - \sqrt{\left(\frac{1.665}{1.6}\right)^2 - \frac{0.665}{0.8}} = 0.538$$

$F'_c = 1,438(0.538) = 774$ psi
$F'_{b1} = F^*_b C_L = F^*_b$ because the roofing
$\qquad$ provides lateral bracing $(\ell_u \approx 0)$
$F'_{b1} = 850(1.15)(1.10) = 1,075$ psi
$F_{cE1} = k_{cE}E'/(L_{e1}/d_1)^2 = 956$ psi

Substituting into the interaction equation,

$$\left(\frac{14.1}{774}\right)^2 + \frac{425}{1,075[1 - 14.1/956]} = 0.401 < 1.0$$

Thus, a 2×10 is certainly adequate; since the interaction value is so low, a 2×8 might also be satisfactory.

Step 7: Properties of a 2×8 from Tables 8.4 and 8.8.

$A = 10.88$ in^2
$b = 1.5$ in.
$d = 7.25$ in.
$S = 13.14$ in^3
$F_b = 850$ psi
$F_c = 1,250$ psi
$E = 1.3 \times 10^6$ psi
$C_D = 1.15$
$C_F = 1.2$ (bending)
$\quad = 1.05$ (compression)
$F^*_b = 850(1.15)(1.2) = 1,173$ psi
$F^*_c = 1,250(1.15)(1.05) = 1,510$ psi

Step 8: Check if a 2×8 satisfies the interaction equation.

$$f_c = (12.5)(15.6)/10.88 = 18 \text{ psi}$$
$$f_b = 24.9(15.6)^2(12)/(8 \times 13.14) = 692 \text{ psi}$$
$$F_c' = F_c^* C_p$$
$$(L_e/d)_x = 15.6 \times 12/7.25 = 25.8$$
$$F_{cE} = 0.3(1.3\times10^6)/(25.8)^2 = 586 \text{ psi}$$
$$F_{cE}/F_c^* = 586/1{,}510 = 0.388$$
$$C_p = 0.351$$
$$F_c' = 1{,}510(.351) = 530 \text{ psi}$$
$$F_{b1}' = F_b^* = 1{,}173 \text{ psi}$$
$$F_{cE1} = 586 \text{ psi}$$

$$\left(\frac{18}{530}\right)^2 + \frac{692}{1{,}173[1 - 18/586]} = 0.610 < 1.0$$

Step 9: Solution suggests that a 2×6 may be satisfactory. Check a 2×6.

$$A = 8.25 \text{ in}^2$$
$$b = 1.5 \text{ in.}$$
$$d = 5.5 \text{ in.}$$
$$S = 7.56 \text{ in}^3$$
$$C_D = 1.15$$
$$C_F = 1.3 \text{ (bending)}$$
$$= 1.1 \text{ (compression)}$$
$$C_r = 1.15 \text{ (repetitive member use; neglected until now)}$$
$$F_b^* = 850(1.15)(1.3)(1.15) = 1{,}460 \text{ psi}$$
$$F_c^* = 1{,}250(1.15)(1.1) = 1{,}580 \text{ psi}$$
$$f_c = 23.6 \text{ psi}$$
$$f_b = 1{,}202 \text{ psi}$$
$$(L_e/d)_x = 15.6 \times 12/5.5 = 34$$
$$F_{cE} = 0.3(1.3\times10^6)/(34)^2 = 337 \text{ psi}$$
$$F_{cE}/F_c^* = 337/1{,}580 = 0.213$$
$$C_p = 0.202$$
$$F_c' = 1{,}580(0.202) = 319 \text{ psi}$$
$$F_{b1}' = 1{,}460 \text{ psi}$$
$$F_{cE1} = 337 \text{ psi}$$

Thus,

$$\left(\frac{23.6}{319}\right)^2 + \frac{1202}{1{,}460[1 - 23.6/337]} = 0.89 < 1.0$$

Therefore, a 2 × 6 is the smallest adequate member for the roof rafter.

Attic Floor/Ceiling Joist Design

Select a No. 2 Hem-Fir member surfaced dry and used at less than 19% moisture content to carry the loads on the attic floor. Assume that the compression flanges of the joists are adequately braced because of the sheathing on the attic floor.

Solution

Step 1: Define the lumber properties and design parameters (assume 2× lumber).

$$C_D = 1.0 \text{ (Fig. 8.18—normal load duration)}$$
$$F_b = 850 \text{ psi (Table 8.4)}$$
$$E = 1.3 \times 10^6 \text{ psi (Table 8.4)}$$
$$F_v = 75 \text{ psi (Table 8.4)}$$
$$F_{c\perp} = 405 \text{ psi (Table 8.4)}$$

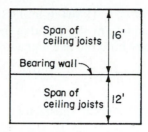

Step 2: Determine the maximum shear and moment.

Assume that the joists are spaced at 12 in. o.c. As noted from the second-floor plan, the controlling span is 16 ft.

$$w = 31 \text{ lb/ft}^2 = 31 \text{ lb/ft}$$
$$V_{\max} = wL/2 = 31(16)/2 = 248 \text{ lb}$$
$$M_{\max} = wL^2/8 = 31(16)^2(12)/8 = 11{,}904 \text{ lb-in.}$$

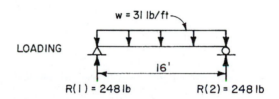

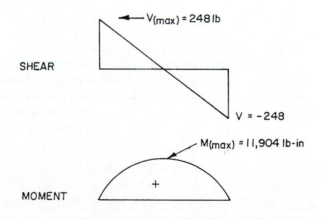

Step 3: Select a joist size that is adequate for flexural stress ($C_L = 1.0$).

Approximate the size using

$$F_b' = F_b C_r = 850(1.15) = 978 \text{ psi}$$
$$S_{\text{est}} = M_{\max}/F_b' = 11{,}904/978 = 12.17 \text{ in}^3$$

A 2×8 is the smallest possible size ($S_{2\times8} = 13.14$ in³).

Now check the adequacy of the 2×8 using 2×8 properties.

$$F'_b = F_b(C_D)(C_M)(C_t)(C_F)(C_{fu})(C_r)(C_f)$$
$$= 850(1.0)(1.0)(1.0)(1.1)(1.0)(1.15)(1.0)$$
$$= 1{,}075 \text{ psi}$$

$S_{req'd} > M/F'_b = 11{,}904$ lb-in./1,075 psi $= 11.07$ in³ $< S_{2\times8} = 13.14$ in³. Thus, the 2×8 is indeed adequate. The dead load of the joist is 2.266 lb/ft (Table 8.8). The required section modulus to carry both the live and additional dead load is

$$S_{req'd} = \frac{31.0 + 2.286}{31} (11.07) = 11.88 < 13.14 \text{ in}^3$$

Therefore, the 2×8 is still satisfactory.

Step 4: Check the horizontal shear stress.

$$F'_v = 75 \text{ psi}$$
$$f_v = 3V/(2A) = 3(248 \text{ lb})/[2(10.875 \text{ in}^2)] = 34 \text{ psi}$$

Since $f_v < F'_v$, the 2×8 is adequate for shear.

Step 5: Check the bearing requirements.

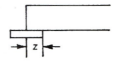

$$F'_{c\perp} = 405 \text{ psi}$$
$$f_{c\perp} = R/bz = 248 \text{ lb}/(1.5 \text{ in.})(z)$$

Since $F'_{c\perp} \geq f_{c\perp}$, it follows that $z \geq 248/(405 \times 1.5) = 0.41$ in. This is not an unreasonable bearing length and is less than 1995 Code minimums.[17] Therefore, bearing is not a problem if the minimum values in the 1995 CABO Code are used.

Step 6: Check the bracing requirements.

It is assumed that the compression edge of the joist will be supported by a plywood sheathing floor in the attic which will provide the lateral stability that is required. Therefore, assume that the plywood sheathing is nailed every 12 in. to the compression edge of the attic joists.

$$\ell_e = 1.63\ell_u + 3d = 1.63(12) + 3(7.25) = 41.31 \text{ in.}$$
$$R_B = (\ell_e d/b^2)^{1/2} = [(41.31)(7.25)/(1.5)^2]^{1/2} = 11.5$$
$$F^*_b = 850(1.10)(1.15) = 1{,}075 \text{ psi}$$
$$F_{bE} = (0.438)(1.3 \times 10^6)/(11.5)^2 = 4{,}305 \text{ psi}$$
$$F_{bE}/F^*_b = 4{,}305/1{,}075 = 4.004$$

$$C_L = \frac{5.004}{1.9} - \sqrt{\left(\frac{5.004}{1.9}\right)^2 - \frac{4.004}{0.95}} = 0.98$$

Thus, $F'_b \approx F^*_b$ and the 2×8 is indeed satisfactory.

Step 7: Check the live load deflection requirement of a uniformly loaded simply supported beam.

$$\Delta = \Delta_L = 5w_L L^4/384EI$$
$$\Delta = [5(20)(16)^4/384(1.3\times10^6)(47.635)] \times 1{,}728$$
$$\Delta = 0.47 \text{ in.}$$

Since $\Delta_{all} = L/240$ (Fig. 8.29) $= 16$ ft $\times$ (12 in./ft)/240 $= 0.80$ in., the deflection requirement is met and a 2×8 is adequate.

Second-Floor Joist Design

Select a No. 2 Hem-Fir member surfaced dry and used at less than 19% moisture content to carry the load created by the second floor. Assume that the compression flanges of the joists are adequately braced because of the sheathing on the second floor.

Solution

Step 1: Define the parameters.

$$C_D = 1.0 \text{ (Fig. 8.18)}$$
$$C_r = 1.15$$
$$F_b = 850 \text{ psi (Table 8.4)}$$
$$E = 1.3 \times 10^6 \text{ psi (Table 8.4)}$$
$$F_v = 75 \text{ psi (Table 8.4)}$$
$$F_{cp} = 405 \text{ psi (Table 8.4)}$$

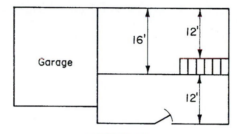

FIRST FLOOR PLAN

Step 2: Determine the maximum shear and moment.

Assume that the joists are spaced 16 in. o.c. As noted from the first-floor plan, the controlling span is 16 in.

$$w = 41 \text{ lb/ft}^2(16 \text{ in.})(12 \text{ in.}) = 54.7 \text{ lb/ft}$$
$$V_{max} = wL/2 = 54.7(16)/2 = 438 \text{ lb}$$
$$M_{max} = wL^2/8 = 54.7(16)^2(12)/8 = 21{,}005 \text{ lb-in.}$$

Step 3: Select a joist size that is adequate for flexural stresses ($C_L = 1.0$).

Approximate the size using $F'_b = 850(1.15) = 978$ psi; thus, $S_{est} = 21{,}005/978 = 21.47$ in^3. So try a 2×10 ($S_{2\times10} = 21.39$ in^3). Now check the 2×10 using the properties of the 2×10.

$$F'_b = 850(1.10)(1.15) = 1{,}075 \text{ psi}$$

$S_{req'd} > M/F'_b = 21{,}005 \text{ lb-in.}/1{,}075 \text{ psi} = 19.5 \text{ in}^3 < S_{2\times10}$. Thus, the 2×10 is indeed adequate. The dead load of the joist is 2.891 lb/ft (Table 8.8). The required section modulus to carry the additional dead load is $S_{req'd} = 43.89/41 (19.5) = 20.88 < S_{2\times10}$. So the 2×10 is still satisfactory.

Step 4: Check the horizontal shear stress.

$$F'_v = 75 \text{ psi}$$
$$f_v = 3V/2A = 3(438 \text{ lb})/2(13.875 \text{ in}^2) = 47 \text{ psi}$$

Since $f_v < F'_v$, the 2×10 is adequate for shear.

Step 5: Check the bearing requirements.

$$F'_{c\perp} = 405 \text{ psi}$$
$$f_{c\perp} = R/bz = 438 \text{ lb}/(1.5 \text{ in.})(z)$$

Since $F'_{c\perp} \ge f_{c\perp}$, it follows that $z \ge 438/(405 \times 1.5) = 0.72$ in.

Step 6: Check the bracing requirements.

It can be assumed that the compression edge of the joist will be supported by the plywood floor sheathing, which will provide the lateral stability that is required. Therefore, assume that the plywood sheathing is nailed every 12 in. to the compression edge of the joists.

$$\ell_e = 1.63\ell_u + 3d = 1.63(12) + 3(9.25) = 47.31 \text{ in.}$$
$$R_B = (\ell_e d/b^2)^{1/2} = [(47.31)(9.25)/(1.5)^2]^{1/2} = 13.9$$
$$F_b^* = 1,075 \text{ psi}$$
$$F_{bE} = 0.438(1.3 \times 10^6)/(13.9)^2 = 2,947 \text{ psi}$$
$$F_{bE}/F_b^* = 2,947/1,075 = 2.74$$

$$C_L = \frac{3.74}{1.9} - \sqrt{\left(\frac{3.74}{1.9}\right)^2 - \frac{2.74}{0.95}} = 0.97$$

Thus, $F_b' \approx F_b^*$, and the 2×10 is indeed satisfactory.

Step 7: Check the live load deflection requirements of a uniformly loaded simply supported beam.

$$\Delta = \Delta_L = 5w_L L^4/384 \, EI$$
$$\Delta = [5(10)(16)^4/384(1.3 \times 10^6)(98.932)] \times 1,728$$
$$\Delta = 0.46 \text{ in.}$$

Since $\Delta_{all} = L/360$ (Fig. 9.29) = 16 ft × (12 in./ft)/360 = 0.53 in., the deflection requirement is met and a 2×10 is adequate. Note that the total deflection would be 0.46 + 11/30(0.46) × 2 = 1.00 in.

First-Floor Joist Design

Select a No. 2 Hem-Fir member surfaced dry and used at less than 19% moisture content to carry the first-floor load. Assume that the compression flanges of the joists are adequately braced because of the sheathing on the first floor.

Solution—16-ft Span Joists

Step 1: Define the parameters.

$$C_D = 1.0 \text{ (Fig. 8.18)}$$
$$C_r = 1.15$$
$$F_b = 850 \text{ psi (Table 8.4)}$$
$$E = 1.3 \times 10^6 \text{ psi (Table 8.4)}$$
$$F_v = 75 \text{ psi (Table 8.4)}$$
$$F_{c\perp} = 405 \text{ psi (Table 8.4)}$$

Step 2: Determine the maximum shear and moment.

Assume that the joists are spaced 16 in. o.c. As noted from the basement plan, the controlling span is 18 ft. However, much of the floor only spans 16 ft. Thus, the size for the 16-ft span will be determined. The 18-ft span will be considered separately.

$$W = 51 \text{ lb/ft}^2(16 \text{ in.})(12 \text{ in.}) = 68 \text{ lb/ft}$$
$$V_{max} = wL/2 = 68(16)/2 = 544 \text{ lb}$$
$$M_{max} = wL^2/8 = 68(16)^2(12)/8 = 26,112 \text{ lb-in.}$$

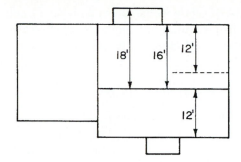

Step 3: Select a joist size that is adequate for flexural stresses.

Approximate the joist size using $F'_b = 850(1.15) = 978$ psi; thus, $S_{est} = 26{,}112/978 = 26.70$ in³. So try a 2×12 with $S_{2×12} = 31.641$ in³. Now check the 2×12 using the properties of the 2×12. Since $C_D = 1.0$ for a 2×12, $F'_b = 978$ psi and the 2×12 is acceptable. The dead load of the 2×12 is 3.516 lb/ft (Table 8.8). Obviously, the 2×12 can carry the additional load.

Step 4: Check the horizontal shear stress.

$$F'_v = 75 \text{ psi}$$
$$f_v = 3V/2A = 3(544 \text{ lb})/2(16.875 \text{ in}^2) = 48 \text{ psi}$$

Since $f_v < F'_v$, the 2×12 is adequate in shear.

Step 5: Check the bearing requirements.

$$F'_{c\perp} = 405 \text{ psi}$$
$$f_{c\perp} = R/bz = 544 \text{ lb}/(1.5 \text{ in.})(z)$$

Since $F'_{c\perp} \geq f_{c\perp}$, it follows that $z \geq 544/(405 \times 1.5) = 0.90$ in.

Step 6: Check the bracing requirements.

Assume that the plywood is nailed every 12 in. to the compression edge of the joists.

$$\ell_e = 1.63\ell_u + 3d = 1.63(12) + 3(11.25) = 53.31 \text{ in.}$$
$$R_B = (\ell_e d/b^2)^{1/2} = [(53.31)(11.25)/(1.5)^2]^{1/2} = 16.3$$
$$F^*_b = 978 \text{ psi}$$
$$F_{bE} = 0.438(1.3×10^6)/(16.3)^2 = 2{,}143 \text{ psi}$$
$$F_{bE}/F^*_b = 2{,}143/978 = 2.19$$

$$C_L = \frac{3.19}{1.9} - \sqrt{\left(\frac{3.19}{1.9}\right)^2 - \frac{2.19}{0.95}} = 0.96$$

Since $F'_b \approx F^*_b$, the 2×12 is indeed satisfactory.

Step 7: Check the live load deflection requirements of a uniformly loaded simply supported beam.

$$\Delta = \Delta_L = 5w_L L^4/384 \text{ EI}$$
$$\Delta = [5(53.3)(16)^4/384(1.3×10^6)(177.979)] \times 1{,}728$$
$$\Delta = 0.34 \text{ in.}$$

Since $\Delta_{all} = L/360$ (Fig. 9.29) = 16 ft × (12 in./ft)/360 = 0.53 in., the deflection requirement is met and a 2×12 is adequate. Again, the total deflection is 0.34 + 11/40(0.34) × 2 = 0.53 in.

The above design indicates that the 2×12 No. 2 Hem-Fir joists must be used for the 16′0″ span on the first floor. It should be recalled from the design calculations in Chapter 9 of this book that the actual Case Study House design (see Sheet A-3, Appendix A) used 2×10 No. 2 Hem-Fir joists. This was possible because the actual design was based upon the

1989 CABO code which allowed higher F_b and E values for No. 2 Hem-Fir. The design calculations presented in Chapter 9, which were based upon the 1995 CABO code, required the use of No. 1 Hem-Fir for the 16'0" span.

Solution—18-ft Span Joists[T]

Step 1: Define the parameters.

Same as for the 16-ft span solution.

Step 2: Determine the maximum shear and moment.

Assume that the joists are spaced 16 in. o.c.

$$w = 51 \text{ lb/ft}^2 (16 \text{ in.})(12 \text{ in.}) = 68 \text{ lb/ft}$$
$$V_{\max} = wL/2 = 68(18)/2 = 612 \text{ lb}$$
$$M_{\max} = wL^2/8 = 68(18)^2(12)/8 = 33{,}048 \text{ lb-in.}$$

Step 3: Select a joist size that is adequate for flexural stresses ($C_L = 1.0$, $C_D = 1.0$).

Approximate the joist size using $F_b' = F_b^* = 850(1.15) = 978$ psi; then $S_{\text{est}} = 33{,}048/978 = 33.79 \text{ in}^3$. A 2×14 with $S_{2\times14} = 43.89 \text{ in}^3$ is satisfactory, but not practical. One solution is to reduce spacing to the next lower standard size, 12 in. Then $S_{\text{est}} = 12/16(33.79) = 25.34 \text{ in}^3 < S_{2\times12}$. The dead load of 3.516 lb/ft (Table 8.8) can easily be carried by the 2×12.

Step 4: Check the horizontal shear stress.

$$F_v' = 75 \text{ psi}$$
$$f_v = 3V/2A = 3(459 \text{ lb})/2(16.875 \text{ in}^2) = 41 \text{ psi}$$

Since $f_v < F_v'$, the 2×12 is adequate in shear.

Step 5: Check the bearing requirements.

$$F_{c\perp}' = 405 \text{ psi}$$
$$f_{c\perp} = R/bz = 459 \text{ lb}/(1.5 \text{ in.})(z)$$

Since $F_{c\perp}' \geq f_{c\perp}$, it follows that $z \geq 459/(405 \times 1.5) = 0.75$ in.

Step 6: Check the bracing requirements.

Assume that the plywood is nailed every 12 in. to the compression edge of the joists.

$$\ell_e = 1.63\ell_u + 3d = 1.63(12) + 3(11.25) = 53.31 \text{ in.}$$
$$R_B = (\ell_e d/b^2)^{1/2} = [(53.31)(11.25)/(1.5)^2]^{1/2} = 16.3$$
$$C_L = 0.6 \text{ (Step 6, 16 ft span solution)}$$

Thus, it is clear that bracing every 12 in. o.c. as provided by plywood nailing is satisfactory.

Step 7: Check the live load deflection requirements of a uniformly loaded simply supported beam.

$$\Delta = \Delta_L = 5w_L L^4/384 \, EI$$
$$\Delta = [5(40.0)(18)^4/384(1.3\text{x}10^6)(177.979)] \times 1728$$
$$\Delta = 0.41 \text{ in.}$$

Since $\Delta_{\text{all}} = L/360$ (Fig. 8.29) $= 16 \text{ ft} \times (12 \text{ in./ft})/360 = 0.53$ in., the deflection requirement is met and a 2×12 at 12 in. o.c. is adequate.

[T]These joists are designed as simply supported beams. The actual design (see Sheet A-3, Appendix A) indicates that they are continuous beams which span over (2)-1 3/4" × 9 1/4" × 10'0" LVL beams.

Bearing Wall Stud Design

Select a Stud Grade Hem-Fir member surfaced dry and used at less than 19% moisture content to carry the wall loads indicated in Fig. 11-12.

Solution: Exterior Bearing Wall between the Attic and the Second Floor

Step 1: Define the parameters.

$$L_e = 1.0\ L\ \text{(Table 11.2)}$$
$$C_D = 1.0\ \text{(Fig. 8.18)}$$
$$C_F = 1.05\ \text{(Compression} - 2\times4)$$
$$F_c = 800\ \text{psi (Table 8.4)}$$
$$E = 1.2\times10^6\ \text{psi (Table 8.4)}$$

Assume a 24-in. o.c. stud spacing.

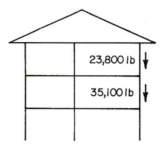

Maximum loads on bearing walls

Step 2: Assume that the stud is a short column ($L_{ey} = 0$) about the weak axis (because of the effect of the wall sheathing).

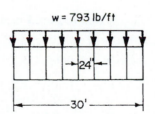

$$P = \text{the load on one stud}$$
$$P = 23{,}800\ \text{lb/(30 ft)(2)} = 793\ \text{lb/ft (2)} = 1{,}586\ \text{lb}$$

Assume a 2×4 and

$$F'_c = F_c(C_D)(C_F) = 800(1.00)(1.05) = 840\ \text{psi}$$
$$\text{Area required} = A = P/F'_c = 1{,}586/840 = 1.9\ \text{in}^2$$

Therefore, 2×4 (5.25 in^2) studs spaced at 24 in. o.c. are adequate to support the load (2×4 is the smallest standard size framing stud; also, this solution presumes $C_p = 1.0$).

Step 3: Check the 2×4 column for buckling.

    Assume that the sheathing is nailed every 12 in. to the studs; thus, $C_L \approx 1.0$ with respect to weak axis of buckling. About the strong axis,

$$(L_e/d)_x = 1.0\ L/d = 1.0(9 \times 12)/3.5 = 30.9$$

C_p calculation:

$$k_{cE} = 0.30; c = 0.8$$
$$F_{cE} = 0.3(1.2 \times 10^6)/(30.9)^2 = 377 \text{ psi}$$
$$F^*_c = 840 \text{ psi}$$
$$F_{cE}/F^*_c = 377/840 = 0.449$$

$$C_p = \frac{1.449}{1.6} - \sqrt{\left(\frac{1.449}{1.6}\right)^2 - \frac{0.449}{0.8}} = 0.397$$

$$F'_c = 840(0.397) = 333 \text{ psi}$$

$$\text{Area required} = P/F'_c = 1{,}586/333 = 4.76 < 5.25 \text{ in}^2$$

So 2×4 studs spaced 24 in. o.c. are still adequate for the second-floor exterior wall framing.

Solution: Exterior Bearing Wall between the Second Floor and the First Floor

Step 1: Define the parameters.

$$L_e = 1.0 \, L \text{ (Table 11.2)}$$
$$C_D = 1.0 \text{ (Fig. 8.18)}$$
$$F_c = 800 \text{ psi (Table 8.4)}$$
$$E = 1.2 \times 10^6 \text{ psi (Table 8.4)}$$

Assume a 16-in. o.c. stud spacing.

Step 2: Assume that the stud is a short column about the weak axis (because of the effect of the wall sheathing).

$$P = \text{the load on one stud}$$
$$P = [35{,}100 \text{ lb}/(30 \text{ ft})](\text{stud spacing}/12)$$
$$= 1{,}170 \text{ lb/ft } (16 \text{ in.})/(12 \text{ in.}) = 1{,}560 \text{ lb}$$
$$F'_c = 800(1.05) = 840 \text{ psi}$$

$$\text{Area required} = A = P/F'_c = 1{,}560/840 = 1.86 \text{ in}^2$$

Therefore, 2×4 (5.25 in^2) studs spaced 16 in. o.c. are adequate to support the load; however, the solution presumes $C_p = 1.0$. About the strong axis,

$$(L/d)_x = 1.0(9 \times 12)/3.5 = 30.9$$
$$C_L = 0.397 \text{ (Step 3, solution for exterior bearing}$$
$$\text{wall between attic and second floor)}$$
$$A_{\text{req'd}} = P/F'_c = 1{,}560/333 = 4.68 \text{ in}^2 < 5.25 \text{ in}^2$$

Thus, 2×4 studs spaced 16 in. o.c. are adequate for the first-floor exterior walls.

NOTE: All interior stud walls have 2×4 studs spaced 24 in. o.c. on the second floor and 2×4 studs spaced 16 in. o.c. on the first floor.

Header Design

The design of the headers over the openings in the bearing walls is not illustrated in this chapter. The structural analysis assumptions regarding the "sharing" of loads between the insulated sheathing and the headers and the subsequent analysis are beyond the scope of this chapter.

Basement Girder Design

Select a No. 2 Hem-Fir member surfaced dry and used at less than 19% moisture content to carry the load on the girder. Assume that the compression flange of the girder is adequately braced by the first-floor joists. Use built-up 2× members.

Solution

Step 1: Define the parameters.

$$C_D = 1.0 \text{ (Fig. 8.18)}$$
$$C_r = 1.0$$
$$F_b = 850 \text{ psi (Table 8.4)}$$
$$E = 1.3 \times 10^6 \text{ psi (Table 8.4)}$$
$$F_v = 75 \text{ psi (Table 8.4)}$$
$$F_{c\perp} = 405 \text{ psi (Table 8.4)}$$

Step 2: Determine the maximum shear and moment.

Assume that the girder is continuous over the three interior supports which are spaced 7.5 ft o.c.

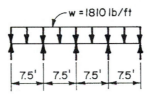

$$w = 54,200/30 = 1,810 \text{ lb/ft}$$
$$V_{\max} = 0.607wL = 0.607(1810)(7.5) = 8,240 \text{ lb (Fig. E.9 in Appendix E)}$$
$$M_{\max} = 0.1071wL^2 = 0.1071(1810)(7.5)^2$$
$$= 130,849 \text{ lb-in. (Fig. E.9 in Appendix E)}$$

Step 3: Select a girder that is adequate for flexural stresses.

$$F_b' = F_b^* = F_b(C_r) = 850(1.15) = 978 \text{ psi}$$

$S_{\text{req'd}} > M/F_b' = 130,849 \text{ lb-in.}/978 \text{ psi} = 134 \text{ in}^3$. Entering Table 8.8, it appears that five 2×12s are the smallest adequate members with $S > 134 \text{ in}^3$. The section properties of the five 2×12s are:

$$A = 84.38 \text{ in}^2$$
$$S = 158.20 \text{ in}^3$$
$$I = 889.89 \text{ in}^4$$
$$\text{Dead load} = 17.58 \text{ lb/ft}$$

Since the beam dead load is only 1% of the live load, it will be neglected in the design.

Step 4: Check the horizontal shear stress.

$$F_v' = 75 \text{ psi}$$
$$f_v = 3V/2A = 3(8240 \text{ lb})/[2(84.38)] = 146 \text{ psi}$$

Since $f_v > F_v'$, the shear strength of five 2×12s is inadequate.
Determine the required area to satisfy the shear requirements.

$$A = 3V/2F_v = 3(8,240)/2(75) = 165 \text{ in}^2$$

Entering Table 8.8, it appears that a built-up girder with ten 2×12s is required, with $A > 165$ in². The section properties of the ten 2×12s are as follows:

$$A = 168.75 \text{ in}^2$$
$$S = 316.41 \text{ in}^3$$
$$I = 1{,}779.79 \text{ in}^4$$
$$\text{Dead load} = 35.16 \text{ lb/ft}$$

Since the beam dead load is only 2% of the live load, it will be neglected in the design.

Step 5: Check the bearing requirements.

$$R \approx 13{,}500 \text{ lb (Fig. 11.13)}$$
$$F'_{c\perp} = 405 \text{ psi}$$
$$f_{c\perp} = R/bz = 13{,}500 \text{ lb/(15 in.)}(z)$$

Since $F'_{c\perp} \geq f_{c\perp}$, it follows that $z \geq 13{,}500/(405 \times 15) = 2.22$ in. This is not an unreasonable bearing length and is less than the minimum value in the 1995 CABO Code.[18] The overall dimension of the ten 2×12s is 15 in. × 11.25 in. Since beam depth is less than the width, lateral buckling is not a factor and $C_L = 1.0$, provided the 2×12s are adequately fastened together to act as a unit.

Step 6: Check the live load deflection requirements of a uniformly loaded simply supported beam.

$$\Delta = \Delta_L = w_L L^4/154EI \text{ (Fig. E.9 in Appendix E)}$$
$$\Delta = [5(1810)(7.5)^4/384(1.3 \times 10^6)(1{,}779.79)] \times 1{,}728$$
$$\Delta = 0.05 \text{ in.}$$

Since $\Delta_{all} = L/240$ (Table 9.8) = 7.5 ft × (12 in./ft)/240 = 0.375 in., the deflection requirement is met and ten 2×12s are adequate.

It is impractical to use ten 2×12 members as the design option for the girder in this house. As noted from the actual basement plan for the Case Study House, an alternative to this design consists of two 1-3/4 in. × 9-1/4 in. laminated veneer (LVL) members. LVL is an engineered product that is available in 1-3/4 in. thicknesses and seven depths from 5-1/2 in. to 18 in. and in lengths up to 60 ft (see Fig. 11.14). Design information would have to be collected from the manufacturer in order to verify the design.

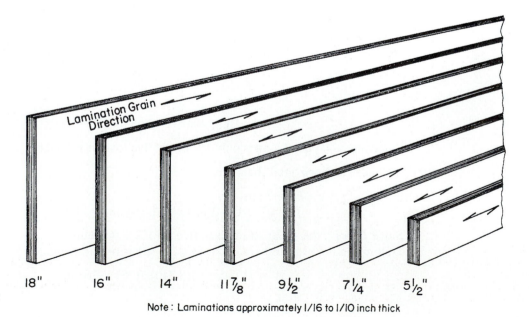

Figure 11.14
Laminated
Veneer Lumber
(LVL)

18"   16"   14"   11⅞"   9½"   7¼"   5½"

Note: Laminations approximately 1/16 to 1/10 inch thick

The LVL member for the Case Study House can be selected using data from Table 11.4.

Given: 1,810 lb/ft load on the girder
Span = 7.5 ft
Allowable deflection = $L/240$

Solution: Using Table 11.4, one 3-1/2 in. × 9-1/2 in. member can carry a total load of 1,862 lbs/ft over an 8-ft simple span. Thus, one 3-1/2 in. × 9-1/2 in. 2.0 ES Parallam® PS2 is adequate. Alternatively, two properly connected 1-3/4 in. × 11-7/8 in. 2.0 ES Microllam™ LVL members are also adequate (Table 11.4).

The PSL and LVL alternatives obviously are more practical than the solid-sawn option. The builder also is assured of minimum performance of the material by the manufacturer.

TABLE 11.4 Allowable Uniform Applied Load for Microllam™ LVL and Parallam® PSL Headers and Beams–Floor Applications (plf)

(a) 1¾-in. 2.0E ES Microllam™ LVL

1¾″×16″ and 1¾″×18″ beams are to be used in multiple-member units only

Span (ft)	One- 1¾″×5½″ Live Load	Total Load	One- 1¾″×7¼″ Live Load	Total Load	One- 1¾″×9½″ Live Load	Total Load	One- 1¾″×11⅞″ Live Load	Total Load	One- 1¾″×14″ Live Load	Total Load	One- 1¾″×16″ Live Load	Total Load	One- 1¾″×18″ Live Load	Total Load
6	305	455	660	763		1,063		1,424		1,795		2,193		2,651
8	134	198	296	440	629	746		979		1,207		1,443		1,701
10	70	102	156	230	338	502	629	745		909		1,074		1,251
12	41	58	92	134	201	297	379	552	599	728		855		988
14	26	36	58	84	129	188	245	361	390	550	566	706	781	816
16	17	23	39	55	87	126	167	244	268	394	390	539	542	672
18			28	38	62	88	119	172	191	279	280	412	390	529
20			20	27	45	63	87	125	141	204	207	303	290	426
22			15	19	34	46	66	93	107	153	157	228	221	322
24					26	35	51	71	83	117	122	175	172	249
26					21	26	40	54	65	91	97	137	136	196
28					17	20	32	43	53	72	78	109	110	156
30							26	34	43	57	64	87	90	126

Table (a) is for one 1¾″ beam. When properly connected together, double the values for two 1¾″ beams, triple for three. See product literature for connection details.

GENERAL NOTES:
1. Values shown are the maximum uniform loads, in pounds per lineal foot (plf), that can be applied to the beam in addition to its own weight.
2. Tables are based on uniform loads and the most restrictive of a simple or continuous span. The shaded areas represent load conditions controlled by a continuous span condition.

(continued)

TABLE 11.4   (continued)

1¾"×16" and 1¾"×18" beams are to be used in multiple-member units only

(b) 3½-in. 2.0E ES Parallam® PSL (for 1¾", 2¹¹⁄₁₆", 5¼", and 7" beams, see multiplier below)

Span (ft)	One- 3½"×5½"		One- 3½"×7¼"		One- 3½"×9½"		One- 3½"×11⅞"		One- 3½"×14"		One- 3½"×16"		One- 3½"×18"	
	Live Load	Total Load	Live Load	Total Load	Live Load	Total Load	Live Load	Total Load	Live Load	Total Load	Live Load	Total Load	Live Load	Total Load
6		2,091		2,163		2,694		2,898		3,652		4,463		5,394
8	1,169	1,470	1,258	1,517		1,862		1,991		2,456		2,935		3,460
10	627	930	676	1,003	1,084	1,421	1,258	1,515		1,848		2,185		2,545
12	372	548	402	592	651	964	758	1,093	1,198	1,480	1,722	1,738		2,010
14	238	347	257	376	420	617	490	722	781	1,093	1,132	1,409	1,561	1,660
16	161	232	174	251	285	416	334	488	535	788	781	1,075	1,084	1,345
18	114	161	123	175	203	292	237	343	382	558	560	822	781	1,058
20	84	115	90	125	149	211	174	249	282	407	414	604	580	850
22	63	84	68	92	112	156	132	185	213	305	315	455	442	643
24	49	63	53	69	87	118	102	140	166	233	244	349	344	496
26	38	47	42	52	69	91	81	108	131	181	194	273	273	390
28	31	36	33	40	55	70	65	84	105	143	156	216	220	310
30	25	27	27	30	45	55	53	66	86	113	127	173	180	250

Table (b) can be used for 1¾", 2¹¹⁄₁₆", 5¼", and 7" width beams. Use the following multipliers to calculate the allowable load for each width: 1¾" beam use values in table × 0.50; 2¹¹⁄₁₆" beam use values in table × 0.77; 3½" beam use values in table; 5¼" beam use values in table × 1.50; 7" beam, use values in table × 2.00.

Table is for one 3½" beam. When properly connected together, double the values for two 3½" beams. See product literature for connection details.

GENERAL NOTES—continued
3. Microllam™ LVL and Parallam® PSL beams are made without camber; therefore, in addition to complying with the deflection limits of the applicable Building Code, other deflection considerations, such as long-term deflection under sustained loads (including creep) and aesthetics, must be evaluated.
4. Lateral support of beam compression edge is required at intervals of 24" o.c. or closer.
5. Lateral support of beams is required at bearing points.
6. Bearing area to be calculated for specific application.

FLOOR BEAM SIZING:
• To size a beam for use in a floor, it is necessary to check both live load and total load. Make sure the selected beam will work in both columns. When no live load is shown, total load will control.
• Total load column limits deflection to L/240. Live load column is based on deflection of L/360. Check local code for other deflection criteria.
• For deflection limits of L/240 and L/480 multiply loads shown in live load column by 1.5 and 0.75, respectively. The resulting live load shall not exceed the total load shown.

Source: *Specifiers Guide: Frameworks Building System*, p. 22. (Truss Joist MacMillan™, 1995).

Columns Supporting the Girder

Select a select structural grade Hem-Fir member used at less than 19% moisture content to carry the load determined above (Fig. 11.13) (assume a post and timber).

Step 1: Define the parameters.

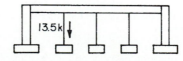

$L_e = 1.0\ L$ (Table 11.2)
$C_D = 1.0$ (Fig. 8.18)
$F_c = 975$ psi (Table 8.4)
$E = 1.3 \times 10^6$ psi (Table 8.4)

Step 2: Assume that the basement column is a pure compression member.
For a first estimate, let $C_p = 1.0$.

$$P = 54{,}200/4 = 13{,}500 \text{ lb}$$
$$F_c' = 975 \text{ psi}$$
$$\text{Area required} = A = P/F_c' = 13{,}500/975 = 13.8 \text{ in}^2$$

Therefore, a 4×6 (19.25 in^2) or a 6×6 (30.25 in^2) wood column is the smallest standard size selection (Table 8.8). Select a square 6×6 section.
Step 3: Check the 6×6 column for buckling.

$$L_e/d = 1.0 \ L/d = 1.0(8 \text{ ft})(12 \text{ in./ft})/5.5 = 17.5$$
$$k_{cE} = 0.3$$
$$c = 0.8$$
$$F_{cE} = 0.3(1.3 \times 10^6)/(17.5)^2 = 1{,}273 \text{ psi}$$
$$F_c^* = 975 \text{ psi}$$
$$F_{cE}/F_c^* = 1{,}273/975 = 1.306$$

$$C_p = \frac{2.306}{1.6} - \sqrt{\left(\frac{2.306}{1.6}\right)^2 - \frac{1.306}{0.8}} = 0.774$$

$$F'c = 975(0.774) = 755 \text{ psi} < f_c = 13{,}500/30.25$$
$$= 446 \text{ psi}$$

Therefore, 6×6 Select Structural columns are satisfactory. Very likely, No. 1 Hem-Fir 6×6 posts are also adequate, and the investigation of this possibility is encouraged.

First-Floor Stairwell Header Design

The joist at the end of the 3.5-ft-wide stairwell (Member $\overline{AB}$ on the Foundation Plan in Appendix A) is only one of several framing members with other than uniformly distributed loads. Member $\overline{AB}$ carries the uniformly distributed first-floor load of 68 lb/ft. It also carries the concentrated load, P, at C. The magnitude of the concentrated load equals the first-floor load of 51 lb/ft^2 times the tributary area for the stairwell header $\overline{CD}$ between points C and E (note that the stairwell header is supported by a column at point E, which is located 6 ft from joist $\overline{AB}$). Thus, the concentrated load at C equals (51 lb/ft)(12.5/2 ft)(6/2 ft) = 956 lbs. The load, shear, and moment diagrams for joist $\overline{AB}$ are sketched below.

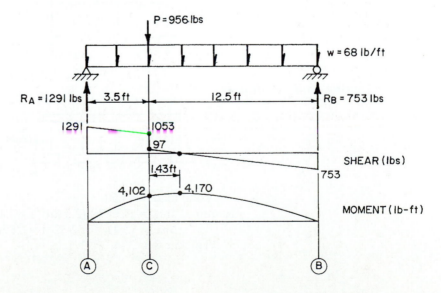

Assuming that the same size (2×12) and grade lumber (No. 2 Hem–Fir) is used for $\overline{AB}$ as for the other first floor joists in this structural analysis, $F_b' = 978$ psi and

$$S_{req'd} = \frac{M_{max}}{F'_b} = \frac{4{,}170(12)}{978}$$

$$= 51.2 \text{ in}^2$$

The section modulus of two 2×12s equals 2(31/641) or 63.28 in³. Thus, two 2×12s are adequate for flexure. The section modulus of two 2×10s is 2(21.39) = 42.78 in³, which is inadequate. Although not very practical, member $\overline{AB}$ would be adequate using one 2×10 and one 2×12 ($S = 53.03$ in³). Of course, shear, bearing and deflection limits of joist $\overline{AB}$ would have to be checked before the final design is selected. These checks are left as an exercise for the reader.

Wall and Column Footing Design

The column footer area is dependent upon the column load and the bearing capacity of the soil. Assume a clayey sand, silty gravel, with an allowable bearing strength of 2,500 lb/ft².

Wall Footings—Bearing Walls

Maximum footer load = 58.3 k
Load/ft = 58.3 k/30 ft = 1.94 k/ft
Required width = (1.94 k/ft)/(2.5 k/ft²) = 0.78 ft = 9.3 in.
Minimum width = 10-in. wall + 2-in. overhang on each side
= 14 in. wide
Minimum thickness = 6 in.

Note that in the actual Case Study house (Sheet A-5), an 8 in. × 18 in. continuous concrete footing was used. The same size footers should be used for the non-bearing end walls for ease of construction.

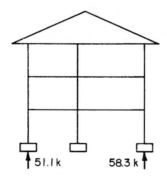

51.1 k   58.3 k

Footing Under Columns

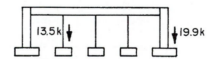

13.5k   19.9k

Total load = 13.5 k
Required area = (13.5 k)/(2.5 k/ft²) = 5.4 ft²
Size = (5.4)^{1/2} = 2.32 ft = 28 in. × 28 in.
Thickness = 8 in.

Note that $2' \times 2' \times 1'$ footings (Sheet A-5) were used under the columns in the actual Case Study House. The soil-bearing capacity assumption that was used in the actual Case Study House design would have to be determined in order to evaluate the adequacy of the actual $2' \times 2' \times 1'$ size footing.

Footing Under Wall (at Girder and Center Line of House)

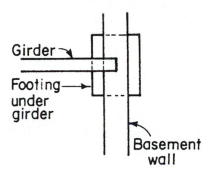

Total load = 19.9 k
Required area = $(19.9 \text{ k})(2.5 \text{ k/ft}^2)$ = 7.96 ft^2
Size = $7.96^{1/2}$ = 2.82 ft = 34 in. $\times$ 34 in. if one assumes the girder
reaction is only transmitted vertically
Thickness = 8 in.

It appears that in the actual Case Study House design the same size footing is used at this location as was used under the wall footings–bearing walls (i.e., 8 in. $\times$ 18 in.). If one assumes the girder load distributes itself at a 45° angle in the basement wall, the load-bearing area becomes $2(8 \times 12)(18)/44$ = 24 ft^2. This is much greater than the required 7.96 ft^2. The required bearing area is exceeded even if the load is distributed at an angle of 20° from the girder reaction point.

Design Summary

A complete design summary of the Case Study House in presented in Table 11.5. The results of the CABO Code analysis in Chapters 9 and 10, the results of the structural analysis in this chapter, and the choices made in the actual house design are all included in the table.

CHAPTER SUMMARY

This chapter presented the structural engineering procedures which can be used to analyze and design the timber tension, compression, and flexural members, as well as the members which are subject to combined loadings. In addition, a detailed structural analysis was presented for sizing the main framing members for the Case Study House.

TABLE 11.5   Design Summary of the Case Study House[1]

Member	CABO Code Design[2]	Structural Analysis	Actual Design[3]
Roof rafters	2×6s at 12 in. o.c.	2×6s at 12 in. o.c.	Roof truss
Attic floor joists	2×8s at 12 in. o.c.	2×8s at 12 in. o.c.	Roof truss
Second-floor joists	2×10s at 16 in. o.c.	2×10s at 16 in. o.c.	2×10s at 16 in. o.c.
First-floor joists			
—16′ span	2×10s at 16 in. o.c.[4]	2×12s at 16 in. o.c.	2×10s at 16 in. o.c.
—18′ span	Not designed	2×12s at 12 in. o.c.	two 2×10s at 16 in. o.c.
Exterior bearing wall between the attic and the second floor	2×4s at 24 in. o.c.	2×4s at 24 in. o.c.	2×4s at 16 in. o.c.
Exterior bearing wall between the second floor and the first floor	2×4s at 16 in. o.c.	2×4s at 16 in. o.c.	2×4s at 16 in. o.c.
Header	two 2×10s	Not designed	two 2×10s
Basement girder	Not designed	ten 2×12s or two 1¾×9½ LVL members	two 1¾×9¼ LVL members
Columns supporting the girder	Not designed	6×6 in. wood	3 in. diameter steel pipe
Wall footings— bearing walls	14 in. × 6 in.	14 in. × 6 in.	16 in. × 8 in.
Wall footings— nonbearing walls	14 in. × 6 in.	14 in. ×6 in.	16 in. × 8 in.
Footings under columns	Not designed	28 in. × 28 in. × 8 in.	24 in. × 24 in. × 12 in.
Footings under wall (at girder and center line of house)	Not designed	34 in. × 34 in. × 8 in.	18 in. × 8 in.

[1]No. 2 Hem-Fir used for wood framing members unless otherwise noted.
[2]The CABO Code Design presented in Chapters 9 and 10 of this book are based upon the 1995 CABO Code.
[3]The actual design of the Case Study House was based upon the 1989 CABO Code, which allowed higher values for Fb and E for No. 2 Hem-Fir than the 1995 CABO Code.
[4]The first floor joists that span 16′0″ require the use of No. 1 Hem-Fir.

It is important to note that this chapter is included to illustrate the types of structural engineering calculations which a licensed engineer or architect would perform in situations where:

1. The actual house design included features that were not covered by the CABO Code structural analysis procedures and therefore could not be performed directly by the homebuilder.

2. The homebuilder wishes to optimize the structural framing phase of the house design by applying acceptable structural engineering principles rather than by using the assumptions built into the CABO Code.

NOTES

[1]*National Design Specification for Wood Construction* (Washington, D.C.: American Forest and Paper Products Association [AF&PA], 1991).

[2]*CABO One and Two Family Dwelling Code, 1995 Edition*, pp. 25–130 (Falls Church, VA: The Council of American Building Officials, 1995).

[3]*National Design Specification for Wood Construction*, p. 6.

[4]Ibid., p. 6.

[5]Ibid.

[6]Ibid., p. 14.

[7]Ibid.

[8]Ibid., p. 102.

[9]Ibid., p. 10.

[10]*1995 CABO Code*, p. 37.

[11]Ibid., p. 35.

[12]Ibid., pp. 89–90.

[13]NDS®, p. 10.

[14]Ibid., p. 15.

[15]Ibid., p. 16.

[16]Ibid., p. 106.

[17]*1995 CABO Code*, pp. 35, 89–90.

[18]Ibid.

BIBLIOGRAPHY

CABO One and Two Family Dwelling Code, 1995 Edition.   Falls Church, VA: The Council of American Building Officials, 1995.

National Design Specification for Wood Construction.   Washington, D.C.: American Forest and Paper Products Association, 1991.

Specifiers Guide: Frameworks Building System.   Boise, ID: Truss Joist MacMillan, 1995.

Supplement to the National Design Specification for Wood Construction.   Washington, D.C.: American Forest and Paper Products Association, 1991.

12

HVAC System Design

INTRODUCTION

This chapter provides an introduction to the design of the energy-related heating, ventilating, and air conditioning (HVAC) systems used in houses. **Heating systems** typically used include: (1) steam distribution/radiator, (2) circulating hot water (hydronic), (3) heat pump/ductwork, (4) gas- or oil-fired furnace/ductwork, (5) electric resistance baseboard or radiant panel, (6) electric warm air furnace, and (7) passive or active solar. **Cooling systems** can consist of: (1) ceiling fans, (2) whole house fans, (3) individual room air conditioners, (4) heat pump/ductwork, and (5) central air conditioner/ductwork types.

Chapters 11–28 of the 1995 edition of the *CABO One and Two Family Dwelling Code*[1] provide guidance about the requirements related to these and other HVAC systems. In addition, Appendix E of the 1995 edition of the *CABO One and Two Family Dwelling Code*[2] addresses the topic of **Energy Conservation** by providing the following note: "Energy Conservation shall be based on the 1995 edition of the *CABO Model Energy Code*.[3] These provisions shall apply to all site-built and prefabricated housing units, with the exception of manufactured (mobile) houses." *Radon Control Methods*[4] are also mentioned for the first time in the 1995 edition of the *CABO One and Two Family Dwelling Code* in Appendix F.

This chapter stresses key fundamental design issues rather than specific references to these CABO documents because of the wide variety of residential HVAC systems that are available.

ENERGY CONSUMPTION IN RESIDENTIAL BUILDINGS

Introduction

Significant cost savings can be realized by constructing energy-efficient houses. These savings will become increasingly important as energy supplies continue to decrease and the prices of raw fuel and electricity continue to rise.

Obstacles to Energy-Efficient Housing

Many techniques and materials that can save energy, without detracting from the quality of the living space, are available to the homebuilding industry. Conservation of energy, however, has not yet become a universally accepted critical issue in the design and construction of houses. This situation is attributable to many factors, one of which relates to the diversity and fragmentation of the homebuilding industry.

Each of the participants in the homebuilding industry brings a different perspective to issues related to energy conservation and consumption requirements. In order for conservation measures to be widely adopted, all participants must understand the energy-related costs and benefits associated with their roles in the homebuilding process.

Another obstacle occurs when some of these participants evaluate energy alternatives solely on an **initial cost basis**. Although initially it may be less expensive to install electric baseboard heat, for instance, a house with a more expensive high-efficiency fuel-fired furnace probably will cost less to operate and, thus, will result in long-term savings to the homebuyer. The procedure of evaluating systems based upon the sum of the initial costs and the present value of lifetime operating costs is known as **life cycle costing**. Thus, decisions based upon a life cycle costing approach to residential energy analysis often produce more cost-effective, energy-efficient housing.

The absence of an effective plan for conducting energy-related inspections during the construction phase and diagnostic measurements of energy performance before actual occupancy also provides an impediment to energy-efficient housing.

Energy Consumption

Energy consumption in houses can be attributed to the following: (1) the building envelope, (2) the heating and cooling equipment, and (3) the internal loads. Improvements made to any or all of these can result in energy cost savings. The overwhelming diversity of energy-related products and construction processes, coupled with the inherently diverse nature of the homebuilding industry, often impede their proper application in practice.

ENERGY MOVEMENT AND THE BUILDING ENVELOPE

Introduction

The words **energy** and **efficiency**, when used to describe the homebuilding process, typically generate thoughts of increased insulation levels. Although adequate insulation levels are a key element in achieving energy efficiency, even very high levels of insulation will not completely prevent heat transfer from occurring.

The **building (i.e., thermal) envelope** provides a boundary between the interior conditions of a house and the exterior environment. This envelope system consists of the following four major subsystems: (1) roof, (2) wall, (3) floor, and (4) foundation or earth. These subsystems, in turn, incorporate numerous elements (sheathing, insulation, windows, and doors) and other supportive subsystems (HVAC, plumbing, and electrical). As noted by Charlie Wing, the thermal envelope (Fig. 12.1) must meet the following conditions in order to be effective: (1) all surfaces of the envelope must be insulated to an appropriate R-value (measure of thermal resistance) for the climate, (2) the entire envelope must be sealed on the warm side by an effective air/vapor barrier, and (3) doors and windows must be weather stripped against infiltration.[5]

As the number of components in the building envelope system increase, energy-efficient design and construction becomes increasingly complicated. From the design perspective, the wide range of building products that can satisfy the building envelope system

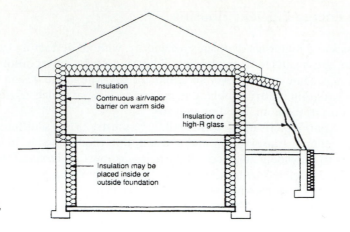

Figure 12.1
The Elements of
the Thermal
Envelope
Source: Charlie
Wing, *The Visual
Handbook of
Building and
Remodeling*
(Emmaus, PA:
Rodale Press, 1990),
p. 282.

requirements for a particular house can be overwhelming. In addition, it also becomes more difficult for the homebuilder to assure that all of the components are assembled correctly and that the interfaces between each component are sealed sufficiently to prevent excessive heat transmission.

Heat Transfer Modes

The three basic modes of heat transfer are: **conduction**, **convection**, and **radiation**. Conduction refers to the transfer of heat through a material from a warmer region to a cooler region. Convection occurs when heat is transmitted by air flow. Radiation refers to the transfer of heat by electromagnetic waves. Heat transmitted by conduction and convection is directly proportional to the temperature difference that exists. In residential applications, it is very common to find at least two, and sometimes all three, of these modes occurring simultaneously.

Each one of these heat transfer modes is illustrated in Figure 12.2, which indicates the energy exchange process that occurs in the wall of a house.

Stein and Reynolds[6] point out that conduction, at varying rates in different materials, is accounted for by the heat flows 1a, 1b, and 1c that are shown in Fig. 12.2. In addition, air rises as it is warmed by the warmer side of the air space. As the air falls down along the cooler side, it transfers heat to that surface (heat flow 2). Radiant energy (heat flow 3) is also transferred from the warmer to the cooler surface. Heat is also conducted from the room air by warm air currents (heat flow 4) that encounter the inside wall, and heat is conducted away from the exterior wall surface by the action of wind (heat flow 5).

Figure 12.2
Nature of Heat Flow through Materials,
Air Spaces, and Assembled Structures.
Thermal action is identified by the following numbers: 1–conduction, 2–convection, 3–radiation, 4–inside surface conductance, and 5–outside surface conductance
Source: Benjamin Stein and John S. Reynolds, *Mechanical and Electrical Equipment for Buildings*, 8th ed, p. 116. Copyright © 1992 John Wiley & Sons, Inc. Reprinted by permission of John Wiley & Sons, Inc.

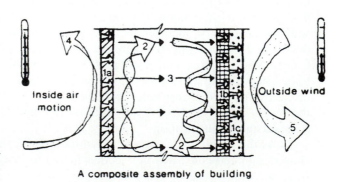

Conductive Heat Transfer

The three major factors determining conductive heat flow into and out of a building are (1) external conditions (temperature, wind speed, and insolation), (2) the area of the building exposed to the elements, and (3) the heat transmission properties of the exposed areas. The general equation that governs one-dimensional, steady-state, conductive heat flow is:

$$Q = U \times A \times \Delta T \qquad \text{(Eq. 12.1)}$$

where

Q = total heat flow (BTU/hr)
U = overall heat transfer coefficient (BTU/hr-ft^2-°F)
A = exposed envelope surface area (ft^2)
ΔT = difference between indoor and outdoor temperature (°F)

Equation 12.1 reveals that a linear relationship exists among the three parameters. Increasing any one of these parameters results in an increase in energy transmission. A more detailed explanation of each of these terms follows.

U-Value and R-Value Determination

In Eq. 12.1, the symbol U, known as the **thermal transmittance** or the "overall heat transfer coefficient," is a measure of the ability of an assembly of materials to allow heat to pass through itself. More specifically, U-values are a measurement of the rate of the heat flow from air to air through 1 sq ft of roof, wall, floor, or other building assembly for a 1° F temperature difference between the air on the inside and the air on the outside. U-values are expressed in BTU/hr-ft^2-°F.

U-values include convective interior and exterior air films if they have an effect on the building assembly. The term **thermal conductance**, or C-value, is also expressed in BTU/hr-ft^2-°F but is associated with the ability of a particular material (or a material assembly) to allow heat to pass through itself. C-values do not include the effect of air films.

A more familiar building material property is the R-value, or **thermal resistance**. The R-value, which is the reciprocal of the C-value, is a measure of a given material's ability to resist heat flow through it. Good conductors of heat, such as metals, have low R-values while good insulators, such as fiberglass, have high R-values. The higher the R-value, the higher is the insulating value. Equation 12.2 expresses the one-dimensional, steady-state, **conductive heat flow equation through a particular building material** as:

$$Q = (A \times \Delta T)/R \qquad \text{(Eq. 12.2)}$$

where

$R = 1/C$ = thermal resistance (hr-ft^2-°F/BTU)

R-values for representative building and insulating materials are shown in Table 12.1. The effect of density on the R-value of insulation is indicated clearly, since density ranges are provided for some of the insulating materials. For example, batt insulation, which is designed to fit into the 3-1/2 in. cavity in a 2 in. × 4 in. stud wall, is typically provided by insulation manufacturers as either R-11 or R-13 (3.14 R/in. or 3.71 R/in., respectively). **High-performance** insulation, which provides such a wall with R-15 (4.29 R/in.), is also available. As also noted in Table 12.1, either R-19 (3.45 R/in.) or a high-performance R-21 (3.82 R/in.) batt insulation can be used to fill the 5-1/2 in. cavity in a 2 in. × 6 in. stud wall. Similar data for representative insulating materials are presented in Table 12.2 in a format which is somewhat more convenient to use for design purposes.

It should be noted that the R-value of insulation is reduced if it is intentionally, or unintentionally, compressed into a smaller cavity than intended by the insulation manufacturer. This effect is shown in Fig. 12.3. Thus, if a 6-in.-thick batt of fiberglass insulation

TABLE 12.1   R-Values of Building and Insulating Materials

Building Materials	R-Value per Inch	R-Value for Thickness Listed
Building Board		
Gypsum or plaster board		
3/8″		0.32
1/2″		0.45
5/8″		0.56
Plywood (Douglas Fir)	1.25	
3/8″		0.47
1/2″		0.62
5/8″		0.77
3/4″		0.93
Hardboard		
Medium density	1.37	
High-density underlay	1.22	
High-density tempered	1.00	
Particleboard		
Low density	1.41	
Medium density	1.06	
High density	0.85	
Underlayment 5/8″		0.82
Finish Flooring Materials		
Carpet and fibrous pad		2.08
Carpet and rubber pad		1.23
Cork tile, 1/8″		0.28
Terrazzo, 1″		0.08
Tile (asphalt, linoleum, vinyl, rubber), 1/8″		0.05
Wood, hardwood finish, 3/4″		0.68
Roofing		
Asbestos–cement shingles		0.21
Asphalt roll roofing		0.15
Asphalt shingles		0.44
Built-up roofing, 3/8″		0.33
Slate, 1/2″		0.05
Wood shingles, plain and plastic film faced		0.94
Siding		
Asbestos–cement, 1/4″, lapped		0.21
Wood, bevel, 1/2×8″, lapped		0.81
Aluminum or steel, over sheathing		
Hollow-backed		0.61
Insulating-board backed, nominal 3/8″		1.82
Insulating-board backed, nominal 3/8″, foil backed		2.96
Woods (12% moisture content)		
Softwoods		
Southern Pine, density=35.6-41.2 lb/ft^3	1.00–0.89	
Douglas Fir Larch, density=33.5-36.3 lb/ft^3	1.06–0.99	
Hem–Fir,Spruce–Pine–Fir, density=24.5-31.4 lb/ft^3	1.35–1.11	
California Redwood, density=24.5-28.0 lb/ft^3	1.35–1.22	

(R-19) is compressed to 4 in. (a 33% reduction in thickness), its R-value is reduced to approximately R-14. Unnecessary compression of insulation into cavities in a wall or floor system, thus, should be a quality management issue during the construction of a house.

Each of the major subsystems of the building envelope consist of a number of different materials. The insulating value of the roof, wall, or floor subsystems can be obtained by summing the R-values of each of the materials. A "convective resistance" factor is added to

TABLE 12.1 *(continued)*

Insulating Materials	R-Value per Inch	R-Value for Thickness Listed
Blanket and Batt		
Mineral fiber, fibrous form processed from rock, slag, or glass		
approx. 3–4″, density = 0.4–2.0 lb/ft^3		11
approx. 3.5″, density = 0.4–2.0 lb/ft^3		13
approx. 3.5″, density = 1.2–1.6 lb/ft^3		15
approx. 5.5–6.5″, density = 0.4–2.0 lb/ft^3		19
approx. 5.5″, density = 0.6–1.0 lb/ft^3		21
approx. 6–7.5″, density = 0.4–2.0 lb/ft^3		22
approx. 8.25–10″, density = 0.4–2.0 lb/ft^3		30
approx. 10–13″, density = 0.4–2.0 lb/ft^3		38
Boards and Slabs		
Glass fiber, organic bonded	4.00	
Expanded pearlite, organic bonded	2.78	
Expanded polystyrene, extruded	5.00	
Cellular polyurethane/polyisocyanurate, unfaced	6.25–5.56	
Cellular polyisocyanurate	7.20	
Loose Fill		
Cellulosic insulation (milled paper or wood pulp), density=2.3–3.2 lb/ft^3	3.70–3.13	
Pearlite, expanded, density =		
2.0–4.1 lb/ft^3	3.70–3.30	
4.1–7.4 lb/ft^3	3.30–2.80	
7.4–11.0 lb/ft^3	2.80–2.40	
Mineral fiber (rock, slag, or glass)		
approx. 3.75–5.00″, density=0.6–2.0 lb/ft^3		11.0
approx. 6.50–8.75″, density=0.6–2.0 lb/ft^3		19.0
approx. 7.50–10.00″, density=0.6–2.0 lb/ft^3		22.0
approx. 10.25–13.75 in., density=0.6–2.0 lb/ft^3		30.0

Source: *1993 ASHRAE Handbook of Fundamentals* (Atlanta, GA: American Society of Heating, Refrigerating and Air Conditioning Contractors, 1993), excerpt from Table 4 in Chapter 22, pp. 22.6–22.9. Reprinted with permission.

the subsystem's total R-value to account for the convective interior and exterior air films if they are present. The U-value of the subsystem is then obtained by determining the reciprocal of the sum of all of the R-values. Thus:

$$U = 1/R_T \qquad \text{(Eq. 12.3)}$$

where

R_T = the sum of the R-values of each of the components of the subsystem

Figure 12.4 illustrates an R-value calculation for a second-floor ceiling/attic subsystem for the section between bottom chords of the attic roof truss. The attic space is assumed to contain 13-1/2 in. of fiberglass blown-in insulation. According to Table 12.2, this would result in an R-30 situation. The convective resistance factor for both the inside air film and the outside (attic) air film is assumed to be R-0.61. It should be noted that this is a simplified R-value calculation because only the heat flow path through the insulation is considered. The heat flow path through the ceiling joists is ignored in this calculation.

The resultant values for this subsystem are:

$$R_T = 31.78 \ (\text{hr-ft}^2\text{-°F/BTU})$$
$$U = 0.0315 \ (\text{BTU/hr-ft}^2\text{-°F})$$

TABLE 12.2   R-Value/Thickness Relationship for Insulating Materials

	Approx. R/In.	Approximate Inches Needed for					
		R-11	R-19	R-22	R-30	R-38	R-49
Loose Fill Machine-Blown							
Fiberglass	R-2.25	5	8.5	10	13.5	17	22
Mineral Wool	R-3.125	3.5	6	7	10.0	12.5	16
Cellulose	R-3.7	3	5.5	6	8.5	10.5	13.5
Loose Fill Hand-Poured							
Cellulose	R-3.7	3	5.5	6	8.5	10.5	13.5
Mineral Wool	R-3.125	3.5	6	7	10.0	12.5	16
Fiberglass	R-2.25	5	8.5	10	13.5	17	22
Vermiculite	R-2.1	5.5	9	10.5	14.5	18	23.5
Batts or Blankets							
Fiberglass	R-3.14	3.5	6	7	10.0	12.5	16
Mineral Wool	R-3.14	3.5	6	7	10.0	12.5	16
Rigid Board							
Polystyrene Beadboard	R-3.6	3	5.5	6.5	8.5	10.5	14
Extruded Polystyrene	R-4.5–R-5.41	3–2	5–3.5	5.5–4	7.5–5.5	9.5–7	12.5
(Styrofoam) Urethane	R-5.4–R-6.2	2	3	3.5	5.0	6.5	8
Fiberglass	R-4.0	3	5	5.5	7.5	9.5	12.5

Source: *Manual J: Load Calculation for Residential Winter and Summer Air Conditioning, 7th Ed.* p. 45
(Washington, D.C.: Air Conditioning Contractors of America, 1986).

Figure 12.3
Effect of
Compression of
Insulating
Materials on
R-Values
*Source: Manual J:
Load Calculations
for Residential
Winter and Summer
Air Conditioning,
7th ed.* (Washington,
DC: Air
Conditioning
Contractors of
America, 1986),
p. 45.

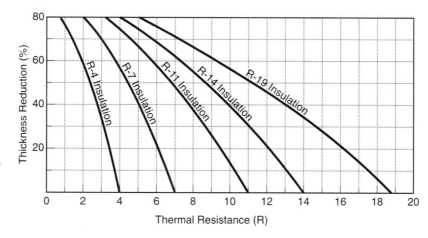

Figure 12.4
R-Value
Calculation for a
Typical
Ceiling/Attic
Subsystem
between the Truss
Lower Chord
Members.

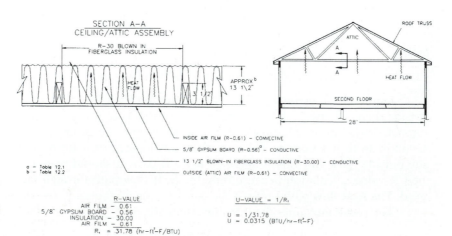

Recommended R-Values

The determination of the R-value of the ceiling assembly in Fig. 12.4 illustrates the calculation procedure that is followed for each of the major subsystems of the building envelope. The R-values that actually are provided by the homebuilder for each of the major subsystems for a particular house depend upon the weather conditions in the geographical area of the United States in which the house is built. Minimum R-values can either be mandated by energy-related legislation in a particular state or be recommended by a particular electric utility that is sponsoring an energy-efficient housing program in its own operating area. Homebuilders can select the minimum R-values themselves if such information is not readily available.

An indication of the severity of the weather in a particular area, and hence the amount of heating required, is the number of degree days (also called heating degree days) which the area experiences. For a particular 24-hour day, the number of degree days recorded for that day is the difference between 65°F and the mean outdoor temperature for the day. It would be expected that the higher the number of degree days, the higher the required or recommended R-value for each of the major subsystems in the house.

A relatively mild location in the southeastern part of the United States, such as Jacksonville, Florida, experiences 1,280 heating degree days while a location such as Augusta, Maine, in the northeastern part of the United States, experiences 7,860 heating degree days. Table 12.3 presents the results of a series of interpretative calculations which determined the minimum thermal properties for several building envelope components that meet the requirements of the *1995 CABO Model Energy Code*.[7]

For the Jacksonville, Florida example, for instance, the indicated R-value for the wall is R-13. This R-value is increased to R-19 for the Augusta, Maine example. It should be noted that the R-values shown in Table 12.3 are for a typical house that has 15% window area on the exterior walls. The indicated insulation values are not the only way to comply with the *1995 CABO Model Energy Code*. Homebuilders are permitted to trade off insulation values among other building components, including equipment efficiencies, and still be in compliance.

Building Envelope Area

The second factor in Eq. 12.1 is the area component. This is simply the exposed surface area on the building envelope. As the exposed surface area is reduced, so too is the heat loss. Thus, for the same square footage of floor space in a house, the optimum energy-related configuration of the "building footprint" is a square because of its minimum resultant wall area. Because single-family residences are very energy-intensive buildings from a heating perspective, a large potential for energy savings exists in house designs which acknowledge this fact.

TABLE 12.3   Recommended Building Envelope R-Values
for Various Heating Degree Day Locations

	Jacksonville, FL	Lexington, KY	Augusta, ME
Heating degree days	1,280	4,710	7,860
Ceiling R-value	R-19	R-30	R-38
Wall R-value	R-13	R-13	R-19
Floor R-value	R-13	R-16	R-19
Basement wall R-value	R-6	R-9	R-12
Slab perimeter R-value	None	R-7	R-18
Glazing U-value	0.65	0.35	0.35

Source: Interpretation of the requirements of Chapter 5 of the 1995 edition of the *CABO Model Energy Code,*™ copyright © 1995, with the permission of the publisher, The International Codes Council, Whittier, CA. Provided by Comfort Home Corporation, Lancaster, PA (August 1995).

Temperature Differential

The final factor in Eq. 12.1 is the temperature differential. This refers to the difference between the inside and outside ambient temperatures. Simply stated, as the winter months become colder, the temperature difference increases (assuming a constant indoor temperature) and a larger heat loss results. In the southern part of the United States, cooling requirements produce the greater energy load during the summer. The outdoor design temperatures are based on the values tabulated for each geographical area of the United States in the *1993 ASHRAE Handbook of Fundamentals*.[8] The temperatures either can represent the 97.5% or 99% design conditions. In other words, colder and hotter temperatures may be expected 2.5% or 1% of the time during the months of December through February and June through September, respectively.

Infiltration and Ventilation

Three convective air exchange mechanisms exist in buildings: forced ventilation, natural ventilation, and infiltration. Each differs in its effect on energy, air quality, thermal comfort, and control. **Forced ventilation** results from air flow driven by a fan system. It typically affords the greatest potential for distribution and flow rate control. **Natural ventilation** occurs through deliberate openings in the building envelope for windows, doors, skylights, venting mechanisms, and other inlets and outlets. It is generally caused by a pressure differential due to the wind or indoor–outdoor temperature differences. Some control capabilities are associated with this mode.

Infiltration is the unintentional and uncontrolled flow of outdoor air through a building due to pressure differentials caused by wind, indoor–outdoor temperature differences, and appliance operation. It is the least reliable in providing adequate ventilation and distribution, because it depends on weather conditions and the location of unintentional openings. It is also the main source of ventilation in envelope-dominated buildings. As noted by ASHRAE, typical infiltration values in housing in North America vary by a factor of about ten, from tight housing with seasonal average air change rates of about 0.2 per hour to housing with air exchange rates as great as 2.0 per hour.[9]

The thermal load of a house is affected by the air exchange rate in three ways. First, there is the need to either heat or cool the outdoor air to an acceptable indoor air temperature as it enters the house. This is called the **sensible component**. Second, the moisture content of the air in a house is changed due to the air exchange. This **latent component** typically requires dehumidification measures in some locations in the summer to counteract the humid outdoor air and humidification measures in some areas in the winter to maintain adequate comfort levels. The third effect on the thermal load of a house due to the air exchange rate is a reduction in the performance of the envelope insulation system.[10]

With regard to the sensible heat loss component, Eq. 12.4, which is taken from the *1993 ASHRAE Handbook of Fundamentals*,[11] can be used to determine the energy required to warm outdoor air entering by infiltration to the temperature of the room:

$$q_s = c_p \, Q\rho(t_i - t_0) \tag{Eq. 12.4}$$

where

q_s = heat flow required to raise temperature of air leaking into building from t_0 to t_i, BTU/h
c_p = specific heat of air, BTU/lb · °F
Q = volume of outdoor air entering building, ft³/h
ρ = density of air at temperature t_0, lb/ft³

Using standard air, $\rho = 0.075$ lb/ft³, and $c_p = 0.24$, Eq. 12.4 reduces to:

$$q_s = 0.018 \, Q(t_i - t_0) \tag{Eq. 12.5}$$

It should be noted that an alternative method for calculating infiltration heat loss is the crack length method. Both that method and the evaluation of the latent heat loss component mentioned are beyond the scope of the presentation in this chapter. The interested reader is referred to the *1993 ASHRAE Handbook of Fundamentals* for additional information.[12]

Representative Contributions to Heat Loss

The relative contributions to the heat loss in a house due to the conductive and convective modes of heat transfer depend upon a large number of interconnected variables related to both the design and the construction of the house. Data of this type are of great interest to the homebuilding industry because they indicate, from a cost/benefit standpoint, the greatest areas for potential energy savings.

Table 12.4 provides estimated heat loss values (in BTU/hr) for a typical 2,000 square foot, two-story house with conventional insulation which is located in an area where the winter design temperature is 0°F outside and 70°F inside (i.e., the temperature difference across each subassembly is assumed to be 70°F). The conductive heat loss estimates are based upon standard industry procedures recommended in *Manual J* of the Air Conditioning Contractors of America (ACCA).[13] It should be noted that the temperature difference across the components, as well as the BTU/hr loss, will vary as the winter design temperature changes for another location. Although the numbers in Table 12.4 will be different for that location, the percentage of heat loss for each component will remain constant.

TABLE 12.4   Representative Heat Losses in a Two-Story, 2,000 Square Foot House[a,b,c]

1. Component Heat Losses

Areas of Heat Loss	Square Feet	BTU/hr loss	% of Total
Windows	235	6,562	17.1
Doors	84	2,342	6.1
Walls	1,725	6,642	17.3
Ceilings	1,033	1,880	5.0
Floor	1,025	1,866	5.0
Subtotal		19,298	50.5

2. System Heat Losses

Areas of Heat Loss	Assumption	BTU/hr loss	% of Total
Infiltration	0.5 AC/hr[d]	10,277	26.8
Ductwork Loss	10%	2,958	7.7
Ductwork Leakage	15%	5,741	15.0
Subtotal		18,976	49.5

3. Building Totals

Total Heat Loss		38,274 BTU/hr	100.0%

[a]Insulation levels: R-38 in attic, R-19 in walls, R-19 in floor.
[b]Double pane windows, insulated doors.
[c]Winter design temperature = 0°F; summer design temperature = 95°F.
[d]Leakage estimate before sealing program is instituted.

Source: Information provided by the Comfort Home Corporation, Lancaster, PA (August 1995).

Losses due to construction practice have also been estimated using standard industry practice. The infiltration, ductwork loss, and ductwork leakage estimates are all deemed to be either reasonable or conservative based upon studies completed in 1994–1995 by the Home Energy Ratings System Council.

The estimated 50.5% of the heat loss through components can be reduced only by increasing the level of insulation in the various subsystems or upgrading the thermal efficiency of the windows. The important thing to note in Table 12.4 is that 49.5% of the heat losses are directly associated with construction practices in the field. A comprehensive total quality management program implemented by the homebuilder during the entire construction process offers the opportunity for considerable energy-related savings for the homebuyer without a great amount of additional cost.

THE KEY ELEMENTS IN THE RESIDENTIAL ENVELOPE

Introduction

The building envelope is a thermodynamic system that, as the name implies, is constantly undergoing change. Energy is constantly transferring across this boundary, attempting to reach a state of equilibrium between the inside of the house and the outside conditions. This dynamic system changes with each season of the year. As indicated in Fig. 12.5, winter is a heating season for most regions of the United States. The sources of heat loss during that season are indicated. Figure 12.6 provides an example of the cooling requirements in most houses in the United States during the summer months.

The objective of the building envelope, from an energy perspective, is to slow down the heat flow process in either direction and thereby reduce the related heating and cooling expenses. This is accomplished by utilizing building materials with high resistance to conductive heat transfer (i.e, a high R-value) in each building envelope subsystem.

The Roof/Ceiling Subsystem

The roof/ceiling subsystem normally accounts for 5–10% of the heat loss in most houses. The insulation in this subsystem is the most significant factor which resists energy transmission since it accounts for approximately 95% of the subsystem composite R-value. High R-values can be achieved because large quantities of insulation can typically be installed in the attic above the ceiling. The most widely used roof/ceiling insulating materials are either loose-fill or batt-style fiberglass and cellulose. The characteristics of each of these are discussed.

Figure 12.5
Winter Heating
Load
Requirements
*Source: Manual J:
Load Calculations
for Residential
Winter and Summer
Air Conditioning,
7th Edition.*
(Washington, DC:
Air Conditioning
Contractors of
America, 1986),
p. 10.

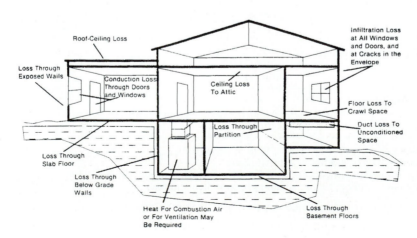

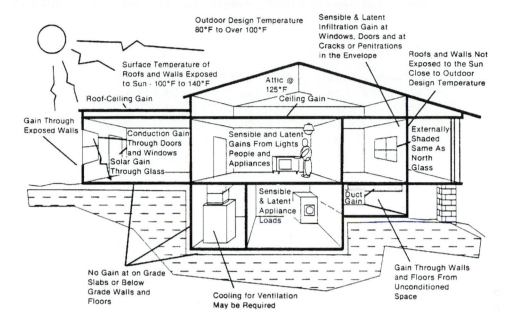

Figure 12.6
Summer Cooling
Load
Requirements
*Source: Manual J:
Load Calculations
for Residential
Winter and Summer
Air Conditioning,
7th Edition.
(Washington, DC:
Air Conditioning
Contractors of
America, 1986),
p. 28.*

Fiberglass

Fibrous glass is a mineral fiber made from molten silica, spun like cotton candy. It is the most widely used insulating material. It has good water absorption resistance and fire resistance properties and excellent resistance to damage by water. Fiberglass insulation is relatively inexpensive and can be installed easily as loose-fill or as batts (blankets), which generally include a vapor retarder facing system.

LOOSE-FILL

Loose-fill fiberglass insulation is typically installed with pneumatic equipment above the ceiling space. Although this method is cheaper than rolling out batts, it does not provide quite the same insulation level (per inch) because of the differences in fiber orientation resulting from the two approaches. Convective heat losses through loose-fill insulation appear to be of concern only at attic temperatures of approximately 0°F or below. Additional thickness can be added to the available attic space to compensate for this situation. As noted in Table 12.2, loose-fill fiberglass provides an approximate insulating value of 2.25 R/in. In order to achieve an R-38 value in the attic, therefore, an approximate thickness of 17 inches of loose-fill insulation is required.

BATTS

Fiberglass batts are available in several variations. Foil-faced fiberglass insulation has an aluminum foil vapor retarder that faces the inside of the building to inhibit the migration of water vapor through the insulation. This is a concern because of the damage which moisture can create inside the cavity space. Kraft paper-faced insulation has an asphalt coating on the paper that acts as a vapor retarder, though it is not as effective as aluminum foil. Unfaced insulation is also available but should be used only where a vapor retarder is not needed or in conjunction with a separate vapor retarder such as polyethylene (plastic) sheets.

Fiberglass batt insulation is typically available in widths of 11 in., 15 in., and 23 in., and in thicknesses of 3-1/2 in., 6 in., 9 in., and 12 in. As noted earlier, the R-value of a specified batt thickness depends upon the density of the insulation. The installation of batts is relatively straightforward, but care must be taken to avoid compressing the blankets and to provide coverage of the entire attic space.

Cellulose

Cellulose is usually made from chemically treated, ground-up, or shredded recycled paper. Installation is straightforward since it is usually blown into the attic space as loose-fill insulation using a blower machine and hose system. Cellulose represents an alternative to fiberglass which provides an R-value of about 3.7 per inch of thickness.

One drawback of cellulose is that some of the chemical treatments associated with the product may react with moisture and form a mixture that can be corrosive to metal water pipes and vents if proper precautions are not implemented. It also readily absorbs and retains water and is very susceptible to water damage. Additionally, as an organic material, it is prone to rot and insect infestation and is flammable.

The Wall Subsystem

The insulation level is also the most significant wall subsystem element affecting energy usage. It is more difficult to estimate its total contribution on wall subsystems, however, because of the variations in wall designs and the influences of windows and doorways.

Insulation

A larger variety of insulation products, including fiberglass, cellulose, and plastic foam, can be used in the wall subsystem. Other materials such as perlite, vermiculite, and rock wool have also been used, but are not as common in contemporary residential construction in the United States.

FIBERGLASS

Fiberglass insulation, as noted under ceiling applications, is the most common cavity-fill insulation. Although blown-in fiberglass is becoming more widely used for walls, kraft paper-faced batts are currently more common. In a 2 in. × 4 in. wall, 3-1/2 in. batts will yield an R-value of R-11 to R-15, while 5-1/2 in. batts can be used in a 2 in. x 6 in. wall cavity to produce R-19 to R-21.

CELLULOSE

Sprayed **cellulose** for walls is a different product than the blown-in loose-fill variety used in attics, but it consists of the same constituent parts. In high-pressure wall applications, cellulose is mixed with a binder and sprayed into the wall cavity prior to installing drywall. A potential disadvantage of this material is that sufficient drying time is required before the wall can be closed in. An R-value of 3.1–3.4 per inch is achieved.

PLASTIC FOAMS

Plastic foams are classified as rigid insulation since these products are typically supplied in 4 ft × 8 ft rigid sheets. They can be installed on either the exterior or the interior faces of the wall, with the former being the most common. Although rigid foams are effective insulating materials, some have secondary side effects such as flammability and off-gassing. The two most widely used varieties of plastic foams are (1) polyurethane/polyisocyanurate and (2) polystyrene.

Polyurethanes and **polyisocyanurates**, available in board or as foam, possess the highest R-value (R-5 to R-7.2 per inch) of the plastic foams. Two disadvantages of these materials are flammability and dimensional stability. The foamed-in-place products may also cause problems due to expansion.

Polystyrene insulation has both high thermal resistance and low moisture absorption. It can be used on walls, on foundations, and under slab-on-grade construction. Rigid sheets are convenient to use as sheathing on exterior walls underneath siding, brick, and stucco.

Since the sheets have little lateral strength, diagonal bracing or plywood must be used at the corners of the house to provide the house with structural resistance to racking. Polystyrene insulation has an R-value that varies from R-3.6 to R-5.0 per inch depending on its density and the process used to produce it.

Windows

Windows represent a major source of energy loss, since they can account for up to 40% of the heat loss from a well-insulated house. As houses become tighter and better insulated, the proportion of heat loss attributable to windows continues to increase, primarily because insulated windows are less efficient than even an uninsulated wall. Thus, a large amount of heat can flow through them. They do allow sunlight to pass through in the winter, thereby adding solar energy to the house. The same sunlight, however, is the major factor in raising summer cooling requirements.

Glass offers very little insulation value (R-0.6 for a 1/8-in. pane). In order to increase the effectiveness of windows, therefore, a number of layers of glazing must be used. Further improvements can be made by adding low-emissivity (low-E) coatings to the glazing surface. R-values can approach R-4.5 for low-E triple-glazed windows.

A final method of improving energy efficiency in windows is to fill the spaces in the window units with a gas other than air, such as argon or krypton, which is less conductive to heat flow than air. These windows can have R-values ranging from R-4 to R-5.

Window frames and jambs are also an important part of an energy-efficient window system. Although they represent a small percentage of the entire window opening, thermally broken (no direct metal-to-metal energy path from inside to outside) framing can improve comfort, increase energy efficiency, and reduce condensation on interior framing members.

Doors

Doors also can be a source of large heat losses. Doors transmit heat by conduction through them and infiltration around them. Standard solid wood doors vary in R-value from R-1.3 for a 1-in.-thick door without a storm door, to about R-3 for a 2-in.-thick door with a storm door.

Insulated steel doors are more energy efficient because they typically use 24-gauge sheet metal to enclose an inner core of solid polyurethane or expanded polystyrene. An insulated metal door with a storm door can provide a thermal resistance of R-8. Adding glass panels to the door increases the aesthetics and installation cost of the door, while also greatly reducing its energy efficiency.

The Floor and Foundation Subsystems

These two subsystems are best described together, since the energy management strategies are dependent upon the combination used. Four of the commonly accepted foundation types are **full basements**, **partial basements**, **crawl spaces**, and **slabs on grade**. The foundation subsystem is in contact with the earth. Therefore, it generally experiences a smaller temperature differential between the interior (the basement) and the exterior surface because the temperature fluctuations of the earth are smaller than those of the air that surrounds the above-grade part of the house. With reasonable insulation levels, these two subsystems should account for no more than 15% of the total heat loss in a house.

Full Basement Foundation Systems

Before any decisions are made about foundation and floor insulation for this system the homebuyer or builder must decide whether the basement space will be conditioned (heated and cooled). If the basement is left unconditioned, then insulation is usually installed underneath the first floor.

If the basement is to be conditioned, however, perimeter insulation must be installed on either the interior or exterior wall faces, or within the cavities of masonry block exterior walls. For exterior applications, only extruded polystyrene or rigid fiberglass is recommended since expanded polystyrene and polyurethane may tend to absorb moisture, which can lead to problems. Foundation insulation levels typically range from R-5 to R-10.

The Floor Slab

The concrete floor slab in a full basement foundation can either be insulated or uninsulated. If the space is to be used as a living area, it is typically recommended that a 1-in. layer of rigid insulation be placed under the floor slab to help keep the floor surface warm.

Diagnostic Testing of the Building Envelope

The thermal integrity of the building envelope may be verified and anomalies located and corrected by one or more diagnostic techniques. Fan pressurization and depressurization of the entire home by use of a **blower door**, for instance, can reveal the location and approximate magnitude of air leakage sites.

Tracer gas dilution may also be employed to estimate natural infiltration. If a sufficient inside-to-outside temperature differential exists, infrared thermography techniques can also be used to indicate areas of the building envelope that experience the highest levels of heat flow. These levels may be due to excessive air leakage, higher conduction rates of elements of the envelope, or improper or missing insulation between the supporting ceiling or the wall framing.

MECHANICAL SYSTEMS

Introduction

The range of heating, cooling, and domestic water heating options that are available to the homebuilder are at least as broad as the diversity of building envelope options just discussed. The advent of improved designs and features such as larger heat exchange surfaces, better materials, and refined condensing technologies has enabled manufacturers to produce appliances that use 60–70% of the fuel they used a decade ago while providing the same amount of space conditioning. Because of the lower operating costs, significant savings can be realized by the homeowner because of such equipment.

Mechanical System Considerations

In most cases, the total operating costs of a mechanical system will far exceed the purchase price over the life of the system. The operating costs are governed by fuel costs, system efficiency, and maintenance costs. Of these three items, fuel costs and system efficiency are usually the primary factors which influence the operating costs of mechanical systems in residential applications. Maintenance is not usually as critical.

Fuel Costs

The heat content of each fuel is used as the common denominator to convert each fuel to the same units. Heat contents are measured in British Thermal Units (BTUs) per unit of fuel consumption. A BTU is the amount of heat needed to raise the temperature of one pound of water 1°F. This is approximately equal to the amount of heat that is released when a match is burned in its entirety. Table 12.5 lists the heat contents of the various fuels that are commonly used in residential systems. Coal is also sold by the ton and natural gas is also sold by the 100 cubic feet and the nearly equivalent **therm** (100,000 BTU content).

TABLE 12.5   Energy Contents of Fuels

Fuel		BTUs	Unit
Coal	Anthracite	15,000	lb
	Bituminous	13,000	lb
	Cannel	11,000	lb
	Lignite	11,000	lb
Electricity		3,412	kwhr
Gas	Natural	1,030	cu ft
	Propane	91,600	gal
Oil	#1 (kerosene)	134,000	gal
	#2 (residential)	139,000	gal
	#4	150,000	gal
	#6	153,000	gal
Wood	Fir, Douglas	21.4×10^6	cord
	Hemlock	18.5×10^6	cord
	Hickory	30.6×10^6	cord
	Maple, red	24.0×10^6	cord
	Maple, sugar	29.0×10^6	cord
	Oak, red	24.0×10^6	cord
	Oak, white	30.6×10^6	cord
	Pine, pitch	22.8×10^6	cord
	Pine, white	15.8×10^6	cord
	Poplar	17.4×10^6	cord
	Spruce	17.5×10^6	cord

Source: Charlie Wing, *The Visual Handbook of Building and Remodeling,* p. 371 (Emmaus, PA: Rodale Press, 1990).

The heat contents in Table 12.5, along with the local costs of each fuel and the seasonal efficiency of the mechanical system, can then be substituted into Eq. 12.6 to compare the relative energy costs for each system:

$$\text{Cost per million BTUs} = (1 \times 10^6 \times P)/(F \times AFUE) \qquad \text{(Eq. 12.6)}$$

where:

P = price per unit of fuel ($)
F = BTU content of a unit of fuel
AFUE = annual fuel utilization efficiency of the mechanical system (%)

EXAMPLE

Determine the cost of 1 million BTUs of heat from a #2 fuel oil at $1.32 per gallon which is used in a warm-air furnace with an AFUE rating of 85%.

$$\text{Cost per } 10^6 \text{ BTU} = (1 \times 10^6 \times \$1.32)/(139,000 \times .85)$$
$$= \$11.17$$

Efficiencies of Mechanical Equipment

The efficiency factor in Eq. 12.6 is an indicator of how effectively a piece of mechanical equipment utilizes a fuel source. Furnace and boiler efficiencies are measured by their Annual Fuel Utilization Efficiencies (AFUE). The higher the AFUE, the more efficient is the furnace or boiler. Energy-efficient furnaces and boilers can have AFUEs as high as 97%.

Heat pumps operating in the heating mode are rated in terms of the Heating Season Performance Factor (HSPF). This is a ratio of the total heat output to the total electrical input. HSPF values typically range from 5.25 to 9.25 BTUs/hr per watt, with a higher HSPF denoting a more efficient machine. Some systems, such as water and ground source heat pumps, are rated using the Coefficient of Performance (COP). The COP is a ratio of the rate of heat removal or heat delivered to the rate of energy input. COP values typically range from 2.0 to 4.0.

Air conditioners and heat pumps operating in the cooling mode are rated by their Seasonal Energy Efficiency Ratio (SEER). This is a ratio of the total cooling output to the total electrical input. SEER values range from 10 to 16 BTUs/hr per watt or more. A higher SEER also denotes a more efficient machine.

As efficiencies increase, proper balance between the sensible and latent capacities and the total cooling become more critical because cooling may occur at a faster rate than dehumidification, resulting in reduced comfort for the occupants of the house.

Contemporary Heating and Cooling Systems

The mechanical systems which provide either heating, cooling, or both consist of the equipment which heats or cools the medium of heat transfer (i.e. steam, water, or air), a distribution network which delivers and returns the heat transfer medium to the equipment, and the device in each room which distributes the energy. Figure 12.7 illustrates some of the typical mechanical systems that are used in residential construction.

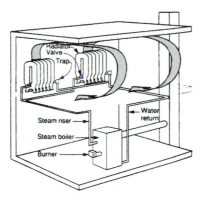

(a) **Steam Boiler**
In a steam system, steam is produced in a steam-rated oil or gas boiler, circulated through insulated pipes to room radiators, and condensed in the radiator, giving up its heat of vaporization. The condensed water then drains back to the boiler for reheating

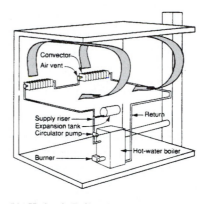

(b) **Hydronic Boiler**
The hydronic, or forced-hot-water, system heats water in a gas or oil boiler and circulates it through loops of pipe to distribute heat to separate heating zones.

Figure 12.7
Typical
Residential
Heating Systems
Source: Charlie
Wing, *The Visual
Handbook of
Building and
Remodeling*
(Emmaus, PA:
Rodale Press, 1990),
pp. 368–370.

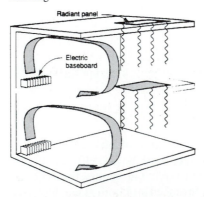

(c) **Electric Baseboard and Radiant Panels**
Electric-resistance baseboards and radiant ceiling panels convert electricity to heat with 100 percent efficiency; no heat goes up a flue.

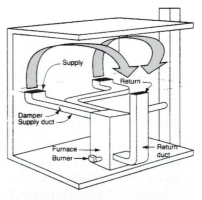

(d) **Warm-Air Furnace**
In a modern warm-air furnace heat is produced by clean and efficient combustion of gas or oil, and the warm air is distributed evenly throughout the building by a blower, supply and return ducts, and registers.

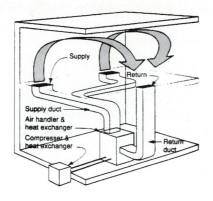

(e) Heat Pump

Heat pumps operate on the same principle as refrigerators. By compressing and expanding a gas (the refrigerant), they reverse nature and pump heat from a cooler "source" to a warmer "sink". By reversing the pump, you can cool as well as heat a house.

(f) Coal/Wood Stove

Solid-fuel (wood or coal) stoves heat both by radiation to the immediate surroundings and by natural convection of warmed air.

Figure 12.7
(continued)

Electric Resistance Systems

Electric resistance systems include baseboard heaters, hot-air electric furnace systems, and radiant panels.

Baseboard heaters, placed in the rooms of the house that are to be heated, consist of electric resistance elements encased in a metal housing. These systems rely on natural convection to circulate heated air and, therefore, require no fan or ductwork. They are usually the least expensive units to purchase and maintain. Operating costs, however, tend to be higher than for any other system except electric resistance furnaces.

Hot-air electric resistance furnace systems also use an electric resistance element to warm the supply air. In this case, however, the heating process is performed at a central furnace location and then the heated air is delivered to the different zones through ductwork.

Electric radiant panels rely on the radiant heat transfer process to supply heat. Electric elements, which can be embedded in ceiling, floor, or wall panels, transfer heat to the surrounding objects and people instead of to the air.

Electric Thermal Storage

Electric thermal storage systems take advantage of discounted electricity rates that are available during off-peak hours. These systems typically operate by heating water or ceramic bricks through the night and then slowly releasing the stored heat during the day to heat the house. Some utilities offer as much as a 40% discount for residential uses during off-peak hours, making these systems advantageous in that respect.[14] However, high installation costs can outweigh the savings in energy usage.

Fuel-Fired Furnaces

These systems typically consist of a furnace, a distribution network, supply and return registers in each heated location, and a thermostat. The furnace, which is the heart of the system, contains a combustion chamber, heat exchanger, flue, blower, filter, and controls. These furnaces use a variety of fuels such as gas, oil, wood, and coal, with gas and oil systems being the most commonly used.

High-efficiency gas-fired furnaces installed in today's new homes typically have an AFUE greater than 90%. They generally are of the sealed combustion type where the appliance obtains 100% of its combustion air from the outside and vents the exhaust gases directly to the outside. Such appliances generally burn more efficiently and offer no risk of introducing the harmful combustion gases into the home.

Hydronic Boilers

Warm water boilers are similar to furnaces except that combustion gases transfer heat to a circulating water medium instead of to air. These systems, which distribute heated water to hydronic convectors located in each room, are attractive because they require little or no power to circulate the heated water. They also can be easily and inexpensively zoned. The system requires a minimum of space, and an even distribution of heat is supplied quietly and comfortably.

Air Conditioners

The majority of new houses currently being constructed include central air-conditioning systems. The air conditioner shown in Fig. 12.8 is of the "split system" type that can be used in conjunction with the gas-fired furnace discussed.

A central air conditioner has two basic parts: the condenser unit and the evaporator coil. The **evaporator** (cooling) coil is located in the supply air plenum of the furnace while the **condenser** is located outside the house, hence the term "split system." Refrigerant gas is compressed, then cooled and condensed in the condenser and throttled to the evaporator coil. A blower circulates air through the evaporator coil, cooling the air, and then blows it into the house through the same duct system utilized by the furnace.

Heat Pumps

Heat pumps operate by extracting heat from a cold medium and transferring it to a warmer medium. **Air source heat pumps** (Fig. 12.9) are one of the most popular systems used by residential homebuilders. When operating in the cooling mode, heat is extracted from the

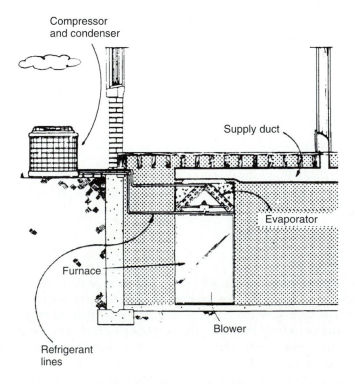

Figure 12.8
Central Air
Conditioner
Source:
Pennsylvania
Heating Systems
Manual
(Harrisburg, PA:
The Pennsylvania
Energy Office,
1989), p. 26.

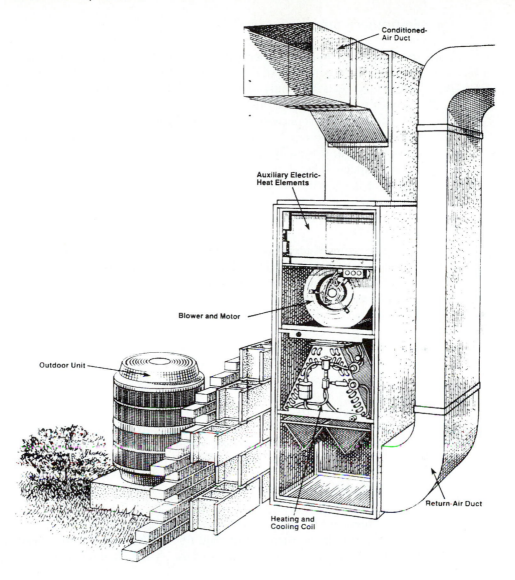

Conditioned-Air Duct

Auxiliary Electric-Heat Elements

Blower and Motor

Outdoor Unit

Heating and Cooling Coil

Return-Air Duct

Figure 12.9
Air Source Heat
Pump

*Source: What is a
Heat Pump and an
Add-on Heat Pump*
(Greensburg, PA:
West Penn Power
Company, 1989),
p. 2.

inside air and released to the outdoors. In the heating mode, heat is absorbed from the out-side air and released to the indoors.

An air source heat pump operates with a high efficiency until the outdoor temperature is in the range from 25–30°F. As temperatures go below this level, the amount of heat which can be extracted from the air decreases at the time that the house's heating requirements are increasing. The point at which the heat pump cannot extract enough energy from the air to heat the house is known as the **balance point**. When this situation occurs, supplemental heat must be provided to make up the difference. Common sources of supplemental heat include electric resistance and gas-fired and propane furnaces.

Water and ground source heat pumps operate by drawing heat from and rejecting heat to the ground, wells, ponds, or streams instead of to the air. Operating costs are about one-half of those for air source heat pumps because ground temperatures are more stable than air temperatures and because water can carry much more heat than air. In addition, waste heat from the system can be used to provide domestic hot water at no cost in the summer and at substantial savings in the winter. Even though these systems are gaining popularity, they are still much less common than air source heat pumps because of their expensive and difficult installation requirements.

Equipment Sizing

In addition to reducing operating costs by selecting high-efficiency heating and cooling equipment, proper sizing of the systems is also critical. A system that is too large wastes fuel and money because it only operates at peak efficiency for short periods of time since it is constantly cycling on and off.

In order to properly size a **gas-fired furnace**, a heat loss analysis should be completed that considers all areas and insulation levels of the walls, ceilings, floors, doors, and windows. The heat loss analysis determines the peak hourly heating demand in BTU/hr on the coldest expected day of the year in the area in which the house will be built. If possible, the output of the furnace should be no more than 25% over this peak hourly heating demand. For example, if the peak hourly heating demand for a particular house is calculated to be 20,000 BTU/hr, a furnace with an output in the range of 20,000–25,000 BTU/hr should be selected.

The sizing of an **air conditioner** follows a procedure which is similar to the sizing of a furnace. A heat gain analysis that considers the same areas and insulation levels noted should be performed. The importance of proper sizing of air conditioners is readily apparent since equipment cost is directly proportional to size: doubling the cooling output nearly doubles the cost. If the system is oversized and therefore operates only for short periods of time (due to constant on/off cycling), the consumption of electricity is increased and the level of dehumidification in the house may be lower. This is extremely critical since cooling comfort in a house is highly influenced by the amount of humidity in the air.

The proper sizing of **heat pumps** can be somewhat difficult since the system provides both heating and cooling capabilities. If the system is sized for heating in a colder climate, it will most likely be oversized for cooling. If the system is sized for cooling in a warmer climate, it will most likely be oversized for heating. A system that is undersized for heating requires electric resistance to provide the supplemental heat needed in the winter. Even though electric resistance is 100% efficient, it can be very expensive to operate.

Duct Systems

Duct systems are an integral part of warm-air furnace and heat pump systems (Fig. 12.10). Ducts are usually constructed of rigid fiberglass board or galvanized sheet metal. A rigid fiberglass board duct system (i.e., a Class 1 rigid air duct) is easier to install and usually

Figure 12.10 Conventional Warm-Air Furnace and Ducts

Source: Benjamin Stein and John S. Reynolds, *Mechanical and Electrical Equipment for Buildings*, 8th ed, p. 365. Copyright © 1992 John Wiley & Sons, Inc. Reprinted by permission of John Wiley & Sons, Inc.

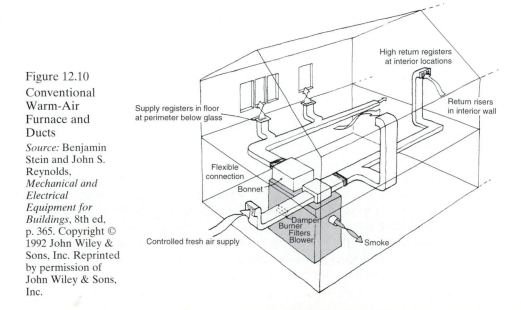

results in a quieter system. Fiberglass ducts, however, are not as durable as galvanized sheet metal ducts. Conventional house construction typically consists of either a fiberglass or sheet metal: (1) fully ducted system or (2) partially ducted system with flexible runouts to individual room locations.

Effective distribution of conditioned air requires careful attention to both the design of the duct system as well as its installation in the house. It is important that accurate design heat loss and heat gain calculations be performed and that the necessary air quantities be established. The performance characteristics anticipated in a satisfactory ductwork design will never be realized, however, unless a total quality management program is instituted by the homebuilder to ensure that the ductwork is properly installed by the subcontractor performing the work.

In order to minimize air leakage, all duct joints between sections should be made as tight as possible, particularly when the ductwork is in unconditioned spaces. In addition, the return portion of the system should receive as high a level of installation quality as the supply portion. Whenever possible, air flow measurements should be taken throughout the entire length of the completed distribution system in order to ensure that the designed supply and return air quantities are actually achieved.

CASE STUDY HOUSE EXAMPLE

Introduction

A design procedure which can be used to select the size of the heating equipment is illustrated in this section of the chapter using the Case Study House plans provided in Appendix A. Only the steps involved in sizing and routing the ductwork are presented, however, because a detailed explanation of the ductwork design procedure is considered beyond the scope of this chapter.

The design procedure which selects the size of the cooling equipment is similar to the procedure used for the selection of the heating equipment. It is somewhat more complicated, however, because the effects of radiation from the sun and internal loads from people and appliances, including dehumidification, must be included in the calculation.

Overview of the Design Procedure

The design procedure which is presented is a modification of the basic procedure recommended in *Manual J: Load Calculation for Residential Winter and Summer Air Conditioning, Seventh Edition*, published by the Air Conditioning Contractors of America (ACCA).[15] It includes the following steps.

Sizing of the HVAC System

1. Description of the house.
2. Description of the assumed envelope features.
3. Intitial selection of the heating and cooling equipment.
4. Design day selection.
5. Calculation of the U-values and areas of the exposed building envelope.
   The U-values for each of the important parts of the exposed building envelope are calculated using the procedures discussed earlier in the chapter.
6. Determination of the conductive design heat losses.
   The total conductive design heat losses are determined by using Eq. 12.1. Outside design conditions for the United States are provided in *Manual J*, published by ACCA.[16] As an example, Table 12.6 presents the winter design day temperatures used to calculate the ΔT term in Eq. 12.1 for 10 representative locations

TABLE 12.6   Representative Design Conditions in Pennsylvania

Location	Winter 97-1/2% Design Temperature	Heating Degree Days below 65°F
Pennsylvania		
Allentown Airport	9	5,810
Chambersburg	8	5,170
Erie Airport	9	6,540
Harrisburg Airport	11	5,280
Johnstown	2	7,804
Lancaster	8	5,560
Philadelphia Airport	14	5,180
Pittsburgh Airport	5	5,950
Reading City Office	13	4,960
State College	7	6,160
Williamsport	7	5,950

Source: *Manual J: Load Calculation for Residential Winter and Summer Air Conditioning, 7th Ed.,* pp. 59–60 (Washington, D.C.: Air Conditioning Contractors of America, 1986).

in Pennsylvania. Also presented are the number of heating degree days for each location. For design purposes, it will be assumed that the Case Study House is being built in State College, Pennsylvania.

7. Determination of the infiltration heat losses.

The heat losses due to infiltration through the envelope are determined using the **air change method**. This method assumes that a building will undergo a number of air changes per hour (ACPH) based upon the building type, construction, and use. It is typically expressed as a percentage of the total conditioned space volume. For example, a 1,000 sq ft one-story house with 8-ft ceilings that experiences 0.5 ACPH loses 4,000 ft^3 of conditioned air each hour as indicated:

$$\text{Volume of air lost} = 1,000 \text{ ft}^2 \times 8 \text{ ft} \times 0.5 \text{ ACPH}$$
$$= 4,000 \text{ ft}^3 \text{ of air each hour}$$

8. Determine the total **design day heat loss** (the sum of the conductive and the infiltration heat losses).

9a. Determine the required size of the heating system—**manual approach**.

An appropriately sized heating system can be selected by multiplying the heat loss obtained in Step 8 by a safety factor of 1.25. This is the heating system output that is required for the house. It should be noted, however, that many heating systems are rated by energy input instead of output. The efficiency of the selected mechanical system can be considered by dividing the required output by the appropriate efficiency factor as shown in Eq. 12.7:

$$Q_\text{sys} = \frac{Q_\text{tot} \times 1.25}{\eta} \qquad \text{(Eq. 12.7)}$$

where:

Q_tot = total design day heat loss
1.25 = safety factor (maximum value)
η = system efficiency

9b. Determine the required size of the heating system—**computer approach**.

The above procedures represent a simplified analytical approach that can be used to determine the approximate heating loads and heating equipment size for a given house situation and configuration. Microcomputer-based energy analysis programs have been developed to expedite this process and take advantage of more sophisticated HVAC analytical procedures. The Carrier Corporation, for instance, has developed the Hourly Analysis Program (HAP)[17] for use in conducting various types of HVAC engineering analyses. Before using any computer program, however, time should be taken to compare its results to a set of manual calculations to assure that it functions properly.

Estimate of the Annual Fuel Consumption

The design heat load value is used to size the heating equipment for a house. In addition, it is often desirable to estimate the total amount of energy required to heat the house throughout the year. Three analytical methods can be used to provide such an estimate: the degree day method, the bin method, and hourly analysis techniques. Each of these will be discussed briefly.

THE DEGREE DAY METHOD

The **Degree Day method** assumes that over a long period of time, solar and internal heat gains in a residential structure will offset the heat lost to the environment until the mean daily outdoor temperature is lowered to 65°F. The number of heating degree days recorded for a particular day, as noted earlier, is the difference between 65°F and the mean outdoor temperature for that day. Databases containing this information for various locations have been developed over the years. For example, the value for heating degree days in State College, PA, is 6,160 (Table 12.6).

The general fuel calculation relationship using the appropriate geographical degree day value is shown in Eq. 12.8:

$$F = \frac{24(DD)Q\, C_D}{\eta(t_i - t_o)H}$$

(Eq. 12.8)

where:

F = the quantity of fuel required for the period
DD = the degree days for the period (F-day)
Q = the total calculated heat loss based on design conditions, t_i and t_o (BTU/hr)
η = an efficiency factor that includes the effects of rated full-load efficiency, part-load performance, oversizing, and energy conservation devices
H = the heating value of the fuel (BTU/unit volume of mass)
C_D = interim correction factor for degree days based on 65°F. $C_D = 0.62$ (Fig. 12.11) for the 6,160 degree days used for State College, PA.

Figure 12.11
Correction Factor
versus Degree
Days
*Source: 1989
ASHRAE
Handbook of
Fundamentals*
(Atlanta, GA:
American Society
of Heating,
Refrigerating and
Air Conditioning
Contractors, 1989),
Chapter 28, Fig. 1,
p. 28–2. Reprinted
with permission.

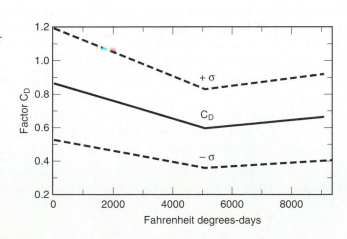

THE BIN METHOD

One of the shortcomings of the degree day method is that it does not consider the effect of temperature when heat pumps are used. As outdoor temperatures go down, so does heat pump performance. Degree day calculations are not based on hourly temperatures, but rather on daily average temperatures.

To account for this effect, the **bin method** can be used. This method groups the same temperatures into bins and performs energy calculations for those hours with the equipment operating under those conditions. This method can be performed manually, but a computer solution is recommended.

THE HOURLY ANALYSIS METHOD

The **hourly analysis method** is the most complicated procedure and typically requires a computer. In this method, heat balances are performed on each room surface on an hour-by-hour basis, resulting in hourly loading conditions. This information is then used to determine the required flow rates, temperatures, humidities, and coil data for the air system. Power consumption and energy use data can then be simulated for the central plant.

COMPARISON OF METHODS

Each of these methods provides energy estimates with different degrees of accuracy. The degree day method can be used to manually produce reasonable estimates in a short period of time or to compare heating requirements from one location to another. Accuracy is increased by using the more complex bin method and hourly analysis procedures, using a computer to perform the simulation calculations.

Sizing and Routing of the Ductwork System

Once the design day heating and cooling load calculations have been completed, the design of the ductwork system can be performed. A design procedure which produces satisfactory results is presented in *Manual D: Duct Design for Residential Winter and Summer Air Conditioning*, published by the Air Conditioning Contractors of America (ACCA).[18] The steps involved include:

1. Calculate the air flow that is required for the entire house based upon both the **design heat loss** and the **sensible heat gain**.

2. Size the fan based on the greater of the two air flows calculated for the design heat loss and the sensible heat gain.

3. Calculate the air flow that is required for each room. The design room air flow is the greater of the two air flows calculated from either the room heat loss or the sensible heat gain.

4. Determine the available static pressure of the selected fan.

5. Determine the total effective length of each of the duct runs, including the return ducts and the supply ducts.

6. Determine the friction rate.

7. After the air flow and friction rate are known, the duct sizes can be determined by using a friction chart or a duct slide rule.

Although the ductwork system for a house can be designed manually, the process can be expedited by using a number of readily available microcomputer-based energy analysis programs.

Application of the Design Procedure to the Case Study House

Sizing of the HVAC System—Manual Approach

DESCRIPTION OF THE CASE STUDY HOUSE

The Case Study House is a 1,680 sq ft, two-story house (Fig. 12.12). It contains three bedrooms, formal living and dining rooms, a large kitchen with a nook, a family room, and a two-car garage. The set of plans for this house is included in Appendix A.

ASSUMED ENVELOPE FEATURES OF THE CASE STUDY HOUSE

A summary of the assumed envelope features for the Case Study House, as well as the source of these features, is included in Table 12.7. The actual components were identified from the drawings when possible; otherwise, a realistic assumption was made. It should be noted that the basement is assumed to be unheated, so the primary heat loss is from the first floor into the basement. If the basement were assumed to be heated, then the primary heat loss would be from the basement through the basement walls and slab into the ground. A design procedure which addresses the heated basement situation is presented in Chapter 25 of the *1993 ASHRAE Handbook of Fundamentals*.[19]

INITIAL SELECTION OF THE HEATING AND COOLING EQUIPMENT FOR THE CASE STUDY HOUSE

It is assumed that a high-efficiency (96%) gas furnace is being used to heat the house. This system, although expensive in initial cost compared to electric baseboard heaters, is relatively inexpensive to operate. It is also assumed that central air conditioning will be provided for the Case Study House. The cooling energy calculations are not shown, however,

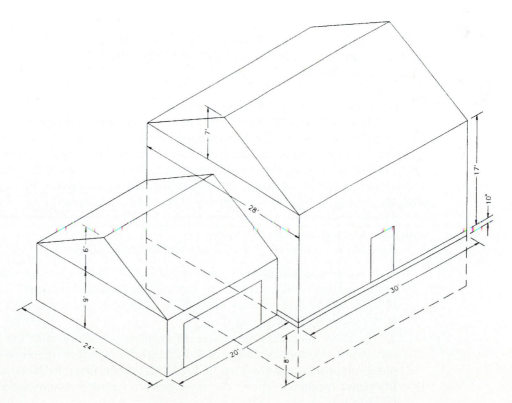

Figure 12.12
Important
Dimensions of the
Case Study House

TABLE 12.7   Envelope Features of the Case Study House

System	Component	Source
Exterior Wall	2×4″ studs @ 16″ o.c. with R-13 fiberglass batt insulation (with vapor barrier), 1/2″ insulated sheathing	House Plans: Appendix A
Second-Floor Ceiling/Attic	Wood trusses @ 24″ o.c. with R-38 blown-in fiberglass insulation (above the second-floor ceiling)	House Plans: Appendix A
Windows	R-2.75 insulating glass	Assumption
First Floor	3/4″ tongue-and-groove flooring supported by 2×10″ floor joists @ 16″ o.c., R-19 fiberglass, kraft paper faced, batt insulation between the basement and the first floor	House Plans: Appendix A
Basement	Unheated basement with uninsulated walls and slab	House Plans: Appendix A
Infiltration	0.5 air changes per hour (ACPH)	Assumption
Doors	R-4 wood doors	Assumption
Interior Wall Finish	5/8″ gypsum board	Assumption
Interior Floor Finish	Carpet	Assumption
Exterior Wall Finish	Vinyl siding	Assumption

because they are more difficult to perform and are beyond the scope of this book. The interested reader is referred to *Manual J: Load Calculation for Residential Winter and Summer Air Conditioning*[20] for an excellent discussion of the calculations required.

DESIGN DAY SELECTION FOR THE CASE STUDY HOUSE

It is assumed that the house will be constructed in State College, PA. Reference to Table 12.5 indicates that the winter 97-1/2% design temperature is 7°F.

**CALCULATION OF THE U-VALUES AND THE AREAS
OF THE EXPOSED BUILDING ENVELOPE FOR THE
CASE STUDY HOUSE**

A. Ceiling between the Second Floor and the Attic

Figure 12.4 presented a simplified approach to analyzing a typical ceiling subsystem. A close examination of Fig. 12.4, however, reveals that there are actually two different paths available for heat to migrate from the second floor up to the attic: (1) entirely through the blown-in fiberglass insulation and (2) through the bottom chord of the truss and the blown-in fiberglass insulation above it. If the combined effects of each of these paths are considered, the overall R-value of the ceiling will be slightly less than if there were only blown-in fiberglass insulation above the ceiling. Figure 12.13 illustrates the R-value calculation procedure for this parallel path situation.

Several points related to the calculation should be noted:

1. It is assumed, based on the data in Table 12-1, that an *average* value of R-1.23/inch can be used to represent Hem–Fir lumber (the indicated range is 1.35–1.11). Thus, the R-value of the 2×4 in. truss bottom chord (actual depth = 3-1/2″) is assumed to be R = (1.23/in.)(3.50 in.) = 4.31.

2. The R-38 value must be reduced in the thermal path through the 2×4 in. truss bottom chord. An approximation of R-2.25/inch for blown-in fiberglass insulation is used based upon the data provided in Table 12.2. Therefore, the R-value of the blown-in fiberglass insulation above the truss bottom chord is assumed to be R = 38.0 − (2.25/in.)(3.5 in.) = 30.13.

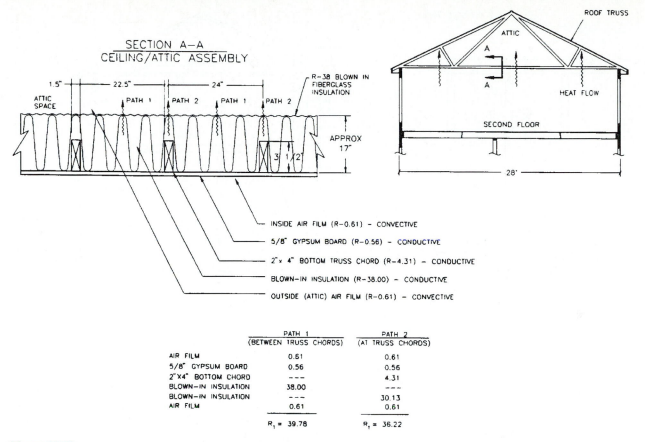

Figure 12.13
Second Floor Ceiling R-Value Calculation

3. In order to calculate the overall R-value for the dual-path condition, the fractional contribution of each path must be considered. As noted in Fig. 12.13, the roof trusses are spaced 24 in. o.c. This means that for every 24 in., there is 1.5 in. of truss and 22.5 in. of insulation.

The appropriate calculations for the overall assembly are:

Component	Between Truss R-Value	At Truss Chord R-Value
Inside Air	0.61	0.61
5/8″ Gypsum Board	0.56	0.56
2×4″ Bottom Chord	—	4.31
Blown-in Insulation (Path 1)	38.00	—
Blown-in Insulation (Path 2)	—	30.13
Attic Air	0.61	0.61
Total:	$R_{ins} = 39.78$	$R_{chord} = 36.22$

$$(\text{ft}^2\text{-hr-}°\text{F/BTU})$$

$$\text{U-value} = 1/\text{R}: U_{ins} = 0.0251 \quad U_{chord} = 0.0332 \ (\text{BTU/ft}^2\text{-hr-}°\text{F})$$

$$U_{clg} = [(U_{ins} \times A_{ins}) + (U_{chord} \times A_{chord})]/A_{tot}$$

$$= [(0.0251 \times 22.5) + (0.0332 \times 1.5)]/24$$

$$= (0.56475 + 0.0498)/24 = 0.6146/24$$

$$= 0.0256 \ (\text{BTU/ft}^2\text{-hr-}°\text{F})$$

$$R_{clg} = 1/U_{clg} = 1/0.0256 = 39.10 \ (\text{ft}^2\text{-hr-}°\text{F/BTU})$$

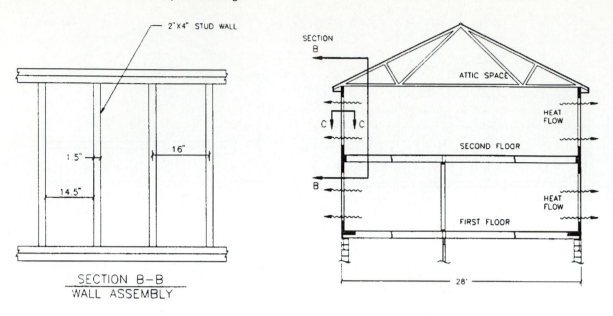

SECTION B-B
WALL ASSEMBLY

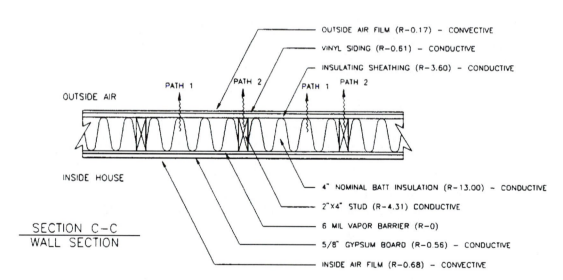

SECTION C-C
WALL SECTION

Figure 12.14
Exterior Wall R-Value Components

This example indicates that when the effects of the two paths for heat migration are considered, there is a slight reduction in the R-value at the truss chord from R-39.78 to R-39.10.

B. Exterior Walls

There are also two paths available for heat to migrate through the exterior walls: (1) through the insulation and (2) through the stud. Since the 2×4 in. studs are spaced 16 in. o.c., for every 16 in. there is 1.5 in. of stud and 14.5 in. of insulation. Figure 12.14 illustrates the R-value calculation procedure for this parallel path situation. Several points related to the calculation should be noted:

1. The value that was assumed for the convective resistance factor for the outside air film of R-0.17 considers the fact that true "outside" conditions with moving air apply. The convective resistance factor for the inside air film is assumed to be R-0.68.

2. Table 12.1 does not provide an R-value for vinyl siding, so the R-value corresponding to hollow-backed aluminum or steel siding (i.e., R-0.61) is used as a reasonable approximation.

3. Table 12.1 indicates that the R-value per inch thickness of cellular polyisocyanurate is 7.20. Thus, an R-3.60 is assumed for the 1/2 in.-insulated sheathing in the Case Study House.

The appropriate calculations for the overall assembly are:

Component	Between Stud R-Value	At Stud R-Value
Outside Air	0.17	0.17
Vinyl Siding	0.61	0.61
1/2″ Insulating Sheathing	3.60	3.60
4″ (nom.) Batts	13.00	—
2×4″ Stud	—	4.31
6 Mil. Vapor Barrier	—	—
5/8″ Gypsum Board	0.56	0.56
Inside Air	0.68	0.68

$$\text{Total} \qquad R_{ins} = 18.62 \qquad R_{stud} = 9.93$$
$$(\text{ft}^2\text{-hr-}°\text{F/BTU})$$
$$\text{U-Value} = 1/R: U_{ins} = 0.0537 \qquad U_{stud} = 0.1007 \ (\text{BTU/ft}^2\text{-hr-}°\text{F})$$
$$U_{ew} = [(U_{ins} \times A_{ins}) + (U_{stud} \times A_{stud})]/A_{tot}$$
$$= [(0.0537 \times 14.5) + (0.1007 \times 1.5)]/16$$
$$= 0.0581 \ (\text{BTU/ft}^2\text{-hr-}°\text{F})$$
$$R_{ew} = 1/U_{ew} = 1/0.0581 = 17.2 \ (\text{ft}^2\text{-hr-}°\text{F/BTU})$$

C. Windows
Double-pane insulating glass:

$$\text{R-value} = 2.75$$
$$\text{U-value} = 1/R = 1/2.75 = .364$$

D. Doors
2-1/2 in. wood door with storm door:

$$\text{R-value} = 4.0$$
$$\text{U-value} = 0.25$$

E. First Floor/Basement
There are also two paths available for heat to migrate from the first floor into the basement: (1) through the insulation and (2) through the floor joists. Since the 2×10 in. floor joists are spaced 16 in. o.c., for every 16 in. there is 1.5 in. of joist and 14.5 in. of insulation. Figure 12.15 illustrates the R-value calculation procedure for this parallel path situation.

Several points related to the calculation should be noted:

1. Table 12.1 indicates that 3/4-in. Douglas Fir plywood sheathing has an R-0.93 and that carpet with a rubber pad has an R-1.23.

2. The R-value for a 2×10 in. Hem-Fir floor joist (actual depth = 9-1/4 in.) is assumed to be:

$$R = (1.23)(9.25) = 11.38$$

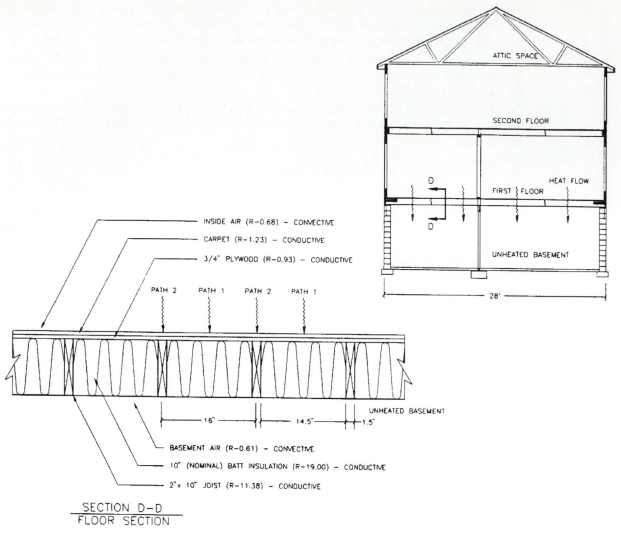

Figure 12.15
First Floor/Basement R-Value Components

The appropriate calculations for the overall assembly are:

Component	Between Joist R-Value	At Joist R-Value
Inside air	0.68	0.68
Carpet	1.23	1.23
3/4″ plywood	0.93	0.93
2×10″ joist	—	11.38
6″ (nom.) batts	19.00	—
Basement air	0.61	0.61

Total: $\qquad$ $R_{ins} = 22.45$ $\qquad$ $R_{joist} = 14.83$
$$(\text{ft}^2\text{-hr-}°\text{F/BTU})$$

U-Value = $1/R_{tot}$: $U_{ins} = 0.0445$, $U_{joist} = 0.0674$ (BTU/ft^2-hr-°F)

$$U_{ff} = [(U_{ins} \times A_{ins}) + (U_{joist} \times A_{joist})] / A_{tot}$$
$$= [(0.0445 \times 14.5) + (0.0674 \times 1.5)]/16$$
$$= 0.0466 \text{ (BTU/ft}^2\text{-hr-}°\text{F)}$$
$$R_{ff} = 1/0.0466 = 21.4 \text{ (ft}^2\text{-hr-}°\text{F/BTU)}$$

TABLE 12.8   Summary of Subsystem U-Values and Areas

Subsystem	R-Value (hr-ft²-°F/BTU)	U-Value (BTU/hr-ft²-°F)	Area (ft²)	U × Area (BTU/hr-°F)
Ceiling/Attic	39.1	0.0256	840[a]	21.5
Exterior Walls	17.2	0.0581	1702[b]	98.9
Windows	2.75	0.364	232	84.4
Doors	4.0	0.25	38	9.5
First Floor/ Basement	21.4	0.0466	840[c]	39.1

[a]Ceiling area = 30′ × 28′ = 840 sq ft (see Fig. 12.18).
[b]Exterior wall area
   = 17′ × [30′+30′+28′+28′] − [232 ft² + 38 ft²]
   = 1702 sq ft (see Fig. 12.12).
[c]First-floor area = 30′ × 28′ = 840 sq ft (see Fig. 12.12).

These calculations of the U-values and the areas of the exposed building envelope are summarized in Table 12.8.

SIZING OF THE HVAC SYSTEM

A. Determination of the Conductive Design Heat Losses for the Case Study House
   The total conductive design heat loss equation becomes:

$$Q = \text{sum of } [U \times A \times \Delta T] \text{ for all of the subsystems}$$

where

   Q = total heat flow (BTU/hr) from the house on the design day
   U = overall heat transfer coefficient (BTU/hr-ft²-°F) for each subsystem
   A = exposed envelope surface area (ft²) for each subsystem
   ΔT = design temperature difference (or an adjustment thereof) for each subsystem

Both the **above-grade** and the **below-grade** situations must be considered when the appropriate values of ΔT are determined. For the above-grade situation, it has been noted previously that the 97-1/2° winter design temperature for State College, PA, is 7°F. If an indoor temperature of 72°F is assumed, then:

$$\Delta T = 72°F - 7°F = 65°F \text{ (the design temperature difference above grade)}$$

When the appropriate ΔT value across floors over unheated basements or enclosed crawl spaces is analyzed, consideration must be given to the design details associated with the foundation wall as well as the insulation value of the backfill around the walls. If less than 50% of the wall height is below grade, the **design temperature difference** is sometimes reduced by 50% or a modified design temperature difference based upon local design considerations is used.[21]

For the Case Study House example, Section AA on Sheet A-5 in Appendix A indicates that all of the basement is below grade. Since the HVAC unit will be located in the basement, it will therefore be assumed that the ambient temperature in the basement will be 50°F on the coldest day of the year. Thus:

$$\Delta T = 72°F - 50°F = 22°F \text{ (the adjusted design temperature difference between the first floor and the basement)}$$

TABLE 12.9   Subsystem Conductive Heat Losses

Subsystem	$U \times A$ (BTU/hr-°F)	ΔT (°F)	$U \times A \times \Delta T$ (BTU/hr)
Ceiling/Attic	21.5	65	1,398
Exterior Walls	98.9	65	6,429
Windows	84.4	65	5,486
Doors	9.5	65	618
First Floor/Basement	39.1	22[a]	860
			14,791

[a]Adjusted temperature difference due to unheated basement assumptions.

The conductive heat losses for each of the subsystems are summarized in Table 12.9. As noted, the total conductive heat loss $Q = 14,791$ BTU/hr.

B. Determination of the Infiltration Heat Losses for the Case Study House

The total infiltration heat losses using the **Air Change Method** are as follows (assuming an air change rate of 0.5 ACPH):

$$\text{First-Floor Volume} = 28 \times 30 \times 8 = \quad 6,720 \text{ ft}^3$$
$$\text{Second-Floor Volume} = 28 \times 30 \times 8 = \quad 6,720 \text{ ft}^3$$
$$\text{Total} \quad 13,440 \text{ ft}^3$$

Thus, Q, the volume of outdoor air entering the house per hour, is (0.5 ACPH) (13,440 ft^3) = 6,720 ft^3/hr. According to Eq. 12.5:

$$Q_{\text{INfl}} = 0.018\, Q(t_i - t_o)$$
$$= 0.018(6720)(72 - 7) = 7,862 \text{ BTU/hr}$$

C. Determine the Total Design Day Heat loss for the Case Study House

$$Q_{\text{tot}} = Q_{\text{cond}} + Q_{\text{infl}}$$
$$= 14,791 + 7,862$$
$$= 22,653 \text{ BTU/hr}$$

D. Determine the Required Size of the Heating System for the Case Study House— manual approach

$$Q_{\text{sys}} = \frac{22,653 \times 1.25}{0.96}$$
$$Q_{\text{sys}} = 29,496 \text{ BTU/hr}$$

Estimate of the Annual Fuel Consumption for the Case Study House—Degree Day Method

To illustrate the process of manually estimating fuel consumption, the degree day method will be used for the Case Study House design. This procedure, as noted earlier, provides only an approximate estimate of the annual fuel consumption because of the inherent shortcomings associated with it. Nonetheless, it is a straightforward analytical technique that can be calculated quickly after the heat loss and mechanical system parameters have been established. For the Case Study House design, the quantity of fuel required for the system is estimated to be:

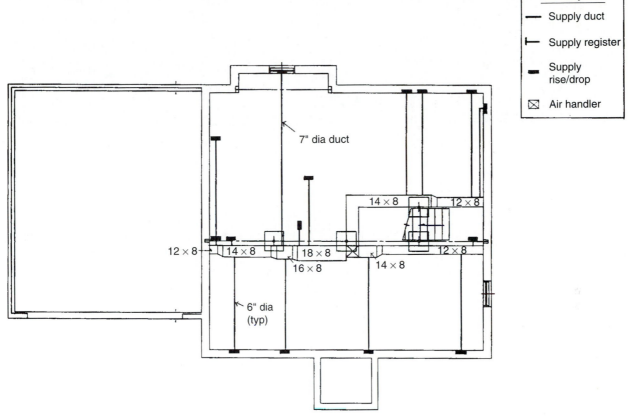

Figure 12.16
Ductwork Design for the Case Study House Basement (Supply)
Source: Information provided by the Comfort Home Corporation, Lancaster, PA (August 1995)

$$F = \frac{24(DD)Q\,C_D}{\eta(t_i - t_o)H} = \frac{24(6,160)(29,496)(0.62)}{1.0(72 - 7)(100,000\ \text{BTU/therm})} = 416\ \text{therms/yr}$$

Table 12.5 can be used to determine the volume of natural gas that will be required. The cost of heating the Case Study House could then be determined by using a locally determined cost of natural gas.

Sizing and Routing of the Ductwork System— Manual Approach

The ductwork sizing and routing calculations for the Case Study House are not shown because, as indicated earlier, they are difficult to illustrate in a concise manner and therefore are considered to be beyond the scope of this chapter. Although a complete ductwork system design is not provided, Figs. 12.16 and 12.17 indicate the final results of the design of the supply and return ductwork in the basement of the Case Study house that were based upon the procedures provided in *Manual D: Duct Design for Residential Winter and Summer Air Conditioning.*[22]

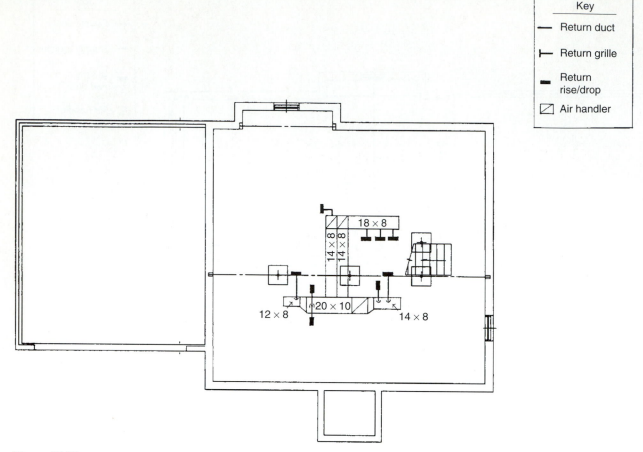

Key

— Return duct

⊢ Return grille

▬ Return rise/drop

▱ Air handler

18 × 8

14 × 8
14 × 8

12 × 8   20 × 10   14 × 8

Figure 12.17
Ductwork Design for the Case Study House Basement (Return)
Source: Information provided by the Comfort Home Corporation, Lancaster, PA (August 1995)

ENERGY-EFFICIENCY ISSUES

With energy efficiency in housing becoming increasingly important, builders and home-owners must be aware of opportunities that exist to help them meet this goal. Electric utilities throughout the United States now offer incentive programs that promote building energy-efficient homes. In addition, building code requirements have become more stringent, and thus more effective, in promoting energy-efficient homes.

Life cycle costing techniques can also provide information regarding the optimum selection of HVAC equipment and building envelope features. Since homebuyers in the United States typically move every 8 to 10 years, an alternative which uses a **monthly cost basis** (i.e., energy savings/payments for improvements per month) has become popular.

Utility/Builder Programs

Many electric utilities throughout the United States offer incentives to builders and/or homeowners in order to promote the construction of energy-efficient homes. Ten utilities, for example, offer a program developed by the Comfort Home Corporation which includes prescriptive and performance standards including 100% final inspection of all homes that are included in the program. The **Comfort Home program** is marketed to ensure comfort and energy savings and consists of the following requirements:[23]

1. Insulation levels exceeding the normal state requirements.

Ceilings:

With unheated space above	R-38
Cathedral/sloped ceilings	R-30

Walls:

Exterior	R-19
Bandjoists	R-19

Floors:

Over vented crawl spaces, garages	R-19
Over unvented crawl spaces	R-19
Over unheated basements	R-19
Concrete slab/perimeter insulation	R-10

Basement Walls:

Conditioned, wall surface greater than 50% above grade	R-19
Conditioned, wall surface less than 50% above grade	R-13

Windows and Doors:

Windows	R-2.17
Sliding glass doors	R-2.17
Exterior doors	R-5
Glass in door	R-1.82

Ducts, Supply and Returns:

All ductwork in unconditioned spaces	R-5

2. Infiltration and thermal integrity control, with an air infiltration rate not exceeding 0.5 air changes per hour. In addition, infiltration is not to be concentrated in any one area but is to be spread throughout the dwelling to ensure comfort.

3. Proper equipment sizing and system design based on room-by-room heat loss and heat gain calculations.

4. Energy-efficient and properly insulated water-heating systems.

5. High-efficiency appliances.

6. Thermal integrity enhancements, during construction, to improve insulation performance and reduce infiltration.

7. Periodic inspection of the building shell and mechanical systems during construction to ensure the quality of the installation.

8. Final inspection of insulation efficiency, mechanical equipment performance, ductwork tightness, and building infiltration to ensure that all standards are met.

9. A warranty that guarantees energy savings and customer comfort.

The Comfort Home Program has been honored by governmental and environmental groups for its proactive role in the incorporation of better building practices in today's residential construction. Utilizing existing materials and subcontractor infrastructures, the Comfort Home approach typically increases the energy efficiency of new houses by 25–40%. This is accomplished by increased quality control of every stage of construction in order to increase ductwork efficiency and thermal integrity.

Code Requirements

The 1995 edition of the *Model Energy Code*[24] (MEC) is an example of a code that contains regulations regarding energy usage in residential buildings. The MEC is promulgated jointly by Building Officials and Code Administrators International, Inc. (BOCA); International Conference of Building Officials (ICBO); and Southern Building Code Congress International, Inc. (SBCCI), under the auspices of the Council of American Building

TABLE 12.10   Minimum Acceptable Subsystem R-Values

| Components | Minimum Total R-Values for Specified Degree Day Locations | | |
	2,000	4,000	6,000
Roof/Ceilings	26.00	29.00	38.46
Walls with less than 14% Windows and Doors	7.02	9.34	13.96
Slab on Grade	6.00	6.00	7.20
Floor over Unheated Spaces	14.29	20.00	20.00
Crawl Space Walls	6.67	11.11	16.67
Basement Walls	6.32	8.92	10.30

Source: Interpretation of the requirements on pp. 19–50 of the 1995 edition of the *CABO Model Energy Code*,™ copyright © 1995, with the permission of the publisher, The International Codes Council, Whittier, CA. Provided by Comfort Home Corporation, Lancaster, PA (August 1995).

TABLE 12.11   Minimum Performance Requirement
of the Common Mechanical Systems

System	Minimum Performance
Air Source Heat Pump	6.8 HSPF
Water Source Heat Pump	3.8 COP (Standard Rating)
Groundwater Source Heat Pump	3.4 COP (High-Temperature Rating)
	3.0 COP (Low-Temperature Rating)
Gas Furnaces	78% AFUE
Central Air Conditioner	10.0 SEER

Source: Interpretation of the requirements on pp. 19–21 of the 1995 edition of the *CABO Model Energy Code*,™ copyright © 1995, with the permission of the publisher, The International Codes Council, Whittier, CA. Provided by Comfort Home Corporation, Lancaster, PA (August 1995).

Officials (CABO). The MEC is updated annually and a new edition is published approximately every three years. The code contains regulations regarding (1) the building envelope, (2) mechanical systems, (3) service water heating, (4) electrical power and lighting, and (5) calculation methodologies for tradeoffs between all components with minimum standards for each.

The minimum subsystem R-values defined by the MEC for locations with approximately 2,000, 4,000, and 6,000 Heating Degree Days, for example, are presented in Table 12.10. The MEC should be consulted to identify the specific requirements in a particular location.

The minimum system efficiencies required by the MEC for the most common types of heating and cooling equipment are shown in Table 12.11.

CHAPTER SUMMARY

This chapter has provided an overview of HVAC system design. It began by briefly discussing energy consumption and the basics of energy movement. The key elements in the residential envelope and some of the alternatives that exist with those elements and mechanical systems were covered next. A set of manual calculations that can be used to determine the heating performance of a given house was provided. Finally, an overview of energy-efficiency issues that promote the building of energy-efficient homes was provided.

NOTES

[1] *CABO One and Two Family Dwelling Code, 1995 Edition*   (Falls Church, VA: The Council of American Building Officials, 1995).

[2] Ibid., p. 327.

[3] *Model Energy Code, 1995 Edition*   (Falls Church, VA: The Council of American Building Officials, 1995).

[4] 1995 CABO Code, pp. 329–330.

[5] Charlie Wing, *The Visual Handbook of Building and Remodeling*, p. 282   (Emmaus, PA: Rodale Press, 1990).

[6] Benjamin Stein and John Reynolds, *Mechanical and Electrical Equipment for Buildings, 8th Edition*, p. 116 (New York: John Wiley & Sons Inc., 1992).

[7] *Model Energy Code, 1995 Edition*, pp. 21–53   (Falls Church, VA: The Council of American Building Officials, 1995).

[8] *1993 ASHRAE Handbook of Fundamentals*, pp. 24.4–24.15   (Atlanta, GA: American Society of Heating, Refrigeration and Air Conditioning Engineers, 1993).

[9] Ibid, p. 23.10.

[10] Ibid, pp. 23.1–23.2.

[11] Ibid., pp. 25-13–25.14.

[12] Ibid.

[13] *Manual J: Load Calculation for Residential Winter and Summer Air Conditioning, 7th Edition*   (Washington, D.C.: Air Conditioning Contractors of America, 1986).

[14] *Pennsylvania Heating Systems Manual*, p. 19   (Harrisburg, PA: The Pennsylvania Energy Office, 1989).

[15] *Manual J*.

[16] Ibid, pp. 24.4–24.15.

[17] *Hourly Analysis Program*   (Syracuse, NY: Carrier Corporation, 1988).

[18] *Manual D: Duct Design for Residential Winter and Summer Air Conditioning, 8th Ed.*   (Washington, D.C.: Air Conditioning Contractors of America, 1995).

[19] *1993 ASHRAE Handbook*, pp. 25.10–25.13.

[20] *Manual J*, pp. 27–42.

[21] Information provided by Comfort Home Corporation, Lancaster, PA (August 1995).

[22] *Manual D*.

[23] Information provided by the Comfort Home Corporation, Lancaster, PA (August 1995).

[24] *Model Energy Code*, 1995 Edition.

BIBLIOGRAPHY

1989 ASHRAE Handbook of Fundamentals.   Atlanta, GA: American Society of Heating, Refrigeration and Air Conditioning Engineers, 1989.

1993 ASHRAE Handbook of Fundamentals.   Atlanta, GA: American Society of Heating, Refrigeration and Air Conditioning Engineers, 1993.

CABO One and Two Family Dwelling Code, 1995 Edition.   Falls Church, VA: The Council of American Building Officials, 1995.

Hourly Analysis Program.   Syracuse, NY: Carrier Corporation, 1988.

Manual D: Duct Design for Residential Winter and Summer Air Conditioning, 8th Ed.   Washington, D.C.: Air Conditioning Contractors of America, 1995.

Manual J: Load Calculation for Residential Winter and Summer Air Conditioning, 7th Edition.   Washington, D.C.: Air Conditioning Contractors of America, 1986.

Model Energy Code, 1992 Edition.   Falls Church, VA: The Council of American Building Officials, 1992.

Model Energy Code, 1993 Edition.   Falls Church, VA: The Council of American Building Officials, 1993.

Model Energy Code, 1995 Edition.   Falls Church, VA: The Council of American Building Officials, 1995.

Pennsylvania Heating Systems Manual.   Harrisburg, PA: The Pennsylvania Energy Office, 1989.

Suchar, Michael, Gren Yuill, Harvey Manbeck, and Jack Willenbrock, *An Optimization Analysis of Residential Energy Systems: Housing Research Center at Penn State Report No. 25*.   University Park, PA: The Pennsylvania State University, 1993.

Stein, Benjamin and John Reynolds, *Mechanical and Electrical Equipment for Buildings, 8th Edition*.   New York, NY: John Wiley & Sons, Inc., 1992.

Stricker, Saul, *Optimizing Performance of Energy Systems*.   Columbus, OH: Battelle Press, 1985.

Wing, Charlie, *The Visual Handbook of Building and Remodeling*.   Emmaus, PA: Rodale Press, 1990.

13

Plumbing System Design

INTRODUCTION

This chapter discusses the design fundamentals related to (1) the pressurized water supply and distribution system and (2) the sanitary drainage system of a house. The latter, which is often called the DWV (drain, waste and vent) system, relies on gravity to make its mixture of liquids and solids flow. Chapters 29 through 38, as well as Appendices B and C, of the 1995 CABO Code,[1] provide guidance about the requirements related to these two systems. An in-depth coverage of these chapters is not provided here, but the requirements are used as a framework for discussing plumbing system design in residential construction.[*]

WATER SUPPLY AND DISTRIBUTION SYSTEM

Water

Water, in relation to its use in residential buildings, can be categorized as either potable or nonpotable. **Potable water**, which is suitable for human consumption, must be available in a house to satisfy the drinking and cooking needs. **Nonpotable water** exists in two forms: either as **gray water** or **black water**. Gray water is not suitable for human consumption, but it may be used for flushing water closets (toilets), watering grass and gardens, washing cars, and for any use other than drinking and cooking. Black water, which contains toilet waste, must be treated once it leaves the house.

Rain, the source of most of the water available for use, replenishes both surface water and groundwater. **Surface water** occurs when rain runs off the surface of the ground into streams, rivers, and lakes. **Groundwater** is the water that percolates through the soil, building the supply of water below the earth's surface. The water level formed below the surface of the earth due to groundwater at a particular location is known as the **water table**.

---

*Sections and tables from the 1995 CABO Code are cited only for instructional purposes, and the treatment of the Code provisions contained in the design solutions illustrated in this chapter is not intended as a substitute for a complete and thorough understanding of all building code provisions to which any particular project may be subject.

Public versus Private Water Systems

The water supply system of any house must be connected to a **public water system** if it is available. Before proceeding with the design of the water supply system, it is important to determine the location of the water main, the charge to connect to the system, and the available pressure in the main at the property line.

When public systems are not available, private water supply systems are installed. **Private water supply systems** usually consist of wells that remove water from the groundwater beneath the water table. The required depth of the well is determined by the depth of the water table below the ground surface and the rate at which water must be pumped.

Components of a Water Supply System in a House

The primary components of the typical water supply system are shown in Fig. 13.1 and discussed next.

Water Meter

Most public water supply systems require the installation of a meter to measure and record the amount of water being used. The meter may be located in the ground, near the street, or inside the house (usually in the basement).

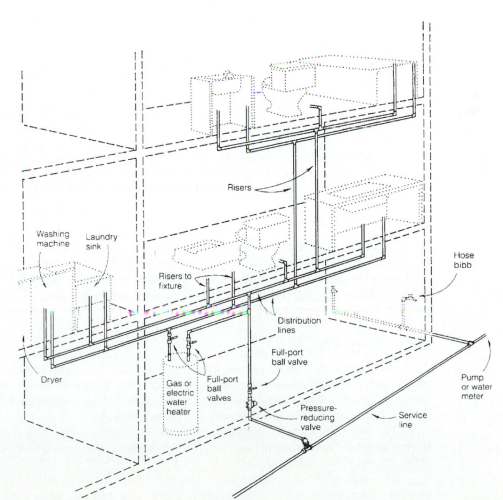

Figure 13.1
A Typical Fresh
Water Supply and
Distribution
System
Source: Peter
Hemp, *Plumbing a
House* (Newtown,
CT: The Tauton
Press, Inc., 1994),
p. 27.

Water Service Line

The outside pipe connecting the water main or other source of potable water supply to the water distribution system inside the house is called the **water service line**. The minimum pressure rating for water service piping, as noted in the 1995 CABO Code, must be 160 psi at 73°F.[2] Sizing requirements, the details of which are discussed later in this chapter, are also mentioned in the 1995 CABO Code:

> The water-service pipe shall be of sufficient size to furnish water to the dwelling in required quantities and pressures, but in no case shall be less than 3/4 in. nominal diameter. Exact sizing to account for total demand and for pressure drop due to friction loss shall be determined in accordance with the procedure outlined in Appendix C or in accordance with the tables in Section 3409.6.[3]

Building Main Water Supply System

The building main connects to the public or private water supply system and extends into the house to the highest riser. The building main is typically located in the basement, in a crawl space, or below the concrete floor slab. Cold water (under pressure) is brought into the house through the water meter and a main shut-off valve (Fig. 13.1). A back flow preventer of some type is usually included in the system at this point.

The cold-water supply system, which sometimes first passes through a water softener, then branches to all fixtures and water-using appliances inside, and to hose bibbs outside, the house. The hot water supply system is created by passing a portion of the cold water through a domestic water heater.

Typical pressure requirements for the water supply system are presented in the 1995 CABO Code, as follows:

> **3403.2 Pressure**. Minimum average static pressure (as determined by the local water authority) at the building entrance for either public or private water service shall be 40 psi.
>
> **3403.2.1 Outlet discharge**. The minimum pressure at the point of outlet discharge shall not be less than a flow pressure of 8 psi for all fixtures except where manufacturers require a higher pressure. In determining the minimum pressure, allowance shall be made for pressure losses during maximum demand periods. Pressure and flow requirements to special fixtures, such as low one-piece toilets, shall be determined prior to sizing the piping system.
>
> **3403.2.2 Pressure-reducing valve**. Maximum average static pressure shall be 80 psi. When main pressure exceeds 80 psi, an approved pressure-reducing valve shall be installed on the domestic water branch main or riser at the connection to the water-service pipe.[4]

Domestic Water Heater

A water heater is required in all houses. Water heaters come in a variety of sizes; can use oil, gas, electricity, or even solar energy as a fuel source; and can be located in the basement or crawl space, in a closet or cabinet, or even under a counter. The 1995 CABO Code[5] provides guidelines for determining minimum water heater tank size. Water heaters must be protected by relief valves to control both excessive pressure and temperature conditions.

Other Components

The other basic components include the various parts of the piping system. A **riser** is a water pipe which extends vertically for one full story or more to convey water to branches or to a group of fixtures. The water supply pipe between the fixture supply and a main water-distribution pipe or group main is known as a **fixture branch**.

The fixture branch is connected to a fixture or fixture fitting by a pipe section called a **fixture supply**. The **fixture fitting** is any device which controls or guides the flow of water into or conveys water from a fixture.

The closing of valves in the water supply system may cause the water to stop suddenly and cause the pipes to rattle (commonly referred to as **water hammer**). **Air chambers/shock**

absorbers are usually installed on all hot and cold water pipes at fixtures with valves such as faucets, hose bibbs, dishwashers, clothes washers, tubs, showers, and water closets to prevent that occurrence. These air chambers typically are twice the diameter of the pipe on which they are mounted and are a minimum of 18 in. in length. They usually are hidden in the wall cavity adjacent to the nearest fixture or device they are protecting from inertial shock.

Plumbing Fixtures

A **plumbing fixture** is a device which requires both a water-supply connection and a discharge to the drainage system. Each fixture is assigned a Water Supply Fixture Unit (WSFU) value when the water supply system is designed. WSFUs represent the contribution of each fixture type to the peak flow rate. The plumbing fixtures commonly used in a house are discussed next.

Showers

Showers are commonly available in porcelain, enameled steel, or fiberglass. Shower components must have at least 900 sq in. of floor area and be of sufficient size to inscribe a circle with a diameter not less than 30 in.[6] If a separate shower compartment is not desired, a shower head may be installed over a bathtub.

Shower heads should be of the water-conserving type. All showers must be equipped with control valves of the pressure balance, thermostatic mixing, or the combination pressure balance/thermostatic mixing types with high limit stops set to limit water temperature to a maximum of 120°F.

Lavatories

Lavatories (i.e., bathroom sinks) are generally available in vitreous china or enameled iron. They are available in a large variety of sizes, and the shape may be square, rectangular, round, or oval. The lavatory may be wall hung, set on legs, or built into a cabinet.

Water Closets

Water closets (toilets) are constructed of solid vitrified china and are usually of the flush tank type. The flush tank water closet has a water tank as a part of the fixture. When the handle is depressed, the valve in the tank lifts and releases water to "flush out" the tank. When the handle is released, the valve drops, and the tank refills with water.

Bathtubs

Bathtubs are usually constructed of enameled iron, cast iron, or fiberglass. They are available in various sizes, the most common being 30 in. wide, 12 to 16 in. high, and 4 to 6 ft long. Fiberglass bathtubs are currently the most common alternative used in houses. If they are cast in one piece, they must be placed in their final location before the walls of the house are framed. Bathtubs must have control valves of the pressure balance, thermostatic mixing, or combination pressure balance/thermostatic mixing type with high limit stops set to limit water temperature to a maximum of 120°F.

Sinks

Sinks usually are made of enameled cast iron or stainless steel, and are available in single- or double-bowl arrangements. A garbage disposal is often connected to the sink in the kitchen.

Water Supply System Materials

A listing of the ASTM standards which define each acceptable type of material that can be used for "Water Service, Supply and Distribution" piping is provided in the *1995 CABO Code*.[7] Included are the various copper and plastic piping options which are used most commonly in present new house construction. Also included is the galvanized piping option which will typically be found in houses that were built more than 20 years ago.

Copper Pipe and Fittings

The three typical types of copper pipe used in the water supply system are: (1) hard supply pipe, (2) soft supply pipe, and (3) flexible tubing.

Hard supply pipe, which is sold in lengths up to 20 ft, comes in three wall thicknesses: M (thin wall), L (medium wall), and K (thick wall), with M usually being adequate for above-ground plumbing. This type of pipe must be cut and soldered with fittings whenever a change in direction is required, since it cannot be bent without crimping. Nominal diameters range from 1/4 to 1 in.

Soft supply pipe, which is sold in 30, 60 or 100 ft coils, can be bent around curves without crimping and therefore does not require change-of-direction fittings. Soft supply pipe is available in two wall thicknesses: L (medium wall) and K (thick wall). Nominal diameters range from 1/4 to 1 in.

Flexible tubing of corrugated or smooth copper or chrome-plated copper is available in short lengths to link supply pipe to fixtures. Nominal diameters of 3/8 to 1/2 in. are available.

The typical types of fittings that are used with copper pipe are shown in Fig. 13.2. A number of these fittings are soldered. This allows all of the pipes and fittings to be set into place before the joints must be soldered, generally allowing faster installation of copper pipe as compared to plastic. Another advantage of copper pipe is that the required spacing of the supports is larger than the required support spacing for plastic pipe.

Plastic Pipe and Fittings

Plastic pipe is available in either flexible or rigid form. The two common flexible types, which are sold as coils of tubing, are PB (polybutylene) and PE (polyethylene).

The three common varieties of rigid plastic pipe are: (1) PVC (polyvinyl chloride), (2) CPVC (chlorinated polyvinyl) chloride), and (3) ABS (acrylonitrile–butadiene styrene). CPVC pipe is more expensive than PVC pipe, but it can be used for both cold-water and hot-water systems. These piping materials must be able to sustain a minimum working pressure of 100 psi at 180°F.[8]

The typical fittings that are used with plastic pipe are shown in Fig. 13.3. The fittings shown in Fig. 13.3 are cemented in place with a permanent solvent—cement. Transition fittings—which join plastic pipe of a different material—often have threads on one end.

Galvanized Steel Pipe

Galvanized steel water supply pipe and fittings are coated with zinc to resist corrosion. Although stronger than either copper or plastic pipe, galvanized steel pipe has a high cost because the joints must be threaded. It is also corroded by soft (acidic) water, and it is susceptible to scale from hard-water deposits. This type of pipe, which currently is not widely used, will not be discussed further.

Valves

Valves (Fig. 13.4) control the flow of water through the water supply system. An accessible main shutoff valve is required, for instance, near the entrance of the water service into the

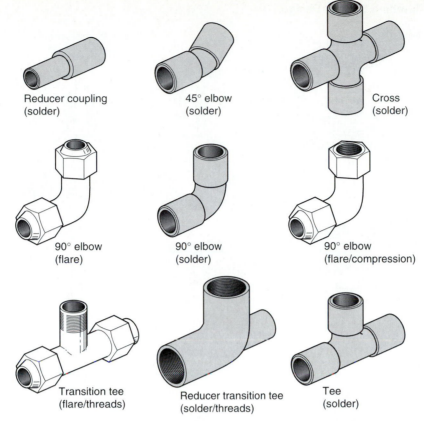

Figure 13.2
Copper Fittings for Rigid Copper Water Supply Pipe

Source: Sunset Basic Plumbing Illustrated (Menlo Park, CA: Sunset Publishing Corporation, 1991), p. 53.

Reducer coupling (solder)

45° elbow (solder)

Cross (solder)

90° elbow (flare)

90° elbow (solder)

90° elbow (flare/compression)

Transition tee (flare/threads)

Reducer transition tee (solder/threads)

Tee (solder)

Note: Transition tees are threaded when they connect copper pipe to galvanized pipe.

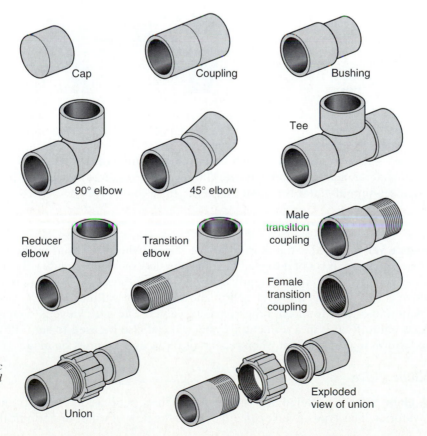

Figure 13.3
PVC and CPVC Fittings for Rigid Plastic Water Supply Pipe

Source: Sunset Basic Plumbing Illustrated (Menlo Park, CA: Sunset Publishing Corporation, 1991), p. 49.

Cap

Coupling

Bushing

90° elbow

45° elbow

Tee

Reducer elbow

Transition elbow

Male transition coupling

Female transition coupling

Union

Exploded view of union

395

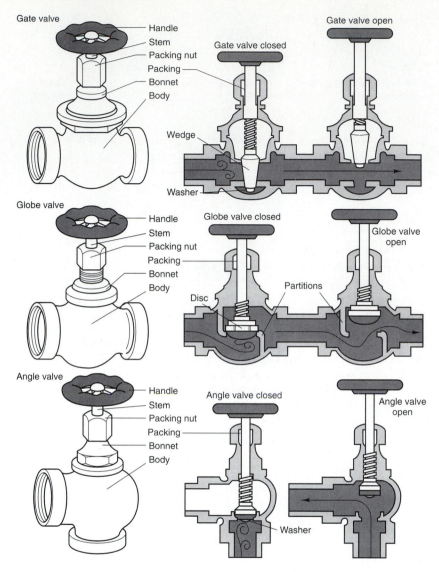

Figure 13.4
Three Common
Valve Types
*Source: Sunset Basic
Plumbing Illustrated*
(Menlo Park, CA:
Sunset Publishing
Corporation, 1991),
p. 22.

house. A shutoff valve must also be installed in the cold-water supply pipe to the water heater, at or near that appliance. Although not required, valves at fixtures, risers, and branches simplify later service to these items.

Gate Valve

A **gate valve** provides the most effective service when it is used in the fully open or fully closed position; it is not designed to adjust the flow of water. A gate valve functions by using a wedge-shaped leaf to seal tightly against two metal seats. An important feature of the gate valve is that there is less obstruction and turbulence within the valve, resulting in a lower friction loss than found in other types of valves. The gate valve is commonly used as the main shutoff valve in a residential system. It can also be used to shut off the flow of water to fixtures and equipment when repairs or replacements must be made.

Globe Valve

A **globe valve** is installed at locations where there is occasional or periodic use, such as for lavatories. This type of valve can either close the flow of water or periodically adjust the flow of water going through the valve. This valve functions by turning the handle, which in turn forces a washer against the metal seat, thus stopping the flow of water. The flow of water through the valve is increased as it is opened by turning its handle. The design of the globe valve forces the water passing through to make two 90° turns, thereby greatly increasing the friction loss as compared to the gate valve.

Angle Valve

An **angle valve** operates similarly to the globe valve since its washer is compressed against a metal seat to stop the flow of water. It is commonly utilized for outside hose bibbs. The angle valve has a much higher friction loss than the gate valve and about half the friction loss of the globe valve, since the water only has to make one 90° turn.

Check Valve

A **check valve** has a hinged leaf which opens to allow the flow of water in the desired direction. The leaf closes if there is any flow of water in the opposite direction. The check valve works automatically, thereby eliminating the need for a handle.

Relief Valve

A **relief valve** is held closed by a spring or some other means and is designed to automatically relieve the equipment pressure or temperature which is in excess of its setting or rating. In general, a relief valve should be installed wherever there is any danger of the **water pressure** or **temperature** rising above the design working pressure or temperature of the pipe fittings or container (such as a water heater or a water pressure tank). The common types of relief valves include: pressure relief valves, temperature relief valves, and combination pressure/temperature relief valves.

Typical Water Supply Layout

The hot-water and cold-water supply piping for the simple bathroom shown in Fig. 13.5 illustrates how the piping and fittings discussed are integrated into the water supply system.

Fire Sprinklers

Even though most building codes require that one or more smoke detectors be installed in new housing construction, many people maintain that fire sprinklers are probably the most effective method of reducing fire-related deaths and damages.

In areas where the code requires a fire sprinkler system, it usually includes a system of pipes (steel, copper, or approved plastic) which distribute pressurized water throughout the house. The piping system terminates in a series of sprinkler heads, which are devices that are activated (broken) by the heat of the fire. The failure of the sprinkler heads allows water to flow through the device to the fire. Residential sprinklers are usually activated at lower temperatures (135°F to 165°F) than the sprinkler heads in commercial structures.

An alternative to a "wet" sprinkler system is a "dry" sprinkler system, which often is used in commercial construction.

Design of the Water Supply System

The 1995 CABO Code presents both a simplified method and a detailed engineered method for designing the water supply system for a single-family house. The simplified method[9] (also called the **velocity limitation method**) may be used where available water supply pressures are at least 40 psi and the elevation of the highest fixture above the service valve does not exceed 25 ft.

1. Reducing tee, 3/4" × 3/4" × 1/2"
2. Reducing elbow, 3/4" × 1/2"
3. 90° elbow, 1/2"
4. Valve body
5. Drop ell with threaded outlet
6. Shower arm
7. Threaded nipple, 1/2"

8. Shut-off valve
9. Supply tube, 3/8"
10. Type L pipe, 3/4"
11. Type L pipe, 1/2"
12. Coupling, 3/4"

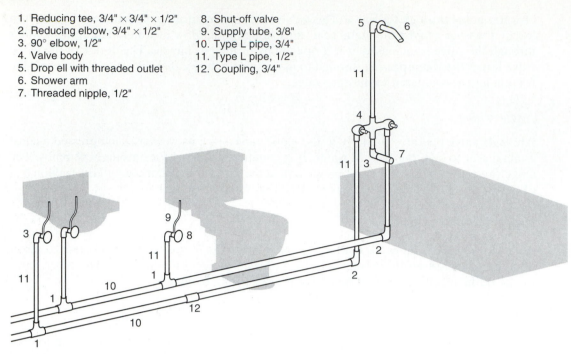

Figure 13.5
A Typical Bathroom Water Supply System
Source: Charlie Wing, *The Visual Handbook of Building and Remodeling* (Emmaus, PA: Rodale Press, 1990),
p. 233.

The detailed engineering method[10] (also called the **uniform friction drop method**) provides greater precision than the simplified method and can be used for determining pipe
sizes for any given supply pressure and for a greater variety of conditions. The steps
involved in each method are described next.

The Simplified Method

Step 1: Prepare an isometric layout (or riser diagram) of the water supply system which indicates all of the fixtures on the layout in their approximate relative locations in the house and
the approximate routing of the hot and cold water lines (see Figure 13.11a for an example).

Step 2a: Determine the total number of water supply fixture units in the house.

The supply load in the building water distribution system is determined by the total
load on the pipe being sized, in terms of WSFUs. WSFUs for a variety of cases are defined
in Table 13.1. The water supply fixture unit is a measure of the probable hydraulic demand
on the water supply by various types of plumbing fixtures. The water supply fixture unit
value for a particular fixture depends on its volume rate of supply, on the time duration of
a single supply operation, and on the average time between successive operations.

Table 13.1 also lists WSFU values for fixture groups such as a full bath or multiple
baths. These values should be used when sizing common piping to a fixture group, instead
of separately accounting for each fixture in the group, since there is only a small probability
that all of the fixtures in a fixture group will be in use at the same time. The fixture group
values should also be used when the main water supply pipe into the house is being sized.

TABLE 13.1 Water Supply Fixture Unit Values for
Various Plumbing Fixtures and Fixture Groups

Type of Fixtures or Group of Fixtures	Water Supply Fixture Unit Value (WSFU)		
	Hot	Cold	Combined
Bathtub (with/without overhead shower head)	1.0	1.0	1.4
Clothes washer	1.0	1.0	1.4
Dishwasher	1.4	—	1.4
Hose bibb (sill cock)[a]	—	2.5	2.5
Kitchen sink	1.0	1.0	1.4
Lavatory	0.5	0.5	0.7
Laundry tub	1.0	1.0	1.4
Shower stall	1.0	1.0	1.4
Water closet (tank type)	—	2.2	2.2
Full-bath group with bathtub (with/without shower head) or shower stall	1.5	2.7	3.6
Half-bath group (water closet and lavatory)	0.5	2.5	2.6
Kitchen group (dishwasher and sink with/without garbage grinder)	1.9	1.0	2.5
Laundry group (clothes washer standpipe and laundry tub)	1.8	1.8	2.5
Multiple-bath groups:			
1-1/2 baths	2.0	3.3	4.4
2 baths	2.6	3.9	5.2
2-1/2 baths	2.8	4.2	5.6
3 baths	3.2	4.7	6.3
3-1/2 baths	3.4	5.1	6.8
Additional 1-1/2 bath if part of a group	0.3	0.6	0.8

For SI: 1 gpm = 3.785 L/m.
[a]The fixture unit value 2.5 assumes a flow demand of 2.5 gpm, such as for an individual lawn sprinkler device. If a hose bibb/sill cock will be required to furnish a greater flow rate, the equivalent fixture-unit value may be obtained from Table 3409.3 or from Figure C201 of Appendix C.

Step 2b (*optional*): Determine the maximum supply demand in the system in gallons per minute (gpm).

The volumetric demand flow rate in the service pipe or in various parts of the water-distribution system can be determined by using Table 13.2 after the fixture unit loads have been determined. For a hose bibb or lawn sprinkler system which imposes a continuous demand, the continuous demand should be estimated separately and added to the gpm demand for fixtures supplied by the pipe being sized.[11]

Step 3: Determine the size of the fixture branches.

The minimum sizes for fixture branches can be determined by referring to Table 13.3.

Step 4: Determine the size of the remaining piping on the basis of connected water supply fixture units.

The size of the water service mains, branch mains, and risers—which could be composed of copper, steel, PB and CPVC, or CPVC and PE pipe—can be determined from the 1995 CABO Code.[12] Table 13.4 illustrates the pipe sizing table for copper tubing. All of the tables assume a nominal static pressure of 40 psi or greater at the main shutoff valve. The following procedure can be used for the pipe sizing step:

a. Working upstream from the farthest fixture or fixture group, add the fixture unit values and note the totals for each riser, branch, or main to be sized. Use fixture group values where applicable.

TABLE 13.2 Demand Flow Rate as a
Function of Fixture Unit Load

Load[a] (WSFU)	Demand[a] (gpm)
2	2.0
3	3.0
4	3.8
5	4.5
6	5.1
7	5.8
8	6.5
9	7.2
10	7.7
12	9.0
14	10.4
16	11.6
18	12.7
20	14.0
25	16.8
30	19.5

For SI: 1 gpm = 3.785 L/m.
[a]Interpolation may be used to obtain intermediate values.

Source: Reproduced from the 1995 edition of the *CABO One and Two Family Dwelling Code,*™ copyright © 1995, with the permission of the publisher, The International Codes Council, Whittier, CA, Table 3409.3, p. 214.

TABLE 13.3 Minimum Size of Fixture Branches[a]
(Fixtures Water Supply Pipes)[b]

Type of Fixture or Outlet	Nominal Pipe Size (in.)
Bathtub (with/without shower head) Clothes-washer supply fitting Dishwasher Kitchen Sink Laundry tub (one and two compartment) Shower head Wall hydrant/sill cock/hose bibb	1/2
Bar sink Bidet Lavatory Water closet (close-coupled tank type)[c]	3/8

For SI: 1 in. = 25.4 mm.
[a]Table not applicable to manifold system. See Section 3409.6 of the CABO Code.
[b]For special fixtures or fittings, size according to the manufacturer's installation instructions.
[c]Also see Section 3403.2 of the CABO Code, or according to the manufacturer's specifications.

Source: Reproduced from the 1995 edition of the *CABO One and Two Family Dwelling Code,*™ copyright © 1995, with the permission of the publisher, The International Codes Council, Whittier, CA, Table 3409.4, p. 214.

TABLE 13.4   Pipe Sizing Based on Velocity Limitation for
Copper Water Tube[a,b,d]

Nominal Pipe Size (inches)	Type K		Type L		Type M	
	Carrying Capacity					
	gpm	WSFU	gpm	WSFU	gpm	WSFU
1/2	5.44	6	5.81	7	6.34	8
3/4	10.9	15	12.1	17	12.9	20
1	19.4	30	20.6	32	21.8	35
1-1/4	30.3	54	31.3	57	32.6	61

For SI: 1 in. = 25.4 mm, 1 gpm = 3.785 L/m, 1 ft/sec = 0.3048 m/s.
[a]The relation between carrying capacities in gpm and WSFU is based on Table C401 and Figure C101 of Appendix C of the CABO Code.
[b]Table based on velocities as follows:
   Copper—9 fps
   See Table 3409.5a for complete footnote.
Where local experience or manufacturer's recommendations specify lower velocities, the carrying capacities shall be reduced accordingly.
[c]See Table 3409.5a for complete footnote.
[d]Values are based on materials which conform to the following standards:
   Copper water tube—ASTM B 88
   See Table 3409.5a for complete footnote.

Source: Reproduced from the 1995 edition of the *CABO One and Two Family Dwelling Code*,™ copyright © 1995, with the permission of the publisher, The International Codes Council, Whittier, CA, Table 3409.5a, p. 214.

b. Determine the pipe size for a piping interval from the water supply fixture unit values using tables such as Table 13.4. It should be noted that Table 13.4 can be used with either WSFU or gpm values.

c. These steps are repeated for each succeeding piping interval where additional loads connect, based on the total WSFU load at that point, until all mains, branch mains, and risers have been sized back to the water heater (hot) and service valve (cold). This procedure may also be used to size the service pipe.

The Detailed Engineered Method

Step 1: Prepare an isometric layout or riser diagram for the water supply system which indicates the location of the fixtures on the layout (see Fig. 13.11(a) for an example).

   The results obtained in each step of the calculations should be marked on the diagram at the appropriate locations.

Step 2: Determine the **equivalent length** of the basic design circuit (BDC).
   As noted in the 1995 CABO Code:

> The basic design circuit (BDC) is the longest run of piping from the water source to the highest and most remote fixture or water outlet on the system. In most systems, the BDC may be assumed to consist, first, of the cold-water supply piping extending from the source to the water heater, and then the hot-water supply piping extending from the water heater to the highest and most remote hot-water outlet on the system (terminal fixture).[13]

Each valve or fitting in the water distribution network must be assigned the "equivalent length" of the pipe which will have the same pressure loss as the pressure loss through the valve or fitting. The 1995 CABO Code provides information for this determination.[14] Alternatively, it allows the assumption to be made that the equivalent length to be added for all of the valves and fittings in the piping system will be 50% of the developed length of the BDC.

The equivalent length due to the valves and fittings is added to the developed length (i.e., the actual length) of the BDC to determine the **total equivalent length** of the BDC. This is the key value that is used to determine the **friction loss design factor**.

Step 3: Calculate the pressure available for overcoming friction.

The pressure available for overcoming friction must also be obtained. It is determined by **subtracting the sum** of the following from the **minimum static pressure**, in psi, which is available at the water main or other source:

a. The required minimum flow pressure at the terminal fixture of the piping. In most residential systems using tank-type water closets, a minimum flow pressure of 8 psi may be assumed.

b. The friction loss, in psi, due to the flow through fixture supply branches. The CABO Code allows a value of 5 psi to be assumed for this loss.

c. The elevation head loss, in psi, from the main or other source up to the terminal fixture. This is calculated as the product of the elevation difference, in feet, times the factor of 0.433 psi per foot.

d. The sum of friction losses, in psi, caused by certain equipment which is to be installed in the system, such as a water meter, water softener, or check valve. The most common source of such losses is the water meter. Appropriate values for this loss can be determined in the 1995 CABO Code.[15] For equipment other than the water meter in the BDC (i.e., water softener, check valve, instantaneous or tankless water heater, etc.), the friction loss should be determined from the manufacturer or some other reliable source.

Thus, if the available static pressure at the main is assumed to be 50 psi and the sum of the pressure losses calculated in 3(a)–3(d) is 20 psi, the **total pressure available to overcome the friction** due to the flow of water in the pipe system is 30 psi.

Step 4: Calculate the friction loss design factor, Δp, for the basic design circuit (BDC).

The **friction loss design factor**, Δp, psi/100 ft, for the BDC is obtained by dividing the total pressure available for friction (see Step 3) by the total equivalent length of the basic design circuit (see Step 2) and multiplying the result by 100:

$$\Delta p = \frac{(\text{psi available for friction})}{(\text{total equivalent length})} \times 100$$

This Δp factor must be known before the hot and cold building mains, as well as the primary branches and rises, can be sized.

Step 5: Determine the total fixture unit load and equivalent demand flow rate.

The total fixture unit load is determined for the service pipe, the hot and cold building mains, and each primary branch and riser using the same procedures described in Step 1 of the Simplified Method, and the data in Table 13.1. The **equivalent demand flow rate** for each of the previously-noted components can be determined by using Fig. 13.6.

Step 6: Determine the size of the main service pipe.

Two sources of information for this calculation are available in Appendix C of the 1995 Cabo Code. One of the sources is a set of tables which provides the appropriate size for different types of pipe material using the values obtained in Steps 2, 4, and 5.[16]

An alternative approach (Table 13.5) uses the Δp value obtained in Step 4 and the demand flow rate obtained in Step 5 for the type of pipe being used in the water distribution system. Values may be interpolated. The size of the main service pipe should be the smallest size which will furnish the required flow rate for the given friction loss design factor.

Step 7: Determine the size of each hot and cold water main, primary branch, and riser.

Table 13.5 also can be used to perform these calculations. For each pipe section, the portion of the table for the appropriate piping material is entered with the friction loss

Figure 13.6

Demand
Flow Rate
Determination

Source: Reproduced from the 1995 edition of the *CABO One and Two Family Dwelling Code,*™ copyright © 1995, with the permission of the publisher, The International Codes Council, Whittier, CA, Fig. C201, p. 317.

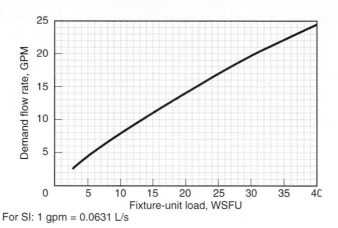

For SI: 1 gpm = 0.0631 L/s

TABLE 13.5   Minimum Flow Rates,[a] Q, for Copper Water Tube[b] which Yields Friction Losses not Exceeding Indicated Values at Velocities not Exceeding Maximum Recommended Values[c] in Hot and Cold Building Mains, Primary Branches, and Risers of the Water Distribution System

Nominal Pipe Size (in.)	Friction Loss Design Factor, Head in Lbs per sq in. per 100 ft of length, Δp													
	2	3	4	5	10	15	20	30	40	50	60	70	80	100
	Flow Rate, gpm[d]													
Copper Water Tube—Type K, ASTM B 88														
1/2	1.2	1.5	1.8	2.0	3.0	3.7	4.3							5.4
3/4	3.1	3.9	4.5	5.1	7.5	9.4								10.9
1	6.6	8.2	9.7	10.9	15.9									19.4
1-1/4	12.0	15.0	17.4	19.8	20.8									30.3
Copper Water Tube—Type L, ASTM B 88														
1/2	1.4	1.7	2.0	2.3	3.3	4.1	4.8							5.8
3/4	3.6	4.4	4.6	5.8	8.4	10.7								12.1
1	7.2	8.9	10.4	11.6	17.1									20.6
1-1/4	12.6	15.8	18.3	20.2	29.1									31.3
Copper Water Tube—Type M, ASTM B 88														
1/2	1.3	1.9	2.2	2.5	3.7	4.6	5.4							6.3
3/4	3.9	4.8	5.6	6.3	9.2	11.5								12.9
1	7.7	9.7	11.3	12.5	18.4									21.8
1-1/4	13.2	16.3	19.0	21.3	30.7									

[a]Flow rates calculated for "Fairly Smooth Condition."
[b]See Table C401 for information on steel pipe, polybutylene (PB) tubing, and chlorinated polyvinyl chloride (CPVC) pipe).
[c]Limiting rates correspond to 8 fps for copper, 10 fps for steel, and 12 fps for plastics. Values may be interpolated between columns.
[d]$Q = 4.57\,\Delta p^{0.0546}\,D^{2.64}$, where Q is the gpm, Δ is in psi/100 ft, and D is in inches, I.D.

Source: Reproduced from the 1995 edition of the *CABO One and Two Family Dwelling Code,*™ copyright © 1995, with the permission of the publisher, The International Codes Council, Whittier, CA, excerpt from Table C401, p. 316.

design factor Δp, obtained in Step 4, and the appropriate demand rate obtained in Step 5. The size of the pipe element obtained should be the smallest size which will furnish the required flow rate for the given friction loss design factor. The set of tables noted in Step 6 also can be used for this calculation.

SANITARY DRAINAGE SYSTEM

Waste Stream

Once the water is used by a particular plumbing fixture, **waste matter** is created. Different quantities and types of waste matter are created by the various fixtures commonly found in houses. The next step in the design process, therefore, is to develop a **sanitary drainage system** which can dispose of the waste matter, both fluid and organic, which is accumulated. The wastes come from almost all sections of the house, including bathrooms, kitchens, and laundry areas. Because all the wastes tend to decompose quickly, one of the primary objectives of the drainage system is to remove the decaying wastes from the house before they cause odors or become health hazards.

Public versus Private Sewage System

The drainage system of a house must be connected to a public sewage system if one is available. The sanitary sewer system which connects the house to either a **public sewage treatment plant** or a **private or individual sewage disposal system** must exclude stormwater, surface runoff, and subsurface water, unless the public sewer has been designed as a "combination" sewer (i.e., one that carries both storm and sanitary drainage).

When a public sewer system is not available, it becomes necessary to install a private on-lot sewage disposal system. Minimum requirements for such a system are provided in the 1995 CABO Code.[17] The type of private sewage disposal system which is selected is often a function of location, soil porosity, groundwater level, and local practice.

The most commonly used private on-lot sewer system is the **septic tank**, which is a water-tight receptor that receives the discharge of a building drainage system. The tank is constructed so that the solids are separated from the liquid. The organic matter is digested through a period of detention, and the liquids are allowed to discharge into the soil outside the tank through a system of open joints or perforated piping to a disposal area or seepage pit.

Components of a Sanitary Drainage System in a House

The primary components of a typical sanitary drainage system (also called the drain, waste, and vent [DWV] system) are shown in Fig. 13.7. The flow in the DWV system is by gravity alone, so there must be slopes in all the "horizontal" pipes. Each of the primary components of the DWV system is discussed next.

Traps

A **trap** is a fitting, either separate or built into a fixture, which provides a liquid seal to prevent the emission of sewer gases back through the fixture into the house. A trap does not affect the flow of sewage or wastewater through it. The trap shown in Fig. 13.8 is commonly a U-shaped section of pipe in lavatories and sinks. Such traps can be made of lead, cast iron, cast or drawn brass (the most common material), or approved plastic. Traps are built directly into water closets, where they are made of vitreous china.

The design criteria for traps are provided in the 1995 CABO Code.[18] The requirements for trap seals (i.e., traps must have a liquid seal of not less than 2 in. or more than

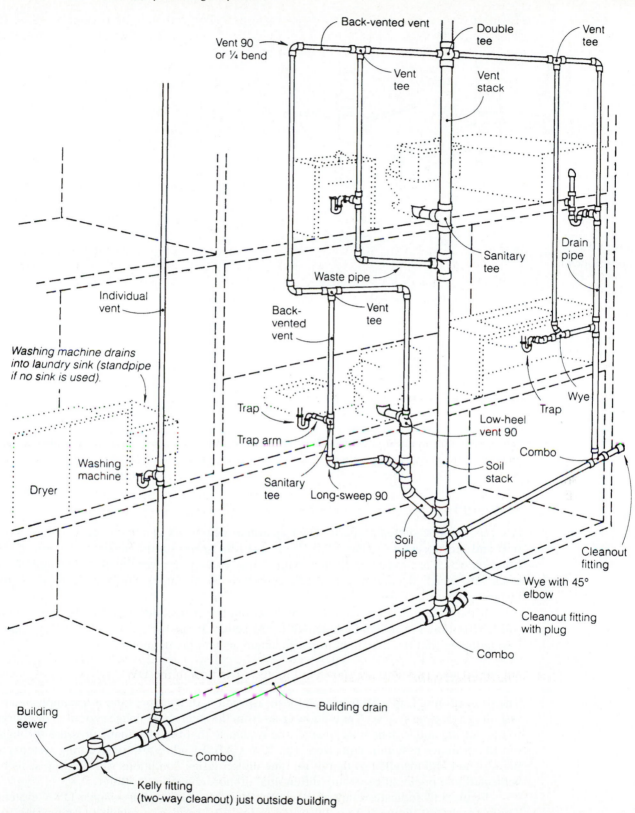

Figure 13.7
A Typical Sanitary Drainage (DWV) System
Source: Peter Hemp, *Plumbing a House* (Newtown, CT: The Taunton Press, Inc., 1994), p. 16.

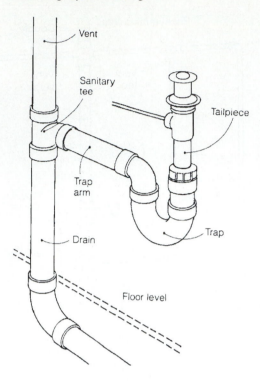

Vent

Sanitary
tee

Tailpiece

Trap
arm

Drain

Trap

Floor level

Figure 13.8

Installation of a
Trap for a
Lavatory

Source: Peter
Hemp, *Plumbing a
House* (Newtown,
CT: The Taunton
Press, Inc., 1994),
p. 106.

4 in.) and the maximum vertical distance from the fixture outlet to the trap weir (i.e., no greater than 24 in.) are specified. In addition, information about the size of traps and trap arms is provided (Table 13.6).

Trap Arms

All traps must be vented properly. The **trap arm** is the portion of a fixture drain between the trap and the closest vent (Fig. 13.8). There are restrictions on the maximum length of the trap arm which is allowed between a fixture's trap and either a vent stack or the main drain stack into which it empties (often called the critical distance). For a 1-1/2 in. pipe, for instance, the critical distance is 6 ft.[19]

If the fixture drain is vented properly within the critical distance, the drain pipe may run horizontally for an indefinite length to the actual drain stack. The 1995 CABO Code[20] also indicates that the minimum trap arm length is two pipe diameters.

The Vent System

The protection of trap seals from siphonage, aspiration, or back pressure is accomplished by installing a venting system. **Vents** allow gases from the sanitary drainage system to discharge to the outside and sufficient air to enter the system so that the pneumatic pressure throughout the drainage system is equalized. The 1995 CABO Code[21] requires venting systems to be designed and installed so that at no time under design conditions will the trap seals be subjected to a pneumatic pressure differential of more than 1 in. of water.

Figure 13.7 indicates some of the trap venting alternatives for a simple DWV system. The type of vent piping system illustrated in Fig. 13.7 services a number of plumbing fixtures, or group of fixtures, in a house. In such a system, the **vent stack** is the vertical vent pipe which provides circulation of air to and from the drainage system and which extends through one or more stories. Each vent stack, as illustrated, will typically have a number of vent pipes connected to it.

TABLE 13.6   Fixture Trap Size/Trap Arm Information

Plumbing Fixture	Trap Size Minimum (in.)
Bathtub (with or without shower head and/or whirlpool attachments)	1-1/2
Bidet	1-1/4
Clothes washer standpipe	1-1/2
Dishwasher (on separate trap)	1-1/2
Floor drain	2
Kitchen sink (one or two traps, with or without dishwasher and garbage grinder)	1-1/2
Laundry tub (one or more compartments)	1-1/2
Lavatory	1-1/4
Shower	1-1/2
Water closet	*

For SI: 1 in. = 25.4 mm.
*Consult fixture standards for trap dimensions of specific bowls.

Source: Reproduced from the 1995 edition of the *CABO One and Two Family Dwelling Code,*™ copyright © 1995, with the permission of the publisher, The International Codes Council, Whittier, CA, Table 3701.7, p. 233.

Each house is required to have at least one **main vent stack** which usually extends through the roof of the house. This vent stack should be installed through the roof behind a ridge line or on the back side of the house whenever possible for appearance purposes.

An **individual vent**, such as the one servicing the laundry sink adjacent to the washing machine on the first floor in Fig. 13.7, is a pipe which is installed to vent a single fixture drain that either connects to the vent system above or terminates independently outside the building. This type of a vent is called a **dry vent** because it leads directly to the roof without servicing any other fixtures above. It does not carry any waste material.

In many houses, especially single-family houses, widely separated fixtures make it impractical to use a single main vent stack. In such a situation, each fixture, or a number of fixture groups, may be connected to a secondary vent stack. It is considered good practice to limit the number of roof penetrations as the DWV venting system is being designed.

Soil and Waste Stacks

The fixture branches feed into a vertical pipe, referred to as a **stack**. When the waste pipe conveys sewage containing fecal matter (human waste) from water closets, as well as liquid free from fecal matter, the stack is referred to as a **soil stack**. (The main soil stack in Fig. 13.7 serves that purpose.) When the stack conveys only liquid sewage not containing human waste, it is referred to as a **waste stack**. (The stack which services the bathtub on the second floor and the sink in the kitchen in Fig. 13.7 serves that purpose.) A **stack vent** is the extension of a soil stack or waste stack above the highest horizontal drain connected.

Stack venting is a method of venting a fixture or fixtures directly through the stack vent instead of using an individual fixture vent. It is a convenient approach if the fixture is close to the soil or waste stack.

If fixtures such as the dry-vented bathtub on the second floor in Fig. 13.7, for instance, are not close enough to the stack vent, then a **branch vent** can be used to connect the vents from these fixtures to the vent stack. This practice, also called **backventing** or **reventing**, involves running a vent loop vertically from the trap arm until it is at least 6 in. above the fixture's flood rim before it is turned back to reconnect with either the main vent stack or a secondary vent stack.

The 1995 CABO Code also allows **wet vents** in certain cases. In wet venting, the venting serves as both a drain for upper fixtures serviced by the vent as well as a vent for fixtures that are tied into the vent at a lower level. Each wet vented section must be at least one pipe size larger than the required pipe under normal conditions. Some building codes do not allow wet vents at all because they can become clogged with waste, while others permit only wet vented toilets.

Fixture Branches/Horizontal Drainage Branch

A **fixture branch** is a drain pipe serving one or more fixtures which discharges into another portion of the drainage system. A **horizontal drainage branch** is a drain pipe which extends laterally from a soil or waste stack or building drain. It receives the discharge from one or more fixture drains. Horizontal drainage piping must be installed with uniform slopes not less than 1/4 in. per ft for 3 in. diameter and less, and not less than 1/8 in. per ft for diameters of 4 in. or more.

Building Drains

The **building drain** (Fig. 13.7) is the lowest piping that collects the discharge from all other drainage piping inside the house and conveys it to the building sewer, which is typically 3 ft (914 mm) outside the building wall. The building drain must slope 1/8 in. to 1/4 in. per ft as it carries waste to the building sewer. A **cleanout** must be provided near the junction of the building drain and building sewer to allow for cleaning. The cleanout can be either inside or outside the building wall, provided it is brought up to finish grade or to the lowest floor level. Location of the building drain is dependent primarily upon the elevation of the public sewer line. Ideally, all of the plumbing wastes of the building will flow into the sewer by gravity. If the height of the sewer requires the drain to be placed above the lowest fixtures, it is necessary for the low fixtures to drain into a sump pit. When the level in the sump pit rises to a certain point, an automatic float or control will activate the pump, which raises the waste out of the pit and into the building drain.

Building Traps

The 1995 CABO Code[22] specifically states that building traps *shall not be installed* except when directed by the administrative authority in those special cases where sewer gases are extremely corrosive or noxious. Other building codes might require a building trap on the building drain near the building wall. This trap acts as a seal to keep gases and animals from entering the sewage system through the sewer line. When a building trap is used, a fresh air inlet is required to allow fresh air into the system. This ensures that the trap seal is not siphoned through.

Building Sewers

The **building sewer** is the part of the drainage system which extends from the end of the building drain and conveys its discharge to the public or private sewer, individual sewage-disposal system, or other point of disposal. The building sewer slopes 1/8 in. to 1/2 in. per foot and should never be smaller in diameter than the building drain.

Sanitary Drainage System Materials

A listing of the ASTM standards which define each acceptable type of material that can be used for the **sanitary drainage** (i.e., DWV) **system** is provided in the 1995 CABO Code.[23] A similar listing of each acceptable type of material for the building sewer piping component is also provided.[24]

The most common types of materials used are two types of **plastic pipe** (PVC–DWV and ABS–DWV), **copper DWV pipe**, **galvanized steel** and **galvanized wrought iron** (which cannot be used underground), and **cast iron**. In addition, extra-strength vitrified clay pipe can be used for the underground building sewer portion of the system.

Plastic Pipe

Both PVC (which is off-white in color) and ABS (which is black) are less expensive and easier to work with than the copper, galvanized, and cast iron alternatives. Plastic pipe is easily cut and assembled (it is easily solvent welded or threaded), and its light weight enables one person to handle long sections of the pipe.

However, some disadvantages exist in using plastic pipe. For instance, it expands and contracts with changes in temperature and has to be supported with approximately twice as many hangers as cast iron pipe in order to keep it from bending. It also generates poisonous gases in a fire, and readily transmits the sound of water running through it when the system is in use.

Of the two types, PVC pipe is less susceptible to mechanical and chemical damage, will not burn, and has a slightly greater variety of fittings available. A cast iron or galvanized pipe system can be extended with plastic pipes.

Other Types of Pipe

Copper pipe with a DWV rating is more expensive than plastic pipe. It is usually available in 20 ft lengths and is relatively economical in nominal diameters of 1-1/2 and 2 in.

Both galvanized and cast iron pipe are strong and are not attacked by chemicals or affected by boiling water. On the other hand, galvanized pipe is expensive, has to be threaded in order for pieces to be joined, and is susceptible to rust on the inside. Cast iron pipe, which was used extensively in the past, is heavy (it requires a stronger structural support system), is difficult to cut, and is very time consuming to install.

DWV System Fittings

Joint connection information for a DWV system is provided in the 1995 CABO Code.[25] Although each type of pipe material has its own set of fittings, the presentation in Fig. 13.9 provides a general indication of the types of fittings that are used in a DWV system.

Design of the DWV System

The simplified design procedure for DWV systems[26] and the basic design information related to the venting system for single-family houses are presented in the 1995 CABO Code.[27] An engineered procedure for sizing plumbing vents is also provided.[28]

The Simplified Method

Step 1: Prepare an isometric layout of the DWV system which indicates all the fixtures on the layout in their appropriate relative locations in the house and the appropriate drain and vent lines (see Fig. 13.12(a) for an example).

Step 2: Determine the total number of drainage fixture units in the house.

There is a similarity between the design calculations for the water supply system presented earlier in this chapter and the design of the DWV system, since both begin with the calculation of the number of "fixture units" (Table 13.7). In this case, however, the load on the DWV piping system is represented in terms of the **drainage fixture units** (d.f.u.). The d.f.u. is a measure of the probable discharge into the drainage system by the various types of plumbing fixtures that are used in the DWV piping system.

The drainage fixture unit value for a particular fixture depends on its volume rate of drainage discharge, on the duration of a single drainage operation, and on the average time between successive operations. It is important to note that Table 13.7 also lists values for fixture groups such as a full bath or multiple baths. These values can be used instead of accounting for each fixture in the group because there is only a small probability that all of the fixtures in a fixture group will be in use at the same time.

CHANGE-OF-DIRECTION FITTINGS

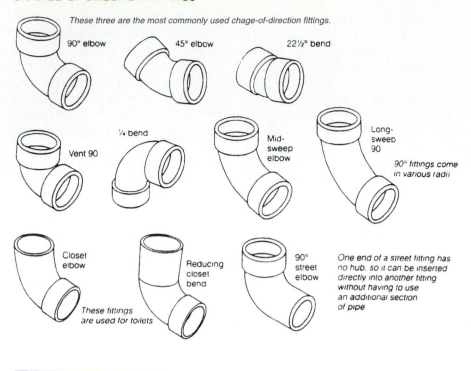

These three are the most commonly used chage-of-direction fittings.

BRANCH FITTINGS

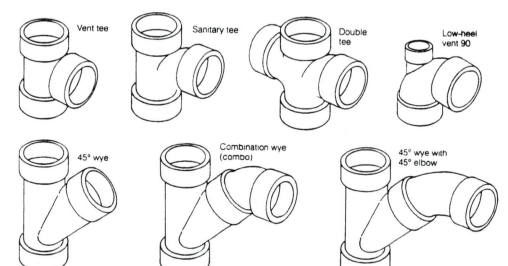

Figure 13.9
The Type of
Fittings Used in a
DWV System
Source: Peter
Hemp, *Plumbing a
House* (Newtown,
CT: The Taunton
Press, Inc., 1994),
pp. 17, 18.

Step 3: Determine the size of the traps and the fixture drains.

The fixture trap sizes are determined by using Table 13.6. The drainage pipe into which the trap discharges should be the same size as its associated trap.

Step 4: Assign accumulated drainage fixture units to the remaining unsized drain and vent pipes.

The sizing of the drain and vent pipes requires an accumulation of the number of drainage fixture units. The calculation procedure starts at the top floor, or the most remote fixture on each branch, and moves downstream toward the building drain. Using the infor-

TABLE 13.7 Tabulation of Representative
Drainage Fixture Units

Type of Fixture or Group of Fixtures	Drainage Fixture Unit Value (d.f.u.)[a]
Bar sink	1
Bathtub (with or without shower head and/or whirlpool attachments)	2
Bidet	1
Clothes washer standpipe	2
Dishwasher	2
Floor drain	0[b]
Kitchen sink	2
Lavatory	1
Laundry tub	2
Shower stall	2
Water closet (tank type)	4
Water closet (flushometer tank)	4
Full-bath group with bathtub (with or without shower head and/or whirlpool attachment on the bathtub or shower stall)	6
Half-bath group (water closet plus lavatory)	5
Kitchen group (dishwasher and sink with or without garbage grinder)	3
Laundry group (clothes washer standpipe and laundry tub)	3
Multiple-bath groups:[c]	
1-1/2 baths	7
2 baths	8
2-1/2 baths	9
3 baths	10
3-1/2 baths	11

[a]For a continuous or semicontinuous flow into a drainage system, such as from a pump or similar device, 1.5 fixture units shall be allowed per gpm of flow. For a fixture not listed, use the highest d.f.u. value for a similar listed fixture.
[b]A floor drain itself adds no hydraulic load. However, used as a receptor, the fixture unit value of the fixture discharging into the receptor shall be applicable.
[c]Add 2 d.f.u. for each additional full bath.

mation provided in Table 13.7, drainage fixture values are accumulated for the individual fixtures and fixture groups.

Step 5a: Determine the horizontal branch drain pipe sizes.

Branch drain sizes are determined by equating the assigned d.f.u. values obtained in Step 4 to the pipe sizes shown in Table 13.8.

Step 5b: Determine the stack sizes.

Stacks can also be sized by equating the assigned d.f.u. values to the pipe sizes shown in Table 13.8. In addition, the 1995 CABO Code specifies the following design requirements for single-stack system venting.

TABLE 13.8   Maximum Number of Drainage Fixture
Units That May Be Connected to Various Sizes of
Branches and Stacks

Nominal Pipe Size (in.)	Any Horizontal Fixture Branch	Any One Vertical Stack or Drain
1-1/4[a]	—	—
1-1/2[b]	3	4
2[b]	6	10
2-1/2[b]	12	20
3	20[c]	48[d]
4	160	240

For SI: 1 in. = 25.4 mm.
[a]1-1/4-in. pipe size limited to a single-fixture drain or trap arm. See Table 3701.7.
[b]No water closets.
[c]Maximum three water closets.
[d]Maximum six water closets.

Source: Reproduced from the 1995 edition of the *CABO One and Two Family Dwelling Code,*™ copyright © 1995, with the permission of the publisher, The International Codes Council, Whittier, CA, Table 3505.4.1, p. 220.

3601.8.1 General. The stack shall be considered a vent for all fixtures discharging to the stack when installed in accordance with the requirements of this section.

3601.8.2 Stack Installation. The stack shall be vertical and shall not be offset. All fixture drains shall connect separately except that fixtures within the same group shall be permitted to be connected to the same drain. See Figures 3601.8.2a and 3601.8.2b for typical stack vent installations.

3601.8.3 Stack vent. A stack vent shall be provided for the stack. The size of the stack vent shall be at least one-half the diameter of the required stack size but not less than 1-1/4.

3601.8.4 Stack size. The waste stack shall be sized based on the total discharge to the stack and the discharge within a branch interval in accordance with Table 3601.8.4. The stack shall be the same size throughout its length.[29]

Table 13.9 indicates the maximum loads that can be placed on single stacks in stack vented systems. Figure 13.10 indicates two of the typical stack vent installations mentioned in the 1995 CABO Code.

TABLE 13.9   Maximum Drainage Loads on Single
Stack Systems

Stack Size (in.)	Total Discharge into One Branch Interval (d.f.u.)[a]	Total Discharge for Stack (d.f.u.)
1-1/2	1	2
2	2	4
2-1/2	NL	8
3	NL	42
4	NL	50

For SI: 1 in. = 25.4 mm.
[a]NL means no limit.

Source: Reproduced from the 1995 edition of the *CABO One and Two Family Dwelling Code,*™ copyright © 1995, with the permission of the publisher, The International Codes Council, Whittier, CA, Table 3601.8.4, p. 227.

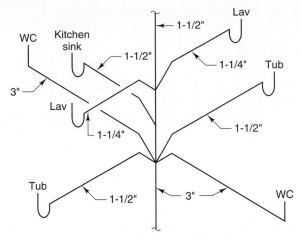

(a) Waste fixtures connected directly to stack

Figure 13.10 Typical DWV Connecting Fixture Drains to Stack in Stack-Vented Systems

Source: Reproduced from the 1995 edition of the *CABO One and Two Family Dwelling Code,*™ copyright © 1995, with the permission of the publisher, The International Codes Council, Whittier, CA, Fig. 3601.8.2a, p. 228.

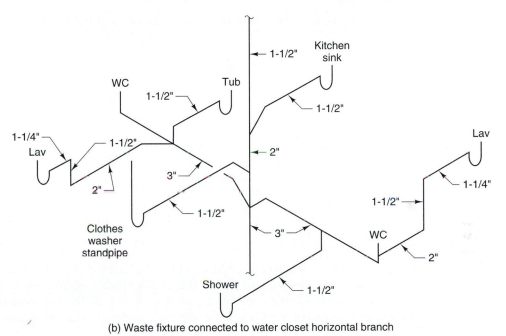

(b) Waste fixture connected to water closet horizontal branch

For SI: 1 inch = 25.4 mm

Step 5c: Determine the size of the vents.

An indication of the type of **dry venting** requirements in the 1995 CABO Code is presented:

> **3601.6 Dry vent sizing.** All dry portions of a venting system including individual and common vents extending from trap arms, stack vents, vent stacks, relief vents, or other dry vents shall be sized based on the accumulated load. The diameter of each vent shall be at least one-half the diameter of the required drainage pipe size computed in accordance with Section 3505.4 except that the vent pipe shall not be less than 1-1/4 in. in diameter. Where vents exceed 40 ft in length, the vent shall be increased by one nominal pipe size. The developed length shall be measured from the farthest point of the vent connection to the drainage system to outside the building.[30]

TABLE 13.10   Maximum Number of Fixture
Units That May Be Connected to the Building
Drain, Building Drain Branches, or
Building Sewer

Diameter of Pipe (in.)	Slope per ft		
	1/8 in.	1/4 in.	1/2 in.
1-1/2[a,b]	—	—[a]	—[a]
2[b]	—	21	27
2-1/2[b]	—	24	31
3	—	42	50
4	180	216	250

For SI: 1 in. = 25.4 mm, 1 ft = 304.8 mm.
[a]1-1/2 in. pipe size limited to a building drain branch serving not more than two waste fixtures, or not more than one waste fixture if serving a pumped discharge fixture or garbage grinder discharge.
[b]No water closets.

Source: Reproduced from the 1995 edition of the *CABO One and Two Family Dwelling Code,*™ copyright © 1995, with the permission of the publisher, The International Codes Council, Whittier, CA, Table 3505.4.2, p. 220.

A number of additional vent size requirements must be met if there is a wet venting situation present in the DWV system. These are also specified in the 1995 CABO Code.[31]

Step 5d: Determine the pipe diameter and slope of the building drain and the building sewer.

This determination can be made using Table 13.10.

Additional Information

The explanation of the Simplified Method has highlighted the important steps and related reference material that are required to design the DWV system. Not all of the details presented in the 1995 CABO Code have been included. Reference should be made to the appropriate sections of the Code for additional clarification of the requirements.[32]

The Engineered Procedure for Sizing Plumbing Vents

An alternative **engineered procedure** for sizing plumbing vents is presented in the 1995 CABO Code.[33] Although not discussed here, this procedure may be of interest during the design process if the objective of achieving a more optimized design is desired.

CASE STUDY HOUSE EXAMPLE

The design procedure for the water supply and sanitary drainage systems which has been presented in this chapter is illustrated by applying it to the Case Study House plans provided in Appendix A.

Design of the Water Supply System

The design of the water supply system for the Case Study house follows the four steps of the simplified procedure that were defined earlier in this chapter. Figures 13.11(a)–13.11(d) are associated with these steps.

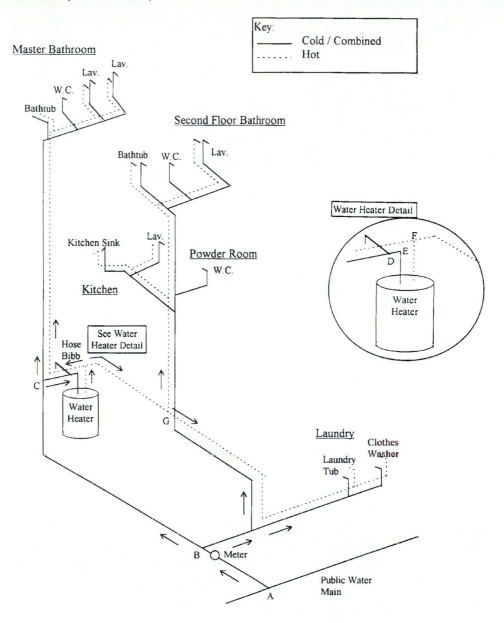

Figure 13.11a
Isometric of the
Water Supply
System: Case
Study House in
Appendix A

Step 1: Prepare the Isometric Drawing of the Hot and Cold Water Supply Systems
    An analysis of the plans of the Case Study House in Appendix A results in Fig. 13.11(a).

Step 2: Determine the Total Number of Water Supply Fixture Units in the House
    A review of the house plans in Appendix A indicates that the Case Study House has: (1) two full-bath groups plus an extra lavatory (i.e., there is a double lavatory in the master bathroom) on the second floor, (2) a half bath and a kitchen sink on the first floor, (3) a laundry group in the basement (it is assumed that it includes a laundry tub and a clothes washer), and (4) an outside hose bibb.

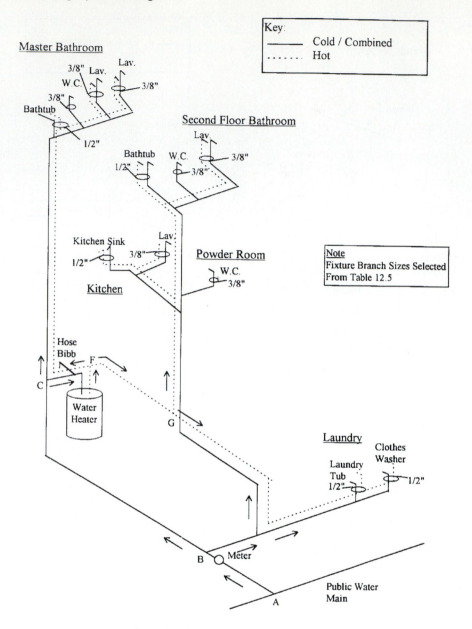

Figure 13.11b
Isometric of the
Water Supply
System with the
Fixture Branch
Sizes Indicated

The total tabulated hot, cold, and combined water supply fixture units (WSFUs) for the house, based upon Table 13.1, are:

Fixture	Number of WSFUs		
	Hot	Cold	Combined
2-1/2 bath groups	2.8	4.2	5.6
Extra lavatory	0.5	0.5	0.7
Kitchen sink	1.0	1.0	1.4
Laundry group	1.8	1.8	2.5
Hose bibb	—	2.5	2.5
TOTAL	6.1	10.0	12.7

These total WSFU values, which incorporated fixture group values as appropriate, are the basis for the design of the water service piping into the building.

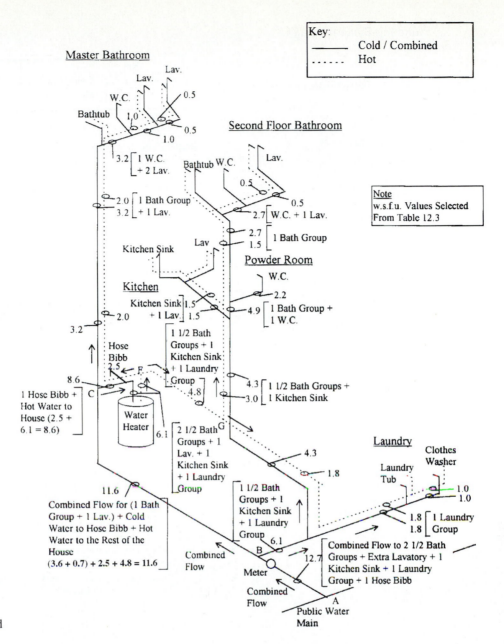

Figure 13.11c
Isometric of the
Water Supply
System with the
WSFUs Indicated

Step 3: Determine the Size of the Fixture Branches

The sizes of the pipes serving the individual fixtures are obtained directly from Table 13.3. The branch piping to the lavatories and the water closets is 3/8 in.; the branch piping to the bathtubs/showers, laundry tubs, clothes washers, kitchen sinks, and hose bibbs is 1/2 in. [Fig. 13.11(b)].

Step 4: Determine the Size of the Remaining Piping

The size of the remaining piping is determined by evaluating the connected water fixture units using Table 13.4. It will be assumed that Type K copper tube will be used for the building service piping and that Type L copper tube will be used for the interior water distribution network.

In order to use Table 13.4, it is necessary to assign a total water-fixture unit value to each unsized hot, cold, and combined pipe (see Table 13.1). The proper approach is to start at the downstream end of each line [i.e., the master bath, the second-floor bath, and the laundry group in Fig. 13.11(b)], and systematically move toward the hot-water heater or the cold-water supply. Advantage should be taken of fixture group values as full fixture groups become part of the "upstream portion" of the line. The results of this procedure are shown in Fig. 13.11c.

417

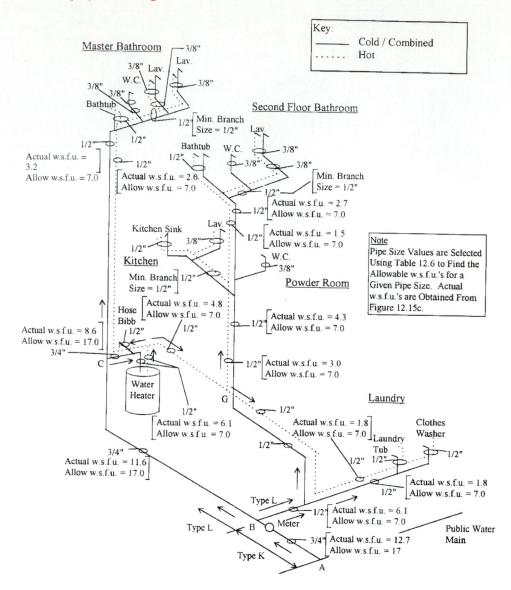

Figure 13.11d
Isometric of the
Water Supply
System with the
Pipe Sizes
Indicated

The tabulation of fixtures, or groups of fixtures, which appears next to the WSFUs assigned to each section of the piping network in Fig. 13.11(c) indicates the process. Section B–C deserves special attention since it carries: (1) the combined flow corresponding to the one bath group and the extra lavatory in the master bedroom, (2) the cold-water supply to the hose bibb, and (3) the hot water supply to the rest of the house [which excludes the master bedroom, which was accounted for in (1)]. The 4.8 WSFU value assigned to Section F–G in Fig. 13.11(c) represents that hot-water supply.

The appropriate pipe size for each section of the piping system is selected from Table 13.4 by verifying that the selected pipe size has an allowable WSFU which is equal to or greater than the actual number of WSFUs which were assigned to the section of pipe. The minimum acceptable pipe size in Table 13.4 is 1/2 in. The completed design, which is shown in Fig. 13.11(d), indicates these actual and allowable WSFU values for each section of the piping system.

The building service pipe extends from the outside water main through the meter to the first branch line as Type K copper pipe. This section of pipe, labeled as Section A–B in Fig. 13.11(c), carries the combined service into the house. The total combined service for all of the fixtures in the house is 12.7 WSFUs (Step 2). For Type K copper pipe and a load of 12.7 WSFUs, the pipe can be sized as 3/4 in., according to Table 13.4.

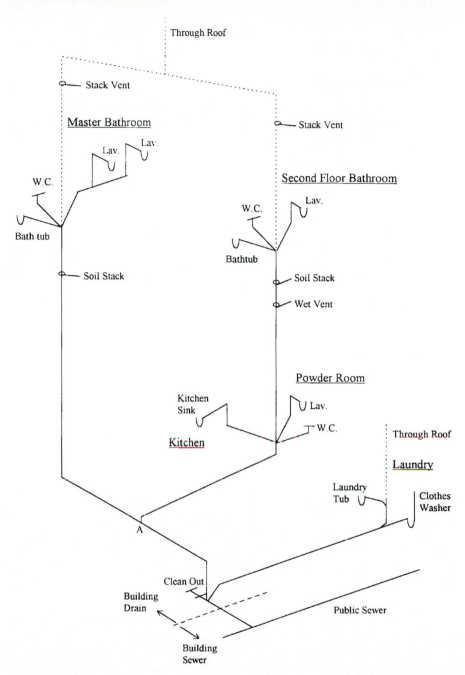

Figure 13.12a
Isometric of the
DWV System:
Case Study House
in Appendix A

Design of the Sanitary Drainage System

The design of the DWV system for the Case Study House follows the five steps of the Simplified Method that were defined earlier in this chapter. Figure 13.12(a)–13.12(d) are associated with these steps.

Step 1: Prepare the Isometric Drawing of the DWV System

    An analysis of the plans of the Case Study House in Appendix A results in Fig. 13.12(a).

Step 2: Determine the Total Number of Drainage Fixture Units in the House

    The total number of drainage fixture units (d.f.u.s), which should incorporate fixture group values as appropriate, is the basis for the design of the building drain which conveys

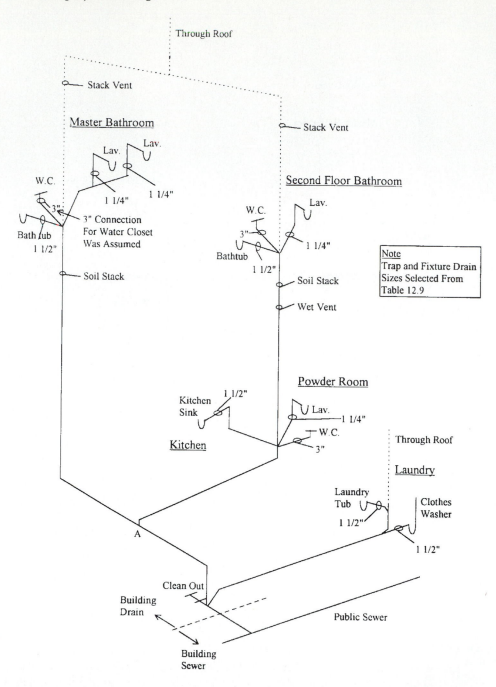

Figure 13.12b
Isometric of the
DWV System
with the Trap and
Fixture Drain
Sizes Indicated

the building discharge to the building sewer. The appropriate d.f.u.s for the Case Study House, based upon Table 13.7, are:

Fixture	Number of d.f.u.s
2-1/2 bath groups	9
Extra lavatory	1
Kitchen sink	2
Laundry group	3
TOTAL	15

Step 3: Determine the Size of the Traps and the Fixture Drains

The sizes of the traps for each fixture are specified in Table 13.6. The size of the fixture drains are the same sizes as their associated traps. The selected sizes [Fig. 13.12(b)] are:

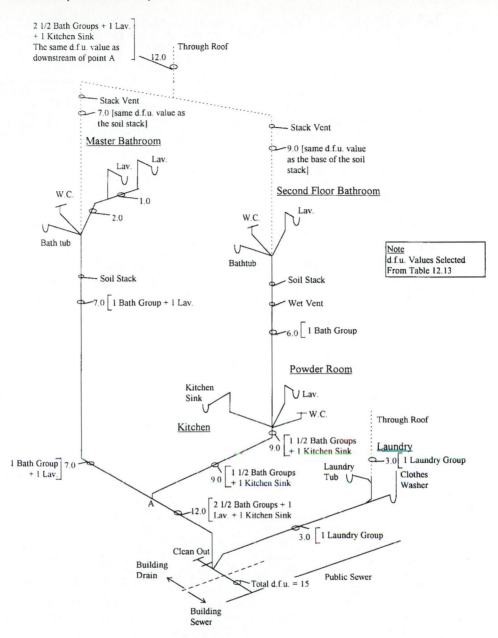

Figure 13.12c
Isometric of the
DWV system with
the d.f.u.s.
Indicated

1-1/4 in. for lavatories and 1-1/2 in. for the tubs, kitchen sink, clothes washer, and laundry tub. It also is assumed that a 3-in. connection to the water closet is required.

Step 4: Assign Accumulated d.f.u.s to the Remaining Unsized Drain and Vent Pipes

The calculation begins at the upstream end of the master bath, second-floor bath, and laundry group fixture branches shown in Fig. 13.12(a), and moves downstream toward the building drain. The d.f.u. values are obtained from Table 13.7 and the results are shown in Fig. 13.12(c). Group drainage fixture unit values are used where appropriate.

The d.f.u. values assigned to the stack vents correspond to the d.f.u. values assigned to the stacks that they are venting [Fig. 13.12(c)]. The soil stack which transports the discharge from the second-floor bathroom, powder room, and kitchen, for instance, is assigned a value of 9 d.f.u.s. As a result, the stack vent for these fixtures is also assigned a value of 9 d.f.u.s.

Step 5a: Determine the Horizontal Branch Drain Pipe Sizes

The branch drains are sized according to Table 13.8. As indicated in Fig. 13.12(d), the branch drain from the lavatories in the master bathroom is sized as 1-1/2 in. This is because

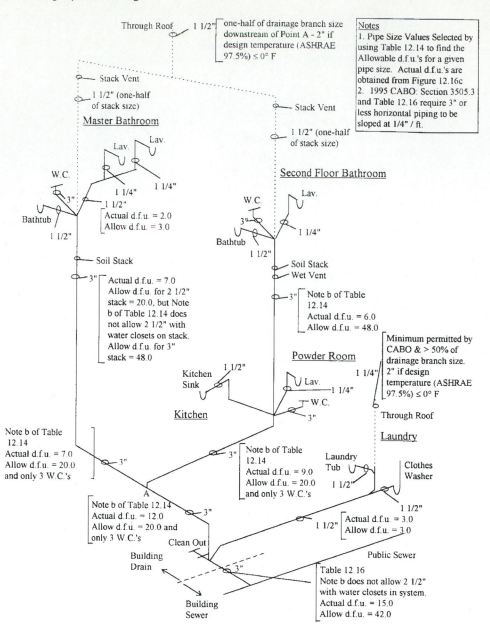

Figure 13.12d
Isometric of the
DWV System
with the Pipe
Sizes Indicated

the 3 d.f.u. allowable value for a 1-1/2 in. line in Table 13.8 exceeds the assigned 2 d.f.u. value in Fig. 13.12(c). The branch drain from the laundry group is also sized at 1-1/2 in., but in this case is at its maximum allowable capacity of 3 d.f.u.s.

Fig. 13.12(d) also indicates that the horizontal branch drains from the soil stacks which start at the master bath and the second-floor bathroom, respectively, are sized at 3 in. because, according to Note b of Table 13.8, this is the minimum allowable size if water closets are being served. Finally, the 1995 CABO Code requires a minimum slope of 1/4 in. per foot for horizontal branch drains 3 in. or less.

Step 5b: Determine the Stack Sizes

The two soil stacks shown in Fig. 13.12(a) are sized using Table 13.8. In each case, the design decision is based not upon the number of d.f.u.s that have been assigned to each stack, but rather upon Note b in Table 13.8 (which states that the minimum soil stack size is 3 in. when water closets are being served).

Step 5c: Determine the Size of the Vents

The vents for the two stacks serving the bathrooms are combined so that only one roof penetration is required [Fig. 13.12(a)]. A separate roof penetration is chosen for the laundry group vent because of its remoteness from the other wet areas of the house.

The 1995 CABO Code, as mentioned earlier in the chapter, states that dry vents must be at least one-half the diameter of the size of drain being served or 1-1/4 in., whichever is greater.[34] As a result, Fig. 13.12(d) indicates that the two vents from the master bath and the second-floor bath stacks are sized at 1-1/2 in. and the vent from the laundry group is sized at 1-1/4 in.

According to the 1995 CABO Code, the minimum vent sizes shown in Fig. 13.12(d) may have to be increased to 2 in. if the house is located in an area with a design temperature (ASHRAE 97.5% condition) less than 0°F.[35]

Step 5d: Determine the Size and Slope of the Building Drain and the Building Sewer

The building drain and the building sewer are sized according to Table 13.10. Since the system serves three water closets, the minimum acceptable drain size is 3 in. At a slope of 1/4 in./ft, a maximum of 42 d.f.u.s can be connected to a 3-in. building drain. A 3-in. size is satisfactory, because as indicated in Step 1 and in Fig. 13.12(c), the total DWV system load is only 15 d.f.u.s.

CHAPTER SUMMARY

This chapter has discussed the fundamentals of the water supply and sanitary drainage systems for a house. The design criteria used in this chapter are specific to the 1995 edition of the *CABO One and Two Family Dwelling Code*. Additionally, the water supply and sanitary drainage systems were designed for the Case Study House to illustrate the application of the CABO Code requirements to a specific house.

NOTES

[1] *CABO One and Two Family Dwelling Code, 1995 Edition*, pp. 195–237, 301–323 (Falls Church, VA: The Council of American Building Officials, 1995).

[2] Ibid, Section 3403.1.2, p. 209.

[3] Ibid, Section 3403.4, p. 210.

[4] Ibid, Section 3403.2, p. 209.

[5] Ibid, Table 3301.2, p. 207.

[6] Ibid, Section 3210, p. 204.

[7] Ibid, Table 3403.1, p. 210.

[8] Ibid, Section 3409.1.1, p. 212.

[9] Ibid, Section 3409, pp. 212–215.

[10] Ibid, Appendix C, pp. 307–323.

[11] Ibid, Section 3409.3, pp. 212–213.

[12] Ibid, Tables 3409.5a–d, pp. 214–215.

[13] Ibid, Appendix C, Section C201, p. 307.

[14] Ibid, Table C101a, p. 308.

[15] Ibid, Fig. C101, p. 317.

[16] Ibid, Tables C301a to C301f, pp. 318–323.

[17] Ibid, Chapter 38, pp. 235–237.

[18] Ibid, Chapter 37, p. 233.

[19] Ibid, Table 3602.1, p. 232.

[20] Ibid, Section 3602.1, p. 232.

[21] Ibid, Chapter 36, pp. 223–232.

[22] Ibid, Section 3701.4, p. 233.

[23] Ibid, Table 3502.1, p. 217.

[24] Ibid, Table 3502.2, p. 218.

[25]Ibid, Section 3503, pp. 217–219.
[26]Ibid, Chapter 35, pp. 217–221.
[27]Ibid, Chapter 36, pp. 223–232.
[28]Ibid, Appendix B, pp. 301–306.
[29]Ibid, Section 3601.8, pp. 224 and 227.
[30]Ibid, Section 3601.6, p. 224.
[31]Ibid, Section 3601.7, p. 224.
[32]Ibid, Chapter 35, pp. 217–221; Chapter 36, pp. 223–232; Chapter 37, p. 233.
[33]Ibid, Appendix B, pp. 301–306.
[34]Ibid, Section 3601.6, p. 224.
[35]Ibid, Section 3601.5.5, p. 223.

BIBLIOGRAPHY

CABO One and Two Family Dwelling Code, 1995 Edition.   Falls Church, VA: The Council of American Building Officials, 1995.

Hemp, Peter, *Plumbing a House.*   Newtown, CT: The Taunton Press, Inc., 1994.

Sunset Basic Plumbing Illustrated.   Menlo Park, CA: Sunset Publishing Corporation, 1991.

Wing, Charlie, *The Visual Handbook of Building and Remodeling.*   Emmaus, PA: Rodale Press, 1990.

14

Electrical System Design

INTRODUCTION

This chapter provides an introduction to the electrical systems used in a house. Some of these are standard in most houses; others are entirely optional. Standard systems include the electrical power distribution and lighting systems, which are governed by the *National Electrical Code*®* (NEC®)[1] and Chapters 39 through 46 of the 1995 edition of the *CABO One and Two Family Dwelling Code*.**[2] Telephone and doorbell systems are also considered to be standard systems. Optional systems include: (1) television, (2) fire alarm, (3) security, (4) intercom/closed circuit TV, and (5) smart house technology. Each of these systems is addressed at various levels of emphasis in this chapter.

FUNDAMENTALS OF ELECTRICITY

In order to discuss and understand residential electrical systems, familiarization with basic electrical fundamentals is necessary.

Ohm's Law

Ohm's Law defines a relationship between the voltage, current, and impedance in an electrical circuit. The **voltage** is the force or pressure difference between two points in an electrical system. The **impedance** is the physical characteristic of an electrical system that "impedes" the flow of current when voltage is applied. The **current** is the resulting electron flow when voltage is applied to the circuit. All substances, to greater or lesser degrees, are resistant to the flow of electrical current. Conductors such as metals have low impedance. Various insulators such as plastics, paper, glass, or rubber have high impedance, but no material exists that has no impedance.

---

*National Electrical Code® and *NEC*® are registered trademarks of the National Fire Protection Association, Inc., Quincy, MA 02269.

**Sections and tables from this Code are cited for instructional purposes only, and the treatment of the Code provisions contained in the design solutions illustrated in this chapter is not intended as a substitute for a complete and thorough understanding of all building code provisions to which any particular project may be subject.

In DC circuits, impedance is equivalent to resistance. However, in AC circuits, the impedance has three components: resistance, inductive reactance, and capacitive reactance. As residential systems move away from traditional loads such as incandescent lighting and resistance heating, which have no reactance component, it is important to discuss electrical systems using proper electrical terms.***

Ohm's Law usually is stated as:

$$V = IZ \text{ (AC circuits)}$$
$$V = IR \text{ (DC circuits)}$$

where: V = electrical force or pressure in volts (V)
  I = electrical current in amperes (A)
  Z = impedance in ohms (Ω)
  R = resistance in ohms (Ω)

Other common arrangements of this equation, depending on the unknown term, include:

$$I = V/R, I = V/Z$$
$$R = V/I, Z = V/I$$

EXAMPLE NO. 1

Determine the current through the load in the circuit shown in Fig. 14.1. The load due to the frying pan in the kitchen is $Z = 12$ ohms.

SOLUTION:

$I = V/Z = 120/12 = 10$ amps

EXAMPLE NO. 2

Find the resistance of the water heater in the circuit shown in Fig. 14.2 if it carries a current $I = 18.75$ amps

SOLUTION:

$R = V/I = 240/18.75 = 12.8$ ohms

Watt's Law

Watt's Law establishes a relationship between power, current, and voltage in the same way that Ohm's Law establishes a fixed relationship between voltage, current, and impedance.

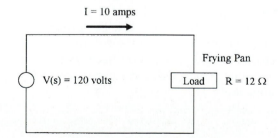

Figure 14.1
Electrical Circuit
for Example
No. 1

***Impedance and resistance are used interchangeably in this chapter recognizing, however, that there is a fundamental difference as explained.

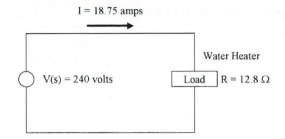

Figure 14.2
Electrical Circuit
for Example
No. 2

Watt's Law usually is expressed as:

$$P = IV$$

where: P = power in volt-amps (VA) or watts (W)
I = electrical current in amperes (A)
V = electrical force or pressure in volts (V)

Other common arrangements of this equation, depending on the unknown term, include:

$$I = P/V$$
$$V = P/I$$

If substitutions are made using the $V = IR$ equation, the following relationships are defined:

$$P = IV = I \times (IR) = I^2R$$
$$P = IV = (V/R) \times V = V^2/R$$

A relationship exists between watts and volt-amps in AC circuits. The total load on an AC circuit is given in volt-amps. This takes into account the inductive and capacitive losses previously mentioned. The relationship is expressed as

$$W = VA \times PF$$

where PF is the power factor, which is expressed as a fraction between 0 and 1.0. PF normally is between 0.5 and 1.0. For a load that is composed of only resistance, such as electric heat, the power factor is 1.0, and **Watts = Volt-Amps**. For other loads, such as motors or fluorescent lighting, the power factor will be less than 1.0, so volt-amps will be greater than watts. All circuits must be sized based on the volt-amp load.

EXAMPLE NO. 3

Determine the power required to operate the water heater in the circuit in Example No. 2 (see Fig. 14.2).

SOLUTION:

$$P = IV = (18.75)(240) = 4,500 \text{ volt-amps or watts,}$$

or

$$P = V^2/R = (240)^2/12.8 = 4,500 \text{ watts or volt-amps}$$

In most cases, it will not be necessary to determine the wattage or volt-amp rating of electrical appliances in the manner indicated in Examples No. 2 and 3 since the rating will be listed on the appliance nameplate. Therefore, the listed wattage or volt-amp rating typically is used to determine the **amperage load** upon the circuit containing the appliance.

Table 14.1 provides wattage and amperage ratings for some of the typical appliances found in the house. Central air conditioners and heat pumps, clothes dryers, baseboard electric heat, kitchen ranges, water heaters, and heat pumps require 240-volt service. Heavy-duty shop equipment may also require that level of service.

TABLE 14.1 Wattage and Amperage Ratings of Typical
Household Appliances

Appliance	Volt-Amps or Watts	Amperes (120 volts)	Amperes (240 volts)
Air Conditioner			
Central	6,720–12,000	—	28–50
Room	630–2,600	5.3–12.1	6.3–10.8
Central Heat Pump			
Heat Pump	6,720–12,000	—	28–50
Resistance Heat Backup	10,000–15,000	—	42–63
Coffee Maker	1,000	8.3	—
Dishwasher	1,030	8.6	—
Clothes Dryer (elec.)	5,600	—	24
Clothes Dryer (gas)	720	6	—
Frying Pan	1,200	10	—
Baseboard Electric Heat	250 per linear ft	—	1 per linear ft
Kitchen Range (elec.)	8,200–13,300	—	34.2–55.4
Oven	3,700	—	15.4
Top	7,300–9,700	—	30.4–40.4
Kitchen Range (gas)	600	5.0	—
Mixer	150	1.3	—
Water Heater (elec.)	4,500	—	18.8
Light			
Incandescent	300	2.5	—
Fluorescent	80	0.7	—
Sump Pump	550–800	4.6–6.7	—
Jacuzzi	1,660–2,400	13.8–20	—
Clothes Washer	960	8	—
Garbage Disposal	500–720	4.2–6.0	—
Bathroom Fan	80–440	0.7–3.7	—
Microwave	950–1,630	7.9–13.6	—
Range Hood	280–720	2.3–6.0	—
Refrigerator	900	7.5	—
Trash Compactor	780	6.5	—

Energy Usage

The unit for measuring the energy that is consumed by a typical electrical appliance or other load is kilowatt hours.

$$E = \frac{PT}{1,000}$$

where: E = energy consumed in kilowatt hours
P = power in watts
T = hours of usage

EXAMPLE NO. 4

Find the energy usage of the water heater in the circuit in Example No. 2 (Fig. 14.2) if it operates for three hours.

$$E = \frac{PT}{1,000} = \frac{(4,500)(3)}{1,000} = 13.5 \text{ kilowatt hours}$$

The homeowner, of course, pays the electric utility company on the basis of the number of kilowatt hours used each month.

Basic Electrical Circuits

The most basic electrical circuit consists, as a minimum, of the following four parts (Fig. 14.3): (1) a source of electrical pressure or voltage, (2) conductors to connect the source to the load, (3) an electrical load, and (4) a switch or other mechanism to control that load. Two conductors are necessary to complete the circuit, since current will flow only when the circuit path is complete from two sides of the source.

Electrical loads can be connected to a power source in either a series or parallel circuit. Each of these types will be discussed.

Series Circuits

In a series circuit, there is only one path for the current to flow. The same current, therefore, flows through all loads of the circuit, and the impedances (i.e., resistances) of each of the appliances or other electric loads are added together to determine the total resistance of the electrical circuit. The wiring in a house typically **is not set up as a series circuit**.

EXAMPLE NO. 5

The series circuit shown in Fig. 14.4 has the following characteristics:

$$V_s = 240 \text{ volts}$$
$$R_1 = 1 \text{ ohm}$$
$$R_2 = 1 \text{ ohm}$$
$$R_{\text{LOAD}} = 22 \text{ ohms}$$
$$R_T = \text{total resistance} = R_1 + R_2 + R_{\text{LOAD}}$$

Determine the current through the 22-ohm load and the voltage across the 22-ohm load

SOLUTION:

$$I = V/R_T = V/(R_1 + R_2 + R_{\text{LOAD}})$$
$$= 240/(1 + 1 + 22)$$
$$= 10 \text{ amps}$$

$$V_{\text{LOAD}} = IR_{\text{LOAD}} = (10)(22) = 220 \text{ volts}$$
or
$$V_{\text{LOAD}} = V_s - IR_1 - IR_2 = 240 - (10 \times 1) - (10 \times 1)$$
$$= 220 \text{ volts}$$

Parallel Circuits

The wiring in a house typically is set up as a **parallel circuit**. In a parallel circuit, there is a separate electrical path through each load, with a part of the total current coming from the source passing through each load. The total current in the parallel circuit, therefore, is the

Figure 14.3
A Basic Electric Circuit

Source: Robert Wood, *House Wiring from Start to Finish, 2nd ed.* (New York, NY: TAB Books, Division of The McGraw-Hill Companies, 1993), p. 6. Reproduced with the permission of The McGraw-Hill Companies.

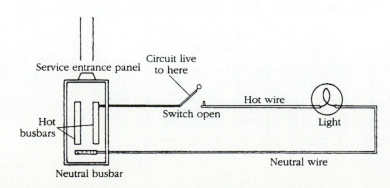

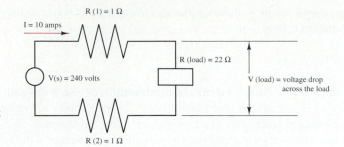

Figure 14.4
Electrical Circuit
for Example
No. 5

sum of the individual currents that pass through each electrical load. The current drawn through each electrical load is proportional to its wattage and inversely proportional to its resistance. The voltage is the same across all loads in parallel circuits.

EXAMPLE NO. 6

The circuit shown in Fig. 14.5 contains the following loads:

$$P_1 = 1,000 \text{ watts (coffee maker)}$$
$$P_2 = 150 \text{ watts (mixer)}$$
$$P_3 = 1,200 \text{ watts (frying pan)}$$

Find the current through each load, the total current through the electrical circuit, and the resistance of each load.

SOLUTION:

$I_1 = P_1/V_s = 1,000/120 = 8.3$ amps
$I_2 = P_2/V_s = 150/120 = 1.3$ amps
$I_3 = P_3/V_s = 1,200/120 = 10$ amps

Therefore:

$I_{total} = I_1 + I_2 + I_3 = 8.3 + 1.3 + 10 = 19.6$ amps
$R_1 = V_s/I_1 = 120/8.3 = 14.5$ ohms
$R_2 = V_s/I_2 = 120/1.3 = 92.3$ ohms
$R_3 = V_s/I_3 = 120/10 = 12$ ohms

With the exception of the special 240-volt circuits shown in Table 14.1 for such appliances as a central air conditioner, clothes dryer, etc., most house wiring is designed to handle no more than 15 or 20 amps. Too many appliances placed on the same circuit draw more amperage than can be carried safely by the circuit. Such uncontrolled amperage could cause excessive heat problems. As will be noted later in this chapter, circuit breakers guard against this type of situation.

Distribution of Current

Electrical current exists in two forms: direct current (DC) and alternating current (AC). **Direct current** is produced by photovoltaic cells, drycells, storage batteries, and generators. Direct current flows steadily in one direction and often, but not always, at constant voltage.

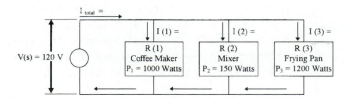

Figure 14.5
Electrical Circuit
for Example
No. 6

The electrical power produced and delivered to houses by utility companies is **alternating current**. In this type of power, the voltage constantly alternates over time from 0 volts to a controlled maximum voltage, back to 0, then down to the same maximum in the opposite direction, and finally back to 0 to start the cycle over again. The number of full cycles per second is referred to as the **frequency**. The current used in the United States cycles at the rate or frequency of 60 times a second. Frequency is expressed in units of Hertz or cycles; therefore, the United States utilizes 60 Hz AC power. Frequencies are different in other countries.

Electrical power is **generated** by utility companies at large generating plants. The voltage is usually generated at approximately 15,000 volts. A step-up transformer then increases it to much higher levels, up to 500–600 kilovolts. This is called **transmission voltage**. The power is transmitted over high-voltage transmission lines to substations within the different subareas of a utility company's service area. There, step-down transformers reduce the voltage to 5,000–35,000 volts for distribution to specific customers. This is called the **distribution system**.

The power generated, transmitted, and distributed by utility companies is **three-phase power**, which uses three wires in the feeder (one for each phase). A neutral may or may not be utilized. The power used in most residences is single phase. This can be obtained by using only one or two of the available three phase wires. Commercial, industrial, and institutional customers mostly use three-phase power.

A single utility transformer typically services several houses and is used to reduce the distributed voltage to the voltage used in homes. This voltage usually is **120/240 volt, single-phase three wire.**

Circuit Types

The following three types of circuits are available with the 120/240 V, single-phase, three-wire system used in houses (Fig. 14.6):

1. a single-phase, two-wire, 120-volt circuit which includes one hot (or phase) wire, one neutral, and one ground wire. This circuit is intended for lights, receptacles, kitchen appliances, and washers.

2. a single-phase, two-wire, 240-volt circuit which includes two hot (or phase wires), no neutral, and one ground wire. This circuit is intended for large loads, such as electric resistance heat or water heaters.

3. a single-phase, three-wire circuit which includes two hot (or phase) wires, one neutral, and one ground wire. It supplies both 120 and 240 volts for loads that require both voltages, such as ranges, clothes dryers, and air conditioners.

For each of these circuit types, the ground wire is not counted as a circuit wire.

A review of Watt's Law will indicate that for a given load, increasing the voltage reduces the current drawn in amps, thus reducing the wire size required to serve a load. Some small loads, such as lights, clocks, fans, etc., are not readily available at 240 volts. This is why some appliances that contain multiple loads, such as ranges, require both 120 and 240 volts.

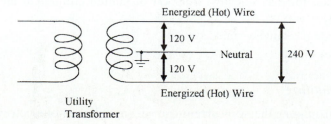

Figure 14.6
Typical
Residential
Voltage System

Grounding

All house electrical systems must be **grounded systems**. This is why each of the three types of circuits just described contains a ground wire. A ground is a conducting connection, either intentional or accidental, between an electrical circuit or equipment and the earth, or to some conducting body that serves in place of the earth. An **accidental ground** is usually referred to as a "ground fault." The practice of **intentional grounding** consists of connecting electrical circuits and equipment (or conducting bodies in close proximity to electrical circuits) to metallic bodies in the earth. There are two types of grounding: (1) **system grounding** and (2) **equipment grounding**.

Grounding Definitions

The definitions below clarify some of the points related to grounding.

Grounded conductor: Neutral. A system or circuit conductor that is intentionally grounded. It is also one of the wires that runs from the electric distribution panel to each 120-volt electrical load. The grounded conductor is considered to be a current-carrying conductor (Fig. 14.6); it carries current under normal conditions.

Ground wire: A green or bare conductor used only for "grounding." It is one of the wires that runs from the electric distribution panel to each electrical load. It is not considered to be a current-carrying conductor and does not carry current under normal operating conditions.

Equipment grounding conductor: The conductor used to connect the metal parts of equipment to the system neutral only at the service entrance. It also does not carry current under normal conditions.

Grounding conductor: A green or bare conductor used to connect equipment or the grounded circuit of a wiring system to a grounding electrode. The grounding conductor begins at the ground/neutral bus bar located in the distribution panel and terminates at the grounding electrode.

System Grounding

With regard to system grounding, power distribution systems fall into two categories: (1) ungrounded and (2) grounded. **Ungrounded systems** do not have an intentional connection between the neutral, or any phase, and ground. **Grounded systems** have both the system neutral and the ground wire connected to the ground. On a grounded system, the earth or ground may be utilized to complete the circuit and allow current circulation from or to its current source. On an ungrounded system, ground cannot be used to complete a circuit.

The primary reasons for system grounding are:

1. Compliance with the minimum safety standards in the 1995 CABO Code[3] and the NEC®;[4]
2. Limit transient overvoltages, caused by restriking ground faults, to the level the equipment is designed to withstand and to levels safe for personnel;
3. To provide for fast, selective isolation of faulted equipment and circuits when a fault occurs;
4. To provide a relatively stable system;
5. Personnel safety.

Equipment Grounding

The practice of grounding the non-current-carrying metal parts of electrical equipment and raceways is known as **equipment grounding**. This includes enclosures and boxes that are metal, but that should not carry current. Equipment grounding also includes the practice of grounding other metallic objects, such as piping systems or structures, that may become energized.

Equipment grounding is accomplished by connecting the green or bare "ground wire" present in each circuit to each metal enclosure or box, and to the green ground terminal of receptacles. The latter serves as a connection point for the ground wire found in many appliance cords. If an accidental connection were to occur between an energized wire and the equipment enclosure, current would then flow safely through the ground wire back to the source.

The primary reasons for equipment grounding are:

1. Compliance with the minimum safety standards in the 1995 CABO Code[5] and the NEC®;[6]
2. Safety;
3. Control of "noise";
4. Transient protection;
5. Reference point for logic and control;
6. Shielding.

Bonding

Bonding is the permanent joining of metallic parts to form a continuous conductive path between enclosures, boxes, etc., and the equipment grounding system. The equipment grounding system is composed of all wires, conduits, enclosures, and boxes that provide for the safe conduction of abnormal current back to its source.

Bonding also refers to the practice of providing a ground connection between the grounded conductor and the grounding conductor (refer to Fig. 14.8). Bonding, which is covered in the 1995 CABO Code,[7] occurs in the distribution panel at the service entrance.

Safety

A word of caution about the safety aspects related to electrical systems is appropriate at this point. *It does not take much electrical current to harm or kill a person.* The average body resistance under normal circumstances (not wet, nor standing in water) is approximately 1,000 ohms. Thus, for a 120-volt current, Ohm's Law indicates that $I = V/R = 120V/1,000$ ohms $= .12$ amps, or 120 mA (10^{-3} amps). Also, for a 240-volt current, $I = 240 V/1,000$ ohms $= .24$ amps or 240 mA. Research has shown the following:

Magnitude of Current	Human Reaction
less than 1 mA	can't feel
1–3 mA	able to sense
5–20 mA	shock, may be painful
20–75 mA	can't let go, loss of muscle control, respiratory distress
75–100 mA	ventricular fibrillation
above 200 mA	flesh burns

Considering this information, it is obvious that residential voltages can cause serious harm. Fatalities are usually caused by the stoppage of breathing or heart function. With these consequences in mind, the following safety rules always should be followed:

- Before work is performed on any wiring, the power should be off. Work should not be performed on energized circuits.
- Tools should have insulated handles.
- Workers should be on a dry surface. Work should not be performed when hands, feet, or the body are wet or when the individual is standing on a wet floor.
- Workers should not work alone.

- Workers should not wear rings, metallic watch bands, etc.
- Workers should wear safety glasses.
- The work area should be well lighted and uncluttered.
- Precautions should be taken to ensure that no part of the body, such as head and hair, or loose clothing can accidentally make contact with the equipment.

A number of codes provide information about safe work procedures on electrical systems. The *NEC*®, National Electrical Safety Code (NESC), and OSHA are only a few. Local jurisdictions may have other requirements. An individual should always be aware of and follow all of the applicable safety codes.

ELECTRICAL POWER AND LIGHTING DISTRIBUTION

Electrical distribution systems for power and lighting consist of several basic components which are typical for every house. The specifics about how these components are arranged, however, differ for each residential design. Each of these will be discussed.

Electric Service

The **electric service** is the part of the electrical system through which the electrical supply from the utility company is delivered and connected to the house. As noted earlier, three-wire service, with one neutral and two hot lines, is typically connected to a house (Fig. 14.7). If the incoming wires pass above ground from a power pole to the house, it is called an **overhead service**. The wires coming to the house are called the **service drop**. These wires termi-

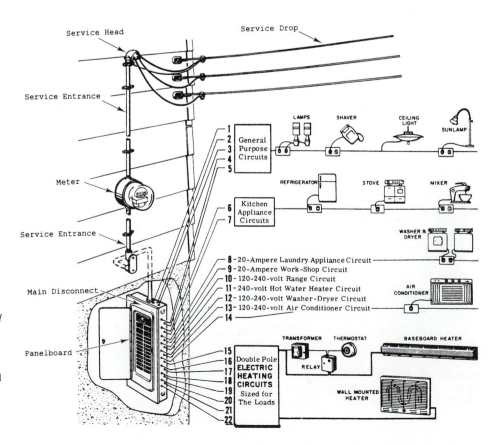

Figure 14.7
Electrical Service
Entrance and
Distribution
Source: William J.
Whitney, *Residential
and Commercial
Electrical Wiring,*
p. 261. Copyright ©
1983 John Wiley &
Sons, Inc. Reprinted
by permission of
John Wiley & Sons,
Inc.

nate in connections to another set of wires that hang down from the service head. The service drop and connection are provided by the power company.

In many residential applications, the choice is made to bring the service into the house underground. The **underground method** allows the power company the option of mounting the final step-down transformer on or under the ground. Power is then delivered to the house through underground wires. The wires that run from the transformer to the house, then up and into the meter, are called the **service lateral**. The service drop and lateral are provided, designed, and owned by the power company.

All electrical energy supplied to power-consuming devices and appliances within the house must pass through the electrical service equipment, where it is metered and protected (Fig. 14.7). The main components of a complete electrical service installation, therefore, include the **service drop** or **lateral**, **meter**, **service entrance**, and **main disconnect(s)**. Figure 14.8 provides a more detailed diagram of these four components and indicates how grounding is accomplished.

Utility Meter Socket and Meter

At the location where the service drop or lateral is attached, power is transmitted through a metering device that records the amount of electrical energy used. It is composed of two parts: a socket or base (provided by the homeowner) and a meter (provided by the power company; Fig. 14.8). The meter is usually located on the outside of the house in order to be

1. Service drop
2. Service head
3. Service entrance conduit
4. Threaded hub
5. Meter base (socket)
6. Service entrance panel
7. Main disconnect (Main Breaker)
8. Grounding bushing
9. Equipment bonding jumper
10. Main bonding jumper
11. Grounding electrode conductor
12. Service entrance conductor
13. Ground / neutral bus
14. Grounding electrode
15. Metalic cold water service

(not required if plastic)

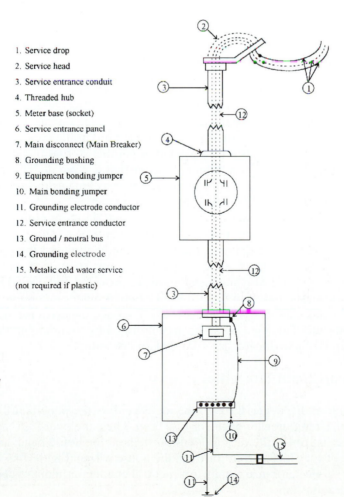

Figure 14.8
Service Entrance
Equipment
Source: William J. Whitney, *Residential and Commercial Electrical Wiring,* p. 262. Copyright © 1983 John Wiley & Sons, Inc. Reprinted by permission of John Wiley & Sons, Inc.

accessible and visible to the utility company meter reader. After the power company installs the meter, the socket is "sealed." Only the power company is permitted access to the meter and socket. An application for service, usually accompanied by a fee or service charge, is required before the power company will extend service to a house.

Service Entrance

The **service entrance**, also called service conductors, is the wiring that extends from the meter socket to the main disconnect (main breaker). For overhead services, the portion from the service head to the meter is also called the service entrance. The $NEC^{®}$[8] and the text and tables in Chapter 41 of the 1995 CABO Code[9] present specific requirements concerning the size and length of the service entrance conductors and all other aspects related to bringing electric service into the house. Service entrance conductors are sized differently than all other conductors. Wire size tables located in other chapters of the 1995 CABO Code should not be applied to service entrance conductors.

Main Disconnect(s)

The **main disconnect**, also called the service disconnect, is the switch or circuit breaker that turns all power to the house off or on. It must be accessible and located as near the service entrance as possible. It may be located outside the house in combination with the meter or inside the house in combination with the distribution panel. Up to six main disconnects are permitted for each individual service entrance.[10]

In normal field practice, particularly when there is only one distribution panel in a house, the main circuit breaker (Fig. 14.8) is located inside the distribution panel.

Service Entrance Size Determination

The procedure for determining the size of the service entrance (i.e., the rating of the main breaker) is described in the 1995 CABO Code.[11] For a typical house, the **main panelboard** is rated at either 100, 200, or 400 amps, with 200 amps being the most common. A **main circuit breaker** of an equivalent or smaller size may be used in the panelboard.

Individual electrical loads are connected in parallel so that the individual currents to each load are additive (Fig. 14.5). A main breaker that is rated at 200 amps can safely supply current to all of the electrical circuits in the house as long as the additive value of all of the currents does not exceed 200 amps per phase conductor.

Once a particular house design has been selected, the steps outlined in Table 14.2 can be used to calculate the minimum size of the service entrance. The 1995 CABO Code provides information about the sizing of the service conductors and the grounding electrode conductor.[12]

The service entrance wire and installation are provided and owned by the homeowner or contractor. Also, the procedure in Table 14.2 only provides *minimum* size requirements. It is advisable, and customary, to include extra service capacity for finished basements, future additions, etc. This usually can be accomplished by rounding up the service capacity to the next standard panelboard size (100, 200, or 400 amps).

Panelboard(s)/Circuit Protection

Electric current, after passing through the meter, enters the panelboard (also called a distribution panel, circuit breaker panel, or fuse box). The panelboard acts as a terminal where branch circuits that run to the various loads throughout the house begin and end. It also prevents the flow of very high current through the house wiring. Figure 14.9 illustrates a panelboard with its service entrance conductors, neutral conductor, main circuit breaker, hot bus bars, and neutral bus bar.

TABLE 14.2   Steps Involved in the Determination of the
Size of the Service Entrance

LOADS AND PROCEDURE
3 volt-amperes per square foot of floor are for general lighting and convenience receptacle outlets
PLUS
1,500 volt-amperes × total number of 20-ampere-rated small appliance and laundry circuits
PLUS
The nameplate volt-ampere rating of all fastened-in-place, permanently connected, or dedicated circuit-supplied appliances such as ranges, ovens, cooking units, clothes dryers, and water heaters
APPLY THE FOLLOWING DEMAND FACTORS TO THE ABOVE SUBTOTAL:
The minimum subtotal for the loads above shall be 100% of the first 10,000 volt-amperes of the sum of the above loads plus 40% of any portion of the sum that is in excess of 10,000 volt-amperes
PLUS THE LARGEST OF THE FOLLOWING:
Nameplate rating(s) of air-conditioning and cooling equipment, including heat pump compressors
Nameplate rating of the electric thermal storage and other heating systems where the usual load is expected to be continuous at the full nameplate value. Systems qualifying under this selection shall not be figured under any other category in this table
Sixty-five percent of nameplate rating of central electric space-heating equipment, including integral supplemental heating for heat pump systems
Sixty-five percent of nameplate rating(s) of electric space-heating units if less than four separately controlled units
Forty percent of nameplate rating(s) of electric space-heating units of four or more separately controlled units
The minimum total load in amperes shall be the volt-ampere sum calculated above divided by 240 volts

Source: Reproduced from the 1995 edition of the *CABO One and Two Family Dwelling Code*,™ copyright © 1995, with the permission of the publisher, The International Codes Council, Whittier, CA; Table 4102.2, p. 247, contains information from the *NEC*®.

As noted in Fig. 14.9, the **phase wire** (i.e., the hot wire) for the branch circuit is connected to the hot bus bar using a current protection (also called overcurrent protection) device called a **circuit breaker**. Fuses were used in older houses to serve the same purpose. Both of these overcurrent protective devices are discussed next.

Overcurrent Protective Devices

The purpose of overcurrent protection is to prevent damage to the conductors or other parts of the electrical system. The damage is typically due to the overheating that would result from an overcurrent. **Overcurrent** refers to the condition resulting from excessive current drawn through a circuit wire designed to carry only a certain amperage. This situation occurs for one of the following two reasons: (1) continuous overloading, or (2) an unintentional and unexpected direct connection of power to ground through little or no resistance. The latter condition is called a **short** or **short circuit**. Overloads are generally much lower values of current than short circuits. Protective devices are required for all phase or hot wires. They are not required for neutral or ground wires. For 120-volt circuits, only one phase has to be protected, so a one-pole (single-pole) protective device is used. For 240-volt circuits, two phases must be protected, so a two-pole (double-pole) protective device must be used.

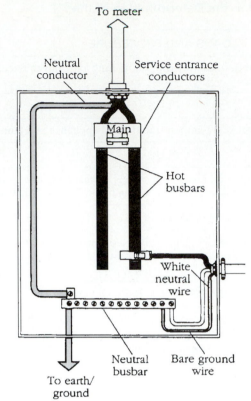

Figure 14.9
Panelboard
Showing Wire
Connections to a
Single-Pole
Circuit Breaker
Source: Robert
Wood, *House
Wiring from Start to
Finish, 2nd ed.*
(New York, NY:
TAB Books,
Division of The
McGraw-Hill
Companies, 1993),
p. 114. Reproduced
with the permission
of The McGraw-
Hill Companies.

Circuit Breakers

The circuit breaker, which is the most commonly used protective device, combines the function of overcurrent protection and a simple power switch. It has an advantage over fuses in that, when tripped by overcurrent, it can be reset almost immediately after the cause of the overcurrent has been corrected.

Fuses

A fuse is an overcurrent protective device with a circuit opening material (element) that is heated and severed (blown) by the passage of excessive current through it. Fuses were commonly used in the electrical systems of older houses. A **plug fuse** has a threaded base. Another common type of fuse is a **cartridge fuse**, which generally looks like a cylinder with two metal ends. It is more difficult to determine if a cartridge fuse has blown.

Grounding Revisited

The detail of the panelboard shown in Fig. 14.10 provides the opportunity to revisit the subject of grounding in order to clarify how it actually occurs. In all new houses built since the early 1960s, a ground wire is attached to every receptacle, switch, and metal junction box. Many appliance plugs have a third prong for grounding which mates to the ground wire connection slot on receptacles. If a fault occurs that would cause sparks, overheating, or the flow of current through an unintended path, the current will instead flow in the path of least resistance—the ground wire—back to the source—the service.

Ground wires from each branch electrical circuit in a house terminate at the panelboard (Fig. 14.10). The ground wires and neutral wires from each branch circuit are attached to the ground/neutral bus. The bus, in turn, and hence the entire electrical system, must be grounded to metal water pipes (if present) and/or to a copper rod driven into the earth outside the house as close as possible to the meter. The ground wire is securely fastened to the rod.

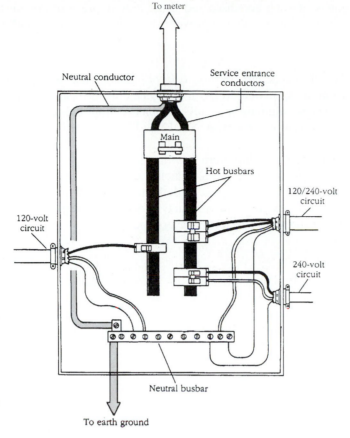

Figure 14.10 Panelboard Showing Wire Connections for 120-Volt, 240-Volt, and 120/240-Volt Branch Circuits

Source: Robert Wood, *House Wiring from Start to Finish, 2nd ed.* (New York, NY: TAB Books, Division of The McGraw-Hill Companies, 1993), p. 117. Reproduced with the permission of The McGraw-Hill Companies.

The length of the grounding rod that is used varies with different types of soils. The length selected must ensure that a proper conductance situation exists.

Conductors

Conductors transmit the current from the panelboard to the various end uses in the house. A conductor that carries current and is not grounded is known as a **phase wire**, or hot wire. A conductor that carries current but is grounded at the service panel is known as a **neutral wire**. Both phase wires and neutral wires are considered to be current-carrying conductors. A conductor that is used to connect the non-current-carrying metal parts of equipment, metal boxes, and other enclosures to the ground bus at the service panel is known as a non-current-carrying conductor or **ground wire**.

Each of these wires, and their connections inside the panelboard, are shown in Fig. 14.9. Figure 14.10 provides an expanded view of a panelboard, which indicates the typical wiring connections for 120-volt, 240-volt, and 120/240-volt branch circuits. As noted, the 120-volt branch circuit consists of one phase wire, which is connected to the hot bus bar using a single-pole circuit breaker, a neutral wire, and a bare ground wire.

The 240-volt branch circuit consists of two phase wires, which are connected to the hot bus bar using a double-pole circuit breaker, and a bare ground wire. The 120/240-volt branch circuit consists of two phase wires, which are connected to the double-pole circuit breaker, a neutral wire, and a bare ground wire.

Materials

The two most commonly used materials for conductors are aluminum and copper, both of which are acceptable according to the 1995 CABO Code and the *NEC*®. The advantages and disadvantages of each are discussed next. Proper precautions must be taken if both types are used in a house in order to prevent the occurrence of **dissimilar metal** problems.

ALUMINUM

Aluminum, over the years, has become very popular for use as a conductor. It is used for most transmission lines, service drops, and service entrance cables. The main reason for its popularity is because aluminum is less expensive and lighter in weight than copper. Aluminum is permitted only for sizes No. 12 and larger, but is typically used only in sizes No. 6 and larger.

The disadvantages of aluminum are that it has lower conductivity than copper, so larger sizes are required, and it has a higher coefficient of expansion, so thermal cycling becomes a concern. As a conductor is loaded and unloaded, its temperature increases and decreases, causing it to expand and contract. This constant expansion and contraction may cause terminations and splices to loosen over time.

COPPER

Copper is an excellent conductor, has a lower resistance and higher conductivity, and is much more widely used than aluminum. It also has better load cycling characteristics than aluminum, thereby lessening the likelihood of loose connections. The main disadvantages of using cooper are that it is more expensive than aluminum and that it is more difficult to bend and handle, especially in larger sizes.

Size

Conductor sizes are commonly based on the American Wire Gauge (AWG). The standard sizes manufactured in accordance with this gauge range from No. 46 up to No. 4/0, where No. 46 is the smallest wire. Sizes larger than No. 4/0 are identified by their cross-sectional areas as measured in thousands of circular mils, abbreviated kcmil. The next larger electrical wire size after No. 4/0 is 250 kcmil.

As indicated by the 1995 CABO Code[13] and the *NEC*®,[14] the **minimum size** of conductors for feeders and branch circuits shall be **No. 14 copper** and No. 12 aluminum. Smaller size wires are permitted for low-voltage signal, alarm, and control circuits such as intercoms, thermostats, bells, buzzers, sprinkler timers, or burglar alarms. Also, smaller sizes are permitted for various exposed flexible cords. Inside individual fixtures may be wired with No. 18 as a minimum.[15]

Insulation

Individual conductors must be covered in order to insulate them from personnel, equipment, and non-current-carrying items, and to protect them from heat and moisture. The 1995 CABO Code[16] specifies the acceptable insulation types as: (1) RH and RHW (rubber); (2) THHN, THHW, THW, THWN, and TW (thermoplastic); (3) UF and USE (underground); or (4) XHHW (moisture and heat resistant). Rubber, the insulation for the types beginning with R, essentially has been replaced in housing by thermoplastic insulation material, with Types THW and THWN being the most popular.

Insulation types are rated according to their maximum allowable temperature, in **degrees Centigrade**. The ratings typically are 60°C, 75°C, and 90°C, as listed in the tables. The ampacity of a conductor is limited to a level which will keep the temperature of the insulation within its temperature rating. The most common wire insulation that is used in housing is a 75°C rated insulation. However, due to limitations in device ratings, even though a 75°C rated insulation is used, the ampacity of the wire must be selected as if a 60°C rated insulation was being used. The 1995 CABO Code[17] provides additional information about this topic.

Color Code

The 1995 CABO Code section indicates that the insulation on individual conductors must be color coded so that conductors being used for specific purposes may be identified easily.[18] The equipment ground or ground wire must be encased in a green insulation or left bare. The neutral wire must be encased in a white or gray insulation, while the hot conductors may be encased in any other color of insulation except green, white, or gray. Typical colors for the hot conductors are black and red.

Wiring Method/Jacket

As noted earlier, the branch circuits in a house provide either (1) 120-volt service in a two-wire format (one hot conductor, one neutral conductor, and one ground wire), (2) 240-volt service in a two-wire format (two hot conductors and one ground wire), or (3) 120/240-volt service in a three-wire format (two hot conductors, one neutral conductor, and one ground wire). The 1995 CABO Code[19] does not permit these conductors to be run individually; they must be grouped together in order to avoid circuit confusion and to provide for the varying degrees and types of physical protection that are required under different operating and environmental conditions. Two or more wires grouped together are called a **cable**. The outer covering is called a **sheath** or **jacket**. Figure 14.11 illustrates the types of cables that provide the levels of service previously noted.

The 1995 CABO Code[20] provides a list of the allowable "wiring methods." Non-metallic sheathed cable (type NM) is the most popular type used for house construction. The group of wires in type NM cable shown in Fig. 14.11 is typically surrounded by both a paper wrapping and a plastic sheathing (i.e., jacket). The 1995 CABO Code[21] also specifies the allowable applications for all of the wiring methods. It indicates, for instance, that type NM cable is suitable for use in feeders, branch circuits, and inside a building. It can also be used

Figure 14.11
Types of Cable
Found in Home
Wiring

Source: Robert Wood, *House Wiring from Start to Finish, 2nd ed.* (New York, NY: TAB Books, Division of The McGraw-Hill Companies, 1993), p. 68. Reproduced with the permission of The McGraw-Hill Companies.

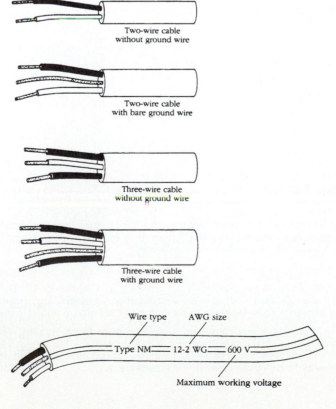

Two-wire cable
without ground wire

Two-wire cable
with bare ground wire

Three-wire cable
without ground wire

Three-wire cable
with ground wire

Wire type   AWG size

Type NM   12-2 WG   600 V

Maximum working voltage

in exposed and concealed conditions that are normally dry. Alternative wiring methods are specified for damp or wet locations.

Ampacity

The purpose of the conductor is to carry current from one place in a circuit to another. **Ampacity** refers to the ability of a conductor to carry current. The ampacity rating of a conductor is dependent upon the following variables: (1) the type of conductor material, (2) the size of the conductor (the larger the conductor, the more current it will carry), and (3) the type of insulation covering the conductor (a conductor with insulation capable of withstanding higher heat will have a higher ampacity rating than a conductor of the same size with a lower insulator temperature rating).

Table 14.3, which presents only the copper wire information provided in the 1995 CABO Code,[22] indicates how the previously noted variables influence ampacity. The ampacity values indicated must be "derated" if the normal ambient temperature is higher than 30°C (86°F). A conductor proximity derating factor may also be necessary when more than three conductors are grouped within a raceway or cable jacket. It should be recalled that a ground wire is not considered to be a conductor, so instances where more than three conductors are grouped within a cable jacket or raceway are rare in residential applications.

TABLE 14.3 Allowable Ampacities for Copper Wire

	Conductor Temperature Rating		
Conductor Size	60°C (140°F)	75°C (167°F)	90°C (194°F)
AWG kcmil	Types TW, UF	Types RH, RHW, THHW, THW, THWN, USE, XHHW	Types THHN, THHW, XHHW
18	—	—	14
16	—	—	18
14	20	20	25
12	25	25	30
10	30	35	40
8	40	50	55
6	55	65	75
4	70	85	95
3	85	100	110
2	95	115	130
1	110	130	150
1/0	125	150	170
2/0	145	175	195
3/0	165	200	225
4/0	195	230	260

In addition, it is important to note that the *NEC*®, through a footnote to this table,[26] limits the allowable ampacities for No. 14 and No. 12 conductors, as follows:

No. 14 — 15 amps
No. 12 — 20 amps

Since the National Electric Code is in effect in the vast majority of jurisdictions, these ampacity limitations should be followed in housing construction. These limitations are also included in the 1995 CABO Code.[27]

Source: Portions of Table 310-6 reprinted with permission from NFPA 70-1996, *The National Electrical Code*®, Copyright© 1995, National Fire Protection Association, Quincy, MA 02269. This reprinted material is not the complete and official position of the National Fire Protection Association on the referenced subject, which is represented only by the standard in its entirety.

Installation Requirements

The requirements for the above-ground installation of the cables constituting the branch circuits in a house are covered in the 1995 CABO Code,[23] which also provides information about the various physical conditions related to cable placement,[24] and about the requirements for underground installation situations.[25]

Feeders and Branch Circuits

Two types of cable circuits typically are found in houses: **feeders** and **branch circuits**. Both are discussed in the 1995 CABO Code.[28] It is important to understand the difference between these two circuits since each has its own set of sizing requirements and criteria.

Feeders[29]

By definition, **feeders** are the conductors between the service equipment (service entrance and main disconnect) and the final branch circuit overcurrent protective device (circuit breaker). In practical terms, a feeder serves a subpanel. It is used when a circuit is extended from a circuit breaker in the main panel, generally a two-pole device, 30 amperes or larger, to another panel. An example of the need for a second panel would be a panel which serves shop equipment, a panel located in a detached garage, or a panel set up to serve only baseboard heating units. It is uncommon to have feeders in residential applications, but they can occur.

Branch Circuits[30]

Most residential wiring consists of **branch circuits** that extend from circuit breakers in the main panelboard to receptacles, lights, motors, heat, and appliances. Branch circuits may serve several loads, such as receptacles and lights, or they may serve only a single load, such as a range or dryer.

It is good practice to separate equipment with motors from branch circuits primarily serving lighting fixtures in order to reduce the tendency of lights to flicker in response to the different phases of motors.

Boxes

The 1995 CABO Code requires that all switches, receptacles, and fixtures, as well as any other splices of conductors, must be installed in electrical boxes (Fig. 14.12).[31] The number of conductors allowed in an electrical box is dependent upon the volume of the box and the wire size of the conductors.[32] The commonly used electrical boxes are composed of either metal or plastic.

Metal Boxes

Metal boxes may be used with any of the wiring systems mentioned. There are two basic shapes of metal boxes: (1) rectangular and (2) octagonal. **Rectangular** boxes are used for switches and receptacles. **Octagonal** or **round** boxes, sometimes equipped with mounting brackets that may be attached directly to ceiling joists, are used for mounting ceiling fixtures or as junction boxes.

Plastic Boxes

Plastic boxes, which are made of either thermosetting polyester or PVC, can be used only with non-metallic (NM) sheathed cable. Plastic boxes are lightweight and easily installed since they do not require grounding, and are supplied in two shapes: (1) rectangular for

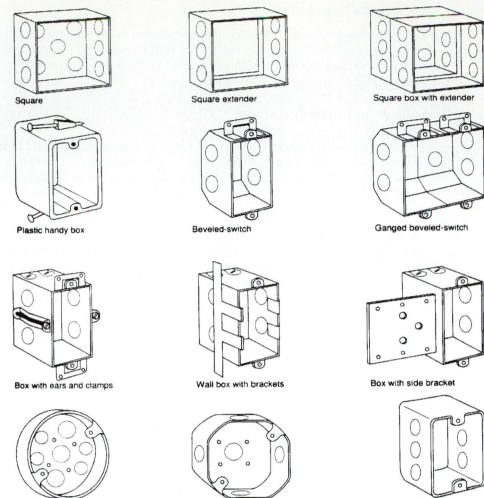

Figure 14.12
Types of Electric
Boxes
Source: Charlie
Wing, *The Visual
Handbook of
Building and
Remodeling*
(Emmaus, PA:
Rodale Press, 1990),
p. 265.

switches and receptacles and (2) round for ceiling fixture mounting. Rectangular boxes are available in 1, 2, 3, and 4 gang sizes. Single gangs vary in depth from 2 in. to 2 1/2 in. and can be mounted directly to the wall studs.

Switches

The basic function of a switch is to open and close a circuit. In order to accomplish this, a switch must contain some form of spring inside to open and close the contacts quickly. If the contacts are slow to open and close, there is danger of **current arcing**—a spark jumping across the contacts as they separate, thus burning the contacts. Switches are available in many types and styles. The most common switch is a single gang-size wiring device, available in ratings from 15 to 30 amperes, with generally one or two poles rated at 120 or 240 volts. These switches include the single-pole, three-way, and four-way varieties.

A **single-pole switch** (Fig. 14.13) is installed when a receptacle, light, or group of lights must be controlled from one switching point. This type of switch is connected in series with the ungrounded (hot) wire feeding the load. It connects/disconnects one conductor, hence the term "single pole".

A **three-way switch** (Fig. 14.14) is used when a load, typically a light fixture or a receptacle, must be controlled from two different switching points. In this case, two three-way switches are used, one at each switching control point.

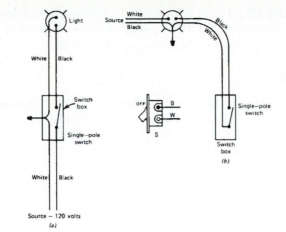

Figure 14.13
Single-Pole
Switch Control
Source: William J.
Whitney, *Residential
and Commercial
Electrical Wiring,*
p. 124. Copyright ©
1983 John Wiley &
Sons, Inc. Reprinted
by permission of
John Wiley & Sons,
Inc.

Notes: (*a*) Typical application of a single-pole switch to control a light from one switching phase.
One hundred twenty volts feeds through switch, black wire is broken at the switch, white wire is fastened
together with wirenut. (*b*) With feed at light. The 120 volts feeds directly to the light outlet; a two-wire
cable with black and white wires is used as a switch loop between the switch and the load. The white wire
is connected to the black wire at the source (NEC 200-7).

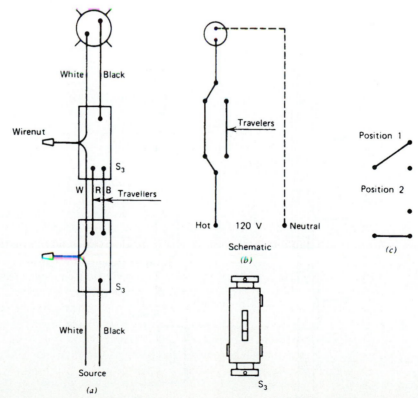

Figure 14.14
Three-Way
Switch
Source: William J.
Whitney, *Residential
and Commercial
Electrical Wiring,*
p. 125. Copyright ©
1983 John Wiley &
Sons, Inc. Reprinted
by permission of
John Wiley & Sons,
Inc.

Notes: (*a*) Two three-way switch control, feed at the first switch control point. (*b*) Schematic wiring
diagram. (*c*) The internal design of the three-way switch allows current to flow through the switch in
either of its two positions.

A **four-way switch** (Fig. 14.15) is used when a load must be controlled from more than two different switching points. Two three-way switches are always used, and four-way switches are added as required for additional switching points.

Other Types of Switches

Several other types of switches are commonly found in houses. A **safety switch** is a heavier, one- or two-pole device used to disconnect equipment, such as a central air conditioner. Such a switch may or may not contain fuses. A **time switch** is used to turn loads, such as lights, on or off at predetermined times. It is available in many sizes and styles. A **motion sensor** is a switch that senses movement and is used primarily to turn lights on. A **thermostat** is also a switch that, in this case, turns loads on and off based on space temperature.

Receptacles

Standard Receptacles

A **receptacle**, or **outlet**, is defined by the 1995 CABO Code as a contact device installed for the connection of a single attachment plug. Since the normal wall receptacle takes two attachment plugs, it is referred to as a **duplex receptacle**. Receptacles are described and identified by poles and wires. The number of poles equals the number of hot contacts. The number of wires includes all connections to the receptacle, including the grounding wire. Receptacles are available in ratings of 10 to 400 amperes, two to four poles, and 125 to 600 volts. The typical duplex wall convenience receptacle in a house is a two-pole, three-wire unit, that has a 125-volt, 15- to 20-ampere rating. Range, dryer, and air conditioning receptacles are exclusively designed for those appliances on 240-volt circuits. The quality or grade of the receptacle can be economy, standard, or specification. Wall duplex receptacles normally are installed vertically, between 12 and 18 in. above the finished floor.

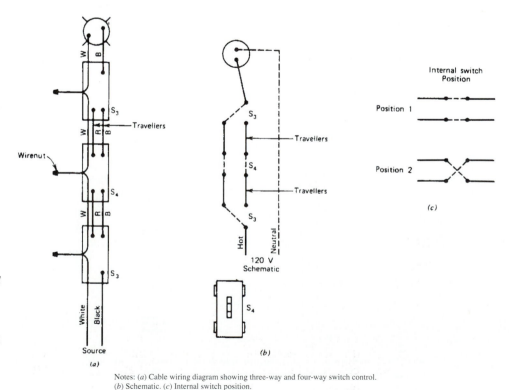

Figure 14.15
Three-Way and
Four-Way Switch
Control
Source: William J.
Whitney, *Residential
and Commercial
Electrical Wiring,*
p. 126. Copyright ©
1983 John Wiley &
Sons, Inc. Reprinted
by permission of
John Wiley & Sons,
Inc.

Notes: (*a*) Cable wiring diagram showing three-way and four-way switch control.
(*b*) Schematic. (*c*) Internal switch position.

Ground Fault Circuit Interrupter (GFCI)

A **ground fault circuit interrupter** (GFCI) is an electronic device designed to protect people against ground fault (current leakage) before it can cause electrocutions, electrical fires, or other serious accidents in a house. A GFCI continuously monitors the current in the two conductors of a 120-volt circuit: the hot wire and the neutral wire. These two currents should always be equal. If the GFCI senses a difference between them of more than 5 ± 1 milliamperes (mA), it assumes the difference is ground fault current and automatically trips the circuit.

Electrical System Design Requirements

After the required floor plans and elevation drawings for a house have been developed, and the projected uses for each room and area of the house have been determined, the design of the electrical system that will meet these use requirements can begin. The first objective of the designer is to make sure that the minimum requirements of the 1995 CABO Code or the $NEC^{®}$ are met. In all likelihood, the final electrical system will be designed to exceed the minimum requirements of the code unless the electrical design is for a basic entry-level home.

The Design Process

Very broadly, the design process includes the following steps:

1. The minimum number of (1) electrical outlets, (2) lighting fixtures, and (3) controls that are required by the code must first be properly located on the floor plan. Figure 14.16 presents a listing of the standard electrical wiring symbols which can be used to assist in this process. Figure 14.17 provides an example of such a design for a family room, bath, and utility room.

2. Pieces of electrical equipment not in the previously noted three categories (i.e., water heater, clothes dryer, etc.), as well as any electrical equipment which is requested by the homeowner but goes beyond the code minimum requirements, should also be added.

3. Any device that serves a piece of end-use equipment, and any directly connected equipment that uses electricity, is considered to be a load. The loads can be divided into **branch circuits** after the **power requirements** of each have been determined. The **maximum number of loads on the same branch circuit is constrained by the total wattage allowed in the Code**.

4. Cable plans and wiring diagrams can be prepared after all the branch circuits have been identified. These documents indicate how the electrical design will be installed in the house.

Minimum Outlet/Lighting Requirements

RECEPTACLES

The following design guidelines, most of which appear in Section 4401 of the 1995 CABO Code,[33] must be considered:

1. *Convenience receptacle distribution*—Every room in the house–other than a hall, stairway, or bathroom–must have receptacle outlets installed so that no point along the floor line in any wall space is more than 6 ft from an outlet. Typically, the 6-ft count starts at a doorway or other break in the wall and is permitted to go around corners, thus including more than one wall. Any separate wall space 2 ft or wider is considered separately.

Figure 14.16 Electrical Wiring Symbols

Source: Ray C. Mullin, *Electrical Wiring Residential* (New York, NY: Van Nostrand Reinhold Company, 1981), p. 7.

⊕ CEILING OUTLET

◖ WALL BRACKET

Ⓛ PS LAMPHOLDER WITH PULL SWITCH

⊙ FLOOR OUTLET

⊡ CEILING OUTLET FOR RECESSED FIXTURE. (OUTLINE SHOWS SHAPE OF FIXTURE)

TV TELEVISION OUTLET

Ⓕ FAN OUTLET

RANGE OUTLET

SPECIAL PURPOSE OUTLET (SUBSCRIPT LETTERS INDICATE FUNCTIONS: DW—DISHWASHER, CD—CLOTHES DRYER, ETC. ALSO a, b, c, d, ETC. SEE SPECIFICATIONS)

DUPLEX OUTLET

DUPLEX OUTLET, SPLIT CIRCUIT

WEATHERPROOF OUTLET

CONVENIENCE OUTLET OTHER THAN DUPLEX. 1 = SINGLE, 3 = TRIPLEX, ETC.

FLUORESCENT FIXTURE (EXTEND RECTANGLE TO SHOW LENGTH)

S   SINGLE-POLE SWITCH

S_D DOOR SWITCH

S_2 DOUBLE-POLE SWITCH

S_3 THREE-WAY SWITCH

S_4 FOUR-WAY SWITCH

S_P SWITCH WITH PILOT

S_{WP} WEATHERPROOF SWITCH

S_{DS} DIMMER SWITCH

*IF THERE IS AN ARROW ON THE CABLE, IT INDICATES A HOME RUN.

TWO-WIRE CABLE OR RACEWAY

THREE-WIRE CABLE OR RACEWAY

FOUR-WIRE CABLE OR RACEWAY

PUSHBUTTON

BUZZER

BELL

CH CHIME (ALSO ⊞)

ANNUNCIATOR

INTERCONNECTING TELEPHONE

OUTSIDE TELEPHONE

Ⓒ CLOCK

Ⓜ MOTOR

Ⓣ TRANSFORMER

Ⓙ JUNCTION BOX

⏚ GROUND CONNECTION

LIGHTING PANEL

POWER PANEL

D ELECTRIC DOOR OPENER

BATTERY

SWITCH LEG INDICATION, CONNECTS OUTLETS WITH CONTROL POINTS

Ⓣ THERMOSTAT

HEATING PANEL

MULTIOUTLET ASSEMBLY ARROWS SHOW LIMITS OF INSTALLATION. APPROPRIATE SYMBOL INDICATES TYPES OF OUTLET. SPACING OF OUTLET IS INDICATED BY X INCHES.

SWITCH AND FUSE

OVERCURRENT DEVICE (FUSE, BREAKER, THERMAL OVERLOAD)

CIRCUIT BREAKER

Figure 14.17 Location of the Required Electrical Outlets and Lighting Fixtures in a Section of a House

Source: Jeff Markel, *Residential Wiring to the 1993 NEC* (Carlsbad, CA: Craftsman Book Company, 1993), p. 146.

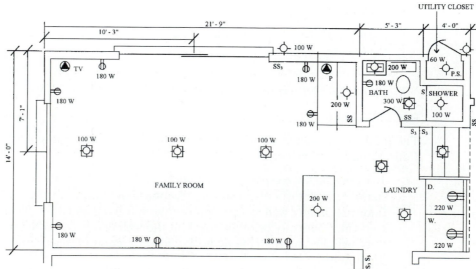

2. *Load rating of a receptacle*—The actual loads on general-purpose receptacle outlets are unpredictable. Some may be blocked by furniture; some may not be placed near a location where the homeowner needs electrical power. As a result, *NEC*® has assigned an **arbitrary load of 180 volt-amperes** to each of these outlets.[34] It is assumed that the average of the loads on all of the outlets will be less than that value.

3. *Small appliance receptacles*—Many small appliances are used in a kitchen. Based upon the indicated power requirements in Table 14.1, it is easy to understand why kitchen currents can be overloaded quickly when a number of the appliances are used at the same time. Therefore, at least two 20-ampere small appliance circuits must be provided to serve all receptacle outlets in the kitchen, pantry, breakfast room, dining room, or similar area. Essentially no other outlets, except for the refrigerator, may be served by these circuits. In larger houses, or where considerable entertaining is done, it is probably advisable to dedicate more than two circuits to serve the kitchen area.

4. *Countertop receptacles*—In kitchen and dining areas, a receptacle outlet must be installed at each wall counter space which is 12 in. or greater in width. Because most countertop appliances have short cords, typically 24 in. or less, no point along the wall line can be more than 24 in. from a receptacle outlet.[35]

5. *HVAC receptacle*—A receptacle outlet is required for servicing any heating, air conditioning, or refrigeration equipment located in attics and crawl spaces.

6. *Additional receptacles*—At least one receptacle outlet is also required in the following locations: (a) bathroom adjacent to the lavatory; (b) laundry area, within 6 ft of the laundry appliance; (c) outdoors, at both the front and back of the house; (d) basement; (e) attached garage; and (f) hallways that are more than 10 ft long. For the basement and garage locations, the receptacle must be in addition to whatever is provided for laundry use.

7. *GFCI receptacles*—According to the 1995 CABO Code,[36] GFCI receptacles are required in all 125-volt, single-phase, 15- and 20-ampere receptacles installed in bathrooms, garages (except those which serve a specific purpose), outdoors (where there is direct grade-level access to the receptacles), crawl spaces (at or below grade), unfinished basements (except those for laundry, dedicated appliances, or sump pumps), and kitchen and bar sink countertops within 6 ft of the sink or wet bar.

The *NEC*® permits the wiring of two additional outlets in series after a GFCI outlet. As a result, all outlets on that circuit are GFCI protected. The *NEC*® also permits placing a GFCI circuit breaker in the panelboard, thus protecting the entire circuit.

LIGHTING FIXTURES

Lighting fixtures are another common residential load. Lighting fixtures are available in many types, styles, and price ranges. Lighting fixtures can be fixed (also called hardwired) or portable, such as table lamps. Although incandescent lamps are traditionally used in residential applications, fluorescent lighting is becoming more popular due to its higher efficiency. For example, a 4-ft fluorescent fixture that uses two 40-watt lamps (80 watts total) creates as much light as one 300-watt incandescent lamp or five 60-watt incandescent lamps. In this case, fluorescent lamps provide an energy saving of 375%.

The following lighting guidelines, based upon requirements in the 1995 CABO Code,[37] should be noted:

1. *Required lighting outlets*—at least one wall switch-controlled lighting outlet shall be installed in every habitable room and in bathrooms, hallways, stairways, attached garages, detached garages provided with electrical power, and at exterior egress doors. A switch-controlled lighting outlet means some type of light fixture. It can be fluorescent or incandescent, ceiling or wall mounted, flush or recessed—as long as it holds a light. A habitable room is a bedroom, family room, dining room, etc.

Exceptions

    (a) In habitable rooms other than kitchens and bathrooms, one or more receptacles controlled by a wall switch shall be considered equivalent to the required lighting outlet.

    (b) In hallways, stairways, and at exterior egress doors, wall-switch control shall not be required where such lighting is remotely, centrally, or automatically controlled.

    (c) Vehicle garage doors.

2. *Stairway lighting control*—Lighting outlets installed at interior stairs shall be controlled by a wall switch at each floor level if the difference between floor levels requires six or more stair risers.

3. *Rooms or spaces for storage and equipment*—In an attic, crawl space, utility room, or basement, some type of lighting fixture is required if that space is used for storage or contains equipment (a gas furnace, water heater, etc.) that needs to be serviced. The light fixture must be controlled by a light switch located at the point of entry to the space.

4. *Load rating of lighting outlets*—The computation of individual lighting loads is based upon known quantities (the wattage listing on the fixture), as well as standardized rule-of-thumb estimates. For example,

    (a) Fluorescent lights—either the volt-amps on the label posted on the fixture may be used; or an estimate of 10 volt-amps per running foot of tube for each tube, plus 15% of the tube volt-amps for the ballast, can be used.

    (b) Recessed incandescent lights—the fixture rating which indicates the maximum allowable lamp wattage for the fixture should be used. If this value is not known, 150 watts safely can be estimated.

    (c) Surface-mounted incandescent lights with a diffuser—A valid rule of thumb for one- and two-bulb units is 200 watts.

MINIMUM OUTLET/LIGHTING LOAD

NEC indicates that the **minimum lighting load** (which includes the receptacle load) for dwelling units is **3 volt-amperes per square foot**.[38] The square footage is computed from the outside dimensions of the house with porches, garages, and unfinished spaces being deducted.

    The minimum requirement of 3 volt-amperes per square foot is primarily used in the calculations to determine the size of the service entrance (see Table 14.2).

Rating of Outlet/Lighting Branch Circuits

As stated earlier, a group of conductors, either two or three, with a ground wire, constitutes a circuit. Individual branch circuits originate at the panelboard and end at a group of loads. The group of loads could be a branch circuit consisting of a number of wall receptacle outlets and light fixtures, or it could be the clothes dryer located in the laundry room. The **maximum number of outlets**, **light fixtures, etc.** that can be grouped into a branch circuit depends upon the **maximum allowable wattage** which a particular branch circuit can carry.

    The following general-purpose branch circuit guidelines, some of which appear as requirements in the 1995 CABO Code,[39] are helpful in rating circuits:

1. *Branch circuit voltage limitations*—A maximum voltage rating of 120 volts can be used on branch circuits that supply lighting fixtures, receptacles for cord-and-plug connected loads of up to 1,440 volt-amperes, and loads of less than 1/4 horsepower.

    A voltage of either 120 volts or 240 volts can be used on branch circuits supplying equipment and appliance loads.

TABLE 14.4   Summary of Branch Circuit Requirements[a,b]

	Circuit Rating		
	15 amp	20 amp	30 amp
Conductors:			
Min. Size (AWG) Circuit Conductors	14	12	10
Maximum overcurrent protection device rating Ampere rating	15	20	30
Outlet devices:			
Lampholders permitted	Any Type	Any Type	N/A
Receptacle rating (amperes)	15 max.	15 or 20	30
Maximum load (amperes)	15	20	30

[a]These gages are for copper conductors.
[b]N/A means not allowed.

Source: Portions of NEC® Table 210-24 reprinted with permission from NFPA 70-1996, *The National Electrical Code*®, Copyright© 1995, National Fire Protection Association, Quincy, MA 02269. This reprinted material is not the complete and official position of the National Fire Protection Association on the referenced subject, which is represented only by the standard in its entirety.

TABLE 14.5   Maximum Allowable Load on General-Purpose Branch Circuits

Amperage	Wire Size	120-Volt Circuit Capacity (Volt-Amps)	240-Volt Circuit Capacity (Volt-Amps)
15	No. 14	$15 \times 120 = 1,800$	$15 \times 240 = 3,600$
20	No. 12	2,400	4,800
30	No. 10	3,600	7,200
40	No. 8	4,800	9,600
50	No. 6	6,000	12,000

2. *Branch circuit amperage rating*—**Branch circuits** shall be rated in accordance with the **maximum amperage rating** or setting on the overcurrent protection device. The rating for other than individual branch circuits shall be 15, 20, 30, 40, or 50 amperes.

3. *Fifteen- and 20-ampere branch circuits*—A 15- or 20-ampere branch circuit can be used to supply lighting units, other utilization equipment, or a combination of both.

4. *Branch circuit requirements*—The requirements for circuits having two or more outlets (other than the receptacle circuits in the kitchen and dining area) are shown in Table 14.4. It should be noted in Table 14.4 that **most of the general-purpose branch circuits in a house will be rated at either 15 amps using No. 14 copper wire conductors or 20 amps using No. 12 copper wire conductors.**

5. *Maximum allowable load on outlet/lighting branch circuits*—The 1995 CABO Code does not directly indicate the maximum allowable load on general-purpose outlet/ lighting branch circuits. However, as shown in Table 14.5, these values can be easily calculated for the permissible branch circuits using Watt's Law.

6. *Allowable number of receptacles per branch circuit*—Based upon the information in items 1 through 5, the previously noted 180 volt-ampere load which *NEC* allows for a single receptacle, and assuming that no additional lighting fixtures are connected to the circuit:

Branch Circuit Rating	Maximum Allowable Number of Receptacles
15 amps	$1,800/180 = 10$
20 amps	$2,400/180 = 13.33 = 13$

It always is acceptable, and many times advisable, to have fewer outlets on a circuit than indicated above.

7. *Example branch circuit calculation*—As an example of the type of branch circuit calculation which is performed based upon the information provided in items 5 and 6, reference is made to the partial houseplan shown in Fig. 14.17.

A first assumption that can be made is that a particular branch circuit will service all of the receptacles and lights in the family room as well as the outdoor wall light. As a result, the calculated loads for Fig. 14.17 would be:

7 electrical outlets (180 watts each)	1,260 watts
3 recessed ceiling fixtures (100 watts each)	300 watts
2 ceiling surface fixtures (200 watts each, estimated)	400 watts
1 outdoor wall light at door (100 watts)	100 watts
	2,060 watts

The special-purpose outlets marked TV, for TV antenna, and P, for telephone, are not loads on the electrical system—they do not connect to the branch circuit.

Clearly this 15-amp, 120-volt branch circuit design would be overloaded since it exceeds the allowable limit of 1,800 watts. The electrical supply to the family room, therefore, must be split between two 15-amp branch circuits or placed on one 20-amp branch circuit. As one alternative, the 700 watts of lighting could be joined to the branch circuit that services the bathroom and utility room. This would reduce the branch circuit in question to 1,360 watts, which is satisfactory for a 15-amp circuit.

The decisions regarding the loads that can be served by each branch circuit for the entire house are performed in a similar fashion.

Other Branch Circuits

Branch circuits also are required for such high-wattage appliances as central air conditioners, clothes dryers, baseboard electric heat, kitchen ranges, and water heaters. Each of these requires an individual 240-volt branch circuit with conductors that are large enough to provide the required ampacity.

For special branch circuits to unique pieces of equipment, the voltage and volt-amp requirements are usually obtained from the nameplate on the equipment. The conductor size is determined to match the current that is being drawn by the equipment.

The 1995 CABO Code[40] provides the following information about some of the specific branch circuits:

1. *Branch circuits for heating*—An individual branch circuit is required for the central heating equipment.

2. *Kitchen and dining area receptacles*—The minimum of the previously mentioned two 20-ampere-rated branch circuits for the receptacles located in the kitchen, pantry, breakfast area, and dining area.

3. *Laundry circuit*—A minimum of one 20-ampere-rated branch circuit must be provided to receptacles located in the laundry area. It can serve only receptacle outlets in that area.

The Rationale for Exceeding Minimum Requirements

The requirements listed in the *NEC* and the 1995 CABO Code, as well as the additional ones introduced herein, are **minimum code requirements** which do not necessarily constitute good design practice. **Minimum requirements** are instituted to provide consumers and the public with **basic and safe electrical service**. In many instances, common sense, in addition to issues such as ease of maintenance, dictate installations that exceed the minimum requirements of the code. Systems should also be designed, for instance, to provide for future flexibility, lowest life cycle cost, and convenience for the occupants of the house.

Some common examples of the types of modifications that can be made to the minimum code requirements include:

1. The use of only 20-amp circuits—The use of No. 14 wire with 15-amp circuit breakers dictates that only 15 amp, not 20 amp, receptacles may be used. Appliances with 20-amp plugs are unusual, but they do exist. Upgrading to 20-amp circuits with No. 12 wire eliminates such conflicts (because a 20-amp receptacle can accommodate a 15-amp plug) while at the same time providing an additional margin of safety against overloads.

2. The kitchen-related electrical loads indicated in Table 14.1 suggest that two electrical circuits may not be enough for large kitchens with significant counter space. Additional circuits may be required. For example, a separate circuit is often added just for the microwave oven.

3. Good design practice dictates that receptacles and lights should be grouped in a logical manner so that if an electrical problem occurs it can be localized and fixed easily. One way to achieve this objective is to provide independent circuits for each room.

4. Bathroom receptacles should not be grouped with other loads. A conventional hair dryer will use as much as 1,800 watts. Overloading is possible if the hairdryer load is combined with other loads. Again, the use of No. 12 wire and 20-amp circuits reduces the possibility of overloads.

5. Areas where large power tools are located—such as shops, garages, and basements—should be provided with No. 12 wire and 20-amp circuits.

6. The use of extra receptacles beyond the minimum requirements will result in a decrease in the use of extension cords and multi-outlet adapters, which are frequent causes of fires.

7. The use of motion sensor switches in frequently unoccupied areas avoids the need to leave lights on for security purposes when the space is not being used.

ADDITIONAL ELECTRICAL SYSTEMS

The discussion thus far has provided an in-depth coverage of the distribution of electrical power and lighting throughout a house. Although these are the primary electrical systems, there are a number of additional ones that should be mentioned because they provide a number of convenience, entertainment, and security services to the homeowner.

Telephone System

Since the 1984 break-up of AT&T™ and subsequent federal legislation, all **telephone wiring** inside a house is the property and responsibility of the homeowner. The local telephone company now provides the wiring and outlets at a cost, just like any other subcontractor.

The installation of telephone wiring and outlets is covered under the *NEC*®.[41] Other standards from the local telephone service provider may also apply.

Doorbell System

Doorbell systems have been a part of residential construction since the 1940s. These systems are simple, inexpensive, and effective. They provide a means to alert an occupant that someone is at the door. Components of a doorbell system include the bell unit, a pushbutton, a transformer which changes the 120-volt line voltage to 24 volts, and a pair of No. 16 or No. 18 low-voltage wires to connect the system components.

Television System

Houses are usually pre-wired for TV usage. The TV signal may originate from an antenna (also called MATV or off air), satellite dish, or cable TV provider. Recent legislation has regulated cable TV companies and their services, and has treated TV wiring the same as telephone wiring. The wiring outside of the home is owned and maintained by the cable TV provider, if any, and the interior wiring (premises wiring) is owned and maintained by the homeowner. The basic parts of the TV wiring system are very similar to those in the telephone system wiring.

The installation of TV wiring and equipment is covered under the *NEC*®.[42] Other local standards may also apply.

Security, Fire Alarm, and Other Systems

Security Systems

Security systems are optional in all houses. Many homeowner insurance companies provide premium discounts if security systems are provided and maintained in working order. Security systems consist of two basic components: detection and alarm. The two are interconnected at a control panel that provides analysis of the detection inputs and determines the appropriate alarm mode.

Wireless security systems, which offer this type of protection for remodeling projects, are also available in the marketplace.

Fire Alarm Systems

Fire alarm systems, in their most basic form, are required in all residential applications by CABO, BOCA, and most other model building codes. Smoke detectors with built-in alarm devices (horns) are required by these codes at various locations in a house. The types specified are either individual battery-powered units or units that are interconnected on a dedicated electric circuit. These types of detectors are called "single station," meaning they provide the detection and alarm function within a single unit.

More sophisticated systems can be provided as an option. They consist of two components similar to security systems: detection and alarm. They can be interconnected through a control panel that provides analysis of the inputs and determines the appropriate alarms. Current systems are microprocessor based.

Intercom and CCTV Systems

Generally, **intercom systems** provide two-way communications from one area to another; **intercoms** can also be used to monitor specific areas, such as a child's nursery. Many telephone systems now provide this function as part of the telephone equipment, so the use of intercom systems has greatly decreased. Current use is primarily for communication to entry locations.

Low-cost Closed Circuit TV (CCTV) systems are available from a variety of manufacturers, usually as part of an entry intercom system. CCTV systems provide increased security through verification of visitors prior to permitting entry.

Computer/Smart House Technology

As the use of computers in houses increases, it is becoming more common to provide dedicated electric circuits for their use when the house is built. The most advanced recognition of the electronic age is represented by what commonly is called the **Smart House** concept.

This concept was first officially marketed in 1991, after many years of research and development by the National Association of Homebuilders and the Smart House Limited Partnership. It provides an owner-friendly electronic management and communications distribution system that permits integrated home automation. Because of the technical sophistication involved, the Smart House concept is practical only in new houses, although application through renovation is technically feasible. The goal of the Smart House is to minimize the time spent on home management; optimize home comfort; maximize energy efficiency, use and cost; and increase owner convenience.

CASE STUDY HOUSE EXAMPLE

Procedures and Assumptions

The design of the electrical system for the Case Study House is provided. The procedures and assumptions related to that design are:

1. The service entrance size and estimated full-load current and wire size will be determined using the method described in Table 14.2.
2. The following design parameters are to be assumed:
   - Separate appliance circuits (with the assumed name plate volt-ampere ratings indicated) are required for the following:
     —Oven/Range–9,000 VA, 120/240 volt, 3 wire
     —Clothes Dryer–5,600 VA, 120/240 volt, 3 wire
     —Dishwasher–1,030 VA, 120 volt, 2 wire
     —Water Heater–4,500 VA, 240 volt, 2 wire
     —Garbage Disposal–1/3 hp, 720 VA, 120 volt, 2 wire
     —Clothes Washer–960 VA, 120 volt, 2 wire
3. Heat and air conditioning will be provided by a central forced-air heat pump system
   - Indoor Heat Pump Fan Unit—1/2 hp, 1,130 VA, 120 volt, 2 wire
   - Outdoor Heat Pump Unit—5 hp, 6,440 VA, 240 volt, 2 wire

4. The requirements of the 1995 CABO Code for lighting and receptacle locations and currents should be used. GFCI receptacles must be specified where they are required.
5. The drawings should indicate:
   a. The main service location for electric, telephone, and CATV. Any reasonable location may be selected. The meter location and the location of connections for telephone and CATV must also be indicated.
   b. The location of all of the electrical components noted below:
      - Main panelboard
      - All fixed lights
      - All receptacles
      - All switches (single-pole, three-way, or four-way)
      - Circuit wiring and homerun

- HVAC equipment
- All telephone outlets (wiring need not be shown)
- All CATV outlets (wiring need not be shown)
- Door chime (wiring and door pushbuttons need not be shown)
- Smoke detectors

6. A branch circuit schedule showing the following information is to be provided:

		Wire			Circuit Breaker	
Ckt. No.	Load Served	No.	Size	Volts	Poles	Size

Design Calculations

Step 1: Compute the **general lighting and receptacle load**.

The floor area of the house is 1,680 ft^2. An initial estimate of the general lighting and receptacle load, using the guidelines in Table 14.2, is:

$$3 \text{ volt-amps/S.F.} \times 1,680 \text{ S.F.} = 5,040 \text{ VA}$$

Step 2: Compute the size of the **service entrance**.

Table 14.2 lists the steps involved as follows:

a. General lighting and receptacle load = 5,040 VA (Step 1)

b. Small appliance and laundry circuits. 20-amp branch circuits (two required in kitchen, one required in laundry) 3 × 1,500 VA = 4,500 VA

c. Individual nameplate volt-amp ratings on dedicated branch circuit supplied appliances

• Oven/range	9,000 VA
• Clothes dryer	5,600 VA
• Dishwasher	1,030 VA
• Water heater	4,500 VA
• Garbage disposal	720 VA
• Clothes washer (considered in 20-amp laundry branch circuit)	—
	30,390 VA

d. Apply the demand factor.

$$10,000 \text{ VA} + (30,390 \text{ VA} - 10,000) \, 0.4 = 18,156 \text{ VA}$$

e. Obtain the total load by adding the largest heating or cooling load.

$$18,156 \text{ VA} + 1,130 \text{ VA} + 6,440 \text{ VA} = 25,726 \text{ VA}$$

f. Size the main service entrance, main service entrance cable, and main disconnect.

$$25,726/240 = 107 \text{ amps}$$

Table 14.6 indicates that a 110-amp circuit breaker or fused switch could be used as the main disconnect. In order to provide for some future expansion, however, a 125-amp circuit breaker will be used. It will be placed in a 200-amp panelboard.

A 125-amp service requires either a No. 2 copper or a No. 1/0 aluminum service conductor and either a No. 8 copper or a No. 6 aluminum grounding electrode conductor (Table 14.6).

TABLE 14.6   Service Conductor and Grounding Electrode Conductor Sizing[a]

Conductor Types and Sizes—THHW, THW, THWN, USE, XHHW (Parallel sets of 1/0 and larger conductors are permitted either in a single raceway or in separate raceways)		Allowable Ampacity	Minimum Grounding Electrode Conductor Size	
Copper (AWG)	Aluminum and Copper-Clad Aluminum (AWG)	Maximum Load (Amps)	Copper (AWG)	Aluminum (AWG)
4	2	100	8	6
3	1	110	8	6
2	1/0	125	8	6
1	2/0	150	6	4
1/0	3/0	175	6	4
2/0	4/0 or two sets of 1/0	200	4	2
3/0	250 kcmil or two sets of 2/0	225	4	2

[a]See Table 4103.1 of the 1995 CABO Code for complete footnote details.

Source: Portions of NEC® Tables 310–16 and 250–94 reprinted with permission from NFPA 70-1996, *The National Electrical Code*®, Copyright© 1995, National Fire Protection Association, Quincy, MA 02269. This reprinted material is not the complete and official position of the National Fire Protection Association on the referenced subject which is represented only by the standard in its entirety.

Step 3: Determine the required minimum number of general lighting and receptacle branch circuits.

Step 1 indicated an estimated general lighting and receptacle circuit load of 5,040 VA. Since all of these are 120V branch circuits, the total amperage required is 5,040 VA/120V = 42 amps. Thus, the minimum number of 15-amp circuits is:

42 amps/15 amps per circuit = 2.8 or 3, 15-amp branch circuits

The minimum number of 20-amp circuits is:

42 amps/20 amps per circuit = 2.1 or 3, 20-amp branch circuits

These minimum requirements typically are exceeded in order to meet the needs of the homebuyers. An example of that situation is illustrated in Step 4.

Step 4: Design and size all of the branch circuits.

The specific calculations for each of the branch circuits that are used in the Case Study House are now presented. Table 14.3, which is excerpted from the 1995 CABO Code[43] is the primary reference for the selection of the type and size of the required conductors. It will be assumed that nonmetallic cable (NM) and 75°C type THW or THWN insulated copper conductors will be used. Although a 75°C rated copper conductor is used, the 1995 CABO Code requires the ampacity ratings for these conductors to be obtained from the 60°C column in Table 14.3. An additional provision related to the 60°C column is that the following allowable ampacities apply:

No. 14 — 15 amps
No. 12 — 20 amps

a. Oven/range—Circuit No. 1 (120V/240V)

The oven-range in the kitchen is the largest single electrical load in a house. It is typically rated at 12 kVA (12,000 VA) and would therefore require its own 240-volt branch circuit rated at 12,000/240 = 50 amps (Table 14.7).

From Column A of Table 14.7, however, the demand that can be assumed for an oven-range is 8,000 VA. The ampacity rating of the conductor for this load is:

8,000 VA/240 V = 33.3 amps

TABLE 14.7   Demand Loads for Electric Ranges, Wall-Mounted Ovens, Counter-Mounted Cooking Units, and Other Cooking Appliances Over 1-3/4 kVA Rating[a,b]

	Maximum Demand[b,c]	Demand Factors (percent)[d]	
Number of Appliances	Column A Maximum 12 kVA Rating	Column B Less than 3- 1/2 kVA Rating	Column C 2-1/2 to 8-3/4 kVA Rating
1	8 kVA	80	80
2	11 kVA	75	65

[a]Column A shall be used in all cases except as provided for in Footnote d.
[b,c]See *NEC*® Table 220-19 for complete footnote details.
[d]Over 1-3/4 kVA through 8-3/4 kVA. As an alternative to the method provided in Column A, the nameplate ratings of all ranges rated more than 1-3/4 kVA but not more than 8-3/4 kVA shall be added and the sum shall be multiplied by the demand factor specified in Column B or C for the given number of appliances.

Source: Portions of *NEC*® Table 220-19 reprinted with permission from NFPA 70-1996, *The National Electrical Code*®, Copyright© 1995, National Fire Protection Association, Quincy, MA 02269. This reprinted material is not the complete and official position of the National Fire Protection Association on the referenced subject, which is represented only by the standard in its entirety.

The 60°C column in Table 14.3 indicates that a No. 8 insulated copper conductor can be selected because it has an allowable ampacity of 40 amps. The 120/240 volt branch circuit for the oven/range will therefore consist of three No. 8 conductors. The current will be protected with a two-pole, 40-amp circuit breaker (CB).

b. Clothes dryer—Circuit No. 2 (120V/240V)
Required ampacity = 5,600 VA/240 V = 23.3 amps
Conductor number and size = three No. 10 (30 amp)
Circuit breaker = two-pole, 30 amp

c. Dishwasher—Circuit No. 3
Required ampacity = 1,030 VA/120 V = 8.6 amps
Conductor number and size = two No. 14 (15 amp)
Circuit breaker = one-pole, 15 amp

d. Water heater—Circuit No. 4
Required ampacity = 4,500 VA/240 V = 18.75 amps
Conductor number and size = two No. 12 (20 amp)
Circuit breaker = two-pole, 20 amp

e. Garbage disposal—Circuit No. 5
Required ampacity = 720 VA/120 V = 6 amp
Conductor number and size = two No. 14 (15 amp)
Circuit breaker = one-pole, 15 amp

f. Indoor heat pump fan—Circuit No. 6
Required ampacity = 1,130 VA/120 V = 9.4 amps
Conductor number and size = two No. 14 (15 amp)
Circuit breaker = one-pole, 15 amp

g. Outdoor heat pump—Circuit No. 7
Required ampacity = 6,440 VA/240 V = 26.8 amps
Conductor number and size = two No. 10 (30 amp)
Circuit breaker = two-pole, 30 amp

h. Clothes washer (laundry)—Circuit No. 8

The 1995 CABO Code[44] requires a minimum of one 20-ampere rated branch circuit for the receptacles located in the laundry area.

Conductor number and size = two No. 12 (20 amp)
Circuit breaker = one-pole, 20 amp

i. Basement—Circuit No. 9

3 receptacles @ 180 watts = 540 watts
4 light fixtures @ 40 watts = 160 watts
Total   = 700 watts
Select a 15-amp circuit, two No. 14 conductors.

The maximum allowable load on a 15-amp/120-volt branch circuit is 1,800 watts (Table 14.5). Circuit No. 9, therefore, is underloaded as designed. Rather than adding additional loads to it from other locations, however, it will be kept as a separate circuit for the basement. This provides flexibility in the future if remodeling to add a playroom in the basement, etc., is anticipated.

j. Living room—Circuit No. 10

7 receptacles @ 180 watts   = 1,260 watts
3 light fixtures @ 100 watts −   300 watts
Total   = 1,560 watts
Select a 15-amp circuit, two No. 14 conductors

k. Kitchen—Circuit Nos. 11 and 12

The 1995 CABO Code[45] requires a total of two 20-ampere rated branch circuits for the "kitchen, pantry, breakfast area, and dining area."
Select a 20-amp circuit, two No. 12 conductors for each circuit.

l. Family room—Circuit No. 13

7 receptacles @ 180 watts   = 1,260 watts
2 light fixtures @ 100 watts =   200 watts
Total   = 1,460 watts
Select a 15-amp circuit, two No. 14 conductors

m. Powder room—Circuit No. 14

1 receptacle @ 1,800 watts = 1,800 watts
(worst case load = 1 blowdryer @ 1,800 watts)
1 light and fan @ 180 watts =   180 watts
Total   = 1,980 watts
Select a 20-amp circuit, two No. 12 conductors

A 20-amp/120-volt branch circuit was selected for the powder room because, as noted in Table 14.5, such a circuit has a capacity of 2,400 watts. A 15-amp/120-volt branch circuit, with a capacity of only 1,800 watts, would not be satisfactory for the indicated powder room load.

n. Garage—Circuit No. 15

A separate 20-amp circuit with two No. 12 conductors is selected for the garage to permit the use of heavy-load power tools.

o. Secondary bedrooms—Circuit Nos. 16 and 17

Circuit 16

6 receptacles @ 180 watts   = 1,080 watts
1 light fixture @ 100 watts =   100 watts
Total   = 1,180 watts

TABLE 14.8   Branch Circuit Schedule for the Case Study House

Ckt. No.	Load Served	Wire			Circuit Breaker	
		No. of Wires	Size	Volts	Poles	Size
1	Oven/range	3	No. 8	120/240	2	40
2	Clothes dryer	3	No. 10	120/240	2	30
3	Dishwasher	2	No. 14	120	1	15
4	Water heater	2	No. 12	240	2	20
5	Garbage disposal	2	No. 14	120	1	15
6	Indoor heat pump fan	2	No. 14	120	1	15
7	Outdoor heat pump	2	No. 10	240	2	30
8	Clothes washer	2	No. 12	120	1	20
9	Basement	2	No. 14	120	1	20
10	Living room	2	No. 14	120	1	15
11	Kitchen	2	No. 12	120	1	20
12	Kitchen	2	No. 12	120	1	20
13	Family room	2	No. 14	120	1	15
14	Powder room	2	No. 12	120	1	20
15	Garage	2	No. 12	120	1	20
16	Bedroom	2	No. 14	120	1	15
17	Bedroom	2	No. 14	120	1	15
18	Bath	2	No. 12	120	1	20
19	Master bedroom	2	No. 14	120	1	15
20	Master bath	2	No. 12	120	1	20

Circuit 17

$$
\begin{aligned}
4 \text{ receptacles @ 180 watts} &= 720 \text{ watts} \\
1 \text{ light fixture @ 100 watts} &= 100 \text{ watts} \\
\text{Total} &= 820 \text{ watts}
\end{aligned}
$$

These two branch circuits can be separated for convenience purposes by selecting two 15-amp circuits and two No. 14 conductors. If it is felt to be an advantage to the homebuyer, the two branch circuits could be combined by selecting one 20-amp circuit and two No. 12 conductors

p. Bathrooms—Circuit Nos. 18 and 20

A separate 20-amp circuit using two No. 12 conductors will be used for each bathroom to permit the simultaneous use of blowdryers in each bathroom.

q. Master Bedroom—Circuit No. 19

$$
\begin{aligned}
6 \text{ receptacles @ 180 watts} &= 1{,}180 \text{ watts} \\
1 \text{ light fixture @ 100 watts} &= 100 \text{ watts} \\
\text{Total} &= 1{,}180 \text{ watts}
\end{aligned}
$$

Select a 15-amp circuit, two No. 14 conductors

Presentation of the Electrical System Design

The results of the design calculations are summarized in Table 14.8 in the form of a **branch circuit schedule**. The results are also shown on the floor plans of the Case Study House in Figs. 14.18 to 14.20.

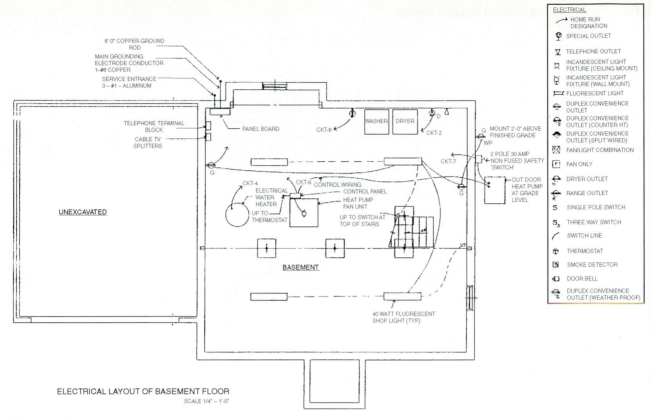

ELECTRICAL

↳ HOME RUN DESIGNATION
⏚ SPECIAL OUTLET
▽ TELEPHONE OUTLET
⌀ INCANDESCENT LIGHT FIXTURE (CEILING MOUNT)
⌀ INCANDESCENT LIGHT FIXTURE (WALL MOUNT)
▭ FLUORESCENT LIGHT
⊖ DUPLEX CONVENIENCE OUTLET
⊖ DUPLEX CONVENIENCE OUTLET (COUNTER HT)
⊘ DUPLEX CONVENIENCE OUTLET (SPLIT WIRED)
▨ FAN/LIGHT COMBINATION
▣ FAN ONLY
⊕ DRYER OUTLET
⊕ RANGE OUTLET
S SINGLE POLE SWITCH
S₃ THREE WAY SWITCH
╱ SWITCH LINE
⊕ THERMOSTAT
⊡ SMOKE DETECTOR
◁ DOOR BELL
⊖ DUPLEX CONVENIENCE OUTLET (WEATHER PROOF)

ELECTRICAL LAYOUT OF BASEMENT FLOOR
SCALE 1/4" – 1'-0"

Figure 14.18
Electrical System Design for the Basement of the Case Study House

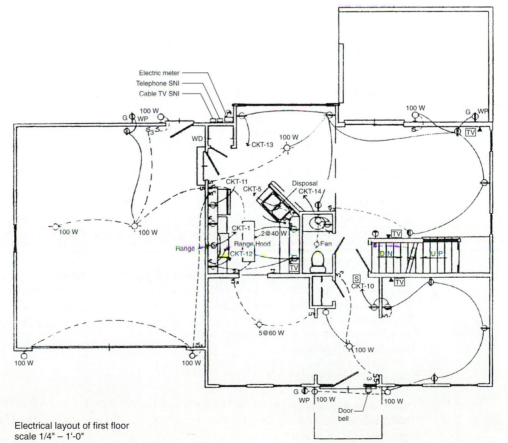

Figure 14.19
Electrical System Design for the First Floor of the Case Study House

Electrical layout of first floor
scale 1/4" – 1'-0"

461

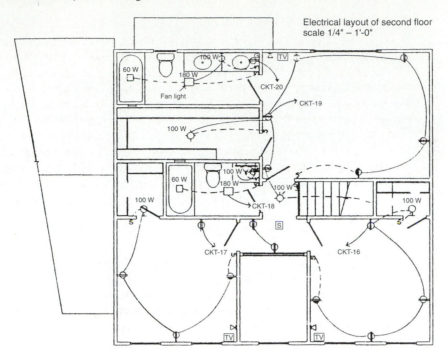

Electrical layout of second floor
scale 1/4" – 1'-0"

Figure 14.20
Electrical System
Design for the
Second Floor of
the Case Study
House

CHAPTER SUMMARY

This chapter has presented the fundamentals of the electric system for a house. The require-ments covered in this chapter are specific to the 1995 CABO Code. The electrical system design of the Case Study House illustrated the application of the CABO Code requirements to a specific house.

NOTES

[1]*NFPA 70-1993-National Electric Code*® (*NEC*®) (Quincy, MA: National Fire Protection Association, 1992).

[2]*CABO One and Two Family Dwelling Code, 1995 Edition* (Falls Church, VA: The Council of American Building Officials, 1995), pp. 239–279.

[3]Ibid, Sections 4107–4111, pp. 252–254.

[4]*NEC*®, Article 250, pp. 70:103–70:139.

[5]*1995 CABO Code*, Section 4408, pp. 271–273.

[6]*NEC*®, Article 250, pp. 70:103–70:139.

[7]*1995 CABO Code*, Section 4107, p. 252; Section 4109, p. 253; Section 4111, p. 254; Section 4408, pp. 271–273.

[8]*NEC*®, Article 230, pp. 70:70–70:90.

[9]*1995 CABO Code*, Chapter 41, pp. 247–254.

[10]Ibid, Sections 4101 and 4102, p. 247; *NEC*®, Article 230, pp. 70:70–70:90.

[11]Ibid, Section 4102, p. 247.

[12]Ibid, Section 4103 and Table 4103.1, p. 248.

[13]Ibid, Section 3906.3, p. 240.

[14]*NEC*®, Article 310–5, pp. 70:160–70:161.

[15]Ibid, Article 410–24(b), p. 70:337.

[16]*1995 CABO Code*, Section 3906.5, p. 242.

[17]Ibid, Section 4205, pp. 257–259.

[18]Ibid, Section 3907, p. 242.

[19]Ibid, Section 4301.2, p. 261.

[20]Ibid, Table 4301.2, p. 261.

[21]Ibid, Table 4301.4, p. 261.

[22]Ibid, Table 4205.1, p. 258.
[23]Ibid, Section 4302, p. 262.
[24]Ibid, Table 4302.1, p.262.
[25]Ibid, Section 4303, p. 263.
[26]*NEC*®, p. 70:174.
[27]*1995 CABO Code*, Table 4205.7, p. 259.
[28]Ibid, Chapter 42, pp. 255–259.
[29]Ibid, Section 4204, pp. 256–257.
[30]Ibid, Sections 4202 and 4203, pp. 255–256.
[31]Ibid, Section 4405, pp. 268–270.
[32]Ibid, Table 4405.12, p. 270.
[33]Ibid, Section 4401, pp. 265–266.
[34]*NEC*®, Article 220–3(c)6, pp. 70:51–70:52.
[35]*1995 CABO Code*, Section 4401.5 and Figure 4401.5, pp. 265–266.
[36]Ibid, Section 4402, pp. 266–267.
[37]Ibid, Section 4403, p. 267.
[38]*NEC*®, Article 220–3(b), pp. 70:50–70:51.
[39]*1995 CABO Code*, Section 4202, pp. 255–256.
[40]Ibid.
[41]*NEC*®, Article 800, pp. 70:791–70:804.
[42]Ibid, Article 810, pp. 70:805–70:810; Article 820, pp. 70:810–70:818.
[43]*1995 CABO Code*, Table 4205.1, p. 258.
[44]Ibid, Section 4203.3, p. 256.
[45]Ibid, Section 4203.2, p. 256.

BIBLIOGRAPHY

CABO One and Two Family Dwelling Code, 1995 Edition.   Falls Church, VA: The Council of American Building Officials, 1995.

Markel, Jeff, *Residential Wiring to the 1993 NEC.*   Carlsbad, CA: Craftsman Book Company, 1993.

Mullin, Ray C., *Electrical Wiring Residential.*   New York, NY: Van Nostrand Reinhold Company, 1981.

NFPA 70-1993—National Electric Code® (*NEC*®).   Quincy, MA: National Fire Protection Association, 1992.

Whitney, William J., *Residential and Commercial Wiring.*   New York, NY: John Wiley & Sons, Inc., 1983.

Wing, Charlie, *The Visual Handbook of Building and Remodeling.*   Emmaus, PA: Rodale Press, 1990.

Wood, Robert W., *Home Wiring from Start to Finish, 2nd Edition.*   New York, NY: TAB Books Division of McGraw Hill, Inc., 1993.

A

The Set of Working Drawings for the "Pauline Model" Case Study House

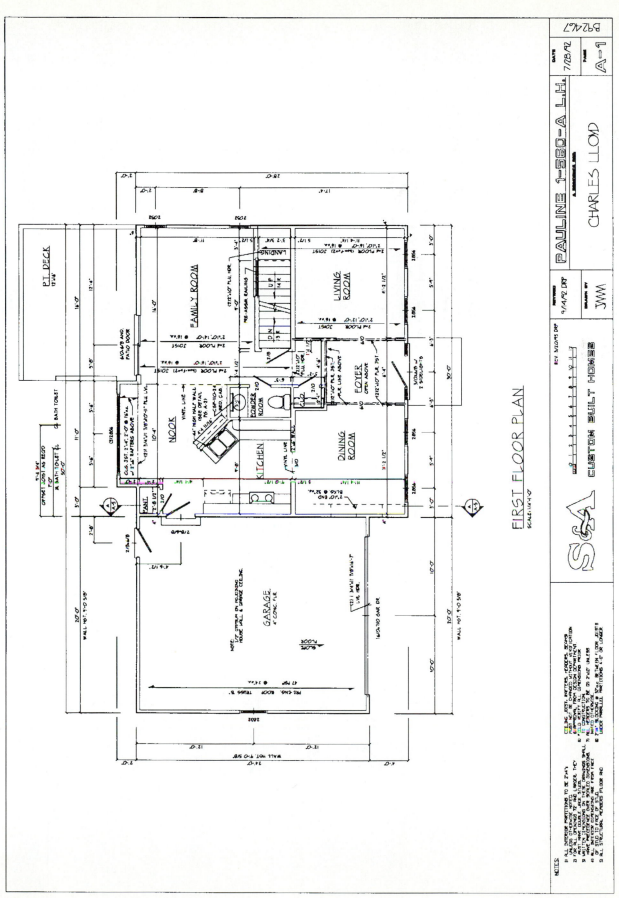

FIRST FLOOR PLAN

FigureA1

465

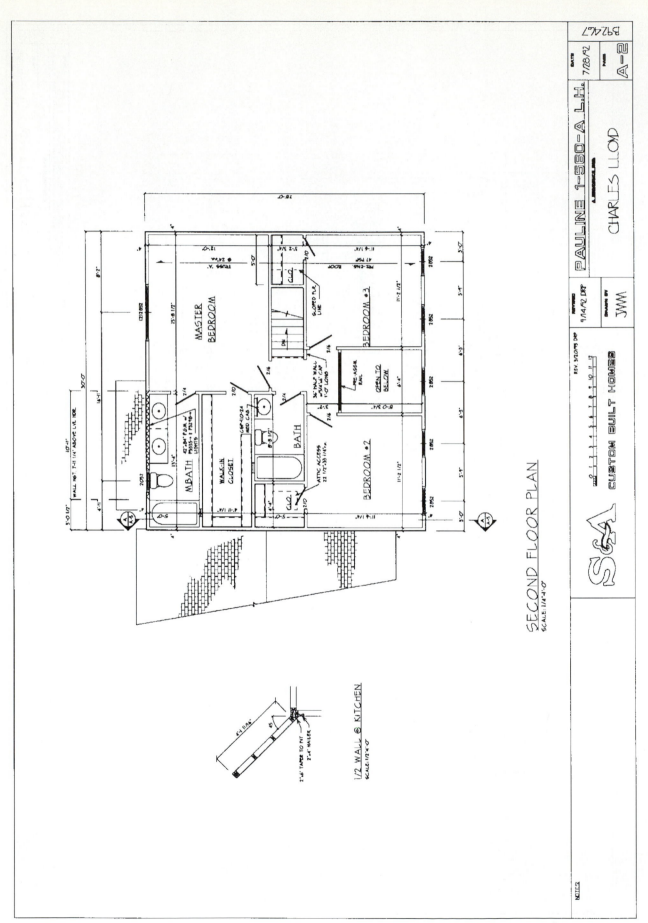

SECOND FLOOR PLAN
SCALE: 1/4"=1'-0"

1/2 WALL @ KITCHEN
SCALE: 1/2"=1'-0"

Figure A2

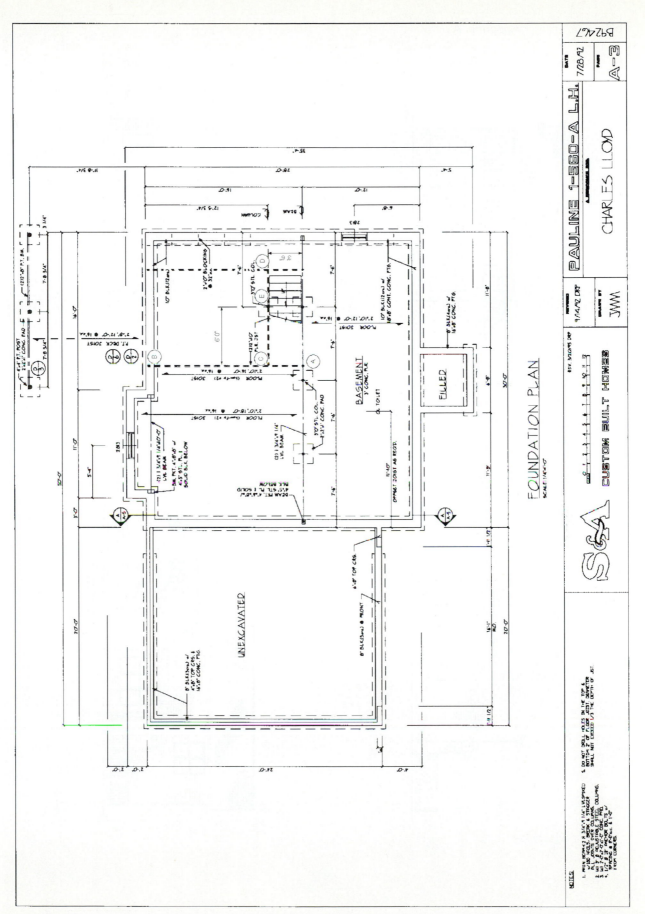

FOUNDATION PLAN
SCALE 1/4"=1'-0"

Figure A3

467

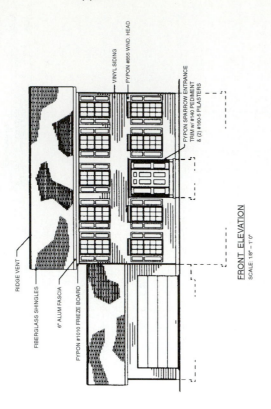

RIDGE VENT

FIBERGLASS SHINGLES

6" ALUM FASCIA

FYPON #1010 FRIEZE BOARD

VINYL SIDING

FYPON #855 WND. HEAD

FYPON SPARROW ENTRANCE
TRIM w/ #140 PEDIMENT
& (2) #160-5 PILASTERS

FRONT ELEVATION
SCALE: 1/8" = 1'-0"

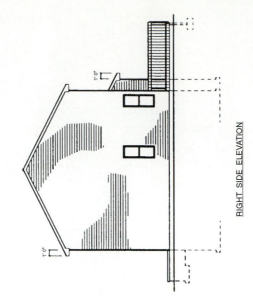

1'-0"

1'-0"

RIGHT SIDE ELEVATION

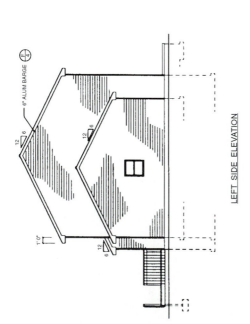

6" ALUM BARGE

12
6

12
6

12
6

1'-0"

LEFT SIDE ELEVATION

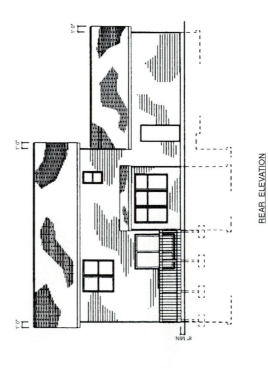

1'-0"

1'-0"

1'-0"

1'-0"

8" MIN.

REAR ELEVATION

Figure A4

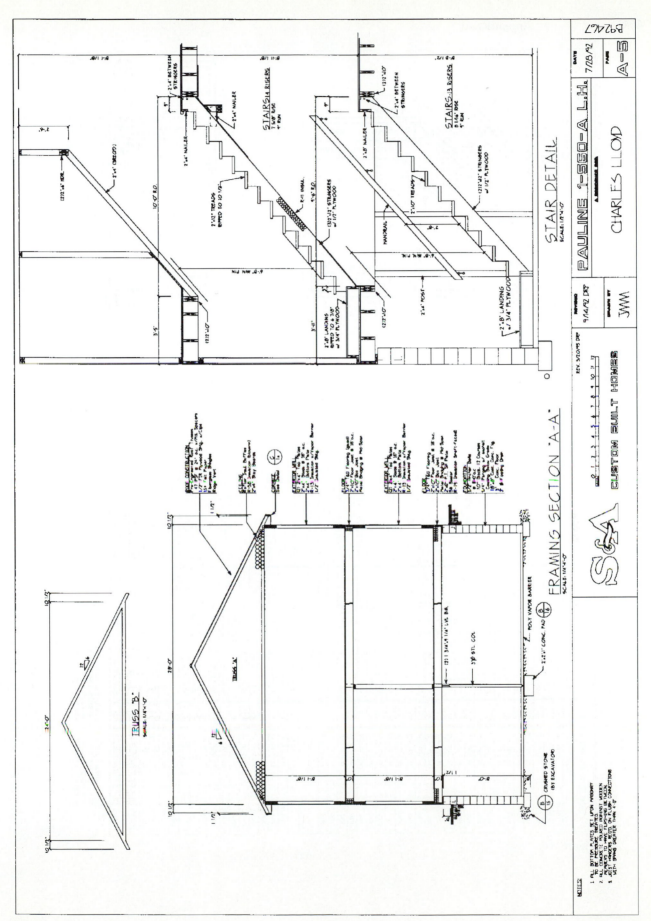

Figure A5

469

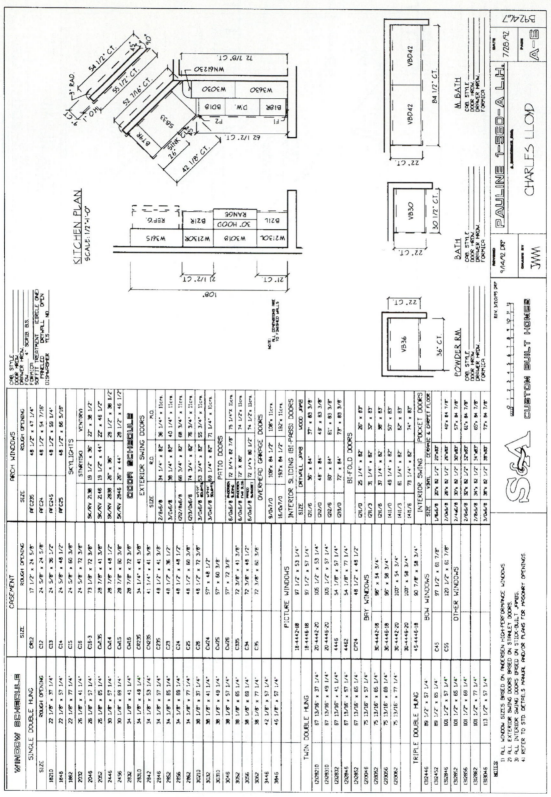

Figure A6

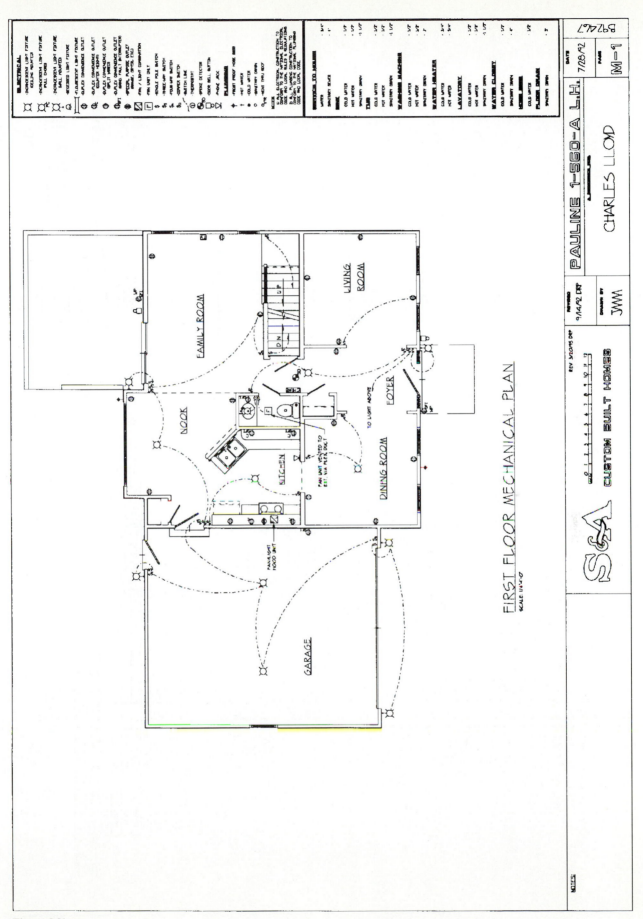

FIRST FLOOR MECHANICAL PLAN
SCALE 1/4"=1'-0"

Figure M1

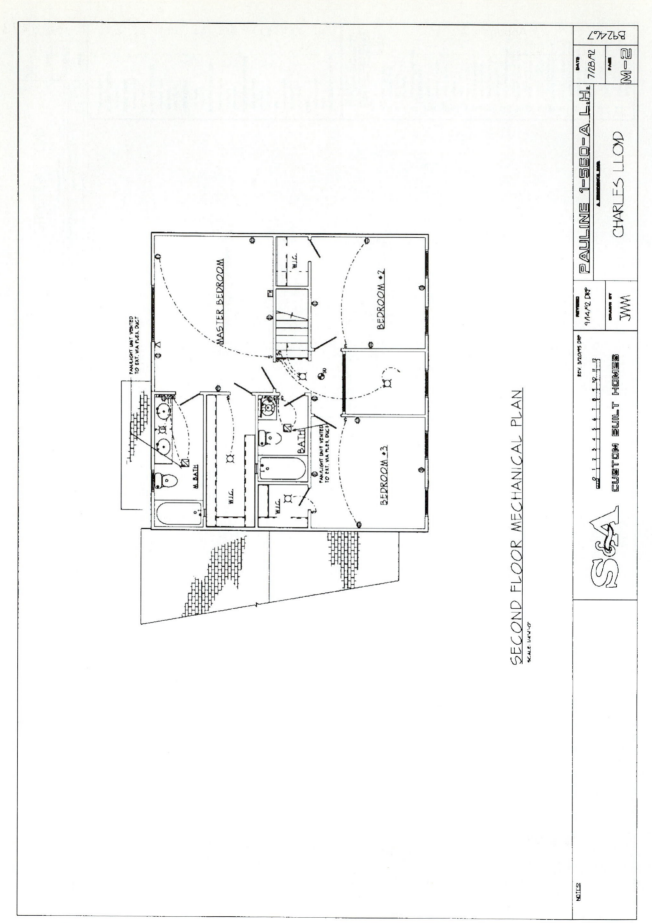

SECOND FLOOR MECHANICAL PLAN

SCALE 1/4"=1'-0"

Figure M2

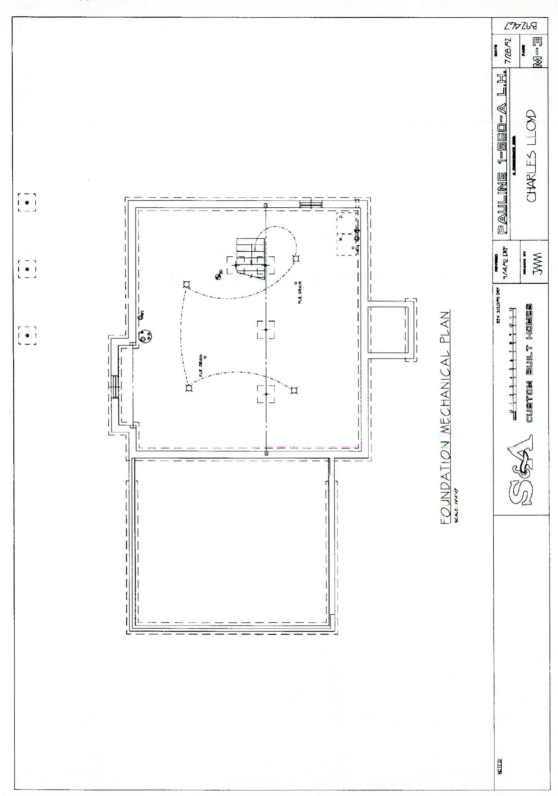

FOUNDATION MECHANICAL PLAN

Figure M3

B

The Residential Construction Contract for the "Pauline Model" Case Study House

Source: Reproduced with the permission of S & A Custom Built Homes, State College, PA (1992).

RESIDENTIAL CONSTRUCTION CONTRACT

AGREEMENT made the _____ day of _____, 19___, by and between S & A CUSTOM BUILT HOMES, INC. (hereinafter referred to as "Contractor" and _____ of _____ (hereinafter referred to as "Buyer").

WITNESSETH that in consideration of the mutual covenants and agreements herein contained, and intending to be legally bound hereby, the parties hereto do agree to the following terms and conditions:

1. Definitions. The following terms shall have the assigned meanings for the purpose of this agreement:

Section 1.1 Contract Documents. The contract documents consist of this Agreement (including specific and general conditions hereto), and drawings, specifications, plans, amendments, and change orders relating to this Agreement. These aforesaid documents constitute the entire contract between the parties, and shall not be modified except by written amendment to the said contract documents. The contract documents do not include, and specifically exclude, any and all oral discussions, understandings, and agreements between the said parties prior to the execution of this Agreement.

2. Contract Work. The Contractor agrees to furnish all the material and labor necessary for the construction of a _____

_____

upon the property of Buyer, which property is located and situated at _____

The said work shall be performed in accordance with certain plans and specifications furnished by Contractor and known and identified as _____ duplicate copies of which have been initialed by the parties hereto and delivered to both Buyer and Contractor pursuant to this Agreement. In consideration for the work to be performed under this contract, Buyers shall pay to contractor the sum of _____ (\$_____) in accordance with paragraph 5 hereinbelow, and subject to additions and/or deductions by change order as provided in paragraph 5.4 of this Contract.

THE ABOVE STATED CONTRACT PRICE IS BINDING UPON CONTRACTOR ONLY IF THE CONTRACT WORK IS COMMENCED WITHIN SIXTY (60) DAYS AFTER THE DATE OF THIS CONTRACT. IN THE EVENT CONTRACT WORK IS NOT COMMENCED WITHIN SIXTY (60) DAYS FROM THE DATE OF THIS CONTRACT, CONTRACTOR, AT ITS SOLE OPTION, MAY TERMINATE THIS CONTRACT AND NEGOTIATE A NEW CONTRACT PRICE WITH BUYER.

3. Obligations of Buyer

3.1. Prior to the commencement of construction, Buyer shall obtain and deliver to Contractor an attorney's certificate of title to the property, which certificate shall be in proper form, or other proof of Buyer's ownership of the premises which is satisfactory to Contractor and its counsel.

3.2. Buyer shall furnish to Contractor all necessary surveys describing the physical characteristics, soils, reports and subsurface investigations, restrictive covenants, building restrictions, legal limitations, utility locations, and a legal description of the building site. Buyer shall review the plans and specifications submitted by Contractor and shall give prompt written notice to the Contractor of any fault or defect in the said plans and specifications.

3.3. Buyer shall secure and pay for necessary governmental approvals or permits, easements, assessments and charges required for the construction, use, or occupancy of permanent structures or for permanent changes in existing facilities. Buyer shall furnish to Contractor reasonable evidence satisfactory to the Contractor, prior to signing this Agreement, that sufficient funds are available and committed to pay for the work to be performed under this Agreement.

3.4. Buyer shall be responsible for obtaining all permits relating to construction of the aforesaid building and for determining how, if at all, the building lot upon which said construction is to occur is encumbered by restrictions, conditions, and/or limitations affecting such construction. Contractor shall not be responsible for determining how, if at all, any laws or ordinances dealing with subdivision, zoning and the like might affect the lot upon which the contract work is to be performed. Buyer shall clearly mark the boundary lines of the building lot prior to commencement of construction thereon by Contractor.

4. Obligations of Contractor

4.1. Contractor will provide all construction supervisions, inspection, labor, materials, tools, equipment, and subcontracted items necessary for the execution and completion of the contract work.

4.2. Contractor will pay all sales, use, gross receipts, and similar taxes related to the contract work to be provided by the contractor, which taxes have been legally enacted at the time of execution of this Agreement.

4.3. The Contractor shall supervise and direct the work, using his best skill and attention.

4.4. The Contractor warrants to the Buyer that all materials and equipment incorporated into the contract work will be new unless otherwise specified, and that all contract work will be of good quality, in conformance with the contract documents.

4.5. Contractor shall be responsible for the acts and omissions of its employees and all subcontractors engaged by Contractor, their agents and employees, and all other persons performing any of the work under this contract on behalf of or with the Contractor.

4.6. The Contractor shall at all times keep the premises reasonably free from the accumulation of waste materials or rubbish caused by the operations of the Contractor. At the completion of the work, the Contractor shall remove all tools, construction equipment, machinery and surplus materials, and shall leave the work "broom" clean or its equivalent, except as otherwise specified.

5. Payment of Contract Price, Progress Payments

5.1. In consideration for the performance of the Agreement, Buyer agrees to pay Contractor, in current funds, as compensation for his service, the Contract Price of _____
_____ ($_____)
(subject to the provisions of paragraph 2 of this Contract).

5.2. The Contract Price shall be paid by Buyer to Contractor based upon written applications for payment which may be submitted by Buyer at the following times:

_____% of the Contract Price upon the execution of this Agreement;

_____% of the Contract Price upon the completion of the foundation;

_____% of the Contract Price when the roof is completed, when framing and sheathing are completed, and when all windows are set in place;

_____% of the Contract Price when the installation of electrical and plumbing systems, in rough form, are completed;

_____% of the Contract Price when the interior drywall is sanded and ready for painting;

_____% of the Contract Price upon substantial completion of the building as defined herein or at the time of occupancy of the building by Buyer.

5.3. The Contract Price is for the materials and labor in construction of the Contract Work ONLY. The Contract Price specifically excludes the cost of excavating (digging, backfilling around foundation) and landscaping and grading. The following items are also specifically excluded from the Contract Price: exterior and interior painting, utility connections, permits, outside concrete sidewalks, installation of septic system, water laterals, driveway, surveying, financing costs, attorney's fees and all other costs not specifically incurred in connection with the installation of the materials specifically provided for in the plans and specifications. The cost of excavating which shall be paid as an addition to the Contract Price includes the cost of rock removal, regardless of whether such removal requires the aid of explosives. The Buyers shall also pay the additional costs involved in providing extra engineering for foundation work which may be required if the Contractor encounters abnormal or unusual subterranean conditions during the course of construction including, but not limited to:

(A) Extra concrete block and labor necessary to install the same for the construction of foundation that may be required in excess of blueprint specifications.

(B) The cost of labor and materials needed to divert or control surface or subsurface water found during the course of construction.

(C) The cost of all fill and topsoil required to backfill and grade the job site.

(D) The cost of removal of all excess dirt and fill from job site.

5.4. The Buyer, without invalidating the Contract, may order Changes in the Work consisting of additions, deletions, or modifications, the Contract Sum and the Contract Time being adjusted accordingly. All such Changes in the Work shall be authorized by written Change Order signed by the Buyer.

5.5. A Change Order is a written order to the Contractor signed by the Buyer or his authorized agent and issued after the execution of this Agreement, authorizing a Change in the Project and/or an adjustment in the Contract Price, or the Performance Time Schedule.

5.6. The Contract Price and the Performance Time Schedule may be changed only by Change Order or as otherwise specified in this Agreement.

5.7. The cost or credit to the Buyer from a Change in the Work shall be determined by mutual agreement.

5.8. Final payment constituting the unpaid balance of the Contract Price as adjusted by Change Order shall be due and payable when the Project is delivered to the Buyer, ready for beneficial occupancy, or when the Buyer occupies the Project, whichever event first occurs, provided that the Project is then substantially completed and this Agreement substantially performed. If there should remain minor items to be completed, the Contractor and the Buyer shall list such items and the Contractor shall deliver, in writing, his guarantee to complete said items within a reasonable time thereafter.

5.9. The making of final payment shall constitute a waiver of all claims by the Buyer except those arising from (1) unsettled liens, (2) faulty or defective Work appearing within one (1) year after Substantial Completion, or (3) failure of the Work to comply with the requirements of the Contract Documents. The acceptance of final payment shall constitute a waiver of all claims by the Contractor except those previously made in writing and still unsettled.

Occupancy of the home by Buyer shall constitute acceptance of same by Buyer and Contractor, except as provided hereinabove, shall thereafter be under no obligation whatsoever to Buyer relative to the construction of said home.

6. Time of Performance

6.1. The Work to be performed under this Contract shall be commenced on or about _____ or fifteen (15) days after the date Buyer has fulfilled all obligations required by Paragraph 3 of this Contract, and except as otherwise provided or permitted by the Contract, shall be substantially completed not later than _____

_____ .

6.2. The Date of Substantial Completion of the Contract Work is the date when construction is sufficiently completed in accordance with the Plan and Specifications so the Buyer can occupy the construction work. Warranties called for by this Agreement shall commence on the Date of Substantial Completion of the construction work.

6.3. If the Contractor is delayed at any time in the progress of the construction work by any act, failure or neglect of the Buyer or by changes in the Project or by labor disputes, fire, unusual delay in transportation, adverse weather conditions not reasonably anticipatable, unavoidable casualties, or any causes beyond the Contractor's control, or a delay authorized by the Buyer, then the Date for Substantial Completion shall be extended for the period of such delay.

7. Insurance

7.1. Indemnity. The Contractor agrees to indemnify and hold the Buyer harmless from all claims for bodily injury and property damage (other than the Work itself and other property insured under Paragraph 7.2) that may arise from the Contractor's operations under this Agreement.

7.2. Contractor's Liability Insurance. The Contractor shall purchase and maintain such insurance as will protect it from claims under workmen's compensation acts and other employee benefit acts, from claims for damages because of bodily injury, including death, and from claims for damages to property which may arise out of or result from the Contractor's operation under this Contract, whether such operations be by it or by any Subcontractor or anyone directly or indirectly employed by any of them. This insurance shall be written for not less than any limits or liability required by law and shall include contractual liability insurance as applicable to the Contractor's obligations under this Agreement.

7.3. Buyer's Liability Insurance. The Buyer shall be responsible for purchasing and maintaining his own liability insurance and, at his option, may maintain such insurance as will protect him against claims which may arise from operations under this Contract.

7.4. Buyer's Property Insurance. Unless otherwise provided, the Buyer shall purchase and maintain property insurance upon the entire Contract Work at the site to the full insurable value thereof. This insurance shall include the interests of the Buyer, the Contractor, Subcontractors and Sub-subcontractors in the Contract Work and shall insure against the perils of fire, extended coverage, vandalism and malicious mischief. Any insured loss is to be adjusted with the Buyer and made payable to the Buyer as trustee for the insured's as their interests may appear, subject to the requirements of any mortgagee clause. The Buyer shall provide a copy of all policies to the Contractor prior to the commencement of the Work. The Buyer and Contractor waive all rights against each other for damages caused by fire or other perils to the extent covered by insurance provided under this paragraph. The Contractor shall require similar waivers by Subcontractors and Sub-subcontractors.

8. Correction of Work.   The Contractor shall correct any work that fails to conform to the requirements of the Contract Documents where such failure to conform appears during the progress of the Work, and shall remedy any defects due to faulty materials, equipment or workmanship which appear within a period of ONE (1) YEAR from the Date of Substantial Completion of the Contract. The provisions of this paragraph apply to Work done by Subcontractors as well as to Work done by direct employees of the Contractor. Contractor does not warrant a waterproof or damp-proof basement. THE CONTRACTOR MAKES AND THERE EXISTS NO OTHER WARRANTIES, WRITTEN OR IMPLIED, CONCERNING THE CONTRACT WORK OR ANY OTHER SUBJECT MATTER OF THIS AGREEMENT.

9. Default by Buyer.   If the Buyer shall default hereunder prior to the beginning of construction work, Contractor shall retain the money paid by Buyer as liquidated damages; and this Contract shall thereupon terminate. If the Buyer fails to make a Progress Payment to Contractor as herein provided through no fault of the Contractor, the Contractor may, upon seven days written notice to the Buyer, terminate the Contract and recover from the Buyer payment for all work completed and for any proven loss sustained upon any materials, equipment, tools, and construction equipment and machinery, including reasonable profit and damages.

Upon default in payment of any installment as provided in Paragraph 5.2, the Buyer hereby authorizes and empowers any attorney of any Court of Record of Pennsylvania, or elsewhere, to appear for and to enter judgment against him (them) for the Contract Price, a sum certain, for all moneys due under this Contract without Defalcation, with costs of suit, release of errors, without stay of execution and with ten percent (10%) added for collection fees; and he also waives the right of inquisition of any real estate that may be levied upon to collect this sum; and does hereby voluntarily condemn the same, and authorizes the Prothonotary to enter upon the fi. fa. his said voluntary condemnation and he further agrees that said estate may be sold on fi. fa. and he hereby waives and releases all relief from any and all appraisement, stay or exemption laws of any state, not in force or hereafter to be passed.

10. NOTICE TO BUYERS REGARDING RADON GAS

10.1. Radon gas is an odorless, colorless gas that is produced naturally by the normal decay of uranium and radium. Uranium and radium are widely distributed in trace amounts in the earth's crust.

Uranium decays into radium, which, in turn, decays into radon gas. Radon gas breaks down into "radon daughters" or "radon progeny."

10.2. Various studies indicate that extended exposure to high levels of radon gas or radon daughters can result in an increased risk of lung cancer.

10.3. Radon gas originates in rock and soil formations, and can be found everywhere in the atmosphere. However, radon is not found in high concentrations outside of buildings because it quickly disperses. Besides being found in the atmosphere, radon can also move into enclosed air spaces, such as basements, crawl spaces, and caves, and can reach high concentrations in such enclosed areas. When radon reaches high concentrations in indoor structures, severe adverse health consequences may result.

10.4. If a house has a radon problem, the problem can usually be remedied by either (a) increasing ventilation and/or (b) preventing the radon gas from entering the house.

10.5. The Environmental Protection Agency ("EPA") has established guidelines on levels of radon in a home that are considered safe. Levels in a residence are measured in terms of pico-curies per liter of air (pCi/l) and the EPA has set its standard at 4 pCi/l. Above this level, the EPA recommends corrective action.

10.6. Further information on radon gas can be secured from the Radon Project Office, 1100 Grosser Road, Gilbertsville, PA 19525, telephone 1-800-23RADON.

REPRESENTATIONS AND AGREEMENTS OF CONTRACTOR AND BUYER

1. The Contractor makes no representations concerning the presence or absence of radon on the Buyer's land.

2. The Contractor's responsibilities under the Residential Construction Contract do not include any responsibility for defects or problems, structural or otherwise, which may result in the accumulation of unacceptable levels of radon gas on Buyer's property or the residence to be constructed pursuant to the Residential Construction Contract. Any and all warranties, expressed or implied, which purport to impose such a responsibility or liability upon the Contractor are hereby specifically disclaimed and excluded.

11. General Provisions

11.1. All rights and liabilities herein given to, or imposed upon, the respective parties hereto shall extend to and bind several and respective heirs, executors, administrators, successors and assigns of said parties; and if there be more than one Buyer, they shall all be bound jointly and severally by the terms, covenants and agreements herein, and the word "Buyer" or "Buyers" shall be deemed and taken to mean each and every person or party mentioned as an owner herein, be the same one or more; and if there be more than one Contractor, they shall be bound jointly and severally by the terms, covenants and agreements herein, and the word "Contractor" and "Contractors" shall be deemed and taken to mean each and every person or party mentioned as a contractor herein be the same one or more.

11.2. Neither the Buyer nor the Contractor shall assign his interest in this Agreement without the written consent of the other except as to the assignment of proceeds.

11.3. This agreement shall be governed by the law in effect at the location of this Project.

Contractor   S & A CUSTOM BUILT HOMES, Inc.

By: _____
                    Authorized Signature

_____
                                                    Buyer
_____
                                                    Buyer

C

The Material
Specifications for the
"Pauline Model"
Case Study House

Source: Reproduced with the permission of S&A Custom Built Homes, State College, PA (1992).

Form FmHA 1924-2
(Rev. 2/87)

USDA - FARMERS HOME ADMINISTRATION
U.S. DEPARTMENT OF HOUSING AND URBAN DEVELOPMENT-FEDERAL
HOUSING ADMINISTRATION

FORM APPROVED
OMB NO. 0575-0042

☐ **Proposed Construction**

☐ **Under Construction**

DESCRIPTION OF MATERIALS

No. _____
(To be inserted by FHA, VA or FmHA)

THE PAULINE "A" - 1248 sq. ft.

Property address _____ City _____ State _____

Mortgagor or Sponsor _____ _____
(Name)   (Address)

Contractor or Builder S & A Custom Built Homes, Inc. _____
(Name)   (Address)

INSTRUCTIONS

1. For additional information on how this form is to be submitted, number of copies, etc., see the instructions applicable to the FHA Application for Mortgage Insurance, VA Request for Determination of Reasonable Value or FmHA Dwelling Specifications, as the case may be.

2. Describe all materials and equipment to be used, whether or not shown on the drawings, by marking an X in each appropriate check-box and entering the information called for in each space. If space is inadequate, enter "See misc." and describe under item 27 or on an attached sheet. THE USE OF PAINT CONTAINING MORE THAN THE PERCENT OF LEAD BY WEIGHT PERMITTED BY LAW IS PROHIBITED.

3. Work not specifically described or shown will not be considered unless

required, then the minimum acceptable will be assumed. Work exceeding minimum requirements cannot be considered unless specifically described.

4. Include no alternates, "or equal" phrases, or contradictory items. (Consideration of a request for acceptance of substitute materials or equipment is not thereby precluded.)

5. Include signatures required at the end of this form.

6. The construction shall be completed in compliance with the related drawings and specifications, as amended during processing. The specifications include this Description of Materials and the applicable Minimum Property Standards.

1. **EXCAVATION:**
   Bearing soil, type _____

2. **FOUNDATIONS:**
   Footings: concrete mix _1-2-4 8" x 18"_ ; strength psi _2500_ Reinforcing _____
   Foundation wall: material _10" Concrete Block_ Reinforcing _____
   Interior foundation wall: material _____ Party foundation wall _____
   Columns: material and sizes _3" adjustable steel colum_ Piers: material and reinforcing _1' x 2' x 2' Concrete_
   Girders: material and sizes _(2)1 3/4" x 9 1/4" LVL beam_ Sills: material _2" x 6" P.T._
   Basement entrance areaway _____ Window areaways _Galvanized metal where required_
   Waterproofing _Asphalt applied under pressure_ Footing drains _3"pipe & 2B stone 6" over top with_ a
   Termite protection _Semi-solid cap block & sill seal_ /positive outlet. Straw or untreated building
   Basementless space: ground cover _____ ; insulation _____ ; foundation vents _paper on top._
   Special foundations _Sump pump if positive outflow cannot be obtained._
   Additional information: _1/4" cement parging to height of foundation. Tarred to finish grade level._

3. **CHIMNEYS:** N/A
   Material _____ Prefabricated (make and size) _____
   Flue lining: material _____ Heater flue size _____ Fireplace flue size _____
   Vents (material and size): gas or oil heater _____ ; water heater _____
   Additional information: _____

4. **FIREPLACES:** N/A
   Type: ☐ solid fuel; ☐ gas-burning; ☐ circulator (make and size) _____ Ash dump and clean-out _____
   Fireplace: facing _____ ; lining _____ ; hearth _____ ; mantel _____
   Additional information: _____

5. **EXTERIOR WALLS:**
   Wood frame: wood grade, and species _Hem Fir #2_ ☒ Corner bracing. Building paper or felt _____
   Sheathing _Insulated_ ; thickness _1/2"_ ; width _4'x8'_ ; ☒ solid; ☐ spaced ____ " o. c.: ☐ diagonal; ____
   Siding _Vinyl_ ; grade _____ ; type _Horiz._ ; size _Double 5_ exposure ____ "; fastening ____
   Shingles _____ ; grade ____ ; type ____ ; size ____ ; exposure ____ "; fastening ____
   Stucco _____ ; thickness ____ "; Lath _____ ; weight ____ lb.
   Masonry veneer _____ Sills _____ Lintels _____ Base flashing _____
   Masonry: ☐ solid ☐ faced ☐ stuccoed: total wall thickness ____ ; facing thickness ____ "; facing material _____
   Backup material _____ ; thickness ____ "; bonding _____
   Door sills _Metal_ Window sills _____ Lintels _____ Base flashing _____
   Interior surfaces: dampproofing, ____ coats of _____ ; furring _____
   Additional information: _____
   Exterior painting: material _Paint & brushes - 2 coats (done by owner)_ ; number of coats ____
   Gable wall construction: ☐ same as main walls; ☐ other construction _____

6. FLOOR FRAMING:

Joists: wood, grade, and species __Hem Fir #2__; other __Hem Fir #2__; bridging __metal__; anchor 6' o.c. metal straps or anchor bolts

Concrete slab: ☒ basement floor; ☐ first floor; ☐ ground supported; ☐ self-supporting; mix _____; thickness __3__".

reinforcing _____; insulation _____; membrane __.006 membrane__

Fill under slab: material __crushed stone__; thickness __2"-4"__ Additional information: _____

7. SUBFLOORING: *(Describe underflooring for special floors under item 21.)*

Material: grade and species __Plywood T & G__; size __3/4"x4'x 8'__ type __Ext. Glue Sheathing__

Laid: ☒ first floor; ☒ second floor; ☐ attic _____ sq. ft; ☐ diagonal; ☒ right angles. Additional information: _____

8. FINISH FLOORING: *(Wood only. Describe other finish flooring under item 21.)*

LOCATION	ROOMS	GRADE	SPECIES	THICK-NESS	WIDTH	BLDG. PAPER	FINISH
First floor							
Second floor							
Attic floor	sq. ft.						

Additional information:

9. PARTITION FRAMING:

Studs: wood, grade, and species __Hem Fir #2__ size and spacing __2"x6" - 16" o.c__ Other _____

Additional information: _____

10. CEILING FRAMING:

Joists: wood, grade, and species __Hem Fir #2__ Other _____ Bridging _____

Additional information: __Pre-Engineered roof trusses at 24" o.c.__

11. ROOF FRAMING:

Rafters: wood, grade, and species __Hem Fir #2__ Roof trusses (see detail): grade and species _____

Additional information: __47 PSF Pre-Engineered roof truss at 24" o.c.__

12. ROOFING:

Sheathing: wood, grade, and species __1/2" Fir plywood__; ☐ solid; ☐ spaced ____" o.c.

Roofing __Fiberglass Shingles__; grade __235#__; size __1 x 3__; type __seal down__

Underlay __15# felt__; weight or thickness ____; size _____; fastening __stapled__

Built-up roofing _____; number of plies ____; surface material _____

Flashing: material __Galvanized__; gage or weight _____; ☐ gravel stops; ☐ snow guards

Additional information: __Drip edge__

13. GUTTERS AND DOWNSPOUTS:

Gutters: material __Aluminum__; gage or weight __.027__; size __5"__; shape __rectangular__

Downspouts: material __Aluminum__; gage or weight __.027__; size __3"__; shape __rectangular__; number __4__

Downspouts connected to: ☐ Storm sewer; ☐ sanitary sewer; ☐ dry-well. ☐ Splash blocks: material and size __Vinyl waterways__

Additional information: _____

14. LATH AND PLASTER

Lath ☐ walls, ☐ ceilings: material _____; weight or thickness _____ Plaster: coats ____; finish _____

Dry-wall ☒ walls, ☐ ceilings: material __gypsum board__; thickness __1/2"__; finish __Painted by owner (2 coats)__

Joint treatment __Tape & drywall compound__

15. DECORATING: *(Paint, wallpaper, etc.)*

ROOMS	WALL FINISH MATERIAL AND APPLICATION	CEILING FINISH MATERIAL AND APPLICATION
Kitchen	Paint - roller & brush	same - latex
Bath	Paint - roller & brush	same - latex
Other		

Additional information:

16. INTERIOR DOORS AND TRIM:

Doors: type __6 panel masonite__; material __masonite__; thickness __1 3/8"__

Door trim: type __Colonial__; material __White Pine__ Base: type __Colonial__; material __White Pine__; size __3 1/4"__

Finish: doors __Paint__; trim __Paint__

Other trim *(item, type and location)* _____

Additional information: __Contractor supplies all materials for painting & varnishing. Customer supplies labor.__

17. WINDOWS: High Performance Insulating Glass
Windows: type __glass__ ; make __Andersen__ ; material __Pine__ ; sash thickness __1 3/8"__
Glass: grade __1/2"insulation__; ☐ sash weights; ☐ balances, type __Spring__ ; head flashing __Vinyl__
Trim: type __Ranch Casing__ ; material __White Pine__ Paint __Paint provided__ ; number coats _____
Weatherstripping: type __Foam & rigid vinyl__ ; material __aluminum__ Storm sash, number _____
Screens: ☒ full; ☐ half; type __Aluminum__ ; number __all__ ; screen cloth material __aluminum__
Basement windows: type __12" x glass__ ; material _____ ; screens, number __*__ ; Storm sash, number __2__
Special windows _____
Additional information: __* Same as windows_____

18. ENTRANCES AND EXTERIOR DETAIL:
Main entrance door: material __Steel insulated__ width __3'-0"__ ; thickness __1 3/4"__ Frame: material _____ ; thickness __3/4"__
Other entrance doors: material __Steel insulated__ width __2'-8"__ ; thickness __1 3/4"__ Frame: material _____ ; thickness __3/4"__
Head flashing __Aluminum__ Weatherstripping: type __rubber&aluminum__; saddles __aluminum__
Screen doors: thickness ____"; number _____ ; screen cloth material _____ Storm doors: thickness ____"; number _____
Combination storm and screen doors: thickness ____"; number _____ ; screen cloth material _____
Shutters: ☐ hinged; ☒ fixed. Railings _____ , Attic louvers __Ridge vent__
Exterior millwork: grade and species __#2 White Pine__ Paint __White provided__ ; number coats __2__
Additional information: __6/0 x 6/8 Andersen Gliding Patio Door at Family Room__

19. CABINETS AND INTERIOR DETAIL:
Kitchen cabinets, wall units: material __Oak__ ; lineal feet of shelves __plans__ shelf width _____
Base units: material __Oak__ ; counter top __Formica or equal__ ; edging __Formica or equal__
Back and end splash __Formica or equal__ Finish of cabinets __Customer's Choice__ ; number coats _____
Medicine cabinets: make __24" x 24" Miami Carey or equal__ model _____
Other cabinets and built-in furniture __21" x 30" vanity in bathroom 21"x36" vanity in pwd rm.__
Additional information: __Cabinets by Yorktowne__   __(2) 21"x42" vanities in m.bath__

20. STAIRS:

STAIR	TREADS		RISERS		STRINGS		HANDRAIL		BALUSTERS	
	Material	Thickness	Material	Thickness	Material	Thickness	Material	Thickness	Material	Thickness
Basement	Hem Fir #2	2" x 10"			Hem Fir #2	2" x 12"	Pine	2"		
Main	Hem Fir #2	2" x 12"			Hem Fir #2	2" x 12"	Pine	2"		
Attic										

Disappearing: make and model number _____
Additional information: _____

21. SPECIAL FLOORS AND WAINSCOT: (Describe carpet as listed in Certified Products Directory.)

	Location	Material, Color, Border, Sizes, Gage, Etc.	Threshold Material	Wall Base Material	Underfloor Material
Floors	Kitchen	Sheet Linoleum		wood	3/4" T&G Plywood
	Bath	Sheet Linoleum		wood	T & G 1/4"
	Bedrooms	Wall to Wall Carpet (UM 44C)			Underlayment
	Liv.Rm & Hall	Wall to Wall Carpet (UM 44C)		wood	Under Vinyl

	Location	Material, Color, Border, Cap, Sizes, Gage, Etc.	Height	Height Over Tub	Height in Showers (From Floor)
Wainscot	Bath	1 pc. Fiberglass tub-shower combination with grab bar.			

Bathroom accessories: ☐ Recessed; material _____ ; number _____; ☐ Attached; material _____ ; number _____
Additional information: __soap dish - 2 towel bars - paper holder - shower rod - tooth brush holder__

22. PLUMBING

Fixture	Number	Location	Make	Mfr's Fixture Identificaiton No.	Size	Color
Sink	1	kitchen	Dayton		20"x30"	S. Steel
Lavatory	3	bath/pwdr	American Standard		16"x19"oval	White
Water closet	3	bath/pwdr	American Standard		12" R.I.	White
Bathtub	2	bath/pwdr	Universal Rundel		5'	White
Shower over tub △	2	bath			1 pc. Fiberglass	
Stall shower △						
Laundry trays						

△ ☒ Curtain rod △ ☐ Door ☐ Shower pan: material _____

Water supply: ☒ public; ☐ community system; ☐ individual (private) system.★

Sewage disposal: ☒ public; ☐ community system; ☐ individual (private) system.★

★ *Show and describe individual system in complete detail in separate drawings and specifications according to requirements.*

House drain (inside): ☐ cast iron; ☐ tile; ☒ other __Plastic__ House sewer (outside): ☐ cast iron; ☐ tile; ☐ other _____

Water piping: ☐ galvanized steel; ☒ copper tubing; ☐ other _____ Sill cocks, number __2 frost free__

Domestic water heater: type __Quick Recovery__ ; make and model __Rheem or equal__ ; heating capacity _____

__18.45__ gph. 100° rise. Storage tank: material __glass liner__ ; capacity __52__ gallons.

Gas service: ☐ utility company; ☐ liq. pet. gas; ☐ other _____ Gas piping: ☐ cooking; ☐ house heating.

Footing drains connected to: ☐ storm sewer; ☐ sanitary sewer; ☐ dry well. Sump pump; make and model _____ ; capacity _____ ; discharges into __Footer drain run from house to surface__ sumphole or sump pump.

23. HEATING

☐ Hot water. ☐ Steam. ☐ Vapor. ☐ One-pipe system. ☐ Two-pipe system.

☐ Radiators. ☐ Convectors. ☐ Baseboard radiation. Make and model _____

Radiant panel: ☐ floor; ☐ wall; ☐ ceiling. Panel coil: material _____

☐ Circulator. ☐ Return pump. Make and model _____ ; capacity _____ gpm.

Boiler: make and model _____ Output _____ Btuh.; net rating _____ Btuh.

Additional information: _____

Warm air: ☐ Gravity. ☒ Forced. Type of system __90% AFUE Condensing Gas Furnace__

Duct material: supply __1"Ductboard__ return __1"Ductboard__ Insulation __Part of Duct__ thickness __1"__ ☐ Outside air intake. N/A

Furnace: make and model __Lenox G-26 Q3-50__ Input __50,000__ Btuh.; output __47,000__ Btuh.

Additional information: __Supply - 30 Ga. S/M above basement insulation/Return - Stud & Joist Cav.__

☐ Space heater; ☐ floor furnace; ☐ wall heater. Input _____ Btuh.; output _____ Btuh.; number units _____

Make, model _____ Additional information: _____

Controls: make and types _____

Additional information: _____

Fuel: ☐ Coal; ☐ oil; ☒ gas; ☐ liq. pet. gas; ☐ electric; ☐ other _____ ; storage capacity _____

Additional information: _____

Firing equipment furnished separately: ☐ Gas burner, conversion type. ☐ Stoker: hopper feed ☐; bin feed ☐

Oil burner: ☐ pressure atomizing; ☐ vaporizing _____

Make and model _____ Control _____

Additional information: _____

Electric heating system: type _____ Input _____ watts; @ _____ volts; output _____ Btuh.

Additional information: _____

Ventilating equipment: attic fan, make and model _____ ; capacity _____ cfm.

"kitchen exhaust fan, make and model _____

Other heating, ventilating, or cooling equipment __Lenox HS29-311 - Condensing Unit Air Conditioner - 2½ Ton Capacity - 10 SEER, with C-23-31 Evaporator Coil (12,000 BTU/ Ton=27,200 BTU Output cooling capacity)__

24. ELECTRIC WIRING: Federal Pacific

Service: ☐ overhead; ☐ underground. Panel: ☐ fuse box; ☒ circuit-breaker; make __or equal__ AMP's __200__ No. circuits __30__

Wiring: ☐ conduit; ☐ armored cable; ☒ nonmetallic cable; ☐ knob and tube; ☐ other _____

Special outlets: ☒ range; ☒ water heater; ☒ other __Dryer__ Exterior, bath, kitchen recpt.

☒ Doorbell. ☐ Chimes. Push-button locations __Front & back doors__ Additional information: __within 6' of sink - GFI__

__Smoke detector each level interconneted with direct wire to box.__

25. LIGHTING FIXTURES:

Total number of fixtures __see plans__ Total allowance for fixtures, typical installation, $ __included in__ price of house.

Nontypical installation _____

Additional information: _____

26. INSULATION:

Location	Thickness	Material, Type, and Method of Installation	Vapor Barrier
Roof	12"	R-38 Fiberglass - blown-in	
Ceiling	3 5/8"	R-13 Fiberglass batts stapled in place 1/2" insulated sheathing	– yes
Wall	4"/6"	R-13/R-19 Fiberglass batts stapled in place	
Floor	4 mill	Vapor barrier	

27. MISCELLANEOUS: *(Describe any main dwelling materials, equipment, or construction items not shown elsewhere; or use to provide additional information where the space provided was inadequate. Always reference by item number to correspond to numbering used on this form.)* Based upon this review, to the best of my knowledge and belief, these documents conform to the "CABO One and Two Family Dwelling Code", 1989 Ed., and the "NFPA 70A Electrical Code for One and Two Family Dwellings", 1990 Ed., which are designated as the development standards for the project. I do hereby certify this drawing or plan and related specifications meet all local code requirements and are in substantial conformity with VA Minimum Property Requirements.

SIGNATURE:

TITLE:

DATE:

HARDWARE: *(make, material, and finish.)*   Brass - Weiser or equal

SPECIAL EQUIPMENT: *(State material or make, model and quantity. Include only equipment and appliances which are acceptable by local law, custom and applicable FHA standards. Do not include items which, by established custom, are supplied by occupant and removed when he vacates premises or chattles prohibited by law from becoming realty.)*

30" Free standing range (Whirlpool)

Electric controlled smoke detector each level wired into an unswitched circuit.

PORCHES:

8" x 16" concrete footer and 8" block. 4" finished slab.

Stoop - minimum 4' x 5'

TERRACES:

GARAGES:

20' x 24' 2 car garage with 16/0 x 7/0 garage door

2/8 x 6/8 service door

Built to CABO One & Two Family Dwelling code specifications.

WALKS AND DRIVEWAYS:

Driveway: width 10' ; base material shale/stone; thickness 2"x4"; surfacing material macadame ; thickness 2"

Front walk: width _____ ; material _____ ; thickness ___". Service walk: width _____ ; material _____ ; thickness_____

Steps: material _____ ; treads _____"; risers _____". Cheek walls _____

Walk required only if driveway not next to house. Stone for approx. 30' driveway included in excavation price.

OTHER ONSITE IMPROVEMENTS:

(Specify all exterior onsite improvements not described elsewhere, including items such as unusual grading, drainage structures, retaining walls, fence, railings, and accessory structures.)

_____
_____
_____

LANDSCAPING, PLANTING, AND FINISH GRADING:

Topsoil _____" thick: ☐ front yard: ☐ side yards; ☐ rear yard to _____ feet behind main building.

Lawns *(seeded, sodded, or sprigged)*: ☐ front yard _____; ☐ side yards _____; ☐ rear yard _____

Planting: ☐ as specified and shown on drawings; ☐ as follows:

_____ Shade trees. deciduous, _____" caliper.	_____ Evergreen trees. _____' to _____', B & B.
_____ Low flowering trees, deciduous, _____' to _____'	_____ Evergreen shrubs. _____' to _____', B & B.
_____ High-growing shrubs, deciduous, _____' to _____'	_____ Vines, 2 year _____
_____ Medium-growing shrubs, deciduous, _____' to _____'	_____
_____ Low-growing shrubs, deciduous, _____' to _____'	_____

IDENTIFICATION.—This exhibit shall be identified by the signature of the builder, or sponsor, and/or the proposed mortgagor if the latter is known at the time of application.

Date _____   Signature_____

Signature_____

D

MASTERFORMAT™ of the Construction Specification Institute

Source: Reproduced from the 1995 Edition of the MasterFormat™—Master List of Numbers and Titles for the Construction Industry, Copyright © 1996, with the permission of the publisher, The Construction Specifications Institute, Alexandria, VA, pp. 25–32.

MASTERFORMAT™—LEVEL TWO NUMBERS AND TITLES

INTRODUCTORY INFORMATION

00001	Project Title Page
00005	Certifications Page
00007	Seals Page
00010	Table of Contents
00015	List of Drawings
00020	List of Schedules

BIDDING REQUIREMENTS

00100	Bid Solicitation
00200	Instructions to Bidders
00300	Information Available to Bidders
00400	Bid Forms and Supplements
00490	Bidding Addenda

CONTRACTING REQUIREMENTS

00500	Agreement
00600	Bonds and Certificates
00700	General Conditions
00800	Supplementary Conditions
00900	Addenda and Modifications

FACILITIES AND SPACES

Facilities and Spaces

SYSTEMS AND ASSEMBLIES

Systems and Assemblies

CONSTRUCTION PRODUCTS AND ACTIVITIES

Division 1—General Requirements

01100	Summary
01200	Price and Payment Procedures
01300	Administrative Requirements
01400	Quality Requirements
01500	Temporary Facilities and Controls
01600	Product Requirements
01700	Execution Requirements
01800	Facility Operation
01900	Facility Decommissioning

Division 2—Site Construction

02050	Basic Site Materials and Methods
02100	Site Remediation
02200	Site Preparation
02300	Earthwork
02400	Tunneling, Boring and Jacking
02450	Foundation & Load-Bearing Elements
02500	Utility Services
02600	Drainage and Containment
02700	Bases, Ballast, Pavements & Appurtenances
02800	Site Improvements and Amenities
02900	Planting
02950	Site Restoration and Rehabilitation

Division 3—Concrete

03050	Basic Concrete Materials & Methods
03100	Concrete Forms and Accessories
03200	Concrete Reinforcement
03300	Cast-in-Place Concrete
03400	Precast Concrete
03500	Cementitious Decks and Underlayment
03600	Grouts
03700	Mass Concrete
03900	Concrete Restoration and Cleaning

Division 4—Masonry

04050	Basic Masonry Materials and Methods
04200	Masonry Units
04400	Stone
04500	Refractories
04600	Corrosion-Resistant Masonry
04700	Simulated Masonry
04800	Masonry Assemblies
04900	Masonry Restoration and Cleaning

Division 5—Metals

05050	Basic Metal Materials and Methods
05100	Structural Metal Framing
05200	Metal Joists
05300	Metal Deck
05400	Cold-Formed Metal Framing
05500	Metal Fabrications
05600	Hydraulic Fabrications
05650	Railroad Track and Accessories
05700	Ornamental Metal
05800	Expansion Control
05900	Metal Restoration and Cleaning

Division 6—Wood and Plastics

06050	Basic Wood and Plastic Materials and Methods
06100	Rough Carpentry
06200	Finish Carpentry
06500	Architectural Woodwork
06600	Plastic Fabrications
06900	Wood and Plastic Restoration and Cleaning

Division 7—Thermal and Moisture Protection

07050	Basic Thermal and Moisture Protection Materials and Methods
07100	Damproofing and Waterproofing
07200	Thermal Protection
07300	Shingles, Roof Tiles, and Roof Coverings
07400	Roofing and Siding Panels
07500	Membrane Roofing
07600	Flashing and Sheet Metal
07700	Roof Specialties and Accessories
07800	Fire and Smoke Protection
07900	Joint Sealers

Division 8—Doors and Windows

08050	Basic Door and Window Materials and Methods
08100	Metal Doors and Frames
08200	Wood and Plastic Doors
08300	Specialty Doors
08400	Entrances and Storefronts
08500	Windows
08600	Skylights
08700	Hardware
08800	Glazing
08900	Glazed Curtain Wall

Division 9—Finishes

09050	Basic Finish Materials and Methods
09100	Metal Support Assemblies
09200	Plaster and Gypsum Board
09300	Tile
09400	Terrazzo
09500	Ceilings
09600	Flooring
09700	Wall Finishes
09800	Acoustical Treatment
09900	Paints and Coatings

Division 10—Specialties

10100	Visual Display Boards
10150	Compartments and Cubicles
10200	Louvers and Vents
10240	Grilles and Screens
10250	Service Walls
10260	Wall and Corner Guards
10270	Access Flooring
10290	Pest Control
10300	Fireplaces and Stoves
10340	Manufactured Exterior Specialties
10350	Flagpoles
10400	Identification Devices
10450	Pedestrian Control Devices
10500	Lockers
10520	Fire Protection Specialties
10530	Protective Covers
10550	Postal Specialties
10600	Partitions
10670	Storage Shelving
10700	Exterior Protection
10750	Telephone Specialties
10800	Toilet, Bath, and Laundry Accessories
10880	Scales
10900	Wardrobe and Closet Specialties

Division 11—Equipment

11010	Maintenance Equipment
11020	Security and Vault Equipment
11030	Teller and Service Equipment
11040	Ecclesiastical Equipment
11050	Library Equipment
11060	Theater and Stage Equipment
11070	Instrumental Equipment
11080	Registration Equipment
11090	Checkroom Equipment
11100	Mercantile Equipment
11110	Commercial Laundry and Dry Cleaning Equipment
11120	Vending Equipment
11130	Audio-Visual Equipment
11140	Vehicle Service Equipment
11150	Parking Control Equipment
11160	Loading Dock Equipment
11170	Solid Waste Handling Equipment
11190	Detention Equipment
11200	Water Supply and Treatment Equipment
11280	Hydraulic Gates and Valves
11300	Fluid Waste Treatment and Disposal Equipment
11400	Food Service Equipment
11450	Residential Equipment
11460	Unit Kitchens
11470	Darkroom Equipment
11480	Athletic, Recreational, and Therapeutic Equipment
11500	Industrial and Process Equipment
11600	Laboratory Equipment
11650	Planetarium Equipment
11660	Observatory Equipment
11680	Office Equipment
11700	Medical Equipment
11780	Mortuary Equipment
11850	Navigation Equipment
11870	Agricultural Equipment
11900	Exhibit Equipment

Division 12—Furnishings

12050	Fabrics
12100	Art
12300	Manufactured Casework
12400	Furnishings and Accessories
12500	Furniture
12600	Multiple Seating
12700	Systems Furniture
12800	Interior Plants and Planters
12900	Furnishings Restoration and Repair

Division 13—Special Construction

13010	Air-Supported Structures
13020	Building Modules
13030	Special Purpose Rooms
13080	Sound, Vibration and Seismic Control
13090	Radiation Protection
13100	Lightning Protection
13110	Cathodic Protection
13120	Pre-Engineered Structures
13150	Swimming Pools
13160	Aquariums
13165	Aquatic Park Facilities
13170	Tubs and Pools
13175	Ice Rinks
13185	Kennels and Animal Shelters
13190	Site-Constructed Incinerators
13200	Storage Tanks
13220	Filter Underdrains and Media
13230	Digester Covers and Appurtenances
13240	Oxygenation Systems
13260	Sludge Conditioning Systems
13280	Hazardous Material Remediation
13400	Measurement and Control Instrumentation
13500	Recording Instrumentation
13550	Transportation Control Instrumentation
13600	Solar and Wind Energy Equipment
13700	Security Access and Surveillance
13800	Building Automation and Control
13850	Detection and Alarm
13900	Fire Suppression

Division 14—Conveying Systems

14100	Dumbwaiters
14200	Elevators
14300	Escalators and Moving Walks
14400	Lifts
14500	Material Handling
14600	Hoists and Cranes
14700	Turntables
14800	Scaffolding
14900	Transportation

Division 15—Mechanical

15050	Basic Mechanical Materials and Methods
15100	Building Services Piping
15200	Process Piping
15300	Fire Protection Piping
15400	Plumbing Fixtures and Equipment
15500	Heat-Generation Equipment
15600	Refrigeration Equipment
15700	Heating, Ventilating, and Air Conditioning Equipment
15800	Air Distribution
15900	HVAC Instrumentation and Controls
15950	Testing, Adjusting and Balancing

Division 16—Electrical

16050	Basic Electrical Materials and Methods
16100	Wiring Methods
16200	Electrical Power
16300	Transmission and Distribution
16400	Low-Voltage Distribution
16500	Lighting
16700	Communications
16800	Sound and Video

E

Load, Shear and Moment Diagrams for Typical Joists and Beams

Source: Robert J. Hoyle and Frank E. Woeste, *Wood Technology in the Design of Structures, 5th ed.* (Ames, Iowa: Iowa State University Press, 1989), excerpt from Appendix C, pp. 381–385.

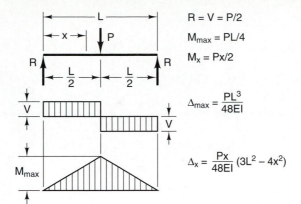

$$R = V = P/2$$
$$M_{max} = PL/4$$
$$M_x = Px/2$$

$$\Delta_{max} = \frac{PL^3}{48EI}$$

$$\Delta_x = \frac{Px}{48EI}(3L^2 - 4x^2)$$

Figure E1

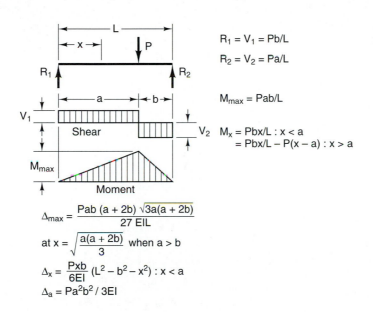

$$R_1 = V_1 = Pb/L$$
$$R_2 = V_2 = Pa/L$$

$$M_{max} = Pab/L$$

$$M_x = Pbx/L : x < a$$
$$\quad = Pbx/L - P(x - a) : x > a$$

$$\Delta_{max} = \frac{Pab(a + 2b)\sqrt{3a(a + 2b)}}{27\,EIL}$$

$$\text{at } x = \sqrt{\frac{a(a + 2b)}{3}} \text{ when } a > b$$

$$\Delta_x = \frac{Pxb}{6EI}(L^2 - b^2 - x^2) : x < a$$

$$\Delta_a = Pa^2b^2/3EI$$

Figure E2

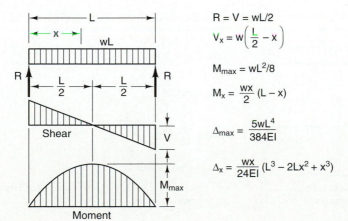

$$R = V = wL/2$$
$$V_x = w\left(\frac{L}{2} - x\right)$$

$$M_{max} = wL^2/8$$
$$M_x = \frac{wx}{2}(L - x)$$

$$\Delta_{max} = \frac{5wL^4}{384EI}$$

$$\Delta_x = \frac{wx}{24EI}(L^3 - 2Lx^2 + x^3)$$

Figure E3

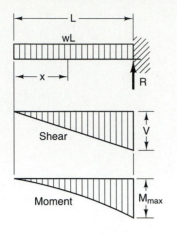

$R = V = wL$

$V_x = wx$

$M_{max} = wL^2/2$

$M_x = wx^2/2$

$\Delta_{max} = wL^4/8EI$

$\Delta_x = \dfrac{w}{24EI}\,(x^4 - 4L^3x + 3L^4)$

Figure E4

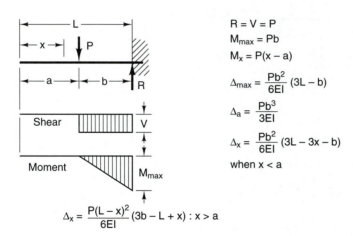

$R = V = P$

$M_{max} = Pb$

$M_x = P(x - a)$

$\Delta_{max} = \dfrac{Pb^2}{6EI}\,(3L - b)$

$\Delta_a = \dfrac{Pb^3}{3EI}$

$\Delta_x = \dfrac{Pb^2}{6EI}\,(3L - 3x - b)$

when $x < a$

$\Delta_x = \dfrac{P(L - x)^2}{6EI}\,(3b - L + x) : x > a$

Figure E5

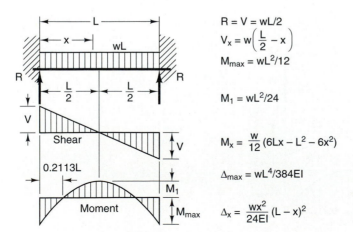

$R = V = wL/2$

$V_x = w\left(\dfrac{L}{2} - x\right)$

$M_{max} = wL^2/12$

$M_1 = wL^2/24$

$M_x = \dfrac{w}{12}\,(6Lx - L^2 - 6x^2)$

$\Delta_{max} = wL^4/384EI$

$\Delta_x = \dfrac{wx^2}{24EI}\,(L - x)^2$

Figure E6

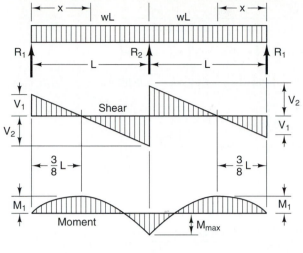

$$R_1 = V_1 = \frac{3}{8} wL \qquad R_2 = 2V_2 \qquad V_2 = \frac{5}{8} wL$$

$$M_1 = \frac{9}{128} wL^2 \qquad M_x = R_1 x - wx^2/2 \qquad M_{max} = wL^2/8$$

$$\Delta_{max} = \frac{wL^4}{185EI} \text{ at } x = 0.4215L$$

Figure E7 $$\Delta_x = \frac{wx}{48EI} (L^3 - 3Lx^2 + 2x^3)$$

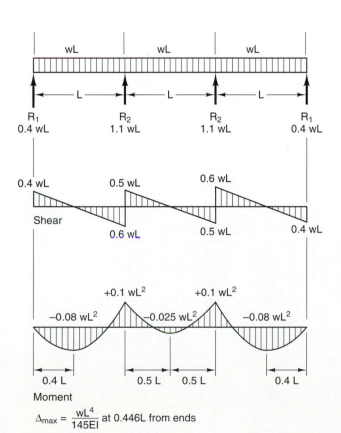

Figure E8 $$\Delta_{max} = \frac{wL^4}{145EI} \text{ at } 0.446L \text{ from ends}$$

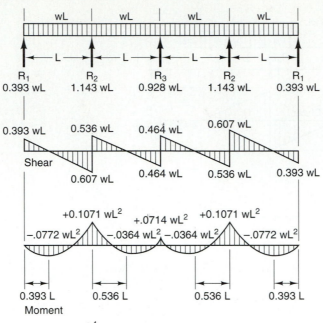

Figure E9 $\Delta_{max} = \dfrac{w}{154}\dfrac{L^4}{EI}$ at 0.44L from ends

Index